Beef Cattle Production Systems

Beef Cattle Production Systems

Andy D. Herring

Department of Animal Science, Texas A&M University

www.cabi.org

CABI is a trading name of CAB International

CABI	CABI
Nosworthy Way	38 Chauncy Street
Wallingford	Suite 1002
Oxfordshire OX10 8DE	Boston, MA 02111
UK	USA
Tel: +44 (0)1491 832111	Tel: +1 800 552 3083 (toll free)
Fax: +44 (0)1491 833508	E-mail: cabi-nao@cabi.org
E-mail: info@cabi.org	
Website: www.cabi.org	

A catalogue record for this book is available from the British Library, London, UK.

Library of Congress Cataloging-in-Publication Data

Herring, Andy D., 1965-
 Beef cattle production systems / Andy D. Herring, Department of Animal Science, Texas A&M University, College Station, Texas USA.
    pages cm. -- (Modular texts)
 Includes bibliographical references and index.
 ISBN 978-1-78064-507-0 (hb : alk. paper) -- ISBN 978-1-84593-795-9 (pb : alk. paper) 1. Beef cattle--Textbooks. 2. Beef--Textbooks. I. Title.

 SF207.H47 2014
 338.1'76213--dc23
                                2014020710
ISBN-13: 978 1 78064 507 0 (hbk)
       978 1 84593 795 9 (pbk)

Commissioning editor: Julia Killick
Editorial assistant: Emma McCann
Production editor: James Bishop

Typeset by SPi, Pondicherry, India

Contents

1 Introduction to Beef Cattle Production Systems

The goal of this book is to provide readers with a fundamental knowledge of beef cattle production. The specific management and genetic considerations are highly variable across geographical regions, cultures and markets, but the fundamentals involved with beef cattle production are (or should be) the same worldwide. In some cases, going through the thought processes involved for a production scenario in an attempt to answer a question is more important than obtaining an individual answer to the question. For so many considerations related to beef cattle production (and livestock production in general), a correct answer to a production question begins with 'it depends.' This is sometimes frustrating for students and managers because they want 'the answer' to the question.

Although there is no perfect breed (or breed combination), no perfect feed or climate, and no perfect management practice, there are many reasonable choices (and many less desirable choices) relative to the production environment and target market; this also means there are numerous desirable combinations of production components – as well as numerous undesirable combinations. Many may view this as problematic because it is human nature to seek easy answers to complex problems; however, a one-size-fits-all mentality is an overly simplistic view of livestock production, and likely most things in life. This chapter sets the stage for evaluation of these beef cattle production combinations, and this general theme is used throughout this book.

1.1 The Importance of Cattle Worldwide

Cattle are important components of agricultural production systems on all continents except Antarctica. Globally there are highly varied cattle types, production environments, market expectations and management strategies. The history of cattle's presence in geographical regions is tied to their genetic history and development (discussed in Chapter 2).

Figure 1.1 shows the relative distribution of cattle worldwide. The density of cattle is strongly related to cultural preferences, management aspects, relationships to farming systems, type of cattle and various environmental resources. We **do not** want the same cattle types or management styles across all environments. Cattle that have historically been present in regions for long periods of time have become adapted to the conditions in which they exist.

1.1.1 Adaptation

There have been many discussions pertaining to adaptation. The various considerations associated with cattle genetic resources are discussed in Chapter 3. The term adaptation refers to a group of individuals that, through genetic mechanisms over time, have become properly accustomed to their environment (a concept related to 'natural selection'). This means that adequate fertility and survivability have been balanced in those particular environments with other traits in an effort to physiologically cope with the circumstances present. From a broad population genetics perspective, over time genetic variability allows for some genes to result in more desirable phenotypes than others relative to the local environmental conditions. Over long times, these genetic changes can lead to sub-species, and speciation (separation of progenitors into new species). The same mechanisms on a smaller scale can lead to breed development, and family groups within breeds being locally adapted. Many traits related to adaptation can be evaluated on live animals through visual signs, many of which are discussed in Chapter 6 and are also related to health and welfare considerations as discussed in Chapter 9.

When animals are relocated from one environment to another and those animals adequately adjust to their new situation/environment, this concept is considered acclimatization (not adaptation). This is where the animal's physiological processes adjust

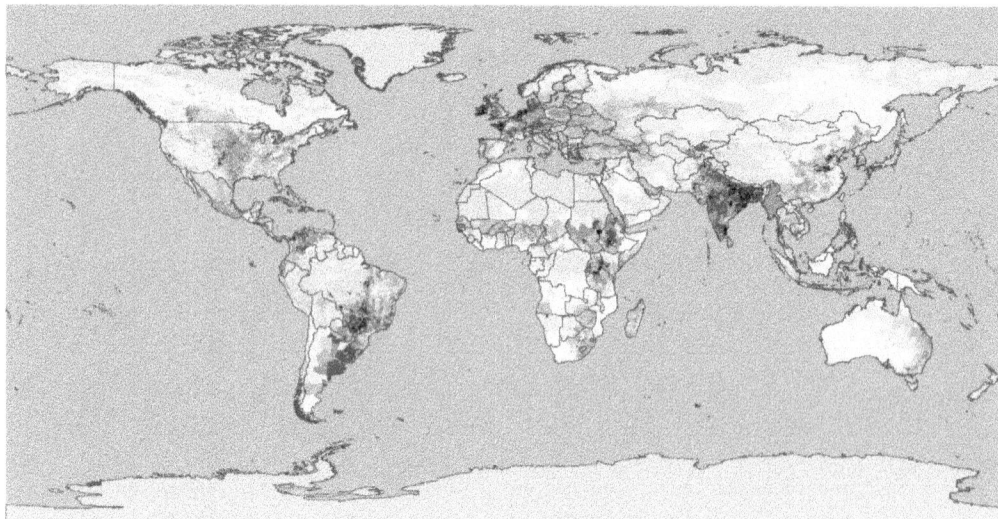

Number per square km

<1	5–10	20–50	100–250	Water
1–5	10–20	50–100	>250	Unsuitable for ruminants

Fig. 1.1. Global distribution and density of cattle. From Food and Agriculture Organization of the United Nations (2012) Gridded Livestock of the World (GLW), Livestock densities page. Available: http://www.fao.org/ag/againfo/resources/en/glw/GLW_dens.html (reproduced with permission).

to the new 'stresses' it is being subjected to. There are several traits of cattle that can be evaluated that provide insight into their relative level of stress, and animals attempt to cope with stressful situations through a variety of mechanisms where they can obtain a state of homeostasis. Groups of animals over time become adapted through genetic changes, but an animal does not become adapted to a new situation in its lifetime. Animals can be adapted to different environments, different environments also produce different phenotypes, and therefore the relative superiority or ranking between groups of animals that may vary across environments is referred to as 'genotype by environment interactions.'

1.1.2 Genotype × Environment Interactions

There are potentially many types of interactions involved with beef cattle production. One particular type of interaction is a genotype by environment interaction. An interaction is a scenario where two (or more) levels of one factor have 'inconsistent differences' depending upon the levels of other factors being considered. The fact that some cattle perform differently in some environments to others is an important fundamental consideration, but this concept does not indicate an interaction. If the difference between breeds is not consistent across the two environments, this is a genotype × environment interaction. Interactions can occur among various levels of genetic background (biological groups, breeds, families within breeds, etc.) and various levels of production environment (natural and artificial) as well as markets. A fundamental understanding of potential interactions for production scenarios leads to a systems-based way of thinking about beef cattle production and resulting sustainability.

1.2 Systems Concepts for Beef Production

The term 'systems' can be defined in several different ways, such as a set or sets of connected things or parts interacting that form a complex whole, sets of things working together as parts of a mechanism or an interconnecting network, functionally related group of elements, organized set(s) of interrelated ideas or principles, etc. A system by definition is complex. The Beef Improvement Federation (BIF, 2010) provided a good summary of the systems approach to beef cattle production, and the discussion below is heavily influenced by its descriptions.

1.2.1 Systems Approach

The systems approach recognizes that interactions among numerous factors influence profitability of beef cattle enterprises. Genetics, production environment and management inputs affect performance, expenses and profit. Managers of beef cattle operations need to recognize that there is more to consider than simply the level of production in one or two traits. What is most important in almost all cases, particularly in regard to sustainability, is the overall efficiency of the operation, which will be dictated by profitability. While the level of production is an important factor affecting profitability, costs of production are equally important, and in many cases reduction of unneeded (or unrecognized) expenses may provide more potential for profit than increased production levels. Another important aspect is that the

same level of production may be desirable in some environments, but undesirable or even detrimental in others.

The 'systems' part of the concept implies that a beef operation is influenced by many components, all of which play a part in determining profitability. Different elements affecting the overall production system include the physical environment, cattle genetic type, mating systems, management philosophy and associated practices, potential for artificial inputs (vaccines, fertilizer, nutrient supplementation, etc.) and market requirements (see Table 1.1). Many of these elements may affect both input costs and market price. As beef production systems are highly complex, there are many individual factors as well as potential interactions to yield successful or unsuccessful outcomes. Understanding of the systems concept and approach is to understand why there is no such thing as a 'one-size-fits-all' answer in most cases.

1.2.2 Livestock Production Systems

Seré and Steinfeld (1995) classified livestock production systems due to soil moisture, daily temperature and how animals were managed, and this classification approach has been widely used in original or modified form in many subsequent reports. The type of livestock production system

Table 1.1. Considerations for matching genetic and environmental resource related to beef cattle production.

Production environment		Production trait considerations[a]					
Feed availability	Stress[b]	Milk production	Mature size	Ability to store energy[c]	Resistance to stress[d]	Calving ease	Lean yield
High	Low	M to H	M to H	L to M	M	M to H	H
	High	M	L to H	L to H	H	H	M to H
Medium	Low	M to H	M	M to H	M	M to H	M to H
	High	L to M	M	M	H	H	H
Low	Low	L to M	L to M	H	M	M to H	M
	High	L	L	H	H	H	L to M
Animal role in terminal breeding system							
Maternal		M to H	L to H	M to H	M to H	H	L to M
Paternal		L to M	H	L	M to H	M	H

[a]L = low, M = medium, H = high.
[b]Heat, cold, parasites, disease, mud, altitude, etc.
[c]Ability to store fat and regulate energy requirements with changing (seasonal) availability of feed; usually indicated by body condition.
[d]Physiological tolerance to heat, cold, internal and external parasites, disease, mud, and other factors; may be considered adaptation to the production environment.
From BIF (2010). The exact values of the traits that correspond to these categories are relative to one another and may be different across production environments.

greatly impacts optimal cattle genetic types, realistic (meaning profitable) management and end products (and by-products).

1.2.3 Beef Cattle Production Systems

Cattle producers can broadly be classified as cow-calf, stocker/grower or finisher (see Table 1.2). These designations may fully describe operations, but combinations are also common. At the cow-calf level, operations (or herds) are seedstock (primarily purebred for breeding purposes) or commercial (primarily not purebred and targeted toward the beef market).

All producers are driven by economics, but commercial producers and seedstock producers can be motivated by different components. Certainly seedstock producers are in the business of selling genetics, and this is largely driven by sales of herd bulls to commercial operations in most cases. Commercial producers must obtain breeding animals from seedstock producers that have complementary production philosophy, and conversely, seedstock producers need to produce and sell animals to commercial producers that have similar philosophy. If these are incompatible, the respective genetic emphases at the seedstock level will be antagonistic to the production environment (which encompasses all non-genetic aspects, including management) at the commercial level. The importation of high growth potential but non-adapted animals into harsh production environments illustrates this concept.

Another consideration in beef cattle production systems is the purpose cattle serve (Fig. 1.2). It is the

Table 1.2. General descriptions of beef cattle production systems based on type of animals produced.

| Seedstock | Commercial (market animal production) | | | Combination seedstock and commercial |
	Cow-calf, production of weaner calves	Stocker, grower, developer	Finisher	
Seedstock herds are mainly purebred herds with animals registered in breed society herd books that produce animals for breeding purposes. They can be classified as elite and multiplier operations. Elite: these produce the highest valued breeding animals in the breed, and sell animals to other seedstock operators. Multiplier: these produce primarily bulls to sell to commercial cow-calf operations.	Cow-calf operations are driven by cow reproductive performance and overall productivity evaluated through calf production to weaning. Commercial cow-calf operations may sell calves close to weaning, or may also/instead sell animals from grower phases, and finishing phase. Different combinations of traits will receive emphasis relative to what age/stage of development animals are sold.	These operations purchase or obtain young animals for growth. This can be grazing-based on rangeland, forage crops, row crop residues or may be feedlot-based. Cattle are grown for some time, but not to completion of market size, weight or fatness. These may utilize calves destined for beef, or replacement heifers or males for breeding purposes.	These operations produce cattle that go directly to slaughter. These can be grazing entirely, grain-assisted with grazing, confined feeding of green crops (such as cut-and-carry), or confined feeding of high grain concentrate diets. These operations can obtain cattle from cow-calf operations or from grower operations.	A commercial herd as component of a seedstock operation adds flexibility for progeny testing, production of recipient females for embryo transfer. It is more common that some seedstock operations also have a commercial herd rather than a primarily commercial herd developing a seedstock portion.

Breeding animals that are sold for breeding purposes have higher values per animal than commercial animals. Utilization of registered purebred cattle for typical non-specialized market uses is undesirable as these cattle will not have higher value carcasses unless there are special circumstances, they may also have less desirable growth or cost of gain as compared to crossbred contemporaries, and they typically have a much higher initial cost per animal. Growing and finishing phases may be stand-alone operations or may be components building upon a cow-calf operation as part of an overall production system. It may also be possible that some systems include ownership at least in part further into the supply chain, such as sales through butcher shops or through specific or specialized beef merchandizing programs.

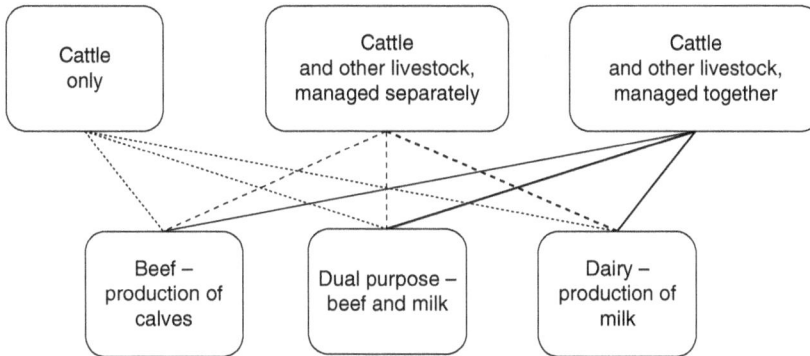

Fig. 1.2. Considerations of the role of cattle in livestock production. In many settings, production of cattle for draft purposes may be a primary use, and this generally aligns more with production of beef calves (as opposed to milk production or dual purpose) considerations.

assumption throughout this book that production of meat or animals that lead to meat is desired. However, the relative importance of beef or the age of animals at which they produce beef can be highly variable. In some contexts the production of beef is the only or primary goal, in some contexts production of beef is equal in importance to production of milk (truly dual-purpose situations), and in other contexts beef production is viewed entirely as a by-product of milk production, draft, or some other purpose. One more beef cattle production system consideration is whether or not cattle are managed solely, with combinations of other ruminants, with other non-ruminant livestock, or in various combinations. Think about grazing and other impacts when cattle, sheep, goats, camels and donkeys are all involved in a livestock enterprise. If these species are mixed together as opposed to being kept separated into different grazing groups, those different management groups also provide for varying production environments as well as resource utilization.

It has been a consideration that seedstock breeders should produce breeding animals with the commercial cattle user in mind; a challenge for the seedstock producer is to: (i) determine what type of cattle fits the commercial customer production system(s) and (ii) produce animals with end products that are acceptable and marketable to consumers. Another big challenge for seedstock producers is to have a well thought-out, long-term breeding system goal where local adaptation is emphasized, particularly for fertility and fitness traits. Under natural selection conditions, groups of animals must first reproduce, then survive, then traits

related to growth and end products can be emphasized adequately. In many cases, seedstock producers may attempt to provide as good a production environment as possible, to allow genetic potential to be expressed. If the environment in the seedstock operation is widely different than in the commercial herds that will utilize those genetics, disastrous results could occur. As a result, it can be a substantial challenge for commercial producers to find seedstock whose offspring will fit their production and marketing systems. Figure 1.3 illustrates the overlap of various factors for managers to consider.

When sub-optimal production environment conditions exist, there will be differences in animal performance expressed, as there will be differences in performance expressed under optimal conditions; however, the same animals or same types of animals may not rank close to the same (the genotype × environment interaction concept). Performance will be lower in less productive conditions (less forage production potential, no supplemental feeding, higher parasite burden, etc.), so it is also a challenge for seedstock producers to market breeding animals to commercial producers that evaluate value based only on performance and live animal condition. Seedstock sales will typically market breeding animals that are excessive in body condition score, because their prices will be lower if the body condition score is lower (even when the lower body condition is optimal for production in the intended environment).

Another point about the systems approach to cattle production is that seedstock and commercial cow-calf producers both need to appreciate and

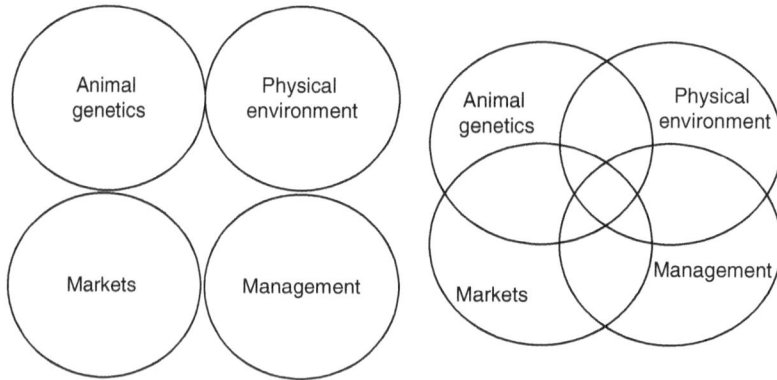

Fig. 1.3. Factors that affect beef cattle production. These broad areas are interrelated – there are many components within each broad area than can overlap and interact with elements of other areas. Understanding these interrelationships is the foundation of systems-based considerations and approaches.

understand the entire production system (supply chain) in which their cattle will be utilized. The particulars of these can be widely different across geographical and political regions. Communal cattle grazing, subsistence livestock production, large cow ranching herds in semi-arid conditions, herds utilized for growing and/or finishing cattle on improved forages, etc. will have highly different goals and resulting emphases on traits of importance for those specific situations. For both the producers and those that make recommendations to producers, it is critical to realize how the cattle and associated production inputs, management production outcomes and valuation of cattle are influenced, and how they could interact.

The fundamental principles of livestock breeding have been based on the concepts that animals tend to resemble their parents and other close relatives for many traits, inbred animals tend to breed better than they perform (and conversely that outbred animals tend to perform better than they breed), and that outbred animals tend to outperform inbred animals for many traits. This is the focus of Chapter 3. As our knowledge improves about the genetics and the underlying biology that affect traits, we become better able to explain what dictates performance and to predict it, but it does not change the fundamentals of livestock breeding; increased knowledge and technologies do, however, greatly affect potential rate of genetic change.

There is continual interest regarding incorporation of genetic information into considerations of cattle production efficiency. Several breed associations currently calculate and report index values on their respective animals based on EBVs (or EPDs), and the potential to incorporate genomics information into genetic evaluations now exists. For instance, Van Eenennaam *et al.* (2011) stated that DNA information could be valuable for seedstock when properly incorporated into genetic evaluations and needs to be incorporated into multiple-trait indices. As more is understood about gene expression and functional gene networks (groups of genes that produce certain related biological function[s]) it is likely that this type of information will add to DNA sequence-based marker utilization and produce a newer type of genomic selection than exists today. Additionally, many reproductive technologies are now available at an economically feasible cost that can alter genetic progress and production system implications. The concepts associated with development of indices in general has largely untapped potential in most beef production scenarios from a general management standpoint (not only genetic emphasis), and developments in many areas of production could benefit greatly from index-type approaches, especially those that combine genetic and environmental information and consider complex relationships among component traits.

1.3 Sustainability Considerations

There is much current discussion about the general concept of sustainability. Local, regional, national and international pressures can all affect the productivity, the profitability, and therefore the sustainability of beef cattle enterprises (Burns *et al.*, 2011).

Much emphasis has been placed on sustainability as it relates to global climate issues and world population growth. Properly designed, implemented and evaluated cattle production systems can be sustainable. Competition for water resources in many areas of the world affects livestock production. Concern about greenhouse gas emissions and relative global climate change has driven policy in some countries to tax estimated greenhouse gas production, including cattle in some cases. The focus of this book throughout is based upon science-based, practical applications. As a non-environmental scientist attempting to review publications associated with environmental impacts, the evidence of global warming from human activities is not conclusive. Undoubtedly the world has experienced much greater climate changes in regard to cooling and warming than what we have experienced in the past 100 to 140 years for which we have climatic data. Policy development always needs to be based on accurate data and to incorporate systems-based considerations.

1.3.1 Societal Demands and Considerations

Several societal issues place pressure on food animal production. Global population is expected to peak at around 9 bn people somewhere around the year 2050. This peak means that global population will then begin to decrease. The majority of this population growth is occurring in developing countries (UNDESAPD, 2013; World Bank, 2013); many developed nations have population rates that cannot sustain their current population numbers and are currently decreasing. Two trends that have occurred in human populations in history are related to personal wealth. As people have accumulated wealth (which has in turn been associated with moving away from subsistence to specialization), there have been historical decreases in birth rates, and there have been increases in meat consumption. This phenomenon of increased meat and other animal foods has been referred to as the 'livestock revolution' by many (e.g. Delgado *et al.*, 2001). A large portion of population growth and increased wealth is occurring in subtropical or tropical areas of the world, many of which have adequate natural resources to produce forage for ruminants but have low forage quality compared to many temperate areas. In societies where there is much wealth, social concerns related to animal welfare and quality can greatly impact demand of products and policy related to production, where production efficiency and/or profitability may be reduced from concern for 'luxury' type issues that are not issues in developing nations.

Livestock have an image problem in developed countries. They are blamed for global warming, for destroying the environment and causing heart disease. Health scares are also a concern – from *E. coli* and other microbes to mad cow disease (bovine spongiform encephalopathy). 'Livestock are seen as wasteful, growing fat on grain that people could eat and polluting the environment with their feces and urine and the gases they give off.' (World Bank, 2013) In developed nations, the vast majority of people are far removed from production agriculture and therefore potentially misled by those that want to do away with farm animal production because of their animal rights beliefs. The truth is that cereal grains and other crops that are fed to livestock are grown specifically for livestock feed; the varieties used for human food are different, and there is no competition for feed for livestock vs food for people. Most of the blame of livestock production worldwide is from a non-systems-based approach and has not considered whole production system evaluations.

Hocquette and Chatellier (2011) recently provided an overview of beef production in the European Union and stated that some ethical (animal welfare, slaughter of animals) and environmental (water quality, biodiversity) considerations were increasingly on the minds of European consumers. The majority of cows in most European countries are dairy cows, and as milk production per dairy cow has increased, fewer dairy cows are needed. A reduction in cow inventory, reliance on forage-only finishing of dairy animals, and increased costs of production have collectively resulted in Europe becoming a net beef importer. European beef is not always available, especially where production is naturally limited by the presence of low-forage areas (Hocquette and Chatellier, 2011).

1.3.2 Reduction of Poverty in Developing Nations Through Livestock Improvements

Although livestock play a vital role in the agricultural and rural economies of many developed countries of the world, they are critical in most developing nations, and have a very different role to that in the developed world. Livestock produce food directly for all peoples, but they also provide for traction (draft power) for farming and transportation in many regions, and the value of manure for fertilization and fire fuel is important in many situations. Most farms in the developing world are too small to justify owning or using a tractor, and the alternatives are animal power or human labor (World Bank, 2013).

The World Bank and International Livestock Research Institute (ILRI) websites (ILRI, 2013; World Bank, 2013) describe how increasing the productivity of livestock-only and mixed crop-livestock systems motivates farmers to protect their grazing lands and use them in potentially sustainable systems for raising livestock rather than simply turning them into cultivated farm ground. This concept is also true in many areas of the developed world. In many situations the use of grazing livestock, including cattle, can maintain and protect natural resources. In many developing areas of the world, livestock provide a regular source of income, and often cattle are a source of capital reserves that may be used similarly to bank accounts by people in developed nations.

In many areas mixed crop-livestock systems are common, but the importance of the livestock component may have been overlooked (World Bank, 2013). Terms used for the non-grain parts of many cereal crops often include things like crop 'residues' or 'by-products', but in many farming systems the farmers may value these 'by-products' as much or more than the grain, and 'improved' varieties or production packages that overlook the feeding value of these 'by-products' could be of little interest and value to the majority of farmers in these regions (World Bank, 2013).

1.3.3 Animal Health and the Global One Health Focus

A greater global recognition that improved study of the interaction of people and animals and resulting health implications in multiple species (including humans) has led to the development of the One Health Initiative. According to its website (One Health, 2013): 'Recognizing that human health (including mental health via the human-animal bond phenomenon), animal health, and ecosystem health are inextricably linked, One Health seeks to promote, improve, and defend the health and well-being of all species by enhancing cooperation and collaboration between physicians, veterinarians, other scientific health and environmental professionals and

by promoting strengths in leadership and management to achieve these goals.' A major goal of this group also stresses the need for joint efforts to inform and educate political leaders and the public sector through accurate media publications (One Health, 2013). These types of efforts are by their nature complicated, but also illustrate the need to understand systems-related considerations for overall human and animal health improvements and their implications.

1.3.4 Climate Change and Environmental Considerations

There has been much discussion in recent years about global and regional climate change. It is beyond the scope of this book to discuss most of the data and considerations associated with these issues. However, there have been reports about the role that livestock production can play in contributing to and mitigating climate change, primarily from greenhouse gas emission considerations. Much of the measurement of greenhouse gas contributions has been reported in carbon dioxide (CO_2) equivalents. Other gases such as nitrous oxide (N_2O) and methane (CH_4) are also involved, and the potential contributions of these gases have much greater greenhouse effects than CO_2 (23 times as much for CH_4 and 300 times as much for N_2O). Carbon dioxide and methane are normal by-products of digestion in all animals, but more of these gases are produced in ruminant animals than in monogastric animals. The FAO report *Livestock's Long Shadow* (Steinfeld *et al.*, 2006) initiated much discussion about the role that animal production can play in regard to climate change, as it reported that 18% of total greenhouse gas emissions from human activities were due to livestock production. Consequently, there is increased pressure to consider potential 'carbon taxes' on livestock production and specifically cattle (United Nations Climate Change Conference, 2009).

More recently, Gerber *et al.* (2013) from the FAO also estimated the potential livestock contribution to environmental impacts and climate change. This report stated that 14.5% of human-induced global greenhouse gas emissions could be attributed to farm animals. This report further stated that meat and milk production from cattle accounted for 41% and 20%, respectively, of farm animal contributions, and that pig meat and poultry (meat and eggs) contributed 9% and 8%, respectively. Table 1.3 shows the estimated emissions from global dairy and cattle meat production, and Fig. 1.4 shows the relative distribution of emissions by production category.

Large variability in total emissions and categories of emissions exist across geographical regions and production systems in both milk and beef production. It may not be obvious why dairy systems are being discussed in a beef production textbook, but globally, many areas have the majority of beef production come about as a by-product of milk production systems, and in some cases no 'beef' breeds may be utilized for beef production. The term beef simply refers to the meat from cattle. Regions of the world Gerber *et al.* (2013) identified as being lowest for greenhouse emissions from beef production were North America, Europe, the Russian Federation, the Near East and Northern Africa, and Oceania (Australia and New Zealand). Other areas that were identified as the highest appear to rely the most for grass-finishing production of beef and produce animals for harvest at later relative ages. This report made an encouraging recommendation in that management techniques that increase efficiency of production (such as increased digestibility of feeds, improved animal husbandry, etc.) also appear to lower carbon footprints. In most instances the optimal production practices that increase efficiency for beef cattle producers can also increase profitability (a major focus of Chapter 12). Many people may assume that grass finishing of beef cattle is more natural and therefore 'more green' than feeding of concentrates, but most of these thoughts do not involve considerations of the whole production system and all its components (for instance see Capper, 2011, and Cooprider *et al.*, 2011). Furthermore, detailed and systems-based considerations are needed before policy considerations that automatically affect grazing-based livestock practices in both developed and developing areas of the world are universally implemented.

1.4 Layout and Organization of this Book

This book is organized in large part by discipline area (breeding and genetics, nutrition, reproduction, meat science, marketing, etc.), but it is the hope of the author that the reader realizes that these disciplines and their associated considerations overlap with one another. These discipline areas are covered in enough detail for the fundamentals of various associated management practices to be understood if the reader is unfamiliar with them, and with enough background information for the reader to understand why recommendations are made. The goal of this book is to be an educational and teaching resource for a wide variety of beef cattle students and producers, and so the material presented is intended to make the reader ponder situations and applications; as a result, study questions are presented at the end of each chapter, and several tables are expressed in both metric units and imperial (English) units or have equivalent tables in the Appendix.

There is a need to understand the underlying biological fundamentals pertaining to traits for improved management and production efficiency. As a result, Chapters 2 through 11 of this book are discipline-oriented. Concepts associated with domestication and genetics of cattle population development (Chapter 2) and genetic changes in traits (Chapter 3) are provided. Concepts pertaining to nutrition (Chapter 4) and grazing management (Chapter 5) are covered next. Many measures of live animal evaluation are related to meat production, typical performance traits, adaptation, functionality and longevity; therefore visual animal evaluation is

Table 1.3. Estimated global production, emissions and emission intensity from cattle milk and meat production.

System	Emissions (million tonnes CO_2 equiv.)		Production (million tonnes)		Emission intensity (kg CO_2 equiv. per kg product)	
	Milk	Meat	Milk	Meat	Milk	Meat
Dairy	1331.1	486.2	508.6	26.8	2.6	18.2
Beef	–	2338.4	–	34.6	–	67.6

From Gerber *et al.* (2013); a metric ton (tonne) is 1000 kg or 2205 lb.

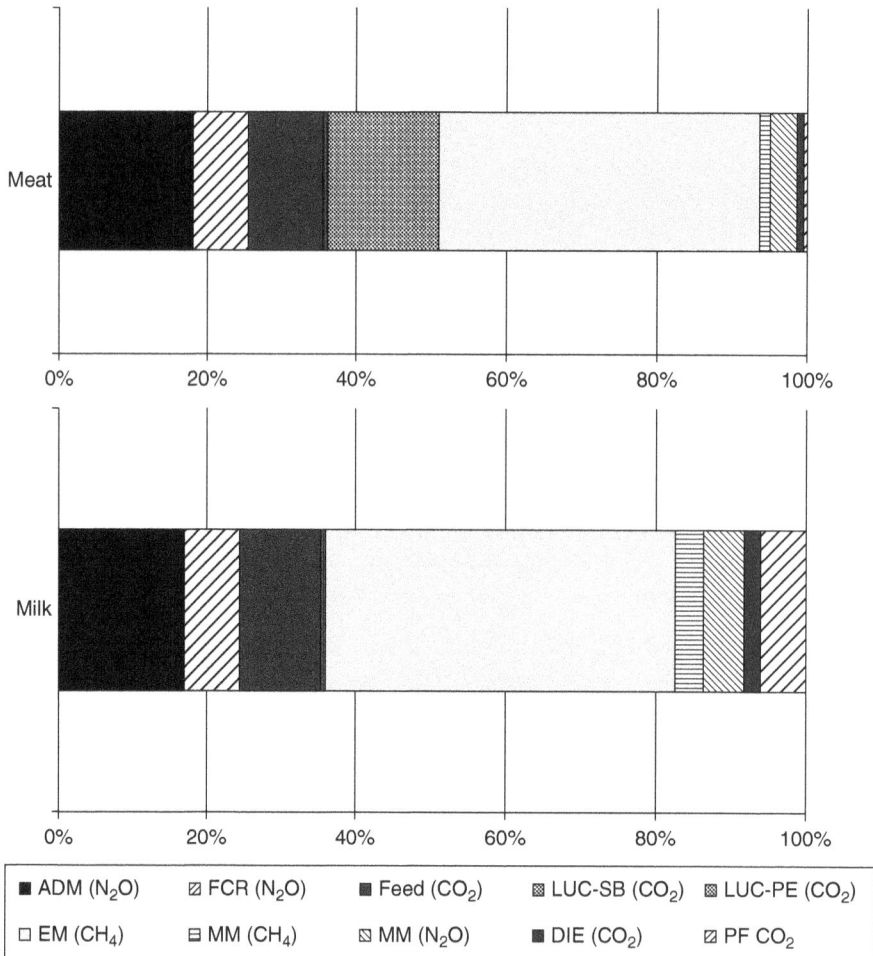

Fig. 1.4. Estimated percentages of global greenhouse gas emission categories from cattle meat and milk production. Data from Gerber *et al.* (2013). ADM = applied and deposited manure, FCR = fertilizer and crop residues, LUC-SB = land use change due to soybeans, LUC-PE = land use change from pasture expansion, EM = enteric methane, MM = manure management, DIE = direct and indirect energy, PF = post-farm. From the Global Livestock Environmental Assessment Model (GLEAM). N_2O = nitrous oxide, CO_2 = carbon dioxide, CH_4 = methane.

the main theme of Chapter 6. Concepts associated with cattle growth and development, from gestation through maturity, are covered in Chapter 7.

Reproduction (Chapter 8) is important for supply and replenishing of herds, excess animals can be sold to increase profit or to (re)build herd sizes, but reproduction is also a direct driver for supply chains (Chapter 13) and the ability to produce meat (Chapter 11) through animal numbers. Survivability and health of animals (Chapter 9), as well as animal stress and behavior (Chapter 10), are also critical for the potential to improve efficiency of production across the industry through better animal well-being and welfare. Principles associated with cattle harvest (slaughter) and meat science (Chapter 11) are critical for efficiency, quality and safety of animal-based human foods, but are also important for development and use of by-products for many industries (agricultural and non-agricultural) that lead to numerous products useful to societies.

Information throughout the book can have variable economic values and implications. Chapter 12 discusses structured ways to assess economic impacts from various production practices. The influence of economic sustainability pertaining to beef cattle production, regardless of the production environment, cannot be overstated. When long-term profitability is the primary focus of a beef cattle operation, the traits of the animals become secondary, and the relative importance of each individual trait also becomes less important. A logical extension of the discipline-oriented information in Chapters 2 through 11, combined with considerations of financial and economic evaluation in Chapter 12, leads to the ideas and discussions regarding supply chain concepts covered in Chapter 13. Although there are specific examples from various countries incorporated throughout the book related to discipline-oriented concepts, Chapter 14 discusses some specific comparisons for environmental and societal conditions, beef production, cattle numbers, herd sizes and livestock production systems across global regions with information from several countries. Hopefully some of these summary tables across countries stimulate readers about how the numerous components of beef cattle production can be varied and contribute to overall systems.

1.5 Summary

All traits of importance in beef cattle production are influenced by both genetics and environment (the concept of environment here means anything that is not of genetic origin). Obviously rainfall, soil type and plant species are environmental influences, but environmental influences may be natural (not influenced by humans) or artificial (influenced by humans), just as there can be natural as well as artificial selection. All of these are important considerations and can potentially interact. In this book production system refers to the total processes and considerations, from breeding decisions to valuation of end product. The end product may be edible beef products, but it could also mean live cattle for beef to a set weight, age or time point, or it could mean breeding animals for commercial and/or seedstock producers.

There are fundamentals that all cattle producers need to be familiar with in order to attain profit, which in most cases involves proper matching of genetic, environmental and economic resources. A foundation in breeding and genetics, nutrition, grazing management, reproduction, health, growth and meat science is critical, and these topics are each addressed in this book as traditional textbook type chapters including study questions. Other chapters include information on domestication and breed development history, supply chain considerations, global comparisons and special situations to provide a broader understanding of whole production systems and the need for managers to use the fundamentals of production disciplines to tailor management for specific scenarios of environments and markets.

1.6 Study Questions

1.1 Explain how it is possible that different types of cattle may be more desirable in some production environments than others.

1.2 Based on the global livestock production system classifications used by Seré and Steinfeld (1995), list four examples of different cattle production systems and develop an example description for each.

1.3 Is it possible that seedstock producers that have both registered and commercial cattle might have some advantages over seedstock producers that have only registered animals? Provide a brief explanation of your thoughts.

1.4 Do you believe that some management practices used in beef cattle production may have varying advantages (or disadvantages) across different environments? Discuss why or why not.

1.5 How is it possible that operations that have (a) breeding cows plus (b) grower and/or finisher cattle have any potential advantages over operations that are exclusively breeding cow or exclusively grower-finisher emphasis?

1.6 Discuss your thoughts on whether or not you believe that it will be important for individual operations to assess and document their greenhouse gas emission status/carbon footprint in the future.

1.7 References

BIF (2010) Chapter 6, Utilization. In: *Guidelines For Uniform Beef Improvement Programs, Ninth Edition.* Beef Improvement Federation, Raleigh, NC. Available at: http://www.beefimprovement.org/content/uploads/2013/07/Master-Edition-of-BIF-Guidelines-Updated-12-17-2010.pdf (accessed 12 May 2014).

Burns, B.M., Herring, A.D., Allen, J.M., McGowan, M.R., Holland, M., Braithwaite, I. and Fordyce, G. (2011) Genetic strategies for improved beef production in challenging environments such as Northern Australia and related implications for the Southern USA. In: *Proceedings of 57th Texas A&M Beef Cattle Short Course*, College Station, TX, USA.

Capper, J.L. (2011) Replacing rose-tinted spectacles with a high-powered microscope: The historical versus modern carbon footprint of animal agriculture. *Animal Frontiers* 1, 26–32.

Cooprider, K.L., Mitloehner, F.M., Famula, T.R., Kebreab, E., Zhao, Y. and Van Eenennaam, A.L. (2011) Feedlot efficiency implications on greenhouse gas emissions and sustainability. *Journal of Animal Science* 89, 2643–2656.

Delgado, C.L., Rosegrant, M.W. and Meijer, S. (2001) *Livestock to 2020. The Revolution Continues.* Ag-Trade, International Agricultural Trade Research Consortium.

Gerber, P.J., Steinfeld, H., Henderson, B., Mottet, A., Opio, C., Dijkman, J., Falcucci, A. and Tempio, G. (2013) *Tackling Climate Change through Livestock – A Global Assessment of Emissions and Mitigation Opportunities.* Food and Agriculture Organization of the United Nations (FAO), Rome.

Hocquette, J.F. and Chatellier, V. (2011) Prospects for the European beef sector over the next 30 years. *Animal Frontiers* 1, 20–28.

ILRI (2013) International Livestock Research Institute. Available at: http://www.ilri.org/ (accessed 16 October 2013).

One Health (2013) One Health Initiative. Available at: http://www.onehealthinitiative.com/ (accessed 8 October 2013).

Seré, C. and Steinfeld, H. (1995) World Livestock Production Systems: Current Status, Issues and Trends. FAO Animal Production and Health Paper No. 127. Food and Agriculture Organization of the United Nations, Rome.

Steinfeld, H., Gerber, P.J., Wassenaar, T., Castel, V., Rosales, M. and de Haan, C. (2006) *Livestock's Long Shadow – Environmental Issues and Options.* Food and Agriculture Organization of the United Nations, Rome.

UNDESAPD (United Nations, Department of Economic and Social Affairs, Population Division) (2013). World Population Prospects: The 2012 Revision, Volume I: Comprehensive Tables, ST/ESA/SER.A/336.

United Nations Climate Change Conference (2009) Climate Change Conference, Copenhagen (COP15), 6–18 December 2009.

Van Eenennaam, A.L., van der Werf, J.H.J. and Goddard, M.E. (2011) The value of using DNA markers for beef bull selection in the seedstock sector. *Journal of Animal Science* 89, 307–320.

World Bank (2013) http://www.worldbank.org/ (accessed 28 August 2013).

2 History and Domestication

The goal of this chapter is to provide the reader with background information and fundamental knowledge regarding basic genetic principles, how different populations of animals relate to one another, and the history of cattle. Elements of speciation, domestication and breed development are discussed. Understanding of population development and relatedness concepts that are highlighted in this chapter provide direct insights into adaptation, underlying breed and biological group differences, matching of genetic resources to environmental conditions, and heterosis considerations that are important points for mating systems, and genetic improvement.

2.1 Genetic Fundamental Concepts

Discussion of the history of cattle needs to begin with an overview of the fundamentals of genetic theory. These basic aspects and associated terminology are discussed here, but other references such as genetics textbooks should be consulted for detailed discussion. Basic aspects are covered so that the reader can have a broad understanding of genetics and understand research results and scenarios discussed in this and subsequent chapters, as well as applications beyond this book.

2.1.1 Cell Replication

Cattle, as are all mammals, are diploid organisms. This means that they have two copies of their genetic material, one half that was inherited from the maternal (female) parent and one half that was inherited from the paternal (male) parent. The entire genetic make-up is generally referred to as the genome, and this indicates the complete DNA (deoxyribonucleic acid) sequence of the organism, housed within the nucleus of cells. In mammals, the DNA sequence within the cell nucleus is generally what is termed the genome, and this is where most

of the DNA exists, however the mitochondria in the cytoplasm also have a small genome as they encode a few proteins related to cellular metabolism. DNA is double-stranded and has a characteristic double helix structure that is a bit like a twisted rope ladder. There are four nucleotide bases (adenine, cytosine, guanine and thiamine, or A, C, G and T, respectively), and it is the order of these bases (or base pairs because DNA is double-stranded) that dictates DNA sequence, which encodes for many different types of products (see Fig. 2.1).

During normal cell division (mitosis), the nuclear genome is replicated and then the resulting daughter cells should be identical to the parent cell they originated from (as well as being identical to each other). When this process does not occur exactly right, a mutation can occur. Mutation typically means a change in the DNA sequence that was not in the original form. The DNA sequence encodes messenger ribonucleic acid (mRNA), and this serves in turn as the template for proteins, enzymes and hormones. There are also several forms of RNA that help with these processes. In general there are specific sequences in DNA that result in different products that affect animal performance. The process by which RNA is made from the DNA template is called transcription, and the process by which resulting gene products such as proteins and hormones are made from mRNA is called translation (Fig. 2.1).

A gene is thought of as a sequence of DNA that results in a specific product such as a protein. Different forms of the gene are referred to as alleles. As a result of cattle and all mammals being diploid, each individual possesses two alleles for each gene (the exception to this is males and genes on the sex chromosomes). There may be many different alleles possible at genes, and populations of animals may differ widely in the frequency of certain alleles. The alleles present in an animal for a certain gene locus is called its genotype. If it has

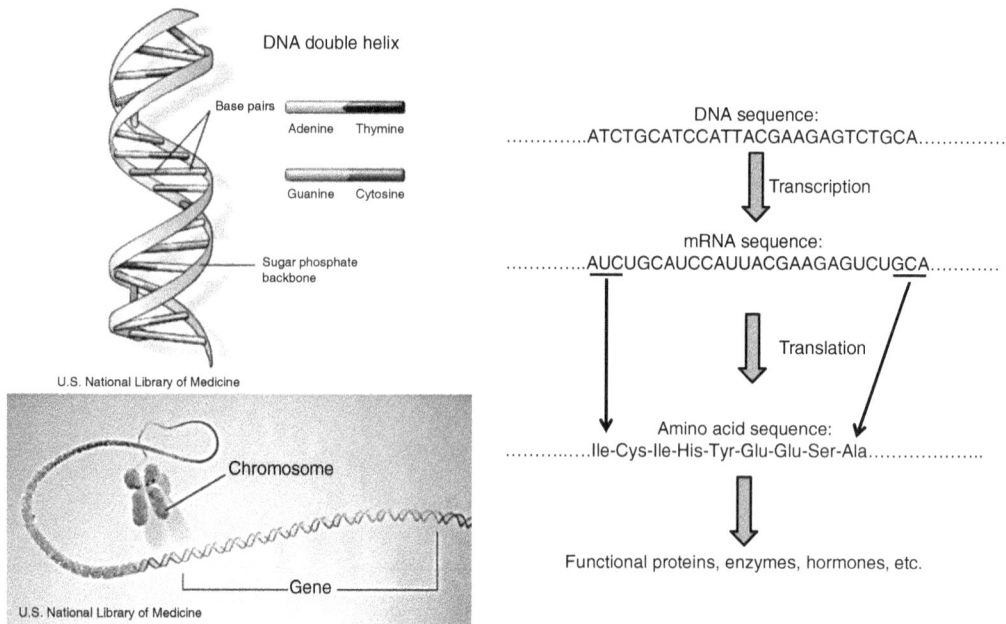

Fig. 2.1. Conversion of DNA into gene products. This is an oversimplification but illustrates the major concepts associated with general genetic theory. Pictorial graphics publically available from http://ghr.nlm.nih.gov

two copies of the same allele it is called homozygous, and if it has two different alleles it is called heterozygous. An animal's genotype influences its phenotype, i.e. the characteristics actually observed. A pattern of alleles across gene loci on the same chromosome is termed a haplotype. Multiple haplotypes that are similar can constitute a haplogroup.

2.1.2 Chromosomes

The DNA is organized into structures called chromosomes. Chromosomes are complex structures of DNA and proteins known as histones. Each species has a characteristic chromosome number. Cattle have 30 chromosome pairs, and therefore 60 total chromosomes (the diploid number is 60, i.e. $2n = 60$). A gene locus refers to a physical location of a gene on a chromosome (the plural version of locus is loci). The chromosomes condense during cell replication and can be seen through a microscope during the metaphase stage of mitosis. The visual representation of the chromosomes is referred to as a karyotype. Chemicals can be used to study the chromosomes as there are specific banding patterns of different chromosomes of each species where the

chemicals bind. Chromosomes that do not differ between the sexes are called autosomes, whereas those that differ between the sexes are called sex chromosomes.

The autosomes are typically numbered from largest to smallest (cattle chromosome 1 is the longest and 29 is the smallest), and the sex chromosomes are the X and Y chromosomes. Male mammals are XY and females are XX. All chromosomes have a specific region called the acrosome which is needed for proper chromosome alignment during cell division, but can also be used to classify and compare chromosomes. The study of chromosomes is known as cytogenetics. The centromere is the place on the chromosomes where maternal and paternal copies align themselves in meiosis. The shapes of chromosomes can be described according to the location of the centromere along the length of the chromosome. The term acrocentric describes chromosomes where the centromere is close to the end of the chromosome, whereas sub-metacentric and metacentric are terms that represent chromosomes with centromeres between the end and the middle or at the middle, respectively, of the chromosome length. Chromosomal rearrangements are possible and have been reported to be involved in speciation processes. A fusion of two acrocentric chromosomes at the centromere is termed a Robertsonian translocation (but may also be referred to as a centric fusion). Robertsonian translocations of specific chromosomes appear to be related to differences in karyotypes among many cattle-related species.

2.1.3 Meiosis and Fertilization

The process to generate reproductive cells is known as meiosis. This leads to production of spermatocytes (sperm cells) in males and oocytes (ova, or eggs) in females. Gamete is a generic term for a reproductive cell. During meiosis the maternal and paternal versions of all the chromosomes replicate and align (maternal and paternal versions of chromosome 1 pair, maternal and paternal versions

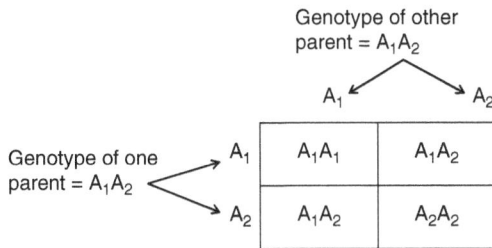

Genotype of other
parent = A_1A_2

A_1 A_2

Genotype of one
parent = A_1A_2

A_1

A_2

	A_1	A_2
A_1	A_1A_1	A_1A_2
A_2	A_1A_2	A_2A_2

Fig. 2.2. Genotypes of parents and progeny. The gametes contain half of the genetic make-up. Genotypes of the parents dictate passing of alleles to progeny, and alleles in gametes dictate genotypes in progeny. In this example both parents are heterozygous (have two different alleles) and are expected to produce three different genotypes among their progeny (two types of homozygotes as well as heterozygotes).

of chromosome 2 pair, etc.). This happens in the reproductive organs (testes of males and ovaries of females). During the time the chromosomes are aligned, there is breakage and reattachment of homologous chromosomes, and these processes are known as crossing over and recombination. Crossing over and recombination provide for new mixtures of paternal and maternal combinations of alleles across gene loci, and therefore provide for new genetic combinations across generations (see Fig. 2.2).

The gametes contain one half of the full genetic make-up, or one copy of every chromosome, and this is referred to as being haploid. At fertilization, the full diploid genetic make-up is re-established (Fig. 2.3). During fertilization, the sperm cell enters the egg, and there will be a new nucleus formed from the combined maternal and paternal genetic information that sets into motion the developmental processes that will impact the individual for the rest of its life. There are mitochondria in the cytoplasm of the egg, and this is passed through maternal inheritance. As the mitochondria are inherited maternally and the Y chromosome is inherited paternally, both have been used to study evolutionary relationships within and across species of animals, and the nuclear DNA sequence of the X chromosomes and the autosomes has been used as well.

2.1.4 History of Genetic Studies

The field of genetics is new compared to many other fields of study. Gregor Mendel, an Austrian monk studying peas and other plants in the mid to late 1800s, is credited with the fundamental discovery

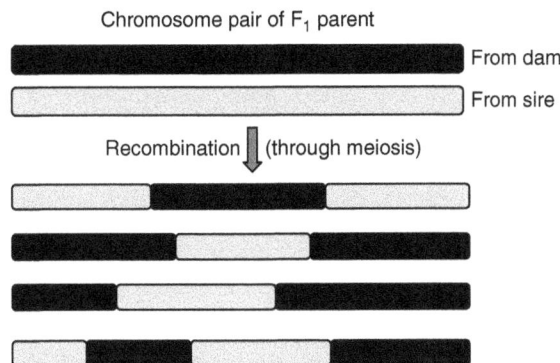

Chromosome pair of F_1 parent

From dam

From sire

Recombination (through meiosis)

Fig. 2.3. Generation of new genetic combinations through chromosomal recombination. Think about the potential variability in gametes (sperm and egg cells) that a parent can produce.

of genetic theory per se, where 'factors' or genes are passed from parent to offspring and that genotypes dictate phenotypes. However, Sir Robert Bakewell of Great Britain is known as the father of animal breeding as he documented bloodlines, pedigrees and planned mating strategies in development of cattle, sheep and horse breeds in the mid to late 1700s. Chromosomes were studied extensively in the first half of the 1900s and were known to be related to inheritance. In the mid-1950s Watson and Crick discovered that DNA was the genetic material carried in the nuclei of cells. It is not possible to see the differences in DNA sequence through a microscope; however, a process called gel electrophoresis was developed where DNA could be carried through a gel with an electrical current, and that patterns of DNA could be seen this way. In the 1970s and 1980s the concept of genetic markers evolved as scientists discovered that patterns of DNA could be visualized, scored and matched with corresponding phenotypes. A process to amplify DNA sequence was invented in 1982, called the PCR (polymerase chain reaction), which allowed extremely small quantities of DNA to be replicated into thousands of copies that greatly aided in the visualization aspects. In the 1980s microsatellite markers were discovered; these are repeats of very short DNA sequences scattered throughout the genome. In the 1990s, single nucleotide polymorphism (SNP) markers were discovered and were thought to be even more prevalent throughout the genome than microsatellites as more DNA sequencing occurred. In the early 2000s widespread DNA sequencing of cattle was under way, and the Bovine Genome Project provided for the first reported DNA sequence of the complete genome in 2006. Following publication and more in-depth study of the bovine genome sequence, many SNP markers were discovered, leading to the development of DNA analysis platforms referred to as SNP chips. Commercially available SNP chips became available with 50,000 SNP, and most recently high-density SNP chips with 700,000 to 800,000+ genotypes have become available (with continuously increasing affordability).

2.1.5 Methods to Study Evolutionary-Genetic Relationships

Relatedness across species has been studied by comparing chromosome gross morphology, chromosome banding patterns, protein amino acid sequence, genetic markers, and DNA sequence. It was studied originally through anatomical evaluation, and this was the process that Sir Charles Darwin used to help formulate his origin of species concept in the mid-1800s, and has been instrumental in archeological studies. In many instances and in many species of mammals, the mitochondria and the Y chromosome have been studied extensively to formulate evolutionary relationships as the mitochondria are inherited almost exclusively through the maternal lineage (as it is found in the cytoplasm of the oocyte and subsequent cells) and the Y chromosome is inherited exclusively in males from their male parent.

Sewall Wright developed the concept of effective population size as the number of parents in a population that contribute genes to the progeny generation. The effective population size dictates the rate of inbreeding even when random mating occurs and is related to both the total numbers of female and male parents as well as the ratio of female to male parents. High degrees of genetic relatedness within a population decrease the effective population size. The genetic effect of inbreeding is to reduce heterozygosity (or increase homozygosity as one is the complementary consequence of the other). The study of distributions of genotype frequencies, allele frequencies, gene flow, and general genetic influences of population dynamics is called population genetics, and Sewell Wright is known as the father of population genetics. Several of these concepts in conjunction with breeding and genetic applications are discussed in more detail in Chapter 3.

The bovine genome (nuclear DNA) was sequenced recently through the Bovine Genome Project, with the official results published in 2009 (Elsik et al., 2009). This report stated that the bovine genome was 2.87 billion base pairs in length and was composed of at least 22,000 protein-encoding genes. There is much utility in comparison of DNA, gene and protein sequences across species to better understand evolutionary biology, and the knowledge obtained in one species helps us learn more about the others. There have also been rapid developments in cattle genetic markers and associated testing because the DNA sequencing has provided for new assays including tests with 50,000 SNP markers, and very recently, high-density assays that may contain 700,000 to 800,000 markers. The cost of this DNA-based technology continues to decline, and some have speculated that individual sequencing of large numbers of animals may be common in just a few years. The statistical analyses and data management of these genotyping activities provide for formidable challenges.

2.2 Species Related to Cattle

All species of living organisms share certain characteristics. This degree of relationship can be compared across species, within species, across and within breeds, and even across and within families. As a result, the degree of relatedness among animals is relative to which base population they are being compared. The term systematics represents the study of the diversification and relatedness of living organisms through time and the term taxonomy represents the field of describing and naming organisms. The field of genetics where relationships among species and other populations is studied from an evolutionary standpoint is referred to as phylogenetics. An example phylogenetic tree is provided and described in Fig. 2.4. Table 2.1 describes the taxonomic groups to which cattle belong. All groups (family, genus, species) share some characteristics in common, but members across the same species have more characteristics in common than members across a genus, which in turn have more characteristics in common than members across a family.

The anatomical similarity among organisms within the same group corresponds to genetic similarity. In evolutionary biology, by studying the differences in DNA that exist today in conjunction with genetic changes assumed to have occurred at some mutation rate (i.e. one base pair change per 1000 years, etc.), it is possible to estimate the time it has taken to yield these differences. There are fossil records of even-toed ungulates (mammals with hooves) from 50 to 60 million years ago (Lenstra and Bradley, 1999), and the family Bovidae probably began its evolution in Africa around 19 million years ago (Huffman, 2014). The subfamily Bovinae has a common ancestor to approximately 14 million years ago, and the Bovini tribe had a common ancestor about 9–10 million years ago. These estimates vary due to the assumptions and techniques employed; as a result reports from individual studies in the literature may deviate from these values (which we cannot know with certainty). However, the concept of speciation is not a process that occurs at a single, short point in time, but is a gradual transition over time (possibly thousands of years). Table 2.2 provides a listing of scientific and common names of species in the Bovinae subfamily. Phylogenetic tree diagrams were mentioned earlier; these can help describe genetic relatedness. Dendrograms or cladograms are also terms that represent these concepts, and these can be expressed in a variety of ways (see Fig. 2.5).

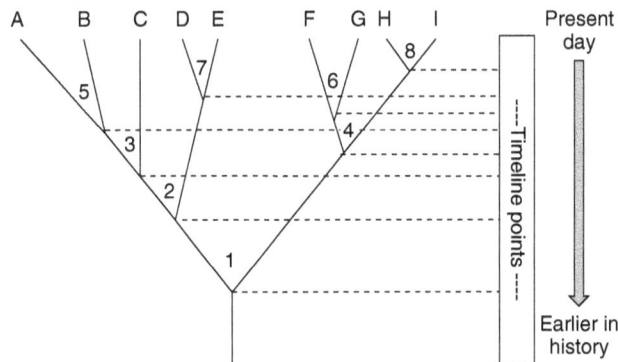

Fig. 2.4. Example of a phylogenetic tree or diagram, illustrating the degree of genetic relatedness across groups of animals. In this generic example groups A through E associate with one lineage and groups F through I associate with another (they diverged at time point 1). Further, groups that show more recent splits in lineage (such as D and E at time point 7) are more similar than groups that show splits earlier (such as D and A at time point 2, F and I at time point 4, etc.).

Table 2.1. Taxonomic levels of domesticated cattle.

Kingdom – Animalia
Phylum – Chordata (internal skeleton and dorsal tubular nerve cord)
Subphylum – Vertebrata (segmented bony backbone or vertebral column)
Class – Mammalia (hair and mammary glands that secrete milk)
Subclass – Theria (marsupials and placentals with nipples on mammary glands)
Infraclass – Eutheria (fetus develops entirely by means of placenta)
Order – Cetartiodactyla (even-toed ungulates, and whales)
Suborder – Ruminatia (stomach with three or four chambers and no upper incisors)
Infraorder – Pecora (stomach with four chambers; true ruminants)
Family – Bovidae (hollow horned; in general cattle-, antelope- and goat-like species)
Subfamily – Bovinae (cattle, spiral-horned antelope and nilgai)
Tribe – Bovini (bison, buffalo and cattle)
Genus – Bos (yak, gaur, banteng, kouprey, domesticated cattle)

Table 2.2. Listing of genus, species and common names of animals in the *Bovinae* subfamily.

Bison
  Bison bison – American bison (sometimes referred to as buffalo in North America)
  Bison bonasus – European bison, or wisent
Bos
  Bos gaurus – gaur, wild form of the domestic gaur (*Bos frontalis*), also known as gayal or mithan
  Bos indicus – the domesticated shoulder-humped cattle from Asian zebu descent
  Bos javanicus – banteng, wild form of Bali cattle
  Bos mutus – wild yak, wild form of the domesticated yak (*Bos grunniens*)
  Bos primigenius – aurochs, wild ancestral form of domesticated cattle, *Bos taurus* and *Bos indicus*
  Bos sauveli – kouprey, wild ox of Cambodia and Vietnam
  Bos taurus – domesticated, non-humped cattle
Boselaphus
  Boselaphus tragocamelus – Nilgai
Bubalus
  Bubalus arnee – wild Asian water buffalo (wild form of the domestic water buffalo, *Bubalus bubalis*)
  Bubalus bubalis – domestic water buffalo
  Bubalus depressicornis – Lowland anoa
  Bubalus mindorensis – tamaraw
  Bubalus quarlesi – mountain anoa
Pseudoryx
  Pseudoryx nghetinhensis – saola
Syncerus
  Syncerus caffer – African or cape buffalo
Tetracerus
  Tetracerus quadricornis – four-horned antelope, chousingha
Tragelaphus
  Tragelaphus angasii – nyala
  Tragelaphus buxtoni – mountain nyala
  Tragelaphus derbianus – giant eland
  Tragelaphus eurycerus – bongo
  Tragelaphus imberbis – lesser kudu
  Tragelaphus oryx – common eland
  Tragelaphus scriptus – bushbuck
  Tragelaphus spekii – sitatunga
  Tragelaphus strepsiceros – greater kudu

2.2.1 Bovini Species

In this section we briefly discuss the different species of the Bovini tribe. The genus of Bovini are the African buffalo (*Syncerus* spp.), the Asian buffalo (*Bubalus* spp.), the bison (*Bison* spp.) and the cattle (*Bos* spp.). Some physical, life cycle and chromosome related traits of these species are provided in Table 2.3. A general description of each species group follows as well.

2.2.2 African Buffalo

African buffalo are found in savannah areas south of the Sahara Desert, and are considered one of the most dangerous game animals. They are large and muscular and can be very aggressive. Forms of the African buffalo have never been domesticated, which may in part be related to its behavior. They have dark hair and light-colored horns that cover the cap (top) of the head. The horns tend to curl up as they proceed laterally away from the head, then upwards and back down toward the top of the head. There are different sub-species of African buffalo. *Syncerus caffer caffer* is the typical wild Cape buffalo (may also be called the savannah

buffalo) that has a characteristic black coat, while *Syncerus caffer nanus* is smaller and red or reddish in color, is much more rare and has been called the forest buffalo. *Syncerus caffer brachyceros* is the Sudan buffalo and is intermediate in size between *S. c. caffer* and *S. c. nanus*. The diploid chromosome number of *S. c. caffer* is 52 ($2n = 52$) whereas that of *S. c. nanus* is 54. There are four biarmed

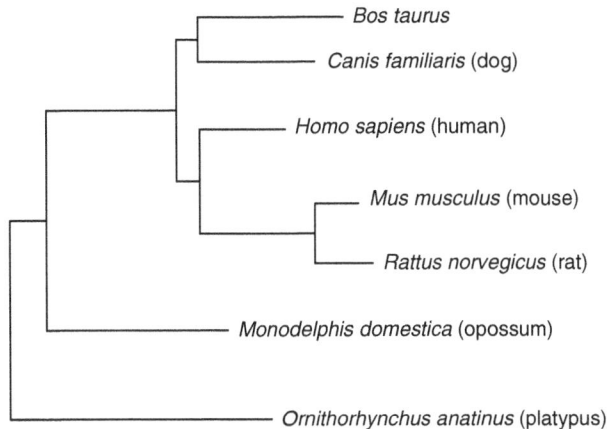

Fig. 2.5. Phylogenetic relationship of cattle with other mammalian species based on data from the bovine genome project. Taken from Elsik *et al.* (2009).

autosomes of *S. c. caffer* and three in *S. c. nanus*. The biarmed chromosomes in *S. c. caffer* correspond to fusions of cattle (ancestral bovid) chromosomes 1;13, 2;3, 5;20 and 11;29 (Gallagher and Womack, 1992). When referring to chromosomes in this manner, 1;13 indicates chromosomes 1 and 13, etc. These two sub-species can interbreed and will produce a viable F1 progeny, but the F1 is infertile due to chromosome imbalance ($2n = 53$).

2.2.3 Asian Buffalo

The wild form (*Bubalus arnee*) of Asian buffalo is known as the arni, but domesticated forms also exist. There are several species of Asian buffalo, with a major distinction between the anoa – a much smaller wild species – and the water buffalo (*Bubalus bubalis*), domesticated water buffalo, which has river and swamp forms. *Bubalus depressicornis* is the lowland anoa, *Bubalus quarlesi* is the mountain anoa and *Bubalus mindorensis* is the tamarao (or tamaraw). Domestication of the water buffalo may have occurred as early as 5000 BC (7000 years BP) in southern China, and/or possibly around 2500 BC (5500 years BP) in Mesopotamia. Water buffalo are known to be extremely calm and docile, with their daily care reportedly left to children in many cases. They are useful species for draft, milk and meat production. There are several breeds of both swamp and river buffalo and the milk from river buffalo is used in Italy to make mozzarella cheese. Porter (1991) stated that the river buffalo has a much higher milk yield, deeper

voice and prefers to wallow in deep, clean water, whereas swamp buffalo like to wallow in mud. River buffalo have longer faces, longer and more muscular legs and heavier heads and horns than swamp buffalo. There is considerable color variation in both river and swamp buffalo and many breeds of both types exist. River buffalo have 50 chromosomes ($2n = 50$) but swamp buffalo have 48 ($2n = 48$). An apparent translocation between river buffalo chromosomes 4 and 9 seems to correspond to swamp buffalo chromosome 1. River and swamp buffalo will interbreed but the F_1 has low fertility. River buffalo have five biarmed autosomes (submetacentric or metacentric), and all the other chromosomes including the X and Y are acrocentric (Iannuzzi and Meo, 2009). The biarmed river buffalo chromosomes correspond to translocations of cattle chromosomes 1;27, 2;23, 8;19, 5;28 and 16;29 (Iannuzzi and Meo, 2009). Tamaraw and mountain anoa have a diploid number of 46 ($2n = 46$); lowland anoa have $2n = 48$. All Asian buffalo have acrocentric X and Y chromosomes.

The buffalo are genetically distinct from the other species in the Bovini tribe. Crosses between African (*Syncerus*) and Asian (*Bubalus*) buffalo are not possible because the hybrid would have an unbalanced chromosome set, particularly because no biarmed chromosomes are shared between these two groups (Iannuzzi and Meo, 2009). Buffalo (African and Asian) cannot interbreed with *Bison* or *Bos* spp. It has been reported that fertilization can occur from water buffalo and cattle crosses but that embryos do not develop past the 8-cell stage.

Table 2.3. Some characteristics of species in the Bovini tribe.

Name	Gestation (days)	Lifespan (years)	Shoulder height (cm)	Adult weight (kg)	Diploid number (2n)	Shape of X chromosome	Shape of Y chromosome
Bison							
American bison	270–300	≤25	≤195	545–820	60	SM	A
European bison	254–272	≤27	180–195	800–1000	60	SM	A
Syncerus							
Cape buffalo (*S. c. caffer*)	340	18–20	130–170	600–900	52	A	A
Forest buffalo (*S. c. nanus*)	340	18–20	100–130	300–600	54	A	A
Bos							
Gaur	275	30	170–220	700–1000	58	SM	SM
Banteng	285	20	160	600–800	60	SM	SM
Kouprey	275	20	170–190	700–900	60?	SM?	SM?
Yak	258	20–25	145–180	300–1000	60	SM	SM
Bos taurus	283	15–25	120–160	350–1200	60	SM	SM
Bos indicus	290	15–25	120–160	350–1200	60	SM	A
Bubalus							
Tamarao	276–315	20–25	95–120	200–300	46	A	A
Lowland Anoa	276–315	≤30	85	≤300	48	A	A
Mountain anoa	276–315	20–25	≤75	≤150	46	A	A
River buffalo	305–315	15–25	130–150	400–900	50	A	A
Swamp buffalo	320–330	15–25	120–145	400–600	48	A	A

Taken from a variety of sources; shoulder height and hip height have a correlation of 0.80 to 0.90. A = acrocentric, SM = sub-metacentric; in some cases metacentric classification of the Y chromosome may have been used. In species where there are domestic and wild types, the wild type is usually larger. There are typically larger ranges in height, weight and color patterns among domesticated species. Lifespan reported is for total expected length of life as would be expected in wild populations (as opposed to age when utility ceases and culling follows as with domesticated species).

It has been estimated that the buffalo and the rest of the Bovini groups diverged approximately 5–6 million years ago.

2.2.4 Bison

There are two species of bison, the American bison (*Bison bison*), sometimes erroneously referred to as buffalo, and the European bison (*Bison bonasus*), also known as the wisent. The American bison was almost hunted to extinction during the last 30 years of the 19th century and therefore experienced a serious population bottleneck, but several thousand individuals exist today; there are also plains and woods varieties of American bison. The European bison went through an even more extreme bottleneck than American bison and only exists today in zoos and wildlife preserves. Physically bison have more massive forequarters (shoulder regions) than hindquarters and small, curved horns. Crosses of American and European bison are reported to be completely fertile. Both species have a diploid number of 60 ($2n = 60$) with a sub-metacentric X chromosome and an acrocentric Y chromosome.

2.2.5 Cattle (*Bos* spp.)

Animals of the genus *Bos* are known as the true cattle and include yak, banteng, gaur, kouprey and domesticated cattle. Yak, banteng and gaur have domesticated populations as well as wild populations. These groups are briefly discussed here individually, and much of the physical description and background discussion relies heavily on information from Porter (1991).

2.2.6 Yak

The scientific name of domesticated yak is *Bos grunniens*, but the scientific name of wild yak is *Bos mutus*; they are the same species as they fully interbreed. Yak may have been domesticated since the dawn of agriculture (10,000 to 12,000 years ago) in the area that corresponds to present-day Tibet, Nepal and Mongolia. They have heavy, long coats and are adapted to the harsh conditions at altitudes of 3000–5000 m. There can be considerable color variation in domestic yak although the wild yak have dark brown coats with a light muzzle ring. In domesticated yak color can range from brown to black to reddish-brown to white. They

can also be roan and can also have white spotting and have white faces. The tails of yak are covered in long hair like the rest of the body, unlike in other *Bos* species. They all appear to have dark pigmented skin. Their main use is as pack animals in many regions, but milk and meat are also important. They are large animals, with bulls much larger than cows, especially in wild yak. Yak make a grunting sound, which led to their species name; wild yak are reported to not make any sounds except during the mating season, and this also influenced their choice of species name. Yak have a hump due to long extensions of the vertebrae, unlike the hump of *Bos indicus* cattle which is mainly muscle. Some scientists (e.g. Nijman *et al.*, 2008) have recently suggested that yak should be removed from the *Bos* genus and given a new genus name because they appear to be more genetically distant to other *Bos* species. Yak have 60 chromosomes ($2n = 60$); the autosomes of yak are acrocentric whereas the X and Y chromosomes are sub-metacentric.

2.2.7 Gaur

Wild gaur have the scientific name of *Bos gaurus* but domesticated gaur were given the scientific name of *Bos frontalis*. Gaur may have been domesticated as recently as 500 BC (2500 years ago) in India, and domesticated gaur may also be known as gayal, mithan (or mithun) or dulong. As with other wild/domesticated forms of the same species, wild and domesticated gaur freely interbreed and suffer no reduction in fertility. Wild gaur are far less common than the domesticated form. Adult wild gaur are dark in color, with males having black hair and younger animals and cows having dark reddish-brown hair; calves may be a light golden color when very young. They also have white stockings on all legs and light color on the forehead. As with other species, there is much more color variation in mithun than gaur and animals may be predominantly white or white spotted with dark ears, eyes and muzzles. Also, the domesticated form has a less pronounced ridge on the front half of its back (due to shorter dorsal vertebral processes) than wild animals. Gaur are said to have a distinct smell from their sweat, which may help ward off parasites. They are reported to seek shade (as opposed to water or mud) to aid against heat stress. The gaur, banteng and kouprey have historically been placed in a separate sub-genus (*Bibos*) of

Bos as they have similarities with one another, and are quite distinct from yak and domesticated cattle. There are three gaur sub-species which show some chromosomal differences. Indian gaur (*B. f. frontalis*) have a diploid number of 58 (2*n* = 58), including 27 pairs of acrocentric autosomes, one pair of sub-metacentric autosomes (due to a fusion between ancestral cattle chromosomes 2 and 28), and sub-metacentric X and Y chromosomes, whereas the Indochinese gaur (*B. f. readei*) and the Malayan gaur (*B. f. hubbacki*) have 56 chromosomes with two pairs of sub-metacentric autosomes (due to an additional fusion between ancestral cattle chromosomes 1 and 29).

2.2.8 Banteng

Banteng have the scientific designation of *Bos javanicus* and the domesticated form is known as Bali cattle. It is believed that Bali cattle may have been domesticated around 3500 BC in Indonesia and/or Indochina. Wild banteng are not common, especially compared to their domesticated form, Bali cattle. Males are typically very dark brown, but in Burma and Indochina are reddish-brown, and this is the typical color in cows and calves. Bali cattle have the same color patterns as banteng, but some animals may also be lighter red or golden. All animals have a light or white muzzle ring, white stockings on their legs and a characteristic white oval patch across their buttocks. They have dark pigmented skin. Bali cattle are noted for having timid temperaments but are easily trained for draft. They are reported to have quite low milk production and very tender and desirable meat. Three sub-species of banteng have been described: *B. j. javanicus* in Java, *B. j. lowi* in Borneo, and *B. j. birmanicus* in Cambodia, Lao PDR, Myanmar, Thailand and Vietnam. Indonesian banteng (*B. j. javanicus*) has a diploid number of 60, but the Cambodian banteng has a diploid number of 56. Ropiquet *et al.* (2008) suggested that Robertsonian translocations (1;29) and (2;28) became fixed in the common ancestor of Cambodian banteng as a consequence of hybridization with the kouprey (*Bos sauveli*) during the Pleistocene epoch.

2.2.9 Kouprey

The kouprey (*Bos sauveli*) is extremely rare, was only first documented in 1937, and has been the subject of debate about whether or not it is a unique species, but recent molecular evidence has been reported that shows it is a unique species; there is not a domesticated version (at least that is known). Several authors have speculated that it may now be extinct. These are the wild ox of Cambodia, reported to be large but very graceful animals, built like a large deer or antelope, and were reported to be able to run quite fast. In color they may be mousy brown or grayish to dark brown, and they have white stockings on the legs. The horns are as long and widespread as any other bovine, and may be corkscrewed in conformation. Reports of any descriptions of the chromosomes of kouprey *per se* cannot be found in the literature.

2.2.10 Domesticated Cattle

Domesticated cattle (*Bos taurus* and *Bos indicus*) were developed from the ancient wild ox (aurochs, aurochsen plural, *Bos primigenius*) that existed across Europe from present-day Britain on the Atlantic coasts to the Asian Pacific coasts. Some of the most recent literature has used the terms *Bos taurus taurus* and *Bos taurus indicus* to refer to the two sub-species, but this is not universally accepted; the traditional terms *Bos taurus* and *Bos indicus* are therefore used throughout this book. The autosomes of cattle (*Bos* spp.) are acrocentric and the X chromosome is sub-metacentric; however, in *Bos taurus* the Y chromosome is sub-metacentric whereas in *Bos indicus* it is acrocentric. Figure 2.6 provides one example from the literature where the evolutionary relationships among Bovini species have been studied and the divergence times estimated.

2.2.11 Interspecies Crosses

Whether or not different populations of animals interbreed has been considered in the study and classification of species. If the degree of divergence and therefore genetic differentiation is great enough species will be reproductively isolated from others, meaning that they cannot interbreed and produce viable offspring. There are many different levels to consider with this concept. Some species can interbreed and produce embryos, but offspring do not survive to birth. Some species can interbreed and produce offspring that are viable and survive, but that are unable to reproduce themselves, as is typical when crossing horses and donkeys. Some species can interbreed where some of the hybrids are fertile, and some

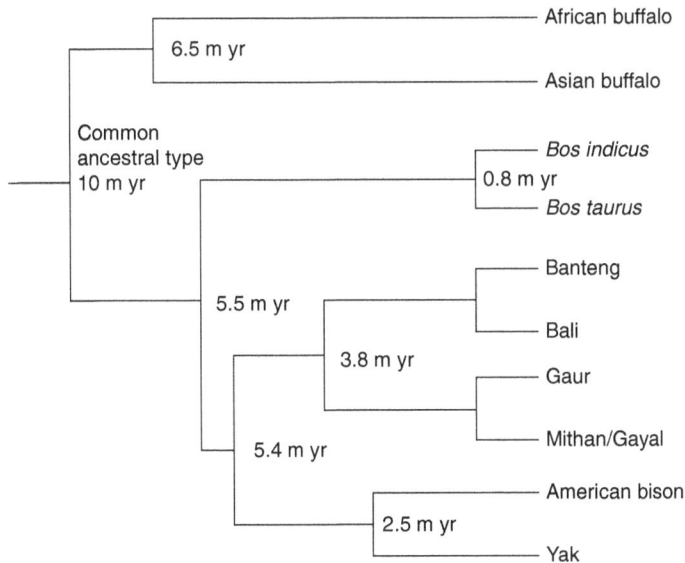

Fig. 2.6. Evolutionary time relationships among some cattle related species. Based on cytochrome b DNA sequence (from Tanaka and Namikawa, 2002).

are sterile. Haldane (1922) proposed that in crosses between species, it is the heterogametic sex that will be sterile; this has come to be known as Haldane's rule, and it is typical of the many interspecies crosses that have been reported among *Bos* and *Bison* species. All types of domestic cattle and related species such as bison, yak, banteng and Bali cattle have the same diploid chromosomal complement ($2n = 60$) and can be inter-crossed (Eldridge, 1985). *Bos taurus* or zebu when crossed with these other species have typically produced fertile F_1 females, but sterile males (Eldridge, 1985). However, the crosses between *Bos indicus* and *Bos taurus* are so fertile that they are superior in performance to either parent type in many environments, which indicates they should not be considered as different species, but as different sub-species; this same phenomenon of full interbreeding is true for crosses of American and European bison.

Some have speculated that the speciation of *Bos* and *Bison* spp. is incomplete because hybridization of these species occurs worldwide and the fertility of male hybrids can be restored by repeated backcrossing. Several species hybrids among *Bison* and/or *Bos* species have functional uses or desirability. The Chinese yakow is a yak–cattle hybrid, and provides an intermediate type between the altitudes

of its parental species. The South-east Asian Selembu cattle are a result of gayal–zebu crossing. The Madura racing cattle of Indonesia are hybrids of banteng and zebu cattle. Beefalo or cattleo are a hybrid of bison and *Bos taurus* cattle.

Yak and bison (and many breeds of *Bos taurus*) are well adapted to cold environments and climates, whereas buffalo, gaur, banteng/Bali and *Bos indicus* (as well as some *Bos taurus*) are adapted to warmer and tropical climates (although some *Bos indicus* cattle adapted to colder environments exist in Nepal and Mongolia). The utility of a population of animals (and/or their hybrids) in an environment outside of what it is adapted to is a critical sustainability (and productivity) consideration and is instrumental in all breeding programs, regardless of the species involved. Concepts associated with interspecies crosses might be considered further regarding terminal crossbreeding systems (Section 3.6) and development of beef supply chains (Section 13.5).

Elements that have contributed to speciation in history are probably still occurring today. Many of these are subtle, such as at the DNA sequence and/or expression level, but some are on a grander scale, such as chromosomal rearrangements. Chromosome mutations are possible and can result from changes within a chromosome (such as inversions, or deletions) and pieces of chromosomes being transferred to other chromosomes (as with insertions, or fusions). Robertsonian translocations, also called centric fusions because the chromosomes become fused at the centromere, seem to be the most commonly reported chromosome abnormality among bovine karyotypes. These genetic rearrangements may still be contributing to potential population differences (Tanaka *et al.*, 2000).

Speciation and domestication are two different concepts, but they are related from the standpoint that over time, genetic differences occur between isolated populations and within populations individuals tend to become more similar to one another. There can be population bottleneck events involved in speciation, domestication and breed development. In 2009, in conjunction with publication of the bovine genome sequence, the International HapMap Consortium assessed genome-wide SNP analyses to investigate structures of cattle breeds. The authors reported that cattle breeds over time have decreased in their effective population sizes (the opposite of what has happened in humans). This is not unexpected as there have probably been multiple population bottleneck events (a severe restriction in the parental population) and favored family lines have been propagated at the expense of others. The story of the domestication of cattle must be gleaned from the archeological and genetic information available at the current time. The history and migration of human cultures are closely tied to the development and use of tools, as well as both animals and plants.

2.3   Domestication

To say the least, the story of cattle domestication is complex, and there are several different speculated timelines and events that occur as a result of the different types of evidence used (archeological observations of animal materials, ancient artistic paintings and carvings, and molecular genetic information). Even among molecular genetic analyses involving DNA sequence, different types of sequence data (autosomal sequence, coding versus non-coding sequence, Y chromosome, mitochondria, etc.) as well as different input assumptions lead to different timeline speculations. It is important to consider all sources of available information simultaneously, but by nature that makes the analyses and their interpretations more complicated.

2.3.1   Domestication Processes

The auroch (*Bos primigenius*) is the direct ancestor of domesticated cattle. It has also been called the wild ox. Several prehistoric cave paintings in Western Europe show artists' interpretations of the appearance of these as well as many other animals such as woolly mammoths and horses. Based on these paintings, the aurochsen had long horns, were lean, apparently light muscled in the hindquarters and had medium-length hair with large skeletal size, perhaps 2 m (6 ft) tall at the shoulder (Porter, 1991). In the frame score discussion in Chapter 7, this would be equivalent to a frame score of 12 or 13, when modern cattle ideally have a frame score of 3 to 7 for most production environments (see Chapter 7 for discussion and calculation of frame score). In ancient times wild cattle were hunted for their hides, horns and meat, as were many other large game species. Aurochs existed from the Atlantic coasts of Europe to the Pacific coasts of Asia. Several populations of aurochs were thought to have existed. How different these auroch groups were genetically as opposed to geographically is not well understood, but large geographical separation over time probably led to the resulting genetic differences.

Exactly what possessed humans to capture some of these animals and try to control them as opposed to strictly hunting them is perplexing, as these were probably very formidable and dangerous animals. The process possibly began with raising some calves that were found or captured. It is possible that animals might have been captured as a sign of virility or as a ceremonial activity. We probably will never know for sure; however, the history of people and their interactions with cattle through time is quite interesting to consider. Bodó (2005) discussed the fact that in an area corresponding to present-day Hungary, transcriptions of ancient documents used different Latin names to describe hunters of wild cattle (*venator buorum*) as compared to hunters of wild calves (*venator bubalinorum*), as these occupations were different – they used different tools (lances and arrows for hunters of wild cattle vs traps for hunters of wild calves), indicating two different intentions, possibly catching and rearing captured calves. In a nice review of cattle domestication knowledge, Ajmone-Marsan *et al.* (2010) discuss some of these ideas and even provide a quote from Julius Caesar about the large size and speed of aurochs, and that even calves could not be tamed. The cattle domestication process is believed to have happened both in the Near East (South-west Asia), corresponding to present-day Iran and Iraq, as well as a separate event in South Central Asia (India). Some have speculated that an additional domestication process may have occurred in North Africa, but this is not generally accepted. Books such as *Cattle: An Informal Social History* (Carlson, 2001) provide interesting and thought-provoking insights as to how cattle have influenced societies (and vice versa) through time.

The domestication process itself is also probably much more complex than most people think at first. For instance, we often assume that when wild animals are captured and bred through human intervention there is a complete split with the wild population. This is probably not the case, however. Just think about how difficult it is at times to keep a neighbor's bull from making unwanted matings when we have fences and well-defined property lines. It is highly likely that there would have been intermixing over time between the 'wild' and the 'domesticated' populations when they both existed in the same environment. Domesticated cattle and their wild ox (aurochs) relatives coexisted for thousands of years in many regions, and the process of domestication was probably much more gradual and transitional than we may think. For instance, Ryder (1984) studied hair samples from 13th century aurochs, medieval aurochs and domesticated cattle, and modern Shorthorn. Ryder (1984) stated that auroch samples had outer and undercoats (as do many wild animals), and this may have been true of the species in general; he also said that the same primitive coat structure persisted in domestic cattle until at least as late as the Middle Ages, but that the typical modern cattle coarse coat pattern was seen in some cattle samples dated as early as 1000 AD. The intermixing of wild and domesticated populations may continue today in cattle-related species of yak, banteng–Bali, gaur–mithan and water buffalo in regions where both types co-exist.

From a genetic conservation point of view, introgression (introduction through breeding) of genetic influence into the wild species from the domesticated type of animal may compromise the genetic integrity of the wild population (Verkaar *et al.*, 2003), but this is probably better than loss of the wild population. Conversely, organized crossing of wild bovines in domestic populations may create animals, populations or even new breeds with

unique properties (Verkaar *et al.*, 2003). Anderung *et al.* (2005) identified auroch mtDNA in a sample of Bronze Age cattle and stated this was probably due to the recent interbreeding of aurochs and domesticated cattle. Bollongino *et al.* (2012) estimated that somewhere between 23 and 452 auroch females were used in the single Near East domestication event based upon DNA samples gathered on ancient and modern cattle in the Euphrates and Tigris valleys. Bollongino *et al.* (2012) also discussed that in regard to a domestication event within a region (not separate domestication events), a large number (such as several hundred or even thousands) of animals would be expected as founders if domestication was technologically straightforward and a non-exact, region-wide phenomenon, but that a smaller number would be consistent with a more complex and challenging process.

Gentry *et al.* (2004) stated that domesticated animals show characteristics of: (i) breeding is under human control or influence, (ii) provides product or service useful to humans, (iii) it is tame, and (iv) it has been selected away from the wild type. In the study of archeological or molecular evidence, these elements are unknown. Large-scale differences in behavior (or other traits) may not relate well to differences in skeletal size, or DNA sequences. It has been assumed that as the domestication process proceeded in cattle, animals became smaller in size up to a point where selection and/or nutritional management provided for an increase in size. In some instances the designation of ancient cattle as domesticated or wild has been based simply on the size of skeletal bones.

Another level of complexity in regard to the study of species related to cattle is that different scientific names have been assigned to the domesticated form versus the wild form even though they fully interbreed. This has occurred with yak, banteng, gaur and water buffalo. It is highly likely that aurochs and domesticated cattle interbred over a considerable time span. Table 2.4 summarizes some of this naming complexity for species of animals under consideration here.

2.3.2 Human History, Migration and Cattle Movement

Some discussion of different ages used to describe human history and time periods is warranted here. There are different naming systems used to describe

Table 2.4. Wild species and their domestic derivatives that traditionally have separate names as they have been referred to in different references.

Wild form	Domestic type
Bos primigenius Bojanus, 1827	*Bos taurus* Linnaeus, 1758
Bos namadicusa Falconer, 1859	*Bos indicus* Linnaeus, 1758
Aurochs of Europe, Asia and North Africa, extinct	Common domesticated cattle
Indian aurochs, extinct	Indian humped cattle or zebu
Bos gaurus H. Smith, 1827	*Bos frontalis* Lambert, 1804
Gaur of India, Burma and Malaya	Domestic gaur, gayal, mithan
Bubalus arnee Kerr, 1792	*Bubalus bubalis* Linnaeus, 1758
Indian water buffalo, arni	Domestic water buffalo
Bos mutus Przewalski, 1883 (yak)	*Bos grunniens* Linnaeus, 1766 (yak)
Yak of mountains of Tibet, Nepal and the Himalayas	Domestic yak
Banteng (*Bos banteng*)	Bali and similar cattle (*Bos javanicus*)

Adapted largely from Gentry *et al.* (2004).

human history as opposed to Earth or geological history. Descriptions of human history relate to types of artifacts found whereas geological history relates to types of rocks and fossils found in different layers of the Earth. Techniques such as radiocarbon dating may be used to estimate times for both of these systems. In geological time, ages, epochs and periods are terms to describe increasingly longer periods. For instance we are living in the Holocene epoch of the Quaternary Period, and within the Holocene epoch there have been different ages such as the Little Ice Age that occurred from the 1200s to 1300s through to the mid-1800s. The Quaternary Period has two epochs: the Pleistocene, which lasted from 2.6 million years ago to approximately 12,000 years ago (i.e. 10,000 BC) and the Holocene, which has lasted from the end of the Pleistocene to the present. Human history has generally been described with the three-age system (Stone Age, Bronze Age and Iron Age) for well over 100 years. These follow chronological order in general, but the degree of overlap and exact times for beginning, transition and ending of these periods vary across different areas of the world. The Stone Age is further broken down into Paleolithic, Mesolithic and Neolithic periods representing early, mid- and late Stone Ages and contain less sophisticated to more sophisticated tools. The development of agriculture in regard to domestication of plant and animal species occurred in the Neolithic period of human history and the Holocene epoch of geological history.

The Stone Age, so called because of the development and use of naturally occurring materials as tools (stone, bone, wood), lasted approximately 3.4 million years, and the degree of sophistication of the tools and their uses progressed through the Paleolithic, Mesolithic and Neolithic periods, respectively. The domestication of cattle (and other animal species) occurred in the Neolithic period, which began somewhere from 10,000 to 8000 BC (10,000 to 12,000 years ago). In general the Holocene epoch and the Neolithic period began at about the same time. This discussion is provided because scientific literature describing aurochs or ancient cattle for instance may refer to time periods with either the human or geological terms, and these terms are unfamiliar to most animal scientists or livestock producers (Fig. 2.7).

Cattle were distributed from points of domestication to all areas of the world by migrating tribes of people. There are no written records of how cattle were dispersed in ancient times, but the appearance of artifacts and fossil evidence in the archeological record has been used to postulate human migration patterns, development of technology and trade among cultures for many years (see Table 2.5). Anderung *et al.* (2005) evaluated mtDNA of Bronze Age cattle in Spain and found African T1 haplotype. They speculated that this occurred through contact of European and African cattle across the Straits of Gibraltar; archeological evidence has demonstrated Iberian–North African interactions of cultures. New DNA-based evidence has provided additional insights into genetic relationship and evolutionary development between and within populations, and this technology itself continues to improve and evolve.

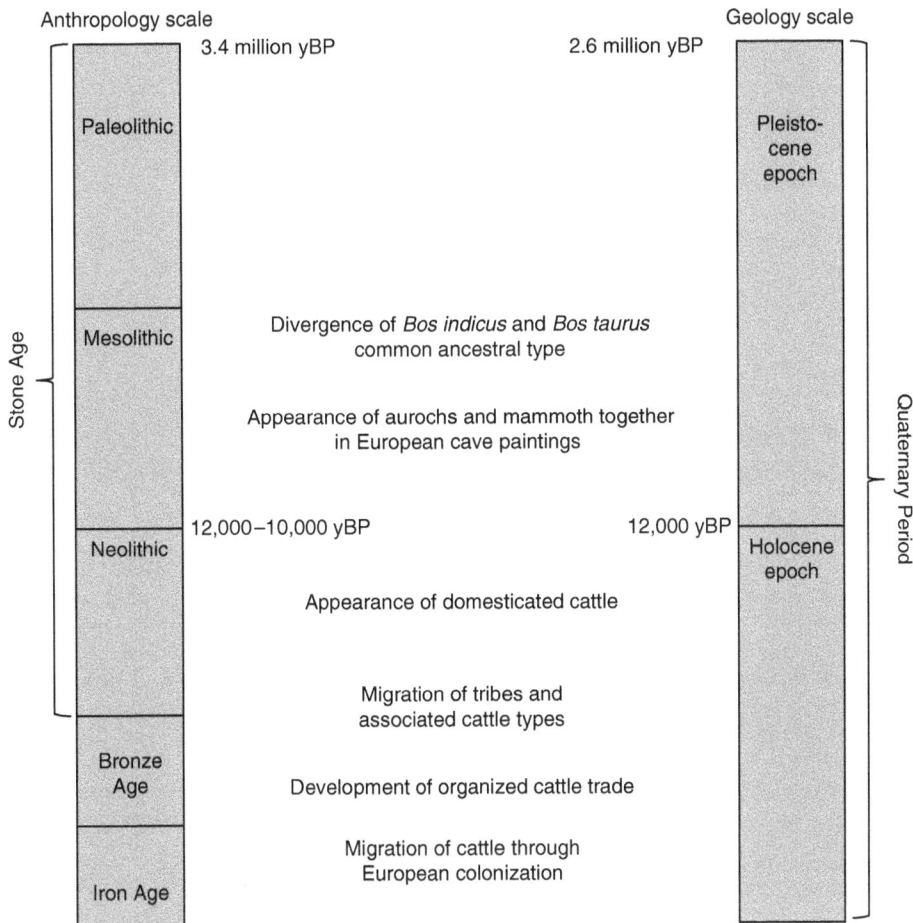

Fig. 2.7. Relative timescales for consideration of cattle domestication and migration history. Graphs are not drawn to scale.

2.3.3 Ancestral Types of Domesticated Cattle

The chronology in regard to the development of different types of ancient cattle had been based on archeological evidence until about 1980. Since then, molecular techniques have evolved and recently produced much new information about genetic relationships. The descendants of these ancient, ancestral types of cattle exist today. The discussions of cattle types based on archeological records are discussed here first, and then new molecular genetic information is incorporated. Because there is more diversity of opinion on the historical aspects of different cattle types, more

detailed citations are provided for much of this section compared to other sections.

The first ancestral type to appear was the long-horned, non-humped cattle. Evidence of these cattle first appears around 6400 BC (or 8400 years BP) in the region of the world that corresponds to the present-day Turkey–Iraq–Iran region, the Fertile Cresent area of the Near East. These were thought to be the most similar to aurochs than later cattle types. Wendorf and Schild (1994) reported archeological records of ancient cattle in the Sahara region of North Africa at several small sites in the Western Desert of Egypt, of large bovid bones thought to be domestic cattle and with radiocarbon

Table 2.5. Major cattle migrations across global regions.

Region	Time frame
India to South-west Asia along spice trade routes	3000 to 2500 BC
Zebu into African Horn region	2000 BC
Etruscan migration from Anatolia to Italian peninsula	900 to 500 BC
Roman cattle to Northern Europe	500 BC to 500 AD
Migration of Bantu tribes from North-eastern Africa to Southern Africa	700 AD
Zebu cattle into Horn of Africa after Islamic conquests	700 AD
Cattle from Spain and Northern Africa (criollo type) to Caribbean, North America, Central America and South America	1500s to 1600s
Portuguese criollo cattle to Brazil	1500s to 1600s
Northern Europe to north-eastern North America	1600s to 1700s
Zebu from India to Brazil	1800s

Some ancient migrations are proposed based on archeological and/or molecular data and remain under debate.
This is not meant to be a complete list, but provides some insight into when cattle may have been moved into global areas that previously did not have cattle.

dates ranging between 7500 and 6000 BC (9500 to 8000 BP), but it was not certain these were actually cattle bones; nonetheless, this has previously been used to speculate about the possibility that a separate, independent center for cattle domestication in north-east Africa could have been possible. This possibility has gone back and forth in its value as a hypothesis, but is not generally accepted. Bollongino *et al.* (2012) discussed, in regard to the Near East domestication event, that after an initial breeding phase that may have lasted 1500 years in an area between the Levant, central Anatolia and western Iran, domestic cattle started to appear in western Anatolia and south-eastern Europe by 6800 BC (8800 BP), in southern Italy by 6500 BC (8500 BP), and in Central Europe by 6000 BC (8000 BP).

The short-horned, non-humped cattle appeared next in the archeological records, at around 5000 BC (7000 BP) in the Iraq–Iran region. These were thought to be smaller in skeletal size, and were known to be used for both draft and milk. Following these were the humped cattle, which appeared in the Iran region around 3000 BC (5000 BP). There are two distinct types of humped cattle in history – the shoulder-humped (zebu, may also be referred to as thoracic-humped) developed in the Indus Valley of Asia (India), and the neck-humped (sanga, may also be referred to as cervico-thoracic-humped) that developed in and are native to Africa.

The Asian shoulder-humped cattle (zebu) had been thought by some to have developed from non-humped ancestral cattle (Epstein, 1971; Epstein and Mason, 1984) in the area now corresponding to Pakistan and India. However, Payne (1970) has stated that the first archeological evidence of zebu appears in Iraq near Mosul. The zebu breeds have physical distinctions from *Bos taurus* in that they possess a hump located over the shoulder, have different head shape, different ear length and set, different horn shape and set and usually more skin through the dewlap and underline (Payne, 1970; Rouse, 1970; Epstein, 1971). Additionally, the zebu breeds are distinguishable from *Bos taurus* by the presence of an acrocentric Y chromosome in the males, as compared to a sub-metacentric Y chromosome (Kieffer and Cartwright, 1968; Potter and Upton, 1979; Halnan and Watson, 1982). Later it was shown that Asian zebu (*Bos indicus*) resulted from a separate domestication event than non-humped (*Bos taurus*) lineage.

There are two distinct types of African humped cattle, the shoulder-humped (zebu) breeds and the cervico-thoracic (neck) humped (sanga) breeds (Epstein, 1971; Epstein and Mason, 1984; Porter, 1991). There have historically been two distinct theories about the development of the humped cattle of Africa. Most authors agree (Epstein, 1971; Epstein and Mason, 1984; Meyer, 1984) that non-humped cattle were domesticated approximately 8000–10,000 years ago (6000–8000 BC) and several had speculated that humped cattle evolved from non-humped ancestors. A difference in opinion has stemmed from archeological records and presumed origins of the sanga African breeds. Epstein and Mason (1984) stated that cervico-thoracic-humped cattle always precede thoracic-humped cattle from a chronological standpoint in archeological evidence, and were probably domesticated in the region that corresponds to modern-day Iran. Phillips (1961) and Meyer (1984) had both postulated that sanga cattle could be descendants of an evolutionary intermediate between non-humped cattle and shoulder-humped zebu. However, males of sanga breeds possess a sub-metacentric Y chromosome, as do males of *Bos taurus* breeds (Meyer, 1984). Epstein and Mason (1984) stated that cervico-thoracic-humped cattle were introduced into Egypt around 1500 BC (3500 BP) and that thoracic-humped cattle did not reach Africa until approximately 400 AD (1600 BP). Conversely, Payne (1970) and Oliver (1983) have stated that shoulder-humped zebu were introduced into northern Africa at roughly 2000–1500 BC. It is possible that in older literature there was some confusion as to what were classified as neck-humped versus shoulder-humped cattle. Both Payne (1970) and Oliver (1983) have stated that African sanga originated in the Ethiopia–Kenya area of East Africa and that the sanga breeds were developed from crossing of shoulder-humped zebu and non-humped cattle. This does not, however, explain the presence of the exclusive sub-metacentric Y chromosome in sanga males. These speculations had all been based on the assumption that a single domestication event (as in the Fertile Crescent region of the Near East) gave rise to all types of domesticated cattle, which has proven to be an incorrect assumption. MacEachern *et al.* (2009) evaluated autosomal DNA sequence and stated that *Bos taurus* breeds and Tuli (African sanga) may have diverged approximately 100,000–200,000 years ago; this result could indicate African and European cattle may have been separately domesticated, or that Tuli may contain genes originating in *Bos indicus*. The full interpretation of this postulation remains to be determined.

2.3.4 Historical Concepts of 'Breeds'

The concept of a 'breed' as we currently know it or think about it, where recorded pedigrees are utilized, did not come about until the mid to late 1700s in Great Britain and Western Europe. It was during this time of industrial revolution that more horses were continually replacing cattle for draft, improved farming techniques were leading to excess grain and forage production, and there was increased urbanization and industrial development (and more meat was desired for these growing urban populations). As a result, an 'all purpose' subsistence-type farm animal was not in as high and widespread demand, and this promoted the use of cattle for specialty traits, and to be thought about from an improved breeding concept. The Shorthorn in Great Britain was the first breed to have a recorded herd book, in 1822.

Many European colonies throughout the world had 'improved' cattle brought into them. However, in many instances the environmental and climatic conditions were extremely hard on the new livestock as they were not adapted to local conditions. In many instances the genetic influence of the local cattle populations was mostly or entirely lost as it was bred out.

2.4 Genetic Techniques to Study Breeds and Groups

Molecular genetic characterization of several types of both ancient and current cattle has been accomplished in recent years. In the 1970s, this was primarily through analyses of blood proteins, but became DNA sequence based in the 1990s. The publication of the bovine genome sequence has accelerated these types of analyses. Baker and Manwell (1980) studied allele frequencies across cattle populations of the world at ten blood protein loci. These authors reported the Asian zebu and *Bos taurus* cattle were very different across most loci, with African zebu intermediate to *Bos taurus* and Asian zebu, and sanga intermediate to African zebu and *Bos taurus*. In a related paper, Manwell and Baker (1980) constructed a phylogenetic tree which, when superimposed upon a map of Europe, Africa and Asia, showed the molecular characterization fitted well with the geographic distribution of the different types of cattle. Manwell and Baker (1980) also stated that the sanga were intermediate to *Bos taurus* and *Bos indicus* for the most part, but did have some unique characteristics that probably developed by adaptation. This process of superimposing genetic data on geographical maps has been employed in many subsequent studies. Loftus *et al.* (1994) utilized DNA sequence data to declare that *Bos indicus* and *Bos taurus* were derived from two distinct domestication events as their genetic differences were so great that they had to pre-date domestication by tens (if not hundreds) of thousands of years.

2.4.1 Mitochondrial DNA

There are two distinct lineages of mitochondrial DNA (mtDNA) in domesticated cattle (taurus and indicus), consistent with two separate domestication events. The haplogroups that have been characterized so far are T (taurus), I (indicus), P (European aurochs) and Q (some native Italian cattle). Among taurus types, these have been designated as T, T1, T2, T3, T4 and T5 haplogroups (see Fig. 2.8). Among indicus types, these have been reported to belong to two major haplogroups (I1 and I2). Within haplogroups, there may be several individual haplotypes. Achilli *et al*. (2008) estimated approximate divergence times from the T haplogroup to be 332,000 BP for I, 75,000 BP for P, and 52,000 BP for Q. Divergence was estimated to have occurred within the last 16,000 years among the T haplogroups and within the last 32,000 years with I haplogroups.

It had been thought that most European cattle resulted from the expansion of a small cattle population from the Near East after domestication, with the T3 haplogroup identified in the Near East and being the most common haplogroup among modern cattle in continental Europe. The T2 haplogroup is also thought to be of Near East origin. African cattle mtDNA lineages are almost exclusively associated with the T1 haplogroup, although it is rare in some cattle of the Near East and Anatolia. T4 has been found only in cattle of north-eastern Asia, and the T5 group has only recently been reported and is not yet well characterized. The recent divergence among the T haplogroup is consistent with the historical domestication timeline, but some have speculated about possible introgression of local wild aurochs into domesticated cattle. The earlier divergence time between the I1 and I2 haplogroups, both of which seem prevalent in India, have led some to speculate that domestication from multiple female lines of influence and/or introgression of Asian aurochs into domesticated populations may have occurred, but this remains unclear.

Lai *et al*. (2006) studied Asian cattle and found those from India, Nepal, the Philippines, and south and south-west China contained lineages belonging to the two haplogroups (I1 and I2) reported in *Bos indicus*. Cattle from north-west, west and central China did not contain any *Bos indicus* lineage assigned to haplogroup I2, but the Mongolian cattle sampled did contain lineage belonging to clade I1. Lai *et al*. (2006) also discussed that although Chinese cattle had both *Bos taurus* and *Bos indicus* influence, the relative levels were regionally distinct. Cattle from south and south-west China showed greater *Bos indicus* influence as compared to cattle from other parts of China. These authors speculated that *Bos indicus* mtDNA was introduced in Chinese cattle after the domestication event in

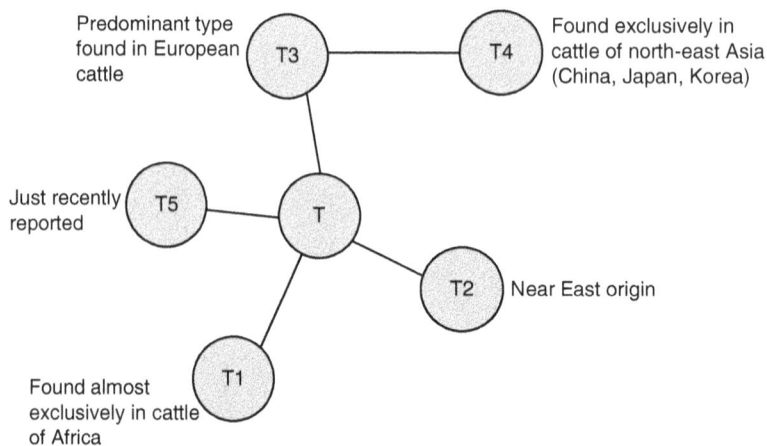

Fig. 2.8. General descriptions of *Bos taurus* haplogroups of mitochondrial DNA. European aurochs and domesticated *Bos taurus* mtDNA may have diverged 50,000 to 80,000 years ago. *Bos indicus* and *Bos taurus* mtDNA may have diverged 200,000 to over 300,000 years ago.

India, *Bos indicus* cattle were gradually introduced into south and south-west China, and then into other regions with those proportionally most distant from India having less influence.

2.4.2 Y Chromosome Analyses

Analyses involving the Y chromosome have been much less common than with mtDNA. Recent analyses have identified informative DNA sites on the cattle Y chromosome allowing the identification of three clusters, or haplogroups (Y1, Y2 and Y3) in domesticated cattle, with Y1 being more frequent in *Bos taurus* from north-western Europe, Y2 being dominant in *Bos taurus* found in southern Europe and Anatolian cattle, and Y3 being exclusive to *Bos indicus* (Götherström *et al.*, 2005; Perez-Pardal *et al.*, 2010). The Y2 haplotypes indicate taurus cattle domesticated in the Near East, and Y1 haplotypes are speculated to indicate European auroch origin, indicating a greater genetic influence of the European aurochs on modern cattle breeds in Europe (Beja-Pereira *et al.*, 2006; Achilli *et al.*, 2008; Perez-Pardal *et al,*. 2010), or an incomplete picture of domestication. Divergence times estimated among Y chromosome haplogroups have been much greater than among mtDNA haplogroups; estimated divergence times between haplogroups have been: Y1 and Y2 – 390,000 BP, Y1 and Y3 – 658,000 BP, and Y2 and Y3 – 376,000 BP (Perez-Pardal *et al.*, 2010). Götherström *et al.* (2005) found introgression of auroch Y chromosome sequence into domesticated cattle in Europe; however, analyses of ancient DNA from aurochs has not been shown to show influence of domesticated cattle back into the wild populations (Bollongino *et al.*, 2006).

Study of Y chromosomes has also been used to clarify speculation about interspecies influences in cattle. For instance, Madura (*Bos javanicus*) cattle in Indonesia have been used for many years in traditional ceremonial cattle racing or 'Karapan Sapi' (bull racing). Madura cattle have been presumed to have some degree of influence of *Bos indicus*. Winaya *et al.* (2008) evaluated Y chromosome microsatellite markers in a small number of animals that showed several alleles known to be of *Bos indicus* origin, but found no *Bos taurus* alleles. The authors concluded that the ancestry of Madura was indeed a blend of *Bos indicus* and *Bos banteng*.

2.5 The Concept, Role and Value of Cattle Breeds Today

The amount and degree of genetic information about existing and earlier types of cattle will continue to increase. Bradley *et al.* (1994) reported a population of N'Dama cattle (non-humped cattle in Western Africa) sampled in The Gambia to exclusively possess an acrocentric Y chromosome, due to crossing with zebu in the region. One N'Dama population in Guinea Bissau contained both sub-metacentric and acrocentric Y chromosome morphologies, whereas a population in Guinea contained only the sub-metacentric form (Bradley *et al.*, 1994). Perez-Pardal *et al.* (2010) stated that four of six Y3 (indicus) haplotypes identified in West African cattle clustered together and were not present in Asian zebu sires and could be evidence of a non-recent introgression of zebu into Africa, such as several thousand years as opposed to several hundred years.

Ginja *et al.* (2009) studied genetic diversity of Creole (criollo) cattle using mtDNA and Y chromosome markers to investigate potential origins of North and South American Creole breeds as well as contributions of breeds from the Iberian Peninsula, the Atlantic Islands and Continental Europe. Some major findings from Ginja *et al.* (2009) about South American cattle follow. African influence to some extent was detected in most breeds corresponding to Iberian and Atlantic Islands groups; African maternal line influence was detected in all Creole populations, whereas African paternal line influence was only found in Caracu and Chiapas. In Continental European and British breeds, African maternal line influence was detected in Charolais, Angus and Hereford, and African paternal line influence was found in British White and Jersey. No *Bos indicus* maternal line influence was detected in any cattle, including the zebu. No European breeds exhibited the *Bos indicus* Y3 haplogroup, but several Creole breeds did, including Texas Longhorn. The occurrence of 100% African-origin alleles on the Y chromosomes of the 20 Jersey bulls sampled is surprising.

Conclusions about breeds based only on molecular (especially a single type such as mtDNA, Y chromosome, autosomal sequence, etc.), on written or archeological evidence, or only on physical appearance should be viewed with caution; however, consideration of multiple types of information provides powerful insights into cattle domestication and breed development history. The underlying genetic differences between and within breeds provide useful genetic information to study evolutionary considerations as well as understanding performance differences. It must also be recognized that breeds and breed differences change over time.

2.5.1 Genetic Conservation

The loss of a breed is not the same concept as the loss of a species, but there are similar considerations. Once the genetic resource is gone, the mixture of genes and gene combinations in those populations are lost. The loss of cattle breeds (and breeds in other livestock species) should be of concern to anyone interested in genetic stewardship. Furthermore several types (or breeds) are simply lost over time due to their real or perceived lower productive value. In many nations, organized genetic conservation programs for domesticated species have been established. The need for this type of activity has become even more obvious as more numerous and detailed molecular genetic characterization has been performed in livestock populations. In many areas, breeds may have already become extinct, and many others (particularly local types in many global areas) are threatened. The banking of semen and embryos of these populations are needed. Simply banking DNA does not allow for new animals to be produced in the future. Simply storing semen and subsequent grading up to create high percentage animals do not provide for resulting animals that would be the same as the foundation population. As more local populations have been studied, many of these have shown large amounts of genetic diversity from others. It is possible that the 'genetic' solution to health problems in some breeds might be found by study of other, maybe even 'non-related' breeds. The unique, historical combinations of climate, pathogens, management, nutritional resources, cultures, etc. where native or long-existing types of animals have developed are likely to produce unique genetic situations.

2.6 Summary

Understanding the fundamentals of genetic theory provides background for understanding family, breed, biological group, and species differences.

Understanding the developmental history of populations of cattle in their native regions should explain to the reader how, over long periods of time, populations of cattle can become 'specialized' to specific environmental conditions. The stories of domestication events in cattle are probably very complex, as is the history and development of many local types of populations, and breeds in general; ongoing characterization of cattle breeds and related species continue to shed light as well as pose questions into the processes. Although interbreeding between groups of cattle is easier today than at any time in the past (due to artificial insemination, transportation, and trade in semen, embryos and live animals), family, breed and biological group differences provide genetic tools to tailor breeding programs in numerous ways. The realization that local populations (which are rare) may provide unique genetic resources is important, and the complete replacement of a local breed or type of cattle simply because it is viewed as 'non-productive' is a decision that needs serious consideration before its implementation is recommended (or accepted).

2.7 Study Questions

2.1 List five types of physical traits that may be useful in distinguishing different breeds or biological types of cattle. Are these similar or different traits that might be useful to distinguish between different species of animals?

2.2 Discuss the two types of genetic information that have been used to infer the maternal line of genetic influence and the paternal line of genetic influence in many genetic studies of populations.

2.3 If a breed of cattle (or any other livestock species) becomes endangered due to very low animal numbers, what are your thoughts on whether or not it should be conserved?

2.4 List four species of animals that exist today that are closely related to domesticated cattle.

2.5 What is a chromosome?

2.6 Do you think it is possible that different types or breeds of cattle today could have differing production characteristics because different cultures of humans with cattle may have had differing animal preferences? Explain why or why not.

2.7 How would you describe the concept of genetic variation?

2.8 References

Achilli, A., Olivieri, A., Pellecchia, M., Uboldi, C., Colli, L., Al-Zahery, N., Accetturo, M., Pala, M., Kashani, B.H., Perego, U.A., Battaglia, V., Fornarino, S., Kalamati, J., Houshmand, M., Negrini, R., Semino, O., Richards, M., Macaulay, V., Ferretti, L., Bandelt, H.J., Ajmone-Marsan, P. and Torroni, A. (2008) Mitochondrial genomes of extinct aurochs survive in domestic cattle. *Current Biology* 18, R157–R158.

Ajmone-Marsan, P., Garcia, J.F., Lenstra, J.A. and the Global Consortium (2010) On the origin of cattle: How aurochs became cattle and colonized the world. *Evolutionary Anthropology* 19, 148–157.

Anderung, C., Bouwman, A., Persson, P., Carretero, J.M., Ortega, A.I., Elburg, R., Smith, C., Arsuaga, J.L., Ellegren, H. and Götherström, A. (2005) Prehistoric contacts over the Straits of Gibraltar indicated by genetic analysis of Iberian Bronze Age cattle. *Proceedings of the National Academy of Sciences USA* 102, 8431–8435.

Baker, C.M.A. and Manwell, C. (1980) Chemical classification of cattle. 1. Breed groups. *Animal Blood Groups and Biochemical Genetics* 11, 127–150.

Beja-Pereira, A., Caramelli, D., Lalueza-Fox, C., Vernesi, C., Ferrand, N., Casoli, A., Goyache, F., Royo, L.J., Conti, S. and Lari, M. (2006) The origin of European cattle: evidence from modern and ancient DNA. *Proceedings of the National Academy of Sciences USA* 103, 8113–8118.

Bodó, I. (2005) From a bottle neck up to the commercial competition: Short history of Hungarian Grey cattle. In: *Proceedings 4th World Italian Beef Cattle Congress, Gubbio, Italy, 29 April–1 May*, pp. 152–154.

Bollongino, R., Edwards, C.J., Alt, K.W., Burger, J. and Bradley, D.G. (2006) Early history of European domestic cattle as revealed by ancient DNA. *Biology Letters* 2, 155–159.

Bollongino, R., Burger, J., Powell, A., Mashkour, M., Vigne, J.D. and Thomas, M.G. (2012) Modern Taurine cattle descended from small number of Near-Eastern founders. *Molecular Biology and Evolution* 29, 2101–2104.

Bradley, D.G., MacHugh, D.E., Loftus, R.T., Sow, R.S., Hoste, C.H. and Cunningham, E.P. (1994) Zebu-taurine variation in Y chromosome DNA: A sensitive assay for genetic introgression in West African trypanotolerant cattle populations. *Animal Genetics* 25, 7–12.

Carlson, L.W. (2001) *Cattle: An Informal Social History.* Ivan R. Dee, Chicago, IL.

Eldridge, F. (1985) *Cytogenetics of Livestock.* Avi Publishing Co., Westport, CT.

Elsik, C.G., Tellam, R.L. and Worley, K.C. (2009) The genome sequence of Taurine cattle: A window to ruminant biology and evolution. The Bovine Genome Sequencing and Analysis Consortium. *Science* 324, 522–527.

Epstein, H. (1971) *The Origin of the Domestic Animals of Africa*, Vol. I. Africana Publishing Corp., New York.

Epstein, H. and Mason, I.L. (1984) Cattle. In: Mason, I.L. (ed.) *Evolution of Domesticated Animals.* Longman, Essex, pp. 6–27.

Gallagher, D.S. and Womack, J.E. (1992) Chromosome conservation in the Bovidae. *Journal of Heredity* 83, 287–298.

Gentry, A., Clutton-Brock, J. and Groves, C.P. (2004) The naming of wild animal species and their domestic derivatives. *Journal of Archaeological Science* 31, 645–651.

Ginja, C., Penedo, M.C.T., Melucci, L., Quiroz, J., Martínez López, O.R., Revidatti, M.A., Martínez-Martínez, A., Delgado, J.V. and Gama, L.T. (2009) Origins and genetic diversity of New World Creole cattle: inferences from mitochondrial and Y chromosome polymorphisms. *Animal Genetics* 41, 128–141.

Götherström, A., Anderung, C., Hellborg, L., Elburg, R., Smith, C., Bradley, D.G. and Ellegren, H. (2005) Cattle domestication in the Near East was followed by hybridization with aurochs bulls in Europe. *Proceedings of the Royal Society B: Biological Sciences* 272, 2345–2350.

Haldane, J.B.S. (1922) Sex-ratio and unisexual sterility in hybrid animals. *Journal of Genetics* 12, 101–109.

Halnan, C.R.E. and Watson, J.I. (1982) Y chromosome variants in cattle *Bos taurus* and *Bos indicus*. *Annales de Génétique et de Sélection Animale* 14, 1–16.

Huffman, B. (2014) Ultimate Ungulate website. Available at: http://www.ultimateungulate.com (last accessed 9 April 2014).

Iannuzi, L. and Meo, G.P.D. (2009) Water buffalo. In: Cockett, N.E. and Kole, C. (eds) *Genome Mapping and Genomics in Domestic Animals.* Springer-Verlag, Berlin, Heidelberg, pp. 19–31.

Kieffer, N.M. and Cartwright, T.C. (1968) Sex chromosome polymorphism in domestic cattle. *Journal of Heredity* 59, 35–36.

Lai, S.J., Liu, Y.P., Liu, Y.X., Li, X.W. and Yao, Y.G. (2006) Genetic diversity and origin of Chinese cattle revealed by mtDNA D-loop sequence variation. *Molecular Phylogenetics and Evolution* 38, 146–154.

Lenstra, J.A. and Bradley, D.G. (1999) Systematics and phylogeny of cattle. In: Fries. R. and Ruvinsky, A. (eds) *The Genetics of Cattle.* CABI Publishing, Wallingford.

Loftus, R.T., MacHugh, D.E., Bradley, D.G., Sharp, P.M. and Cunningham, P. (1994) Evidence for two independent domestications of cattle. *Proceedings of the National Academy of Sciences USA* 91, 2757–2761.

MacEachern, S., McEwan, J. and Goddard, M. (2009) Phylogenetic reconstruction and the identification of ancient polymorphism in the Bovini tribe (Bovidae, Bovinae). *BMC Genomics* 10, 177.

Manwell, C. and Baker, C.M.A. (1980) Chemical classification of cattle. 2. Phylogenetic tree and specific status of the Zebu. *Animal Blood Groups and Biochemical Genetics* 11, 151–162.

Meyer, E.H.H. (1984) Chromosomal and biochemical genetic markers of cattle breeds in Southern Africa. In: *Proceedings. 2nd World Congress on Sheep and Beef Cattle Breeding*. Pretoria, Republic of South Africa.

Nijman, I.J., van Boxtel, D.C.J., van Cann, L.M., Marnoch, Y., Cuppen, E. and Lenstra, J.A. (2008) Phylogeny of Y chromosomes from bovine species. *Cladistics* 24, 723–726.

Oliver, J. (1983) Beef cattle in Zimbabwe, 1890–1981. *Zimbabwe Journal of Agricultural Research* 21, 1–17.

Payne, W.J.A. (1970) *Cattle Production in the Tropics. Vol. I. General Introduction and Breeds and Breeding.* Western Printing Services Ltd, Bristol.

Pérez-Pardal, L., Royo, L.J., Beja-Pereira, A., Chen, S., Cantet, R.J.C., Traoré, A., Curik, I., Sölkner, J., Bozzi, R., Fernández, I., Alvarez, I., Gutuérrez, J.P., Gómez, E., Ponce de León, F.A. and Goyache, F. (2010) Multiple paternal origins of domestic cattle revealed by Y-specific interspersed multilocus microsatellites. *Heredity* 105, 511–519.

Phillips, R.W. (1961) World distribution of the major types of cattle. *Journal of Heredity* 52, 207–213.

Porter, V. (1991) *Cattle: A Handbook to the Breeds of the World.* Facts on File, Inc., New York, NY.

Potter, W.L. and Upton, P.C. (1979) Y chromosome morphology of cattle. *Australian Veterinary Journal* 55, 539–541.

Ropiquet, A., Gerbault-Seureau, M., Deuve, J.L., Gilbert, C., Pagacova, E., Chai, N., Rubes, J. and Hassanin, A. (2008) Chromosome evolution in the subtribe Bovina (Mammalia, Bovidae): The karyotype of the Cambodian banteng (*Bos javanicus birmanicus*) suggests that Robertsonian translocations are related to interspecific hybridization. *Chromosome Research* 16, 1107–1118.

Rouse, J.E. (1970) *World Cattle*. Vol. II. University of Oklahoma Press, Norman, OK.

Ryder, M.L. (1984) The first hair remains from an Aurochs (*Bos primigenius*) and some medieval domestic cattle hair. *Journal of Archaeological Science* 11, 99–101.

Tanaka, K. and Namikawa, T. (2002) Genetic Diversity of Native Cattle in Asia in Present Status and Genetic Variability of Animal Genetic Resources in Asian Region. In: *Proceedings of the 10th NIAS International Workshop on Genetic Resources*. Tsukuba, Japan, pp. 53–62.

Tanaka, K., Yamamoto, Y., Amano, T., Yamagata, T., Dang, V.B., Matsuda, Y. and Namikawa, T. (2000) A Robertsonian translocation, rob (2:28), found in Vietnamese cattle. *Hereditas* 133, 19–23.

Verkaar, E.L.C., Vervaecke, H., Roden, C., Romero Mendoza, L., Barwegen, M.W., Susilawati, T., Nijman, I.J. and Lenstra, J.A. (2003) Paternally inherited markers in bovine hybrid populations. *Heredity* 91, 565–569.

Wendorf, F. and Schild, R. (1994) Are the early Holocene cattle in the eastern Sahara domestic or wild? *Evolutionary Anthropology* 3, 97–123.

Winaya A., Herwintono and Amin, M. (2008) Y chromosomal microsatellites polymorphism in Madura cattle (*Bos javanicus*). *Journal of Biotechnology Research in Tropical Region* 1, 1–4.

3 Breeding and Genetics

The fundamental principles on which genetic change are based include (i) that there are underlying genetic differences among individuals in a population and (ii) that these underlying genetic differences influence the phenotypes (performance) of individuals. The performance (or phenotype) influences the value of animals. Traditional breeding strategies have relied on gathering pedigree information and performance data.

Traits can be classified into two categories. For Quantitatively inherited traits there are many different genes influencing the trait, the individuals show a continuous range in phenotypes, and environment/management influences the phenotype. Most traits of primary economic importance in cattle production fall into this category (weight and size, muscle expression, behavior, fat deposition, etc.), although certain traits like female fertility may not be measured on a continuous scale (open vs pregnant, etc.). Qualitatively inherited traits are influenced by a small number of genes and have distinct categories that individuals can be placed into (red vs black, horned vs polled, spotted vs solid color, etc.). Typical environmental influences do not alter the phenotypes of these traits.

Cattle producers have many tools and approaches for genetic improvement although not all are available in all populations or regions. These tools include breed choices, family choices, breeding systems, genetic predictions such as estimated breeding values, collection of pedigree and progeny information, breed improvement programs through breed societies, government or industry agencies, and genetic tests. This chapter discusses the fundamentals of these concepts that can be helpful for genetic improvement; however, applications of several genetics concepts discussed in this chapter are also incorporated in other chapters.

3.1 Breeds and Biological Types

There are around 1000 different breeds of cattle globally. Most breeds are historically thought to have been deliberately developed with a specific purpose in mind and have evolved from various 'local' cattle types. The term 'breed' is used in this book to represent a population of animals that share a common ancestry, possess certain common physical characteristics (i.e. color, horn shape and size, presence and location of hump, etc.) and also share similar production characteristics related to body composition, size and growth potential, milk production potential, etc. In many cases, there may be as much or more variation in production characteristics within a breed versus between two breeds, especially if they are breeds that belong in the same biological type. Table 3.1 summarizes some biological types of cattle, and Table 3.2 provides example breeds for some of the biological types. Breeds within a biological type share many similar production characteristics, and the proper choice of biological type or types should be the primary genetic foundation for a beef cattle production system, with breed choice within biological type a secondary consideration. If a genetic type is utilized in a production environment to which it is not suited, the input costs are likely to be very high in order to obtain a desired level of production, or the level of production will be quite disappointing relative to the lower level of inputs. The concept that breeds (or biological types, different bloodlines, etc.) do not have consistent performance advantages (or disadvantages) across all environments is referred to as a genotype × environment interaction. There is obviously no one single breed of cattle that is superior for all traits of importance across all production environments.

Not all of the terms used in Table 3.1 to describe cattle groups are exclusive of others. It is likely that combinations of these terms often provide the most meaningful descriptions for several breeds or types (*Bos indicus* dairy, British beef, criollo dual purpose, etc.). Many breeds have been taken from their area of origin and been further developed or changed for local conditions and it is possible that there may be differences among animals of the

Table 3.1. Classification categories of domesticated cattle.

Term	Description and background
Classification based on sub-species	
Bos taurus	Non-humped breeds of the world, originally from Middle East domestication event.
Bos indicus	Shoulder-humped breeds of the world, also known as zebu (or cebu) originally from Indian subcontinent of Asia and generally viewed as highly tropically adapted.
Classification based on geography	
British	Breeds from British Islands off European continent developed for meat production.
Caucasian	Short-horned ancient type of *Bos taurus* cattle from Mediterranean, Middle East, Eastern and South-eastern Europe and Northern Africa.
Chinese yellow cattle	Indigenous domestic cattle of China, not a reference to coat color. Common yellow cattle are typically non-humped; yellow cattle further south have some evidence of a hump.
Continental European	Breeds from Western and Central Europe used for meat production but many had ancestors also used for draft; several are dual-purpose meat and milk.
Criollo	Tropically adapted *Bos taurus* cattle of North and South America whose ancestors originally came from the Iberian peninsula of Spain and Portugal.
Gray Steppe	Long-horned, light haired and dark pigmented *Bos taurus* cattle of Eastern Europe considered an ancient type.
Podolian	Breeds of Central Europe, primarily Italy, that were derived from eastern steppe cattle origin and include Apulian, Chianina, Maremmana, Marchigiana, Piedmontese and Romagnola.
Sanga	Cattle native to Africa that possess a neck hump, originally in Eastern and Southern African continent, some have classified sanga as *Bos taurus* because these breeds possess a sub-metacentric Y chromosome.
Western African humpless	Non-humped breeds that are indigenous and survive in the tsetse fly region of West Africa and include Baoulé, N'Dama, Kuri and Lobi.
Japan/Korea	Wagyu/Hanwoo East Asian breeds known for extremely high levels of intramuscular fat deposition ability.
Classification based on purpose, use or ability	
Beef	Cattle produced for the primary purpose of meat production.
Dairy	Cattle produced for the primary purpose of milk production.
Draft	Cattle used primarily for farm-related work and transportation.
Dual purpose	Cattle intended for two major purposes such as milk and beef, milk and work, etc.
All purpose	Cattle used for all (general use) purposes as historically with subsistence agriculture.

same name that rival differences across several breeds. Furthermore, the same breed in name at an earlier point in time (1880s, 1950s, etc.) may not be the same type of animal as at a later point in time, particularly if the breed is responding to some form of selection pressure (natural and/or artificial). Table 3.2 provides some examples of breeds in relation to various biological cattle types.

There are additional biological groups that could be included here, but many of these would have similar production and adaptation characteristics to the ones used here. There are many ways that breeds and types could be grouped. Understanding the basic aspects of the characteristics of (i) biological groups, then (ii) characteristics of individual breeds, and subsequently (iii) evaluation of family lines within breeds is needed to best match cattle

types with resources and obtain production goals – they provide the foundation of cattle breeding and crossbreeding programs.

Selection across cattle breeds as well as selection within breeds is important to efficiently utilize genetic resources in beef production (Cundiff *et al.*, 1986). Across breed selection becomes increasingly important as different production environments are considered (Frisch and Vercoe, 1978; Vercoe and Frisch, 1992).

3.2 Mendelian Genetic Concepts

Many of the concepts of quantitative genetics are extensions of the single locus (physical location on a chromosome; loci plural) model first proposed by Gregor Mendel (1865). We know now that many biological traits are influenced by several gene

Table 3.2. Example grouping of breeds according to biological group (excluding recent composites).

British beef	Zebu[a]	Continental European dual purpose	Sanga	European dairy (Bos taurus)
Angus/Red Angus, Hereford	American Brahman, Indu-Brazil	Braunvieh, Pinzgauer	Africander (Afrikaner), Mashona	Ayrshire, Milking Shorthorn
Belted Galloway, Highland	Boran, Kankrej (Guzerá)	Gelbvieh, Salers	Ankole, Nguni	Guernsey, Norwegian Red
British White, Shorthorn	Fulani, Ongole (Nelore)	Maine Anjou, Simmental	Barotese, Nkone	Holstein–Friesian, Swedish Red and White
Devon, South Devon	Gir, Red Sindi	Normande, Tarentaise	Bapedi, Tuli	Jersey, Swiss Brown
Dexter, Galloway	Hariana, Sahiwal		Dinka, Tswana	Ayrshire, Milking Shorthorn

Criollo[b]	Continental European beef	Common Chinese yellow	Caucasian	Eastern European Gray Steppe
Blanco Orejinegro, Mexican Corriente	Belgian Blue, Marchigiana	Jinnan, Nanyang	Anatolian Black, Buša	Hungarian Gray, Turkish Gray
Caracú, Romosinuano	Blonde d'Aquitaine, Piedmontese	Kazakh, Qinchuan	Baladi, Greek Shorthorn	Mursi, Ukrainian Gray
Chaqueño, San Martinero	Charolais, Romagnola	Luxi, Tibetan	Brown Atlas, Kurdi	Romanian Steppe
Chusco, Texas Longhorn	Chianina	Mongolian, Yanbian		
	Limousin			

[a]Some of these would be considered dual purpose or dairy type.

[b]Although the term criollo (or criolo) generally refers to cattle developed in the New World from imported Spanish and Portuguese cattle, several breeds in present-day Spain and Portugal such as Barrosã, Morucha and Retinta, etc. would also be in this category.

Examples of some composite breeds regarding country of origin (and component breeds) include: Australia – Belmont Red (Africander, Hereford, Shorthorn); Brazil – Canchim (Charolais, Indu-Brazil), Girolando (Holstein–Friesian, Gir), Ibagé (Angus, Nelore), Tabapuã (Indu-Brazil, Nelore); South Africa – Bonsmara (Africander, Hereford, Shorthorn), Drakensberger (Dutch Groningen, indigenous black cattle); Ukraine – Red Steppe (Red East Friesian, Ukrainian Gray), Southern Ukrainian (Charolais, Hereford, Red Steppe); USA – Barzona (Africander, Angus, Hereford, Santa Gertrudis), Beefmaster (Brahman, Hereford, Shorthorn), Braford (Hereford, Brahman), Brangus (Angus, Brahman), Santa Gertrudis (Shorthorn, Brahman), Senepol (N'Dama, Red Poll), Simbrah (Simmental, Brahman).

loci, with the potential of several alleles per locus. The concept of what a gene is has changed considerably since its 'smallest unit of inheritance' idea was established, and continues to evolve. Many of the concepts of animal breeding and quantitative genetics are abstract in that we cannot actually observe them in animals (such as with a breeding value) or because they are properties of populations (as with heritability and variation) and not individuals.

At any single locus, there is a possibility of n different alleles, but any given individual will possess two alleles (one from each parent). An allele is an alternate form of a gene. The individual may have received the same allele from both parents, making it homozygous, or two alternate forms, making it heterozygous. If there are n alleles in the population, there will be $n(n + 1)/2$ different possible genotypes, of which n will be homozygotes, and $n(n - 1)/2$ will be heterozygotes (see Table 3.3).

This also assumes that genotypes A_1A_2 (A_1 inherited from male parent, A_2 inherited from female parent) and A_2A_1 (A_2 inherited from male parent, A_1 inherited from female parent), etc. are the same, which may not always be the case; genomic or parental imprinting is when there is a different degree of expression for an allele when it is inherited from the sire or dam exists for certain gene loci. The number of genomic regions that show these 'parent-of-origin' effects seem to be in the minority of loci, but more are being reported in most mammal species. The impact of these 'non-Mendelian' effects on most production traits in cattle is unknown, but is likely to be fairly small because many principles such as estimation and application of breeding values have been shown to change breeds (more on this in Sections 3.3 and 3.4).

3.2.1 Mendel's Laws of Segregation and Independent Assortment

Mendel proposed that traits were controlled by factors (genes) that remain discrete across generations. In order for this to happen, genes (alleles) must segregate (separate) from each other from genotypes to individual alleles (i.e. going from diploid concept to haploid concept), so that traits in parents could be preserved in progeny, and later generations. Mendel also proposed that genes controlling different traits segregated independently. The presence of one allele at one locus was not related to the presence of another allele at a second locus. It is now known that this may not be true in many instances because alleles that are in close proximity on the same chromosome have a higher chance of being inherited together (i.e. of being packaged together in the same gamete) compared to loci that are further apart on the same chromosome. It is assumed that alleles from different loci on different chromosomes are always inherited independently.

When thinking about multiple gene loci, the concept of genotype (i.e. $A_1A_2B_3C_2C_2$) takes on a modified meaning. It is easy to see how the possible number of genotypes can become huge numbers when we consider multiple loci and multiple alleles. If there are k genotypes per locus and t loci involved, the total number of possible genotypes is k^t. Earlier it was stated that the number of genotypes per locus was $k = n(n + 1)/2$, and so the number of genotypes across loci = k^t. Table 3.4 shows how the total number of genotypes can approach very large numbers.

Although both the number of loci and number of alleles per locus contribute to the concept of quantitative

Table 3.3. Number of genotypes, heterozygotes and homozygotes at a single gene locus with n alleles.

n # alleles	$n(n + 1)/2$ # genotypes	$n(n - 1)/2$ # hets	n # homs
1	1	0	1
2	3	1	2
3	6	3	3
4	10	6	4
5	15	10	5
6	21	15	6
10	55	45	10
20	210	190	20
50	1,275	1,225	50
100	5,050	4,950	100

Table 3.4. Examples of number of genotypes possible with multiple gene loci.

Number of loci (t)	Number of alleles per locus (n)	Number of genotypes per locus (k)	Total number of different genotypes
1	2	3	3
2	2	3	9
3	2	3	27
4	2	3	81
2	4	10	100
3	3	6	216
3	4	10	1,000
4	3	6	1,296
5	3	6	7,776
4	4	10	10,000
5	4	10	100,000

genetic considerations, it is primarily the multi-locus concept that is thought to be responsible for the continuous genetic variation associated with quantitatively inherited traits. Figure 3.1 illustrates the concept of the normal distribution, and both environmental as well as genetic variation sources lead to continuous distributions of phenotypes.

3.2.2 Gene and Genotype Frequencies

Because we are dealing with effects of genes in populations, the frequencies referred to will be population frequencies, which in fact are estimated from samples in many instances. The population frequencies can then be viewed as probabilities used to define distributions of genes and genotypes. Gene frequency (also called allele frequency) is the common, historical term used to denote the frequency of an individual allele in a population. Gene frequencies are based on the genotype distributions in the population and can be calculated on genotype frequencies directly or on genotype counts (which is the same concept). Table 3.5 illustrates calculations of genotype and allele frequencies.

A genotype frequency is the frequency of individuals possessing a certain genotype in a population, and therefore it is also the probability that a randomly sampled individual from that population will possess that genotype. As with the genotype frequency concept, the gene (or allele) frequency is the frequency of all alleles in a population that are that particular allele, and therefore it is also the

probability that a randomly sampled gamete from that population will possess that allele.

3.2.3 Hardy–Weinberg Equilibrium

Although the gene frequencies can always be calculated from genotype frequencies, only under certain specific circumstances can the genotype frequencies of a population be calculated directly from its gene frequencies. In a random mating, diploid population of large size, a relationship can exist between the gene frequencies and the genotype frequencies, and this is referred to as the Hardy–Weinberg equilibrium (HWE). The assumptions for the HWE are: (i) no selection, (ii) no migration, (iii) no mutations, (iv) autosomal inheritance, (v) infinite population size, (vi) random mating. Under these conditions, the genotype frequencies are a direct function of the allele frequencies, and are called Hardy–Weinberg proportions, and the gene and genotype frequencies remain constant across generations.

Consider a single locus with two alleles where p is the frequency of one allele (A), and q is the frequency of the other allele (a). The Hardy–Weinberg genotype frequencies will be p^2 ($p \times p$) for one homozygote, q^2 ($q \times q$) for the other homozygote and $2pq$ ($p \times q + q \times p$) for the heterozygote. The letters p and q are analogous to f(A_1) and f(A_2) in Table 3.5. Think about all the gametes that come from the male parents, all the gametes from the female parents, and all the ways these alleles could get paired up at fertilization. Here $P(A)$ refers to the probability of

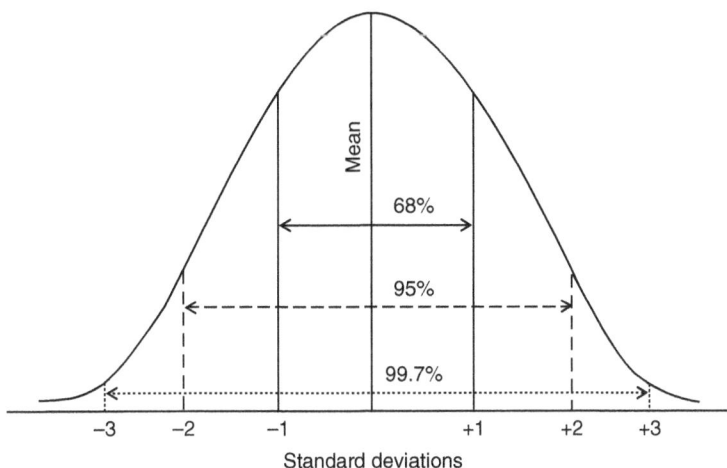

Fig. 3.1. Concept of normal distribution. The normal distribution is continuous, bell-shaped, centerd at the mean, and has defined proportions of the population within 1 (68%), 2 (95%) and 3 (99.7%) standard deviations of the mean.

Table 3.5. Genotype and allele frequency concepts.

Genotype considerations

Genotypes:	A_1A_1	A_1A_2	A_2A_2	Sum
Conceptual frequencies:	$f(A_1A_1)$	$f(A_1A_2)$	$f(A_2A_2)$	1
Count (n):	50	75	32	157
Actual frequencies:	50/157 = 0.3185	75/157 = 0.4777	32/157 = 0.2038	1

The genotype frequencies must total 1 (100%) as the sum includes all the genotypes.

Allele considerations

Allele designations:	A_1	A_2	Sum
Conceptual frequencies:	$f(A_1)$	$f(A_2)$	1
Allele frequencies as function of genotype frequencies:	$f(A_1A_1) + 0.5\, f(A_1A_2) =$ 0.3185 + 0.2389 = 0.557	$f(A_2A_2) + 0.5\, f(A_1A_2) = 0.2038 +$ 0.2389 = 0.443	1
Allele frequencies as function of genotype counts:	(2*50 + 75)/2*157 = 175/314 = 0.557	(2*32 + 75)/2*157 = 139/314 = 0.443	1

There are 314 alleles among the 157 animals because each individual animal has two alleles.
The frequency of the A_2 allele therefore must be 1 – 0.557 = 0.443 because the frequencies of all the alleles must sum to 1.

a parent (or group of parents) passing along the A allele, and P(a) refers to the probability of a parent or group of parents passing along an a allele. The frequency of the allele in the parental group is the same as the probability that a randomly sampled gamete from the parent (or parent group) contains that allele. Only one generation of random mating is required to establish HWE proportions in the resulting progeny generation.

3.3 Forces that Change Allele Frequency

The fundamental processes that have been thought to produce genetic change within and across populations over time give rise to changes in allele frequencies, and therefore other subsequent genetic differences. The four processes recognized to change gene (allele) frequency are genetic drift, migration, mutation and selection. These are briefly discussed below. Parts of this discussion rely heavily on material in Falconer and MacKay (1996).

3.3.1 Chance or Genetic Drift

Chance or genetic drift is associated with small population size. By chance, in populations containing few individuals, females may be mated by a sample of males that possess gene frequencies that differ markedly from those of the population and as a result, their progeny will possess gene frequencies that differ from the parental population. As a result, the allele frequencies can fluctuate randomly across generations. Change in gene frequency from genetic drift is not directional (has equal probability of fluctuating upward or downward in each generation).

3.3.2 Migration

Migration is the introduction of individuals from a second population into the original population to define a new composite population. It should be recognized that for gene frequency to change in one population, there must be a difference in gene frequency between the two populations. And, if there is a difference, the magnitude of gene frequency change due to migration is determined by the rate of exchange as well as the difference in gene frequency between the two populations. Movement of breeding animals between populations that do not differ in gene frequency does not result in genetic migration. In cattle breeding, the effect of migration is achieved through the use of crossbreeding, or through introductions of individuals or family lines of the same breed, which may have been selected for the same (or other) traits in another population. Cattle that moved with settlers and explorers throughout human history resulted in genetic migration into indigenous populations.

More recently, this has been accomplished mainly through artificial insemination and embryo transfer. Using imported bulls that have come from a population selected for performance in a temperate location has been attempted to facilitate rapid genetic change in other 'unimproved' populations.

3.3.3 Mutation

Mutation is the spontaneous change of one allele to another form, and this process results in a new genotype. Mutation may be induced by a number of factors, including certain chemicals and radiation. Mutation rates are so small that they are usually of very little importance to quantitative geneticists. Gene frequency changes so little because of mutation that it can be safely ignored, at least when it first arises. Most new mutations are undesirable (and may even be called deleterious) because they disrupt an already existing and functioning gene; as a result new mutations are probably removed from the population (or kept at very low levels) by selection (natural or artificial). Important exceptions to this are major genes that are deemed as desirable. This is thought to have happened with the mutation in the horned allele that gave rise to polled cattle. The mutated form of the myostatin gene produces the double-muscled phenotype seen in Belgian Blue and Piedmontese (and other) breeds, and there are many different mutated alleles for this gene locus. Whether or not the double-muscled phenotype is desirable depends upon many management and market factors.

3.3.4 Selection

Selection can be thought of as the process that determines which individuals produce progeny and how many progeny individuals produce, and this concept is expanded on below because of its importance both within purebred and crossbred situations. Both migration and selection are very important tools that cattle breeders possess to alter population gene frequencies. Selection may be the only tool in many circumstances, as with a closed, purebred herd. It should also be remembered that there are both natural (imposed by nature) and artificial (imposed by humans) selection occurring within cattle populations, but the degree of both of these elements can be highly variable.

The impact of selection is to alter gene frequencies (and therefore genotype frequencies). In many cases, selection may be to promote a desirable qualitative

trait such as horned or polled, or a reduction in incidence of a genetic disease. The rate at which the frequency of a recessive gene can be reduced in a population depends on the initial frequency of the gene and the selection pressure that can be applied to remove the recessive homozygotes. It is for this reason that it is critical that great care be exercised in the implementation of breeding programs to ensure that deleterious recessive alleles do not accidentally (due to use of carrier individuals in a large proportion of matings) rise to unacceptably high frequencies in a population. Selection of individuals from new populations warrants considerable investigation beforehand, as there can be resulting large (unintended) negative as well as positive effects. The concept of selection at a single gene locus is extended to multiple loci for quantitatively inherited traits.

3.4 Selection for Quantitatively Inherited Traits

The concept of variation is very important, and variation simply means differences. Some traits such as color show distinct values (black, red, white on the face, striped, gray, white, etc.) while other traits such as weight show continuous values where the level of precision dictates the scale that we record that trait (per 10 kg, per 1 kg, per 0.1 kg, etc.). For convenience certain continuous traits may be grouped into classes (small vs moderate vs large mature size; body condition score, sheath score, etc.). Variation in continuous traits can be caused by genetic differences, environmental differences, and interaction between genetics and environment.

These types of traits have been referred to as quantitatively inherited traits, and the classical quantitative genetics model has been: $P = G + E$, where P = performance or phenotype, G = genetics or genotype, and E = environment. So, performance differences are due to genetic differences and environmental differences (and possibly interactions of $G \times E$); furthermore, anything not classified as a genetic influence by default is classified an environmental influence. It is critical to be able to separate environmental from genetic influences for making genetic changes, as with the contemporary group concept and its proper designation; this is discussed in detail with examples in Chapter 13 as genetic management without genetic information is not possible. The G concept also needs to be expanded because genetic influences may be predictably passed from parents in one generation to progeny in the next generation, or they may be due to genetic combinations that depend upon the combination of parents. The predictable part has been referred to as breeding value, additive genetic value or gene content, whereas the genetic combination part has been referred to as hybrid vigor or heterosis. Figure 3.2 illustrates the components of phenotypic variation and its underlying components.

3.4.1 Heritability

The term heritability (represented by notation h^2) describes the percentage of the phenotype differences (phenotypic variation) that is attributed to breeding value (gene content) differences (called the genetic variation). The heritability value is directly related to how quickly a trait may be changed from selection because the phenotypes of animals are more closely related to their breeding values as the heritability increases. The heritability value can vary substantially across traits, and can vary across breeds for the same trait. Tables 3.6 through 3.8 show some heritability values and ranges that have been reported over the years for several important beef cattle production traits. The potential range in heritability for any trait is 0 to 1 (0% to 100%).

Selection has both 'natural' and 'artificial' considerations. Natural selection refers to the survival of the fittest concept and is generally recognized as being associated with wild populations, whereas artificial selection refers to the process where humans make breeding and culling decisions. An effective cattle breeding program must consider both of these aspects relative to the production environment. Animals that are quite extreme and have no practical applications in many settings may be produced through selection, although they may be specialized and sought after in a select few applications (double-muscling, miniatures, extremely large animals, etc.), and these types of animals will be at a disadvantage from a natural selection standpoint under most extensive production environments.

3.4.2 Genetic Prediction Values (Estimated Breeding Values, or EBVs)

The goal of a selection program is frequently to change the breeding values (gene content) of a population, which will be reflected in improved performance. This strategy led to the development of our traditional breeds of cattle. The traits that have been emphasized the most have been quite different across breeds and types around the world. Many cattle breeders may also feel that they need to 'improve' their herd or breed of cattle because their cattle are not desirable for certain traits.

The change in a trait resulting from the imposed selection practice is called selection response. Think about what can impact how much a group of

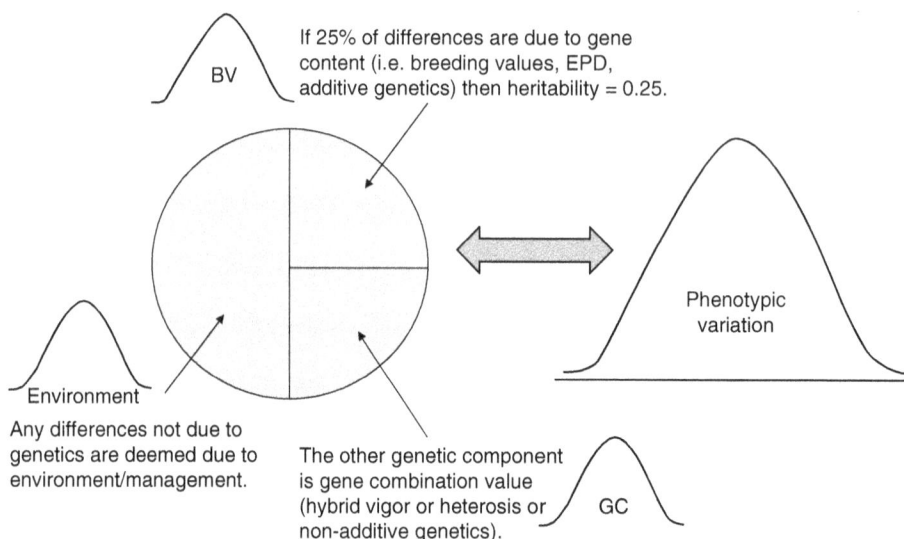

If 25% of differences are due to gene content (i.e. breeding values, EPD, additive genetics) then heritability = 0.25.

BV

Environment
Any differences not due to genetics are deemed due to environment/management.

The other genetic component is gene combination value (hybrid vigor or heterosis or non-additive genetics).

GC

Phenotypic variation

Fig. 3.2. Partitioning of phenotypic variation into its components. The components of phenotype all have continuous variation and follow a bell curve (normal) distribution.

Table 3.6. Heritability estimates for some reproductive traits in cattle.

Trait	Estimate
Age at first calving	0.10–0.25
Age at puberty	0.20–0.40
Calf survival	0.10
Calving interval	0.10
Calving date	0.10–0.20
Calving rate	0.10
Calving to first insemination	<0.10
Days to calving	<0.10
Dystocia	0.30
First service conception rate	0.05–0.20
Heifer pregnancy rate	0.15–0.20
Pregnancy rate	0.05–0.10
Ovulation rate	0.10–0.40
Twinning rate	0.10
Scrotal circumference	0.30–0.50

Table 3.7. Heritability estimates for some size and growth traits in cattle.

Trait	Estimate
Birth weight	0.40
Pre-weaning weight gain (birth to weaning)	0.30
Weaning weight	0.35
Weaning shoulder height	0.80
Weaning hip height	0.80
Feedlot gain	0.35
Pasture gain	0.30
Final feedlot weight	0.45
Final yearling pasture weight	0.45

Table 3.8. Heritability estimates for some growth and end-product traits in cattle.

Trait	Estimate
Carcass weight	0.42
Dressing percent	0.28
External fat thickness	0.39
Longissimus muscle area	0.41
Kidney, pelvic and heart fat	0.48
Marbling score	0.45
Estimated retail product percent	0.28
Retail product weight	0.51
Fat weight	0.52
Bone weight	0.51
Actual retail product percent	0.54
Fat percent	0.51
Bone percent	0.45

animals can be changed genetically due to selection from the parent population to the progeny population. The degree of superiority of the parents relative to their peers, and the heritability of the trait, are the two factors that dictate selection response when the individual's own performance is the basis for selection (and therefore estimated breeding values, EBVs). If there is little genetic variation among the group, there is little potential for genetic change from selection of parents from the group. If most of the performance superiority is not due to breeding value (i.e. heritability is low), utilization of phenotypically superior animals as parents will not result in much genetic change.

Because the heritability is the degree of correspondence of an animal's breeding value to its phenotype, it can be used as the degree of weighting on the animal's phenotypic superiority to predict its breeding value (calculate an EBV) as:

$$EBV = h^2 \times (\text{animal's performance} - \text{peer group average performance}) \quad (3.1)$$

The amount of difference of the individual from its peer (contemporary) group is referred to as the selection differential. The degree of correspondence between the individual's EBV and its true breeding value (which is referred to commonly as accuracy) is also limited by the heritability of the trait from this approach; the accuracy will be the square root of h^2. High levels of accuracy for EBV only occur when large numbers of progeny are evaluated. Many breed societies calculate EBVs on animals that have performance records submitted through their herd improvement programs. The calculation of EBV (or expected progeny differences, known as EPDs) involve incorporation of pedigree information, animal performance records and use of contemporary group designations in an analysis where all information within the breed society's database is used. EBVs are to be compared among animals within the same breed (or same analysis if a multi-breed analysis has been conducted) and therefore EBV cannot typically be compared across different breeds. Recent developments in genomic technology have initiated the calculation of genomic-based EBVs, and genomically-enhanced EBVs that incorporate known genetic markers into the traditional EBV calculation. The general concept of response to selection is shown in Fig. 3.3.

Response per generation can be standardized to an annual rate of response (R) by dividing the number

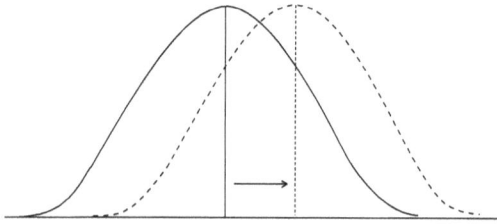

Fig. 3.3. Concept of selection response. Selection response for a quantitatively inherited trait has implied shifting the mean of a population over time as a reflection of which animals are allowed to become parents.

of years separating the turnover of generations (referred to as generation interval). Selection response per year can be expressed in the general format of:

$$R = \frac{\text{Intensity x Accuracy}}{\text{Generation Interval}} \quad (3.2)$$

In this formula, intensity represents the selection differential of selected parents over their peers and is commonly based on EBVs, accuracy refers to accuracy of EBVs and generation interval indicates the time for generation of replacements in years. Breeding values are generally estimated with higher accuracies on sires as they are able to produce more progeny per year.

The accuracy (ACC) values and the estimated breeding values (EBVs, or expected progeny differences, EPDs) are calculated on animals in the performance database of breed societies and provided as a service to their member breeders. The EBVs take into account all available performance information on animals and their relatives, so it is a much more complicated procedure than simply multiplying the heritability by the performance deviation as discussed earlier in the chapter. The accuracy value is the correlation between the animal's calculated genetic value and its true genetic value and can range from 0 (no information and therefore no confidence in its estimate) to 1.0 (where the true breeding value is known without any uncertainty, which is almost impossible). Each time a breed society conducts a genetic evaluation of its animals, new EBVs are calculated for every animal in the database. The higher the ACC value, the less the EBV will deviate from a new estimate; this value is referred to as a possible change value or a confidence interval or value. Table 3.9 provides some of these values across some accuracy values in Australian Brahman cattle.

Table 3.9. Confidence intervals (possible change values) associated with EBVs for a range of accuracy values.

Trait	Accuracy (ACC) values				
	60%	75%	85%	90%	99%
200-day milk	4.4	3.6	2.9	1.7	0.8
200-day weight	7.8	6.4	5.1	3.0	1.4
400-day weight	11.8	9.8	7.8	4.6	2.1
600-day weight	16.4	13.6	10.7	6.4	2.9
Carcass eye muscle area	2.9	2.4	1.9	1.1	0.5
Carcass P8 fat	1.8	1.5	1.2	0.7	0.3
Scrotal size	1.3	1.1	0.8	0.5	0.2
Days to calving	8.7	7.2	5.7	3.4	1.5

Values taken from the Australian Brahman Breeders' Association Sire Summary for June 2011.

3.4.3 Genetic Trend Over Time

The change in breed-average breeding value over time is a reflection of the change in underlying gene frequencies over time. This implies that at some point, response to selection has a set, maximum level. If selection response across generations appears to not be slowing much, then that also probably indicates that the maximal end-point is probably not close. Breed societies often report average EBVs over years and refer to this concept as genetic trend.

3.5 Correlated Traits

In addition to selection for an individual trait, we also need to consider how two phenotypes of two traits of the same individual may be related. As a result, phenotypic correlation, genetic correlation and environmental correlation concepts exist. Phenotypic correlation refers to the degree of association between two phenotypes on the same individual; genetic correlation refers to the degree of association between the breeding values for two traits in the same individual; environmental correlation indicates the degree of association between environmental (and non-additive genetic) deviations for two traits on the same individual. It is important to realize that phenotypic correlations arise from both genetic and environmental causes, but simply knowing the phenotypic correlation does not indicate much about the genetic or environmental correlations. Genetic correlations need to be understood because selection for one trait can also change other traits (Fig. 3.4).

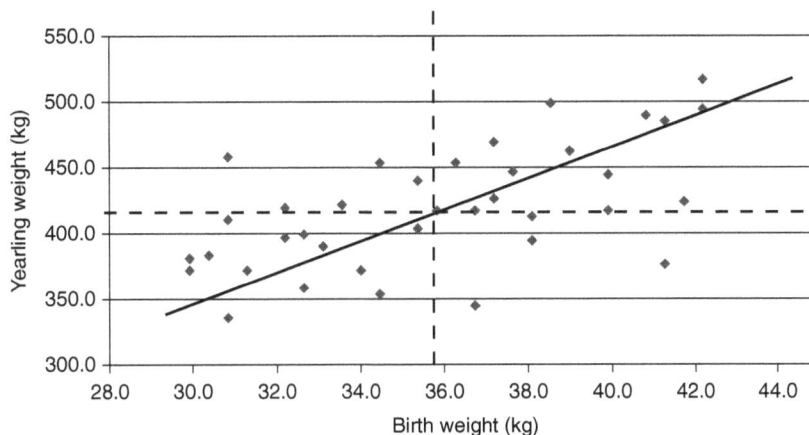

Fig. 3.4. Concept of traits that are correlated. In this example birth weight and yearling weight are positively correlated because as one increases, the other also increases on average. The average degree of relationship is indicated by the solid line. However, there are individuals that deviate from the average pattern, and these can be used to change the population over time (i.e. ability to increase one trait with the reduced increase in another trait). In this example, the average birth weight is 35.8 kg and the average yearling weight is 419.3 kg (broken lines).

3.5.1 Correlated Response to Selection

If the two characteristics are genetically correlated, there will be a corresponding response in characteristic Y due to selection for characteristic X. If the genetic correlation is zero, there will be no correlated response to selection in trait Y regardless of what happens in trait X; conversely, the higher the genetic correlation, the higher corresponding change in one trait from selection on the other trait. The main thing that cattle breeders need to be aware of in selection programs is that many types of traits have varying degrees of correlation, and that simply changing one trait may have consequences on others. Cattle breeders should always avoid single trait selection where only one trait is emphasized. Consequently, when many traits are under selection the rate of improvement in any one trait will be low.

3.5.2 Multiple Trait Selection

It was mentioned earlier that cattle breeders should avoid 'single trait selection'. This concept refers to overly placing selection pressure on some trait while disregarding emphases on other traits. Cattle populations are inherently under both natural selection (imposed by natural environmental influences and constraints) and artificial selection (criteria imposed

by people). It should seem obvious that as more traits are involved in selection programs, there will be less expected progress per trait, but simultaneous changes in all traits are possible. Because many traits are genetically correlated, and the fact that numerous traits affect the fertility, survival, growth and profitability in cattle herds, over-emphasis on any one trait could lead to reduced performance in others.

Different approaches to multiple trait selection have been discussed in animal breeding textbooks. These can be categorized as (i) tandem selection, (ii) use of independent culling levels, and (iii) selection index approaches. Tandem selection simply means selection emphasis on one trait until a desirable level of performance is obtained in that trait, but then selection emphasis switches to another important trait. This approach is very ineffective as genetic progress attained in one trait may be lost in part or entirely when other traits are subjected to selection. Use of independent culling levels refers to the situation where a minimum level performance is established in all traits the breeder wants to emphasize, and animals that do not meet all these levels are not used as parents. The selection index approach utilizes the degree of genetic correlations among all traits, places a specific amount of weighting on each EBV for each trait, and calculates a single index value for each animal, which

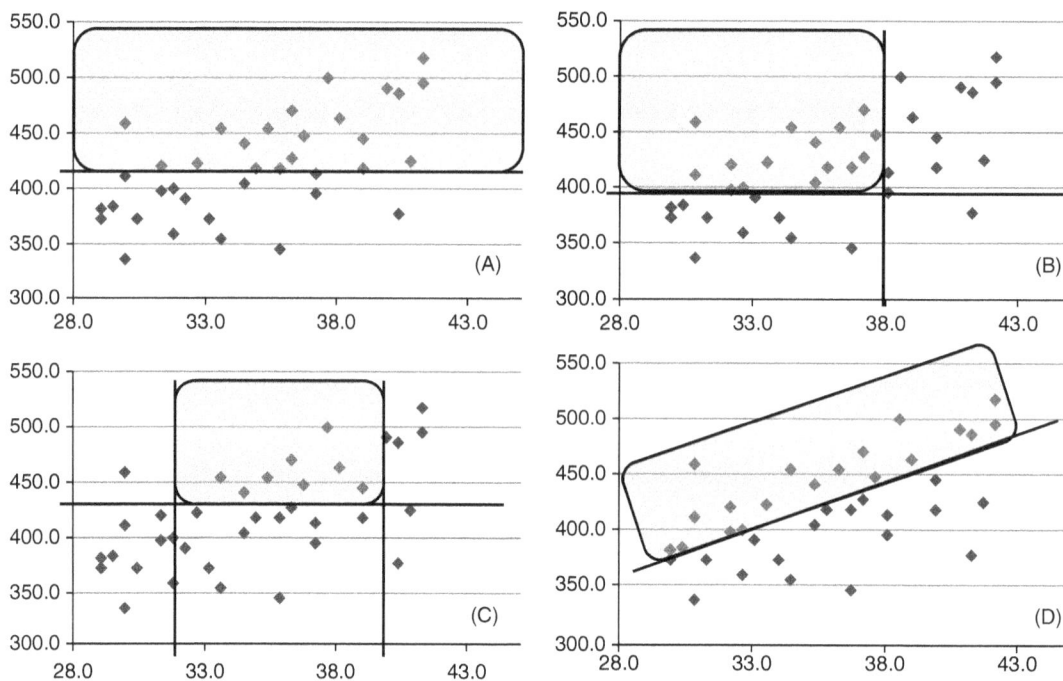

Fig. 3.5. Concepts of multiple trait selection. The same data from Fig. 3.4 is used in these graphs. Shaded areas indicate animals that would be kept for breeding under the following conditions: (A) animals with above average yearling weight (this is single-trait selection), (B) independent culling levels for BWT less than 38 kg and YWT above 400 kg, (C) culling levels for BWT between 31 and 40 kg and YWT over 440 kg, and (D) a selection index approach. Think about which animals would be selected (and culled) and what would happen to YWT and BWT over time with these different approaches.

can be thought of as an aggregate breeding value. Animals can then simply be ranked by the single index values, but information about all traits has been incorporated. Figure 3.5 provides some examples of animals that would be selected as parents under some of these multiple trait selection approaches when two traits are involved.

3.6 Breeding and Mating Systems

In the broadest sense, breeding systems can be thought of as straightbreeding (use of purebred animals to produce the same purebred offspring) and crossbreeding (use and/or production of crossbred animals). Within crossbreeding there are numerous options. The primary points associated with these different types of systems are briefly discussed. Throughout this section, where individual examples

of matings are used, they are always presented as the male parent first as crossed with the female parent (i.e. A × B indicates A bulls bred to B females, B × A indicates B bulls bred to A females, etc.).

3.6.1 Straightbreeding

When populations of animals are straightbred, there are more limitations on making genetic improvements than when crossbreeding is an option. Obviously in maintaining purebred populations, crossbreeding is not an option. In commercial beef production, the true value of having purebred animals is the ability to produce F_1 animals where heterosis is maximized. There are options that should be considered within straightbred populations, however, such as linebreeding and breeding of animals that are not closely related in the pedigree sense.

3.6.2 Grading Up

Grading up refers to the breeding strategy where a single parental breed (or cross) is continually used as a parental type, such as the sire across several generations. Consider the following crosses where animals of breed A are used as sires for each successive generation (see Table 3.10). Beginning with the F_1 generation, the amount of Breed B is halved in each successive progeny generation. This process starts off as a crossbreeding system, but turns into a straightbreeding concept over time. This has been the process used in many regions to propagate introduced breeds. Many breed societies allow for grading up to produce high 'grade' animals of that breed, with some specified grade level classified as 'purebred'. This is the process that has allowed black purebred animals of several European breeds (Simmental, Limousin, Gelbvieh, Braunvieh, etc.) to be produced in North America. The term 'fullblood' is sometimes used in breeds such as these to distinguish animals with only pure ancestry vs those that have been graded up (referred to as 'purebred').

In the past many people have associated the grading up to a new breed as genetic improvement because less productive local or native types of cattle were replaced over time by the genetic influence of a higher producing animal type. The term grading up as used here has no associated value in its overall desirability, but simply means increasing the percentage or influence of one breed type with another through a designed breeding strategy.

3.6.3 Crossbreeding Programs

There are three distinct advantages that crossbreeding strategies offer over purebred (also called straightbreeding) strategies: (i) the ability to blend desirable breed characteristics, (ii) the more desirable performance of crossbred animals relative to the average of the purebred parental types, also known as heterosis or hybrid vigor, and (iii) the ability to use specialized sire and dam parental types, referred to as complementarity. Many people think that heterosis is the primary advantage of crossbreeding (Fig. 3.6), and it may be the main advantage for crosses of particular breeds or types, but the choice of parental breeds should be the foundation of the crossbreeding program. Product-type considerations should be based on gene content. Just as various traits have different heritability values, traits also exhibit different heterosis levels (typically inversely related to heritability). Crossbreeding systems are typically referred to as terminal, continuous or combination systems, and this is based on the ability to produce replacement breeding animals to sustain the systems.

Terminal crosses

A cross that produces progeny that are different in gene content than either parent can generally be thought of as a 'terminal' cross. The cross is terminal in that no replacements are automatically produced for the system as the progeny are genetically different from both parent types. Terminal systems have the potential to produce the highest possible levels of heterosis, as well as the maximum potential to utilize specialized parental types. However the fact that they are 'terminal' provides a serious limitation in regard to sustainability because they rely on other systems or programs or herds to produce replacements.

For the following examples, consider two breeds referred to as A and B. In all crosses presented, the sire breed type is presented first, then the dam breed type, followed by the progeny breed type.

Table 3.10. Concepts of grading up.

Sires	Dams	Progeny	% A in progeny	% B in progeny
A	B	1/2A 1/2B	50	50
A	1/2A 1/2B	3/4A 1/4B	75	25
A	3/4A 1/4B	7/8A 1/8B	87.5	12.5
A	7/8A 1/8B	15/16A 1/16B	93.75	6.25
A	15/16A 1/16B	31/32A 1/32B	96.9	3.1
A	31/32A 1/32B	63/64A 1/64B	98.4	1.6
A	63/64A 1/64B	127/128A 1/128B	99.2	0.8

Repeatedly breeding successive crossbred generations back to the same breed (breed A in this example) over several generations produces animals that are essentially purebreds.

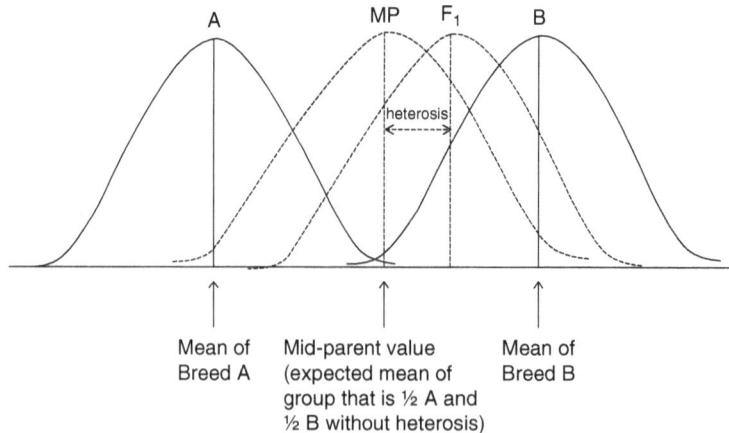

Fig. 3.6. When crossing two breeds (A and B), the mid-parent value is halfway between the purebred means. The amount of deviation of the crossbred population mean from the mid-parent value is the amount of heterosis.

- Production of F_1 progeny: A × B => ½A ½B

  The fraction of any breed in the progeny is the average amount of that breed in the two parents. The sire is 100% Breed A, and the dam is 0% Breed A; therefore the progeny will be 50% Breed A in gene content. The same is true in regard to Breed B. Although half of the alleles in the progeny are A, and the other half are B, all (100%) of the gene loci in the progeny will be heterozygous between Breeds A and B (i.e. at every gene locus the progeny will have an A allele from the sire and a B allele from the dam). The degree of heterozygosity has been used as a predictor of the level of heterosis between breeds that should be expressed for any cross involving those breeds (the F_1 progeny should exhibit 100% of the heterosis between Breeds A and B). This is referred to as the dominance model of predicting heterosis levels in crosses following the F_1 generation with reasonable correspondence to observed performance differences.

- Production of F_2 progeny: F_1 (½A ½ B) × F_1 (½A ½B) => F_2 (½A ½B)

  When F_1 parents are mated, the resulting progeny are termed F_2. They have the same gene content as the F_1 individuals, but not the same expected gene combinations. It is expected that ¼ of the gene combinations in the F_2 are homozygous Breed A (AA) and another ¼ are homozygous Breed B (BB), with ½ heterozygous (¼ AB and ¼ BA, usually assumed to be equivalent). As a result, the expected heterosis in the F_2 generation when only two breeds are involved is 50%.

Additionally, an F_3 cross is when two F_2 parents are mated, and the expected heterosis in the F_3 is also 50%. If there is inbreeding, then the level of heterosis will decrease from these expectations. Barring inbreeding, heterosis is only lost from the F_1 to the F_2 generation; however, that does not guarantee that the F_1 will have superior performance to the F_2. If the trait of interest is maternally influenced, such as birth weight and weaning weight, then the value of having an F_1 dam may be more influential on the trait as compared to the level of heterosis in the progeny. The amount of heterosis in the progeny is referred to as direct heterosis, and the amount of heterosis in the dam is referred to as maternal heterosis.

At first it may seem confusing where these fractions of expected heterozygosity come from, but it is the same genetic concept discussed earlier in the chapter (Sections 3.2.2 and 3.2.3). Think about the breed make-up in the F_1 parents and the possibilities in the F_2 offspring.

If the breed make-up is ½ Breed A, the percentage of alleles in the animal that are Breed A is 50%; additionally, the chance that a gamete contains an allele from Breed A origin is 50% (the expected percentage of alleles in each gamete that are from Breed A is also 50%). So in fact the chance that a Breed A allele is passed from the parent to its offspring is 50%. The approach is the same no matter the fractions of the breeds, or the number of breeds involved. The breed heterozygosity resulting from any cross is the basis of predicting heterosis because the genetic cause of heterosis has been thought of as being due to dominance effects, and when there is dominance, the heterozygotes will have performance similar to the favorable homozygous individual. Results from many cattle crossbreeding projects have shown that heterozygosity closely predicts observed heterosis levels, but that other genetic mechanisms may not be ruled out, particularly in composites where *Bos indicus–Bos taurus* crosses are involved.

- Backcross progeny
  A backcross is when an F_1 is bred back to one of the parental types, and the F_1 can be the sire or the dam. There are two potential backcrosses that can be produced within a given progeny type, one where the sire is a purebred (and the dam is an F_1) or the other where the sire is an F_1 (and the dam is purebred).

In both of these backcrosses, there is ½ or 50% direct heterosis because half of the gene loci in the progeny should be homozygous (AA) and the other half should be heterozygous (AB). However, the first backcross listed is produced from an F_1 dam (with 100% heterosis) whereas the second backcross is from a purebred dam (0% heterosis). These two backcrosses have the same gene content (¾A ¼B), the same direct heterosis (50%), but differing maternal heterosis, which means that calves produced from the first backcross are likely to have higher weaning weights than the second backcross when all other factors are equal.

- Three-breed terminal cross
  A three-breed terminal cross typically refers to a cross where purebred bulls are bred to F_1 cows where the breeds in the cow are different than the breed in the sire (i.e. C × F_1 (½A ½B) => ½C ¼A ¼B). There would be 100% direct heterosis in calves produced in this cross, but the direct heterosis is due to both CA and CB gene combinations, each with 50%. This system allows for maximal direct and maternal heterosis as well as high complementarity potential. A potential disadvantage could be the use of purebred sires that are not adapted to a particular environment to introduce new genetic influence (although this can be avoided through use of semen).
- Four-breed F_1 cross
  A four-breed F_1 cross is where F_1 bulls are bred to F_1 females, and there is no breed in common

between the male and female parents (for example an F_1 bull made from crossing breeds C and D is bred to an F_1 cow generated by crossing breeds A and B, or F_1 (½C ½D) × F_1 (½A ½B) => ¼A ¼B ¼C ¼D).

There would be 100% direct heterosis for this type of progeny, but the sources of the direct heterosis would come from four breed combinations (CA, CB, DA and DB), each occurring at 25%. This type of cross allows for maximum direct and maternal heterosis and also takes advantage of heterosis in sires for traits such as mating behavior, semen quality, etc., and allows for maximum complementarity potential. A potential disadvantage could be lack of available sire types as crossbred sires are not commonly available in many regions, or for many breed crosses; semen availability is very limited on hybrid bulls in most cases.

Rotational crossbreeding systems

Rotational systems represent those where two or more parental types are used. The most common is the two-breed rotation where two different breeds of bulls are used. In rotational systems, the breed least represented in the females is the breed of bull to breed her to.

- Two-breed rotation
  A two-breed rotation of Angus and Hereford means that Angus bulls are bred to Hereford-sired females, and Hereford bulls are bred to Angus-sired females. All dams and progeny are crossbred and have ⅔, ⅓ breed influence. The females produced from one cross generate the type of female that needs to be bred to the other type of bull for the other type of cross (see Table 3.11).
- Three-breed rotation
  A three-breed rotation follows the same logic, but some of the crosses or the expected fractions at equilibrium may not be obvious at first. Suppose that we have Breeds A, B and C. We will utilize these breeds as sires in that same order (A, then B, then C) over time. If F_1 ½A ½B females are bred to C sires, and those female progeny were bred to A sires, and then the resulting females were bred to B sires, and this strategy was followed, there would eventually be stabilization at a 4:2:1 ratio among the breeds in the progeny produced. This is laid out in Table 3.12.

Table 3.11. Initiation and development of a two-breed rotation.

Sires	Dams	Progeny	% A in progeny	% B in progeny
A	B	1/2A 1/2B	50	50
B	1/2A 1/2B	3/4B 1/4A	25	75
A	3/4B 1/4A	5/8A 3/8B	62.5	37.5
B	5/8A 3/8B	11/16B 5/16A	31.25	68.75
A	11/16B 5/16A	21/32A 11/32B	65.6	34.4
B	21/32A 11/32B	43/64B 21/64A	32.8	67.2
A	43/64B 21/64A	85/128A 43/128B	66.4	33.6
B	85/128A 43/128B	171/256B 85/256A	33.2	66.8

This example illustrates how a two-breed rotation (Breeds A and B) starts as a cross of pure breeds and progresses over several generations toward a 2/3, 1/3 ratio at equilibrium.

Table 3.12. Initiation and development of a three-breed rotation.

Sires	Dams	Progeny	% A in progeny	% B in progeny	% C in progeny
C	1/2A 1/2B	1/2C 1/4A 1/4B	25	25	50
A	1/2C 1/4A 1/4B	5/8A 1/8B 1/4C	62.5	12.5	25
B	5/8A 1/8B 1/4C	9/16B 5/16A 1/8C	31.25	56.25	12.5
C	9/16B 5/16A 1/8C	9/16C 9/32B 5/32A	15.6	28.2	56.25
A	9/16C 9/32B 5/32A	37/64A 9/32C 9/64B	57.8	14.1	28.2
B	37/64A 9/32C 9/64B	73/128B 37/128A 9/64C	28.9	57.0	14.1
C	73/128B 37/128A 9/64C	73/128C 73/256B 37/256A	14.4	28.5	57.0
Therefore at equilibrium:					
A	4/7C 2/7B 1/7A	4/7A 2/7C 1/7B	57.1	14.3	28.6
B	4/7A 2/7C 1/7B	4/7B 2/7A 1/7C	28.6	57.1	14.3
C	4/7B 2/7A 1/7C	4/7C 2/7B 1/7A	14.3	28.6	57.1

The order of the sire breeds dictates the fractions of breeds in the progeny generations. For instance, the sire breed order of A, B, C as above will not produce exactly the same breed fractions as the sire breed order of A, C, B, etc.

This sire breed rotation concept can be extended to a four-breed rotation, five-breed rotation, or any number of breeds. In any rotational crossbreeding system the breed least represented in the progeny females is the breed those females need to be bred to. However, the level of complexity and management also increases as the number of breeds increases because to ensure the correct matings for the system, the different types of females must be kept separate from others.

Composites

Composites are crossbred populations of animals that are managed as straightbreds, meaning that all animals are crossbreds, but all sires, dams and progeny are the same crossbred composition (i.e. 5/8 Angus 3/8 Brahman as with Brangus, etc.). As a result, composites by design produce their own replacement females and males. They also do not take advantage of any parental type complementarity. Composites will have heterosis retained that is proportional to their level of breed heterozygosity. As a result, composites that have more breeds and have a more equal representation of component breeds will retain the most heterosis (can you verify why?).

For the example illustrated in Fig. 3.7, four of the 16 possible breed combinations that can result in the progeny are homozygous and 12 of these breed combinations are expected to involve two breeds. As a result, the breed heterozygosity is 12/16 or 75%.

The main reason to form composites is to blend desirable characteristics of parental breeds into a single animal type, but then to also allow for heterosis to contribute to performance. Most composite breeds mentioned in Table 3.3 were probably developed without a recognized attempt to retain heterosis, but in the past 25 years composites have become recognized as useful crossbreeding systems for commercial producers with more emphasis on heterosis.

Sire and dams are the same type of cross and produce progeny that are the same type of cross

		Gametes from dam:			
		$P(A) = 1/4$	$P(B) = 1/4$	$P(C) = 1/4$	$P(D) = 1/4$
Gametes from sire:	$P(A) = 1/4$	$P(AA) = 1/16$	$P(AB) = 1/16$	$P(AC) = 1/16$	$P(AD) = 1/16$
	$P(B) = 1/4$	$P(BA) = 1/16$	$P(BB) = 1/16$	$P(BC) = 1/16$	$P(BD) = 1/16$
	$P(C) = 1/4$	$P(CA) = 1/16$	$P(CB) = 1/16$	$P(CC) = 1/16$	$P(CD) = 1/16$
	$P(D) = 1/4$	$P(DA) = 1/16$	$P(DB) = 1/16$	$P(DC) = 1/16$	$P(DD) = 1/16$

Fig. 3.7. Expected levels of breed combinations in progeny of a composite that has equal amounts of four breeds.

Many 'terminal' crosses may be F_1 generations of potential composites. For instance someone could interbreed progeny from a three-breed terminal cross to produce the second generation of a composite. Individuals produced as progeny in a four-breed terminal F_1 could be interbred as parents to produce the second generation of a composite. Many people's 'terminal' calves are potential replacement animals in several other breeding programs, but they are terminal relative to the cross that produced them because they are a different gene content than either parent and therefore cannot be used as replacements for the parent types because they are genetically different.

3.6.4 Use of Different Biological Types in Crossbreeding Systems

Some biological types of cattle show more heterosis when crossed compared to others. The most commonly known scenario involves *Bos indicus–Bos taurus* crosses. Most of the published recommendations on heterosis levels and heterosis retention for different crossbreeding systems are based on crosses only involving *Bos taurus* breeds. These figures severely underestimate heterosis levels in systems involving *Bos indicus* and *Bos taurus* breeds because approximately twice as much heterosis (sometimes considerably more) is usually seen in the F_1 generation of *Bos indicus–Bos taurus* crosses as compared to F_1 generations of *Bos taurus–Bos taurus* or *Bos indicus–Bos indicus* crosses. As a result, to fairly compare all possible crossbreeding systems, we need to express heterosis levels on a *Bos taurus* equivalent basis, as illustrated in Table 3.13.

Several studies have shown that there may be more heterosis expressed in more challenging environments (probably because one or more of the purebred types show decreased performance in more challenging conditions). It is realistic to expect that there will be more heterosis expressed when breeds of different biological groups are crossed as compared to two breeds within the same biological group, and the more genetic divergence between breeds, the more potential heterosis that may exist. It must be remembered that heterosis can be a desirable and very advantageous benefit to crossbreeding, but for many scenarios, the choice of breeds and the fractions of the breeds involved may far outweigh the potential heterosis benefits if the type of crossbred animal is not a logical fit to the production environment.

It has been known for a long time that heterosis is important for increased beef cattle productivity in *Bos taurus* crosses (Mason, 1966; Warwick, 1968; Cundiff, 1970; Dillard *et al.*, 1980) and that *Bos indicus–Bos taurus* crosses exhibit much higher levels of heterosis in the F_1 generation (for instance Damon *et al.*, 1959, 1960; Cartwright *et al.*, 1964; Franke, 1980; Peacock *et al.*, 1981; McElhenney *et al.*, 1986; Sacco *et al.*, 1989) than do F_1 crosses among *Bos taurus* breeds. Furthermore, the F_1 *Bos indicus–Bos taurus* female has been a very productive cow type under widely varying production environments (Damon *et al.*, 1961; Mason, 1966; Peters and Slen, 1967; Cundiff, 1970; Gregory *et al.*, 1985; Sacco *et al.*, 1989; Vercoe and Frisch, 1992; Cundiff *et al.*, 1993).

Another aspect of heterosis that deserves some discussion is heterosis retention among *inter se*

Table 3.13. Relative heterosis levels (%) for direct and maternal effects in some crossbreeding systems expressed on a *Bos taurus* equivalent basis.

Type of system	Example	Direct	Maternal
Terminal F₁ cross	Charolais × Hereford	100	0
Terminal F₁ cross	Brahman × Hereford	200	0
Two-breed rotation	Angus, Hereford	67	67
Two-breed rotation	Angus, Brahman	133	133
Three-breed terminal	Charolais × F₁ Angus–Hereford	100	100
Three-breed terminal	Charolais × F₁ Brahman–Hereford	150	200
Three-breed rotation	Angus, Simmental, Gelbvieh	86	86
Three-breed rotation	Angus, Brahman, Charolais	143[a]	143[a]
Two-breed composites	50% Hereford, 50% Simmental	50	50
	50% Angus, 50% Brahman	100	100
	5/8 Charolais, 3/8 Red Angus	47	47
	5/8 Shorthorn, 3/8 Brahman	94	94
Four-breed composites	25% Angus, 25% Hereford, 25% Simmental, 25% Charolais	75	75
	25% Angus, 25% Brahman, 25% Charolais, 25% Gelbvieh	112.5	112.5
	25% Angus, 25% Brahman, 25% Limousin, 25% Nelore	125	125

[a]Some crosses in the system will exhibit more heterosis than others. This is an over-simplification of 'real-world' results as different breed combinations can have different amounts of heterosis, and heterosis levels can vary across production environments.

matings (crosses where both parents are the same fractions of breeds) involving *Bos indicus–Bos taurus* combinations. Among *Bos taurus inter se* matings, the heterosis retention seems to be proportional to the expected level of breed heterozygosity, indicating that the dominance model based on breed heterozygosity has done an adequate job in explaining heterosis retention in composite populations involving *Bos taurus* breeds (Gregory *et al.*, 1993), and therefore might be assumed to explain heterosis retention in any breed cross. Other models for heterosis explanation have been offered such as overdominance (Pirchner, 1969), which is really an extension of the dominance theory, parental epistasis and F₁ epistasis (Sheridan, 1981), and interaction of additive loci (Minvielle, 1987).

The increased levels of heterosis observed in crosses between *Bos indicus* and *Bos taurus* are well documented for the F₁ animals. However, the few studies that have evaluated heterosis retention past the F₁ have shown very variable results. In some cases there has been heterosis in the F₂ that seems proportional to decreased heterozygosity from the F₁, but hardly any heterosis apparent in the F₂ generation in some other studies (Rendel, 1980; MacKinnon *et al.*, 1989; Sanders, 2005). As a result the genetic mechanisms that explain heterosis retention in *Bos indicus–Bos taurus* crosses is still under investigation.

3.6.5 General Management Considerations for Crossbreeding Systems

When deciding between rotational, terminal, composite and combination systems as options, several management considerations must be taken into account, several of which are discussed briefly below.

Production of replacement heifers

With rotational or composite systems replacement heifers can be automatically produced as they are of the correct type of female parent used to generate their type, but this is not so with terminal systems. Combination systems allow some replacements and some terminal calves to be produced and provide for an optimal intermediate type of system.

Number of pastures and breeding groups needed

The number of pastures is related directly to the number of dam types in the system, so with terminal and composite systems only one breeding pasture is needed (excluding separation of heifers from mature cows, etc.). Two- and three-breed rotations require a minimum of two and three breeding pastures, respectively. Combination systems can allow for multiple types of calves being produced without increasing the number of breeding groups when multiple sire types are placed with a single cow type, or a single sire type is placed with multiple cow types.

Amount of time required to produce desired type of offspring

When starting a crossbreeding system, whether or not the desired type(s) of dams and/or heifers are available from other sources should be a consideration. Another factor to consider is whether or not the right type of heifers will need to be produced in the intended system. Whether or not an existing market exists for the animals produced should be determined before the final breeding system structure is in place. Another major consideration is whether or not a reliable and sustainable source of herd sires exist, or if they will need to be produced in the system at some point.

Amount of flexibility in marketing or desirability of offspring

The potential of multiple markets for the excess heifers should be thought through; if heifers will be desirable for a wide variety of uses as replacements for someone else, as well as fitting into a growing or finishing grazing or feedlot setting, this provides for numerous marketing opportunities. Also, would the males be desirable as stocker and feeder calves, or as breeding bulls for other operations? Whether or not the offspring can be valuable on either a carcass-based market or a live-basis market also adds flexibility.

Variation in expected color patterns among calves

This may be important only if animals are sold on a live basis (i.e. feedlot cattle sold live or calves marketed through sale barn, potential replacement heifers, etc.) or if other value is determined by color. Color will not affect sale price if animals are sold on a grid carcass system unless there is a specification for particular programs, such as Certified Angus Beef, etc. However, environmental considerations may influence desired color patterns in the cow herd. There can be quite a bit more color uniformity with a terminal system.

3.6.6 Combination Crossbreeding Systems

The challenge for many crossbreeding systems is long-term sustainability, which mandates vision, discipline and flexibility. Programs that combine terminal aspects with rotational and/or composite aspects can be very flexible to target different markets as well as allowing production of replacement heifers. Combination crossbreeding systems are those that have both terminal and continuous (replacement) components. These allow more flexibility in management and marketing opportunities. With combination systems, more types of progeny are produced, but this does not have to be the result of more complex management. If a two-breed rotation of Angus and Hereford was being used, and the operation decided to also produce some Continental European sired calves, then these types of bulls could be used in breeding pastures along with the Angus and/or Hereford bulls.

3.7 Genetic Markers and Genomic Information

Traditional breeding strategies have relied on gathering pedigree information and performance data. These concepts will always be important, but as new genomic information (genetic markers, gene expression and protein formation) for traits of economic importance becomes more available, as source verified programs become more popular, and as managers desire more accurate information for precision animal management, use of genetic markers offers cattle producers new tools to complement traditional approaches.

Genetic markers and tests can be useful for both qualitative and quantitative traits. However, for several qualitative traits, there may be a possible test for 'the gene' that causes the trait; this is not the case for quantitatively inherited traits because several genes influence the trait, and the degree of genetic influence on the trait must also be considered. There is not a single growth gene, a single marbling gene, or a single fertility gene; but it is certainly possible that some genes may have a greater influence than others for these types of traits, or that some genes/genetic markers may influence several traits. Genetic markers offer the potential to get an idea about an animal's genetic make-up before a lot of time and expense are invested in that animal. Traits that are hard to measure, such as feed intake and efficiency, or traits measured later in life, such as carcass quality and reproductive performance, offer a lot of promise from the use of genetic markers. It may also be very useful to have genetic markers to improve traits where the phenotype may be hard to accurately measure, such as health aspects where subclinical illness may go unrecognized. However, just about any economically important quantitative trait could benefit from useful genetic tests.

Several groups around the world have been and are currently working on cattle genetics projects to map (assign to chromosome sequence) genes. The term mapping refers to establishing a relationship of a gene or a genetic marker to another already documented group of genes/markers on a specific chromosome. Certain DNA sequences scattered throughout the genome may show differences across animals, and any DNA sequence that is associated with a particular phenotype or performance level can be a genetic marker. An area on a chromosome that influences a quantitatively inherited trait is called a quantitative trait locus (QTL).

As more research is conducted, more genetic tests become commercially available, and the information upon which these tests are based becomes more informative. Several companies offer genetic testing services in cattle. Many companies list the costs and descriptions associated with their genetic tests on their websites. In early years of genetic tests, individual genotypes of animals for each marker were provided; however, as the number of markers in genetic tests have increased, an index-type or an EBV-type value has become common to report to producers. The concept of genomic selection stems from the potential to screen and identify young animals to become parents earlier in their lives through DNA-based technology. The dairy industry has placed considerable emphasis on genomic selection in recent years, and several beef breeds are investigating the potential of genomically-enhanced EBV (or EPD).

3.8 Genotype × Environment Interactions

The concept that certain biological groups, breeds and even families of cattle have inconsistent performance across varying climatic and management-related environments was introduced in Chapter 1 and really needs to be familiar to people involved in cattle production as this concept is a fundamental consideration for any production system. This is an environmental effect when only considering one genetic group across different scenarios. Likewise, the differences among two or more genetic groups may not be consistent across environments, and when this is true it is referred to as a genotype × environment interaction. Understanding these types

of interactions is crucial for making breeding decisions, particularly when animals are to be produced in environments drastically different than conditions in which they were developed as a population. Figure 3.8 illustrates a classic example of genotype × environment interaction with growing cattle under different management environments. Lack of knowledge or appreciation for this concept has led to detrimental and expensive mistakes when non-adapted cattle with high production potential were brought into harsh environments and performed no better (or worse) than native 'unimproved' cattle. Numerous examples of genotype × environment interactions in cattle have been reported.

3.9 Summary

Understanding the fundamental concepts of inheritance, sources of genetic variation and breeding systems is crucial for effective and profitable beef cattle production systems. There are many opportunities for selection across breeds, and across family lines within breeds. There can be huge differences in how various traits respond to selection, as well as numerous ways to alter selection response. Traits in cattle never need to be changed without considerations of potential correlated changes in other traits. Commercial producers can gain large advantages in fertility and calf survival through heterosis when crossbreeding is used. As new genomic-based technologies become increasingly available, producers need to be informed about the possible ramifications of their use. A well-balanced selection program where cattle are fitted to their production environment(s) and their potential market(s), and where reliable sources of replacement animals are available provide the most potential for sustained success.

3.10 Study Questions

3.1 Briefly explain and compare the concepts of single trait selection and multiple trait selection.
3.2 List five general constraints or concepts that might limit the degree of selection intensity.
3.3 Based on the concepts and values in Tables 3.6 to 3.8, list two traits that should respond to selection fairly rapidly and two traits that should respond to selection slowly; briefly explain your answers.
3.4 If the purebred animals are considered generation 0, what percentages of breeds are expected in

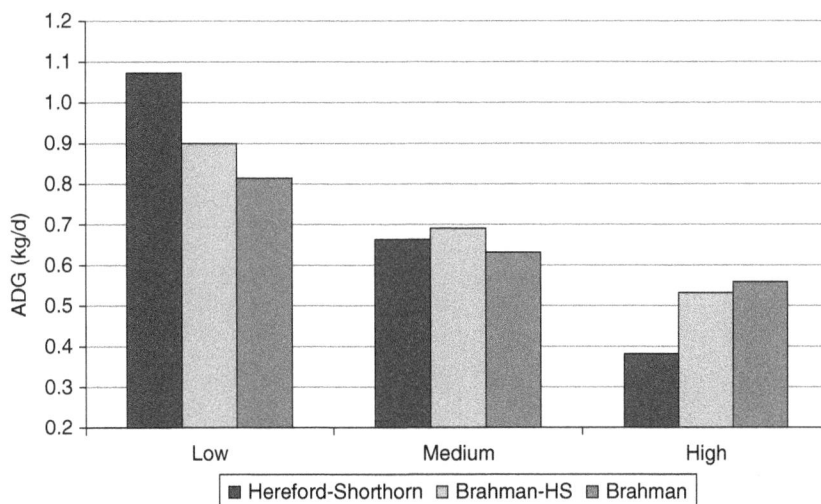

Fig. 3.8. Growth rate (ADG, kg/day) of breed types under different levels of stress in Queensland. The stress level was regulated by degree of cattle tick control (low was complete tick control, medium was moderate degree of control and high was no treatment for ticks). Both composites were multi-generation (as opposed to F_1 generation). Under these same conditions the F_1 Brahman-HS had equal performance to the Hereford-Shorthorn (HS) under low stress and equal performance to straight Brahman under high stress. From Vercoe and Frisch (1992).

the calves from the third generation in a grading up program?

3.5 Scenario

You have just purchased a ranch in (you choose the location) with 400 pregnant F_1 cows (these could be Angus–Hereford, Nellore–Angus, Simmental–Hereford, Tuli–Africander, etc.); half of the cows are bred to bulls of one breed, and half of the cows are bred to bulls of the other breed, and you also have the purebred bulls (for example if you are considering F_1 Angus–Hereford cows, then half of the cows are bred to Angus, half are bred to Hereford, and you also have the Angus and Hereford bulls, etc.).

3.5a Explain how you could use this cow herd to start a two-breed rotational crossbreeding system.

3.5b Could you use this cow herd to start your own two-breed composite? Give one possible example.

3.5c Briefly discuss whether you would consider breeding some or all of these cows to bulls of a third breed.

3.6 Do cow-calf producers with 25 cows have as many realistic options for utilizing crossbreeding systems as producers with 200 cows? Explain.

3.7 Explain the major differences between a two-breed rotational crossbreeding system and a two-breed composite crossbreeding system.

3.8 Provide examples of three different two-breed composites that are expected to have different levels of heterosis retained and why they should be different.

3.11 References

Australian Brahman Breeders' Association Limited (2011) The Brahman Sire Summary. Available from: http://www.brahman.com.au/ (accessed 20 October 2011).

Cartwright, T.C., Ellis, Jr, G.F., Kruse, W.E. and Crouch, E.K. (1964) Hybrid vigor in Brahman–Hereford crosses. Technical Monograph No 1, Texas Agricultural Experimental Station.

Cundiff, L.V. (1970) Experimental results on crossbreeding cattle for beef production. *Journal of Animal Science* 30, 694–705.

Cundiff, L.V., MacNeil, M.D., Gregory, K.E. and Koch, R.M. (1986) Between- and within-breed genetic analysis of calving traits and survival to weaning in beef. *Journal of Animal Science* 63, 27–33.

Cundiff, L.V., Koch, R.M., Gregory, K.E., Crouse, J.D. and Dikeman, M.E. (1993) Characteristics of diverse breeds in Cycle IV of the cattle germplasm evaluation program. Progress Report No. 3. R.L. Hruska US MARC, USDA-ARS.

Damon, Jr, R.A., McCraine, S.E., Crown, R.M. and Singletary, C.B. (1959) Performance of crossbred beef cattle in the Gulf Coast region. *Journal of Animal Science* 18, 437–447.

Damon, Jr, R.A., Crown, R.M., Singletary, C.B. and McCraine, S.E. (1960) Carcass characteristics of purebred and crossbred beef steers in the Gulf Coast region. *Journal of Animal Science* 19, 820–844.

Damon, Jr, R.A., Harvey, W.R., Singletary, C.B., McCraine, S.E. and Crown, R.M. (1961) Genetic analysis of crossbreeding beef cattle. *Journal of Animal Science* 20, 849–857.

Dillard, E.U., Rodriguez, O. and Robison, O.W. (1980) Estimation of additive and nonadditive direct and maternal genetic effects from crossbreeding beef cattle. *Journal of Animal Science* 50, 653–663.

Falconer, D.S. and Mackay, T.F.C. (1996) *Introduction to Quantitative Genetics*, 4th edn. Benjamin/Cummings, Menlo Park, CA.

Franke, D.E. (1980) Breed and heterosis effects of American Zebu cattle. *Journal of Animal Science* 50, 1206–1214.

Frisch, J.E. and Vercoe, J.E. (1978) Utilizing breed resources in growth of cattle in the tropics. *World Animal Review* 25, 8–12.

Gregory, K.E., Trail, J.C.M., Marples, H.J.S. and Kakonge, J. (1985) Characterization of breeds of *Bos indicus* and *Bos taurus* cattle for maternal and individual traits. *Journal of Animal Science* 60, 1165–1174.

Gregory, K.E., Cundiff, L.V., Koch, R.M. and Lunstra, D.D. (1993) Germplasm utilization in beef cattle. Beef Research Progress Report No. 3. R.L. Hruska US MARC, USDA-ARS.

Hammack, S.P. (2009) Texas Adapted Genetic Strategies for Beef Cattle V: Type and Breed Characteristics and Uses. *Texas AgriLife Extension Publication* E-190.

MacKinnon, M.J., Hetzel, D.J.S. and Taylor, J.F. (1989) Genetic and environmental effects on the fertility of beef cattle in a tropical environment. *Australian Journal of Agricultural Research* 40, 1085–1097.

Mason, I.L. (1966) Hybrid vigour in beef cattle. *Animal Breeding Abstracts* 4, 453.

McElhenney, W.H., Long, C.R., Baker, J.F. and Cartwright, T.C. (1986) Production characters of first generation cows of a five-breed diallele: Reproduction of mature cows and preweaning performance of calves by two third-breed sires. *Journal of Animal Science* 63, 59–67.

Mendel, G. (1865) Experiments in Plant Hybridization (reproduced). In: Peters, J.A. (ed.) *Classic Papers in Genetics* (1959), Prentice-Hall, Inc., Englewood Cliffs, NJ, pp. 1–20.

Minvielle, F. (1987) Dominance is not necessary for heterosis: a two-locus model. *Genetic Research* 49, 245–247.

Peacock, F.M., Koger, M., Olson, T.A. and Crockett, J.R. (1981) Additive genetic and heterosis effects in crosses among cattle breeds of British, European and Zebu origin. *Journal of Animal Science* 52, 1007–1013.

Peters, H.F. and Slen, S.B. (1967) Brahman-British beef cattle crosses in Canada. I. Weaned calf production under range conditions. *Canadian Journal of Animal Science* 47, 145–151.

Pirchner, F. (1969) *Population Genetics in Animal Breeding*. Plenum Press, New York.

Rendel, J.M. (1980) Low calving rates in Brahman cross cattle. *Theoretical and Applied Genetics* 58, 207–210.

Sacco, R.E., Baker, J.F., Cartwright, T.C., Long, C.R. and Sanders, J.O. (1989) Lifetime productivity of straightbred and F_1 cows of a five-breed diallel. *Journal of Animal Science* 67, 1964–1971.

Sanders, J. (2005) Evaluation of heterosis retention for cow productivity traits in *Bos indicus/Bos taurus* crosses. In: *Proceedings of 51st Annual Texas A&M University Beef Cattle Short Course*, 1–3 August, College Station, Texas.

Sheridan, A.K. (1981) Crossbreeding and heterosis. *Animal Breeding Abstracts* 49, 131.

USDA-ARS (2013) GPE Progress Reports from U.S. Meat Animal Research Center. United States Department of Agriculture, Agricultural Research Service. Available at: http://www.ars.usda.gov/Main/docs.htm?docid=6238 (accessed 13 December 2013).

Vercoe, J.E. and Frisch, J.E. (1992) Genotype and environment interaction with particular reference to cattle in the tropics. *Australian Journal of Agricultural Science* 5, 401–409.

Warwick, E.J. (1968) Crossbreeding and linecrossing beef cattle experimental results. *World Review of Animal Production*, Vol. IV, pp 37–45.

4 Nutrition

The single largest input in regard to annual cow maintenance cost is forage/feed. It is almost always cheaper to let cows graze growing forage as opposed to transporting forage or other feeds to them. It has been estimated that 50–70% of the cost of weaned calf production is due to feed costs to maintain the cow herd. The challenge for most producers is that adequate amounts of green, growing forage are not available throughout the year. If cows are located where there is never typically any growing forage, then it is not a suitable location for cows unless extremely cheap purchased feeds are available (which is unlikely in most cases). In many situations supplemental feeding of cattle is needed to meet production goals. In some cases, cattle may be fed under confined or semi-confined settings, as in a feedlot. A considerable amount of information in this chapter has been taken from the classic textbooks *The Ruminant Animal*, edited by Church (1988), and *Nutritional Ecology of the Ruminant* by Van Soest (1994). The goal of this chapter is to discuss the fundamental concepts associated with beef cattle nutrition so that general nutrition and associated management decisions can be understood and resulting optimal production can be obtained.

4.1 Classes of Nutrients

It is important to recognize that there are six classes of nutrients: (i) water, (ii) carbohydrates, (iii) protein, (iv) fats or lipids, (v) minerals and (vi) vitamins. Of these nutrients, water is required in the greatest weight per day. Carbohydrates, fat and protein supply energy. The majority of the daily dry matter intake (DMI) is carbohydrates, followed by protein and then fat. Minerals and vitamins are crucial components of the diet, but are supplied in very small amounts. General discussion about each of these nutrients follows.

4.1.1 Water

Cattle perform the best when they have unlimited access to clean water. They typically will not drink more water than they need, but their daily water intake can vary substantially even when climatic and dietary conditions are stable. Both stage of production and environmental temperatures greatly influence water consumption. Table 4.1 shows typical water requirements and consumption for a variety of sizes and types of cattle.

From an environmental aspect, unlimited access to water is critical when very high temperatures and heat stress exist. Cattle can probably only go without water for 24–36 hours before death occurs if ambient temperature is high. Also, cattle that have been severely restricted from water may overconsume it when they are given free access to it again, and death can result from this as well, probably due to resulting electrolyte imbalance. The factors that probably influence water consumption the most are physiological state of the animal (stage of production), the climatic conditions, the type of diet, and the quality of the water. Roughly 60–70% of cattle live weight is water, and it takes roughly 5 liters of water consumption of a beef cow to produce 1 liter of milk.

Different feeds and forages can vary greatly in water content. As a result, feeds are compared on a dry matter basis (0% moisture) for standardized evaluation. Feed consumption is also evaluated on a dry matter basis, and is commonly referred to as dry matter intake (DMI). Not only does this help standardize nutritional comparisons across feeds, but also the economic value. For instance if feed A costs $200 per ton, and feed B costs $300 per ton, which feed is the best buy? Obviously the answer depends on many factors, including the nutrient content of the feed as well as the type of animals it will be fed to. These two feeds may be identical in nutritional content on a dry matter basis and simply have different water content.

Table 4.1. General guidelines for daily water consumption of beef cattle relative to stage of production and temperature.

Weight, kg (lb)	Temperature °C (°F)					
	4.4 (40)	10 (50)	14.4 (60)	21.1 (70)	26.6 (80)	32.2 (90)
	Growing animals, liters (gal)					
182 (400)	15 (4)	16 (4)	19 (5)	22 (6)	25 (7)	36 (10)
273 (600)	20 (5)	22 (6)	25 (7)	30 (8)	34 (9)	48 (13)
364 (800)	23 (6)	26 (7)	30 (8)	35 (9)	40 (11)	57 (15)
	Feedlot animals, liters (gal)					
273 (600)	23 (6)	25 (7)	28 (7)	33 (9)	38 (10)	54 (14)
364 (800)	28 (7)	30 (8)	34 (9)	41 (11)	47 (12)	66 (17)
454 (1000)	33 (9)	36 (9)	41 (11)	48 (13)	55 (15)	78 (21)
	Pregnant cows in winter, liters (gal)					
409 (900)	22 (6)	25 (7)	28 (7)	33 (9)	Not reported	
500 (1100)	25 (7)	27 (7)	31 (8)	37 (10)	Not reported	
	Lactating cows, liters (gal)					
409 (900)	43 (11)	48 (13)	55 (15)	64 (17)	68 (18)	96 (20)
	Mature bulls, liters (gal)					
636 (1400)	30 (8)	33 (9)	38 (10)	44 (12)	51 (13)	72 (19)
727 (1600)	33 (9)	36 (9)	41 (11)	48 (13)	55 (15)	78 (21)

From NRC (2000) with values rounded to nearest liter or gallon. Adequate daily supply of clean water is crucial, and cattle are expected to consume approximately 1 liter of water per every 12 kg of live weight (1 gal per 100 lb) daily under non-extreme conditions; these general guidelines can be used and modified accordingly for many planning purposes.

If feed A was 80% dry matter, its cost would be $0.20 per kg ($200/1000 kg) of feed as purchased (as fed), but its cost per kg of dry matter would be: $200/800 kg = $0.25. Moreover, if feed B was 90% dry matter, its cost per kg as fed would be $300/1000 kg = $0.30, but its cost per kg of dry matter would be $300/900 kg = $0.33.

Water quality considerations

Water quality may be related to environmental conditions such as reduced water flow in creeks or stock ponds that become low due to drought conditions. However, many water quality issues may simply be characteristic of the local geographical conditions or other factors. Water quality is based on what type(s) of substances are in the water that may be harmful to the animals. Most common water quality concerns relate to salinity (dissolved salts), nitrogen-containing compounds such as nitrates/nitrites, biological entities such as bacteria, chemicals from artificial contamination, and possibly blue-green algae. The pH range of water that is acceptable for cattle is typically 6.0 to 8.5.

Stagnant water sources that are small in size pose the highest risk from water quality issues. If cattle have access to more than one water source at a time, they will usually avoid one that has water quality problems, but they may also avoid water sources that taste not as good as others simply from preferred palatability. If cattle are to be located in pastures that have a single water source for extended periods, and the historical use of the water source is unknown, it would be advisable to have the water tested if possible at a local laboratory.

4.1.2 Energy

Energy in feeds comes from carbohydrates, fats and protein, and can be evaluated and expressed under different systems. In regard to cow herd nutritional management, use of total digestible nutrients (TDN) is adequate, and TDN will be the basis of most of the energy discussion in this chapter. However, some discussion of the net energy system is also needed. The net energy system is based on the portioning of the total energy in a feed into different components that result from the digestion steps (Fig. 4.1). Use of the net energy system is more commonly used to formulate diets for intensively managed feedlot and dairy cattle.

The net energy system measures energy in values of calories (or some other measure), and this system is also used in human nutrition to present nutritional values and requirements. A calorie is the amount of energy needed to raise the temperature

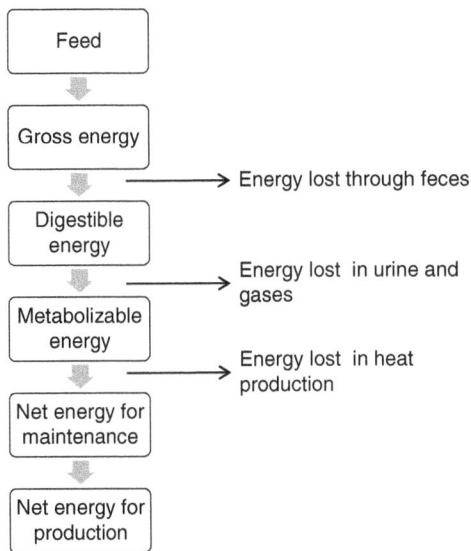

Fig. 4.1. Fractions of dietary energy and energy utilization.

Grains have a high digestibility value because most of the weight is NFE, and mature grass, straw, etc. have a much lower digestibility because the majority of their weight comes from structural carbohydrates. As fiber content increases in feeds, their digestibility decreases. The general breakdown of feeds into these components is referred to as a proximate analysis (Fig. 4.2).

Cattle receive most of the energy from their diets from carbohydrates, but also require fat (lipids) to be present in their feed. Although energy can be obtained from protein, this is not desirable as this is an inefficient process, and protein is usually much more expensive per unit of weight as compared to carbohydrates. Approximately 65–75% of the dry matter weight of grains and forages are carbohydrates. However, not all carbohydrates have the same nutritional value and forages will have higher fiber content than grains; likewise, more mature forages will have more fiber content and less digestible fiber portions than young, immature forages. Although not exact, a general relationship exists between CF and TDN, where if the CF is known, the TDN can be approximated as shown in Table 4.2. In many countries, the CF is required to be published on commercially available feeds.

Although fat is not needed in great amounts in the diets of cattle, it is a necessary component. Fats are needed for absorption of certain vitamins (A, D and E), are needed particularly in neonatal animals for development of the nervous system, and are needed throughout life for proper skin integrity. Supplemental fat in the diet can increase palatability, but levels of 10% or above can reduce palatability and therefore reduce feed intake.

4.1.3 Protein

Protein can be expressed in many ways just as energy can. For most cow herd situations, use of crude protein (CP) is adequate for nutritional management. The CP content is based on the nitrogen percentage in the feed. Because most proteins are approximately 16% nitrogen, the inverse of 0.16 (6.25) is used as a multiplier to yield the CP value. For instance, if we knew a feed was 2.10% nitrogen, its CP value would be: 2.10% × 6.25 = 13.1%, etc. The CP content alone does not indicate its digestibility. Most feeds traditionally fed to cattle are adequate in protein digestibility, but if some non-traditional feed is being considered, the digestibility would need to be evaluated. Feather meal is extremely high in CP, but very

of 1 g of water by 1°C (specifically from 16.5°C to 17.5°C). Consequently, a kilocalorie (kcal) is the amount of energy needed to raise the temperature of 1000 g of water by 1°C, and a megacalorie (MCal) is 1000 kcal, or 1,000,000 calories. In many regions, joules (J) may be used instead of calories. One calorie is equal to 4.184 J (and 1 J is equal to 0.293005736 calories). As with calories, the concepts of kilojoules (kJ) and megajoules (MJ) are more appropriate measures of energy in regard to beef cattle nutritional management. The unit of energy used does not change the concepts employed and are completely interchangeable; the energy requirements of a beef cow might be 11.5 MCal (or 48 MJ) per day, or some other expression, but this is the same concept. In regard to feedstuffs, a feed that has 5.12 MJ of energy per kg has 1.22 MCal of energy per kg.

The TDN of a feed is based on the digestible fractions of crude fat content (measured as ether extract), the crude protein (CP) content and the carbohydrate content that comes from crude fiber (CF) and nitrogen free extract (NFE), and is generally comparable to digestible energy (although it is not exactly the same). The CF comes from structural carbohydrates and generally refers to the weight of the feed that is due to plant cell walls. The NFE component is based on the plant cell interior, which has starches and sugars and other highly digestible carbohydrates.

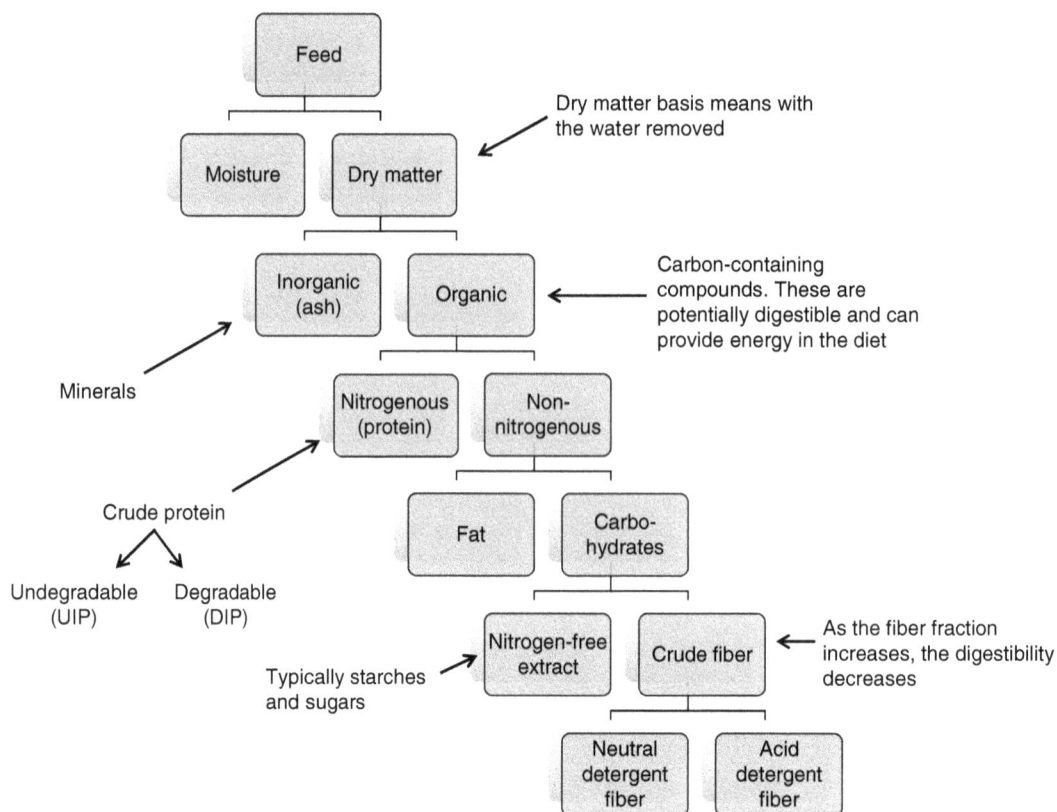

Fig. 4.2. Proximate analysis of feed.

low in digestibility. It costs much more money per kg of protein than compared to per kg of energy, and although energy can be provided to cattle from protein sources, this is to be avoided as much as possible due to its economical inefficiency.

4.1.4 Minerals

Minerals can be classified as macro and micro minerals relative to the amounts that are required in the diet. Minerals whose requirements are measured in units of weight (i.e. grams) are macro minerals, whereas those required in minute quantities (i.e. parts per million) are called micro minerals. The two minerals required in the greatest amounts by cattle are calcium and phosphorus. Cattle required macro minerals are calcium, phosphorus, sodium, magnesium, chloride, sulfur and potassium. Cattle required micro minerals are copper, iron, selenium, manganese, zinc, iodine, cobalt and molybdenum.

Minerals are needed for skeletal formation, various metabolic processes and proper digestion. Mineral deficiencies may result in reduced fertility, retained placentas, weight loss (or improper weight gain) and rough hair coats. A cattle nutrition text or reference such as the beef cattle NRC should be consulted for specific symptoms of various mineral deficiency and toxicity symptoms. References such National Research Council (2000) provide values for the maximum tolerable dietary level of several minerals (the level that when fed for a limited period will not hurt animal performance and should not produce unsafe residues for consumers of products from the animal).

4.1.5 Vitamins

Vitamins are required in very small amounts for various cellular and metabolic functions. Vitamins are broadly classified as water soluble and fat soluble.

Although there may be up to 30 required vitamins, all but three can be synthesized in cattle (or are not known to be needed in the diet). The fat-soluble vitamins A, D and E cannot be synthesized by cattle, but will be present in adequate amounts in many feeds and/or forages. Vitamin A is needed to maintain proper integrity of skin, the lining of the gastro-intestinal tract, and the reproductive tract, eye and mouth membranes and is probably the most important vitamin consideration for cattle. Technically, it is not found in feedstuffs, but its precursors (carotenes or carotenoids) are. It is found in leafy green plants. Vitamin A is needed for membrane function and thought to potentially reduce stress-related illness. Vitamin D is needed for calcium and phosphorus absorption, skeletal formation and calcium mobilization from the skeleton. Vitamin D is found in sun-cured roughages, and may be synthesized to some degree in the body. Vitamin E is usually found in most grains and forages. It is needed for proper

Vitamin A absorption. Vitamin E may also influence the onset of stress-related illness. Animals grazing growing pastures should have adequate levels of these vitamins. Animals in confinement must be given feeds (or supplements) with these vitamins. Cattle grazing dormant forage or fed old hay for extended periods of time (3 months or more) are likely to become deficient in these vitamins. Specific information about deficiency and toxicity symptoms and concerns should be viewed in beef cattle nutrition reference textbooks such as *The Ruminant Animal* (Church, 1988), *Nutritional Ecology of the Ruminant* (Van Soest, 1994), etc.

The purpose of this section has been to give the reader enough fundamental knowledge to understand the underlying principles of nutrition. A summary of some dietary proximate analysis components and the resulting products from rumen fermentation is given in Table 4.3. Understanding some of these basic concepts such as that protein can be used as an energy source gives insight into recommended feeding and supplementation strategies.

4.2 Ruminant Digestion and Digestive System

Ruminants have the ability to survive and thrive on forage-based diets because they have a digestive system where a diverse population of microbes break down cellulose, and utilize high-fiber diets. There are four distinct compartments to the ruminant stomach: reticulum, rumen, omasum and abomasum, each with distinct functions. The rumen is the fermentation vat for the ruminant and by far the largest compartment. The reticulum forms boluses of forage to be regurgitated and further chewed in

Table 4.2. Approximate values of total digestible nutrients (TDN) relative to crude fiber (CF) content.

CF (%)	TDN (%)
4	79
6	76
8	73
10	70
12	67
14	64
16	61
18	58
20	55
22	52
24	49

Table 4.3. General overview of rumen digestion products.

Feed proximate component	Chemical constituent (polymer)	Chemical component (monomers)	Ruminal fermentation products
Nitrogen-free extract	Carbohydrates (hexosan)	Glucose and other hexoses	Acetate, propionate and butyrate
Crude fiber	(Pentosan)	Pentoses	Acetate, propionate and butyrate
Crude protein	True protein and non-protein nitrogen	Amino acids	Acetate, propionate, butyrate, isobutyrate, isovalerate, ammonia
Crude fat	Triglycerides and galactosides	Glycerol and fatty acids	Propionate and saturated fatty acids
Crude ash	Minerals	Elements	Reduced elements, microbial cells, carbon dioxide and methane

From Owens and Goetsch (1988).

the mouth. The omasum is important for water absorption, and the abomasum is analogous to the stomachs of monogastric animals.

The microflora of the rumen is influenced by the type of diet cattle are consuming (see Table 4.4). A rapid change in diet can produce digestive upset, and in certain cases can produce life-threatening issues, which are discussed in detail in Chapter 9. The types of microbes present in the rumen include bacteria, protozoa, fungi and yeast; bacteria are by far the most prevalent. There are certain species of microbes that are starch digesting and others that are fiber digesting; as a result, the entire population of microbes in the rumen is influenced by the diet, and the microbial population dictates which nutrients are available to the animal. The normal pH of the rumen is 5.5 to 7.0. Starch-digesting microbes typically prefer a pH of 5.5 to 7.0, and fiber-digesting microbes prefer a pH of 6.2 to 7.0. When the pH drops below the preferred level of for microbes, they cease to grow. When the pH drops below 5.5, there is potential for acidosis, an undesirable condition than can affect the health and livelihood of the animals. This is discussed in more detail later in the chapter.

4.2.1 Forage Utilization

Ruminants have the ability to consume high roughage diets that cannot be utilized by most monogastric animals. This is a great advantage, allowing livestock production from grazing-based forage resources. However, the conversion of fibrous forages to meat and milk production is not a very efficient process, with something like 10 to 35% of the energy ingested being converted into net energy to the animal, because 20 to 70% of the cellulose in roughages may not be digested (Varga and Kolver, 1997). Figure 4.3 illustrates plant cell components relative to digestibility.

Table 4.4. Grouping of rumen bacteria types according to substrates fermented.

Cellulolytic	Sugar utilizing
Hemicellulolytic	Acid utilizing
Pectinolytic	Lipid utilizing
Amylolytic	Ammonia producing
Ureolytic	Methane producing
Proteolytic	

From Yokoyama and Johnson (1988).

When there is increased forage digestibility more microbial protein is supplied to the small intestine, which in turn means that there will be a reduced need for protein supplementation. There will also be increased volatile fatty acid (VFA) production, which results in energy that can be used by animals. The small intestine is the primary site for protein absorption.

Cattle, like all ruminants in grazing situations, need to maximize forage digestion in order to meet their energy and protein requirements. Factors that limit grazing cattle's ability to meet their nutritional requirements include: forage species, degree of maturity, lignin concentration, and the ammonia requirements of the cellulose-digesting bacteria in the rumen. Forage-based diets (unlike grain-based diets) result in a time period (referred to as the lag phase) required for cellulose-digesting bacteria to attach to forage particles, and therefore the energy available from forages is directly related to surface area available to the rumen microbes. Grazing cattle need to maximize forage digestion for increased performance in traits such as growth (average daily gain) or milk production.

4.2.2 Rumen Microflora

The bacteria and other microbes in the rumen are responsible for digesting feed. The microbes digest feeds by attaching to the feed particles and releasing enzymes. There may be 10 to 100 bn bacteria cells per gram and 100,000 to 1 million protozoa per ml of rumen contents. The environment within the rumen only allows survival of microbes that are suited for the specific conditions found there. The typical gas mixture composition in the rumen may be 65% carbon dioxide, 27% methane, 7% nitrogen, 0.6% oxygen, 0.2% hydrogen and 0.01% hydrogen sulfide, and the temperature is maintained at 38–42°C. Protozoa comprise about 2% of the weight of the rumen contents, 40% of the microbial nitrogen and 60% of the microbial fermentation. Bacteria are responsible for the majority of microbial nitrogen. It is typical that with a forage-based diet, there may be approximately 1 to 3 bn bacteria per ml of rumen contents, but with a grain-based diet there can be as many as 8 to 10 bn bacteria per ml. When ruminant animals are born, they are not yet ruminants because they have no rumen microbial population, and it takes several (9 to 12) weeks for the rumen to become fully functional even when microbes are introduced. Young

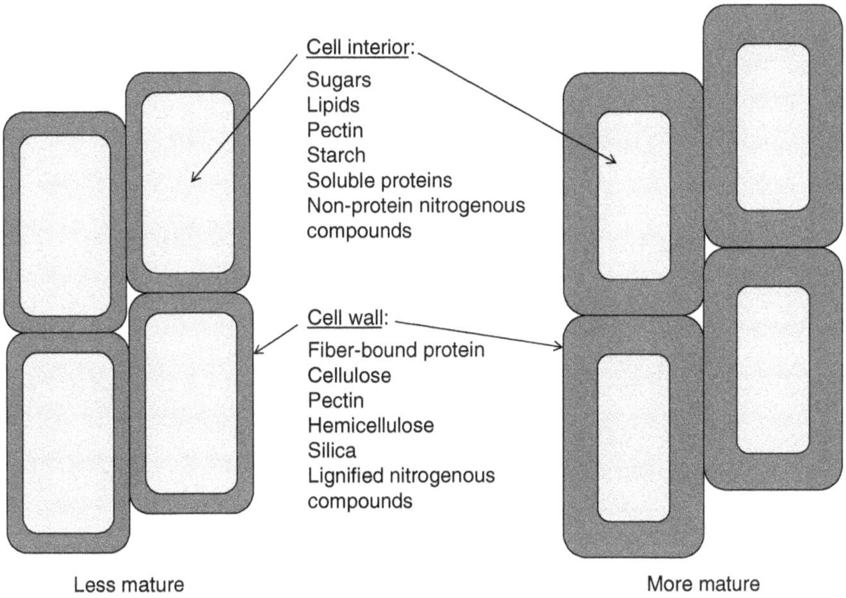

Fig. 4.3. Plant cell components as related to nutritional considerations (adapted from Van Soest, 1982). As plants mature, a higher percentage of their weight becomes structural carbohydrates (and an increased percentage of cell wall constituents).

animals naturally develop a population of microbes from exposure to those from other animals through shared feeding, water sources and other environmental contact.

The surface area of feeds influences how well the microbes can digest them. Processing of roughages and grains increases the surface area of these feedstuffs because it reduces particle size, and then can increase the rate of digestion by allowing more bacteria to attach. This can be done mechanically by the animal through chewing as well as through milling and processing techniques. Roughages are broken down with the rumination process (chewing, swallowing, then later regurgitation and re-chewing) until they are either digested, or become small enough in particle size that they can pass from the rumen into the omasum. Particles leaving the rumen are typically smaller than 1 mm, although particles as large as 50 mm may leave the rumen. Reducing the particle size of many mature forages can reduce maintenance energy expenditures due to a reduction in visceral organ mass and reducing the energy expenditure of rumination and re-chewing, and studies that have evaluated chopped hay as compared to non-processed hay have shown increased digestibility and animal performance; this same phenomenon

has been repeatedly shown with grain processing. Feed particles must be broken down to 20–50 mm to leave the rumen. In some cases, as with maize (corn) meal, the grain can become so processed that it is too readily available for rumen fermentation and presents a nutritional management challenge. Digestion normally occurs from the inside layers of the forage to the outer layers. Limitations to the speed at which this occurs include the physical and chemical properties of the forage, the moisture level of the forage, time for penetration of the waxes and cuticle layer, and the extent of lignification (Varga and Kolver, 1997).

Essentially, all feed proteins are degraded in the rumen to become ammonia. Protein is needed in the diet of ruminants for a nitrogen source for microbial growth and fermentation in the rumen. This is why a non-protein nitrogen source such as urea can be fed to ruminants. This is also the principle behind why protein supplementation of cattle grazing low-quality forage (low protein forage) increases the efficiency of the forage utilization. Microbial protein produced in the rumen typically provides 50–80% of the amino acids needed by the animal. Many proteins are degraded in the rumen and are referred to as degradable intake protein (DIP), but some proteins escape degradation in the rumen and are referred to as undegradable intake protein (UIP). Amino acids that escape the rumen are absorbed in the small intestine. DIP has been reported to be the first-limiting nutrient for beef cattle grazing low-quality forages.

Cellulose-digesting bacteria prefer ammonia (NH_3) as their N source, and this has led to the potential and application of substituting non-protein nitrogen (NPN) for a portion of the degradable true protein in supplements for grazing cattle. This has become a widespread practice in many regions and settings. Care must be used when feeding NPN sources in pastures or pens that have potential for non-ruminant animals to consume the feed it contains because NPN is toxic to non-ruminants.

In production situations where energy is limiting due to relatively low-quality forage and/or reduced DMI, microbial protein reaching the small intestine may be insufficient for desired animal growth or milk production, and combinations of DIP and UIP (also called bypass protein or escape protein) are probably needed. Many feeds that have been used as protein supplements such as cottonseed meal have desirable UIP.

4.2.3 Volatile Fatty Acid Production and Use

With forage-based diets, the VFAs may provide 50–85% of the metabolizable energy used by cattle. Acetate (CH_3–COOH) is primarily used in fatty acid synthesis for subcutaneous and seam fat as well as milk fat. Propionate (CH_3–CH_2–COOH) is converted to glucose in the liver. Butyrate (CH_3–CH_2–CH_2–COOH) is converted to ketones during absorption through the rumen wall and then goes to fatty acid synthesis in adipose tissue and mammary gland tissue. Absorption of VFAs occurs directly from the rumen through the skin cell layer into the blood supply. Ammonia is also absorbed directly from the rumen into the bloodstream, and the amount absorbed is directly related to the amount of ammonia and the pH in the rumen. The majority of energy absorption in cattle occurs through the rumen, although some occurs through the small intestine, and a very small amount occurs through the large intestine.

The diet impacts the end products of rumen fermentation, and therefore subsequent growth rate and milk production. High-concentrate (grain-based) diets result in increased propionate production relative to acetate. Propionate is the only glucose-producing (glucogenic) fatty acid (the only VFA converted to glucose in the liver). Glucose is required for the functioning of the nervous system tissue, muscle, adipose tissue, mammary glands and the fetus. Higher levels of glucose production in the liver have resulted in greater average daily gain, more lean tissue growth per day, and more intramuscular fat (marbling) deposited. Table 4.5 shows some VFA proportions reported for various forage to concentrate ratios in cattle diets.

Table 4.5. Effect of forage to concentrate ratio (F : C) in diet on rumen volatile fatty acid (VFA) ratio production in cattle.

F : C ratio	Acetate (%)	Propionate (%)	Butyrate (%)
100 : 0	71.4	16.0	7.9
75 : 25	68.2	18.1	8.0
50 : 50	65.3	18.4	10.4
40 : 60	59.8	25.9	10.2
20 : 80	53.6	30.6	10.7

From Annison and Armstrong (1970).

4.3 Cattle Energy and Protein Requirements

When the nutritional requirements of cattle are not met, there will typically be a corresponding decrease in performance, and the more severely the nutrient is compromised, the more severe the reduction will be. If deficiencies are severe and prolonged, animal health will be threatened. On the other hand, overfeeding of nutrients beyond what the animals need is wasteful from an economic standpoint and potentially environmentally harmful. In regard to meeting energy and protein requirements it is important to recognize what the general levels are, and that these fluctuate with the physiological state (stage of production) of the animals. Table 4.6 illustrates the general, cyclical fluctuation for energy and protein requirements in a herd of mature cows with an annual (365-day) production calendar.

Early lactation is when the energy, protein, calcium and phosphorus, and other nutrient requirements will be highest; on the other hand, during late lactation and mid-gestation is when nutrient requirements will be lowest. It must be recognized that cows on a target 365-day calving interval will be in more than one stage of production at certain times of the year (i.e. they will be lactating with this year's calf while being pregnant with next year's calf), and for a significant amount of time such as 4–6 months. It is also important to realize that animals in different stages of development have different nutrient requirements, with younger animals needing better nutrition than mature animals. Table 4.7 shows general guidelines for TDN and CP requirements of different cattle ages and production stages.

A challenge in management of grazing beef cattle herds is that both the nutritional requirements of the animals, as well as the nutritional characteristics of the plants, change throughout the year and corresponding production cycles (see Fig. 4.4), and to achieve 'consistent' results regarding animal performance, varying nutritional management is probably needed.

Precise knowledge of nutritional requirements (i.e. exactly how many MJ of energy or g of CP or Ca, etc.) is not needed for typical daily management or to achieve overall production goals. On the other hand, a lack of knowledge about general nutritional requirements of cattle in different stages of production and weights and a lack of general knowledge about the nutritional characteristics of feed resources can be detrimental. Table 4.8 illustrates more precise nutritional requirements of beef cows during stages of gestation and lactation. Tables

showing nutritional requirements for yearling cattle and bulls are provided in the appendix in both metric and imperial units (yearling metric see Table A4, yearling imperial see Table A5, bulls metric see Table A6, bulls imperial see Table A7). Tables 4.9 and 4.10 provide nutritional characteristics of a variety of feedstuffs to be fed to cattle.

The nutritional characteristics or 'book values' of feeds assumes that proper storage is in place. Many problems of stored feeds can be avoided by keeping rodents, insects, moisture/mold and sources of artificial contamination away from stored feeds. The nutritional content of spoiled, old or contaminated feeds can be drastically different from book values and typical situations. Using ingredients that have no problems such as molds and fungi that may occur under typical environmental conditions should be a priority (Chapter 9 discusses health problems resulting from moldy feeds). As a result, periodic testing of feeds for both nutritional content (which can vary), as well as potential toxins that may be common in a region or situation, may be warranted.

The ideal time to harvest forages is before they are mature enough to produce seed heads. Hay quality is directly related to timing of harvest as well as harvest and storage conditions. In some cases if more mature, lower quality hay is being fed it may be a good management strategy to roll out the hay bales along the ground to let animals eat as compared to feeding in a hay ring; if there is a lot of seed in the hay, this can help spread seed in areas of pasture that have suffered from overuse. If hay can be stored long term without exposure to rain, this will greatly prolong its 'shelf life' as well as its nutritional quality.

There are several factors that influence the daily feed intake of cattle. In general cattle will consume (if given the opportunity) somewhere from 1.5 to 3.0% of their body weight per day (dry matter basis). Factors influencing feed intake can be thought of as those of the animal, those of the feed or diet, and interactions. Animal factors include age, stage of production, sex class, and body capacity, as well as differences among individuals, whereas feed or diet factors include TDN, CP, DM%, salt content, plus several types of feed additives (ionophores, hormone and hormone-like products, etc.). The term used to describe when animals have free choice access to feed and water (all they want) is *ad libitum*. Although the nutritional quality affects the palatability of the feedstuffs in

Table 4.6. Production cycle/calendar of the cow herd thought of in 3-month increments relative to protein and energy requirements.

Stage	Crude protein, CP (%)	Total digestible nutrients, TDN (%)
Early lactation	Highest	Highest
Mid-lactation to early pregnancy	Intermediate	Intermediate
Late lactation to mid-gestation	Lowest	Lowest
Late gestation	Intermediate	Intermediate

Table 4.7. General guidelines for crude protein (CP) and total digestible nutrient (TDN) requirements for cattle.

Class of cattle	CP (%)	TDN (%)
Growing animals	12–13	60–65
Lactating mature cows	10	55–60
Dry mature cows	7–8	50–55

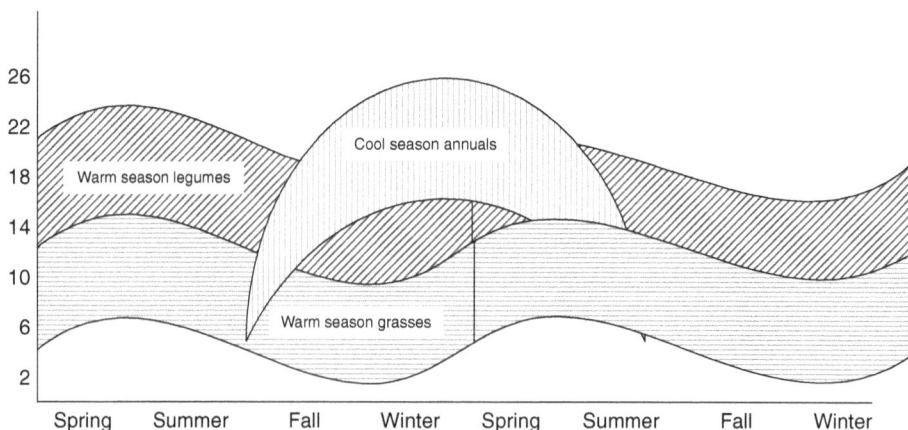

Fig. 4.4. Relative crude protein (CP) levels expected in different classes of forage across seasons. Much variability can occur across and within regions, years and management scenarios.

general, palatability alone (taste preference) does not dictate nutritional content or usefulness of the feed.

4.3.1 General Protein and Energy Supplementation Considerations

Supplementation is the addition of feeds or nutrients to the diets of animals. For many situations, the most typical types of supplementation needed to meet the minimum nutritional requirements of grazing animals are protein and minerals. The degree of mineral supplementation and the level of minerals to be supplemented are direct results of the soil characteristics of the geographical region. For instance in many areas, sodium, chlorine and phosphorus are deficient in the soils, and therefore are deficient in the plants that grazing animals consume in those areas. Other areas may have other minerals that are deficient, and/or minerals that are well above recommended levels, with potential toxicity issues. If possible soil tests of the geographical area provide valuable reference information. Figure 4.5 illustrates some protein and energy supplementation scenarios.

For most cost-effective nutritional management it is very important to match the production calendar to the environmental resource calendar as much as possible. This in many cases will help minimize supplementation feed costs and/or minimize reduction in animal performance. As with genetic considerations, producers should attempt to match the requirements of the cows to pasture conditions as much as possible for optimal use of

forage resources. Although Fig. 4.6 illustrates the expected TDN requirement and forage content fluctuations, a similar trend exists for protein as well.

If the peaks (or valleys) of the requirements and the availability can be matched up, there is more potential to meet the nutritional demands from the forage. For this particular forage species (as is the case with most forages), there is no time of the year that will completely meet the energy and protein requirements of the cow. It is possible that having multiple forage species in the pasture (or in different pastures) will allow for all requirements to be met, and this is discussed in more detail in Sections 5.1.6, 5.4.1 and 5.4.2. In regard to TDN for these scenarios, if the cow calves on 1 March, there will be 5 months where the forage does not meet the requirement, with a severe deficiency in three of those months; however, if the cow calves on 1 October, there will be 6 months with a TDN deficiency, five of which will be severe. This is a simplified example as there is no guarantee that expected nutritional levels will be the same as book values in any given year, or that values will be similar from one year to the next. However, this concept will be true on average across time.

It is apparent that both energy and protein are lacking for at least part of the year, but what is not apparent is that energy does not have to be supplemented to meet the animal's requirements. This is one of the major advantages of utilizing ruminants such as cattle. Two concepts here need to be explained before energy and protein supplementation considerations can be fully understood: forage

Table 4.8. Nutrient requirements of beef cows during gestation and lactation.

Weight (kg)	Expected calf BWT (kg)	DM intake (kg/day)	DM intake (% of BW)	TDN (% DM)	NEm (MJ/kg)	CP (% DM)	Ca (% DM)	P (% DM)	TDN (kg)	NEm (MJ)	CP (g)	Ca (g)	P (g)
Gestating cows, middle third of pregnancy													
408	28.6	7.7	1.9	50	4.06	7.1	0.17	0.14	3.8	30.543	544	13	10
454	31.3	8.2	1.8	50	4.06	7.1	0.17	0.14	4.1	33.054	590	14	11
499	34.0	8.6	1.8	50	4.06	7.1	0.17	0.14	4.4	35.564	635	15	13
544	36.3	9.5	1.7	50	4.06	7.1	0.18	0.15	4.7	38.074	680	17	14
590	39.0	10.0	1.7	50	4.06	7.1	0.18	0.15	5.0	40.585	726	18	15
635	41.3	10.4	1.7	50	4.06	7.1	0.19	0.15	5.3	42.677	726	20	16
680	43.5	11.3	1.6	50	4.06	7.1	0.19	0.15	5.5	45.187	771	21	17
Gestating cows, last third of pregnancy													
408	28.6	8.6	2.1	54	4.61	7.9	0.25	0.16	4.7	40.166	680	21	14
454	31.3	9.5	2.1	54	4.61	7.9	0.25	0.16	5.1	43.514	726	24	15
499	34.0	10.0	2.0	54	4.61	7.9	0.25	0.16	5.5	46.861	816	26	17
544	36.3	10.9	2.0	54	4.61	7.9	0.26	0.17	5.9	50.208	862	28	18
590	39.0	11.3	2.0	54	4.61	7.9	0.26	0.17	6.2	53.555	907	30	20
635	41.3	12.2	1.9	54	4.61	7.9	0.26	0.17	6.6	56.484	952	32	21
680	43.5	12.7	1.9	54	4.61	7.9	0.27	0.17	6.9	59.413	998	34	22

Weight (kg)	Peak milk (kg)	DM intake (kg/day)	DM intake (% of BW)	TDN (% DM)	NEm (MJ/kg)	CP (% DM)	Ca (% DM)	P (% DM)	TDN (kg)	NEm (MJ)	CP (g)	Ca (g)	P (g)
Lactating cows													
408	4.5	10.0	2.5	56	4.89	8.7	0.24	0.17	5.6	48.953	862	24	17
	6.8	10.9	2.7	57	5.07	9.6	0.27	0.18	6.2	55.647	1043	29	20
	9.1	11.8	2.9	59	5.35	10.4	0.30	0.20	6.9	62.342	1224	35	23
454	4.5	10.9	2.4	55	4.80	8.5	0.23	0.17	5.9	51.463	907	25	18
	6.8	11.8	2.6	57	5.07	9.4	0.27	0.18	6.6	58.576	1088	31	21
	9.1	12.2	2.7	59	5.26	10.2	0.29	0.20	7.3	65.270	1270	36	24
499	6.8	12.2	2.5	57	4.98	9.2	0.26	0.18	6.9	61.086	1134	32	22
	9.1	13.2	2.6	58	5.17	10.0	0.29	0.19	7.6	68.199	1315	38	25
	11.4	14.1	2.8	59	5.35	10.6	0.31	0.21	8.3	74.894	1497	43	29
544	6.8	13.2	2.4	57	4.98	9.0	0.26	0.18	7.3	64.015	1179	34	23
	9.1	13.6	2.5	58	5.17	9.8	0.28	0.19	8.0	70.710	1361	39	27
	11.4	14.5	2.7	59	5.35	10.5	0.31	0.21	8.6	77.822	1542	44	30

Continued

Table 4.8. Continued.

Weight (kg)	Peak milk (kg)	DM intake						Lactating cows						
		(kg/day)	(% of BW)											
590	6.8	13.6	2.3	56	4.89	8.9	0.26	0.18	7.6	66.944	1224	35	24
	9.1	14.5	2.4	57	5.07	9.6	0.28	0.19	8.2	73.638	1406	40	28
	11.4	15.4	2.6	59	5.26	10.3	0.30	0.20	8.9	80.333	1542	46	31
635	9.1	15.0	2.4	57	5.07	9.5	0.28	0.19	8.6	76.149	1406	42	29
	11.4	15.9	2.5	59	5.26	10.1	0.30	0.20	9.3	82.843	1587	48	32
	13.6	16.8	2.6	59	5.35	10.6	0.32	0.21	9.9	89.956	1769	53	35
680	9.1	15.9	2.3	57	5.07	9.3	0.28	0.19	8.9	78.659	1451	43	30
	11.4	16.8	2.4	58	5.17	9.9	0.30	0.20	9.6	85.772	1633	49	33
	13.6	17.2	2.6	59	5.35	10.5	0.31	0.21	10.2	92.466	1814	54	37

Original values from NRC (2000) and adapted from Oklahoma State University (2004).
This table is presented in imperial units in Table A3.

Table 4.9. Expected composition of some selected roughage feeds (dry matter basis).

Feed	DM %	NDF %	eNDF % NDF	CP %	DIP % CP	TDN %	NEm MJ/kg	NEg %	EE %	Ca %	P %	K %	S %	Cu ppm	Mn ppm	Zn ppm
Alfalfa hay, early bloom	90	39	92	25	88	60	5.44	3.04	2.9	1.41	0.22	2.51	0.30	13	36	30
Alfalfa hay, mid bloom	90	47	92	22	84	58	4.89	2.86	2.6	1.37	0.22	1.56	0.28	11	28	31
Alfalfa hay, full bloom	90	49	92	17	82	55	4.80	2.40	2.3	1.19	0.24	1.56	0.27	10	28	26
Alfalfa cubes (pellets)	91	46	40	18	70	57	5.07	2.68	2.0	1.30	0.23	1.90	0.35	9	32	18
Alfalfa dehydrated 17% CP	92	45	6	19	41	61	5.63	3.23	3.0	1.42	0.25	2.50	0.24	9	34	21
Bermuda hay, vegetative	90	69	80	15	80	57	5.07	2.68	2.3	0.59	0.28	1.90	0.30	12	170	36
Bermuda hay, early bloom	90	75	90	10	72	53	4.52	2.21	1.9	0.51	0.20	1.60	0.25	10	140	31
Bermuda hay, full bloom	90	79	98	8	68	47	3.60	1.38	1.8	0.43	0.18	1.40	0.21	8	110	26
Corn silage	35	46	70	8	72	72	7.10	4.52	3.1	0.28	0.23	1.10	0.12	4	24	22
Cotton seed hulls	90	87	100	4	55	45	4.15	0.28	1.9	0.15	0.09	1.10	0.05	13	119	10
Fescue hay, early bloom	87	68	98	13	72	57	5.07	2.68	4.8	0.45	0.37	2.50	0.21	11	200	34
Fescue hay, full bloom	88	73	98	9	68	50	4.80	1.48	3.5	0.40	0.26	1.70	0.17	7	100	23
Peanut hulls	91	74	98	8	40	22	3.32	0.00	1.5	0.20	0.07	0.90	–	–	–	–
Prairie hay	91	73	98	6	63	52	4.61	1.11	2.0	0.40	0.15	1.10	0.06	4	59	34
Rice hulls	92	81	90	3	45	13	3.23	0.00	0.9	0.14	0.07	0.50	0.08	3	320	24
Sorghum silage	32	59	70	9	71	59	5.35	2.95	2.7	0.49	0.22	1.72	0.12	9	69	30
Sudan grass silage	31	64	61	10	72	58	5.17	2.86	3.0	0.58	0.27	2.40	0.14	37	99	29
Sunflower seed hulls	90	73	90	4	35	40	3.87	0.00	2.2	0.00	0.11	0.20	0.19	–	–	200
Wheat silage	33	62	61	13	79	59	5.35	2.95	3.2	0.40	0.28	2.10	0.21	9	80	27
Wheat straw	91	81	98	3	40	42	3.97	0.00	1.8	0.16	0.05	1.30	0.17	5	35	6
Wheat straw, ammoniated	85	76	98	9	75	50	4.61	1.11	1.5	0.15	0.05	1.30	0.16	5	35	6

eNDF = % of NDF, DIP = degradable intake protein. Original values from NRC (2000) and adapted from Oklahoma State University (2004).

Table 4.10. Expected composition of some selected feedstuffs.

Type of feed	DM%	NDF%	ENDP[a] % of NDF	CP%	DIP[b] % of CP	TDN%	NEm MJ/kg	NEg MJ/kg	EE%	Ca%	P%	K%	S%	Cu ppm	Mn ppm	Zn ppm
By-product feeds																
Barley malt pellets with hulls	90	50	34	18	64	68	6.55	4.06	1.4	0.19	0.68	0.27	0.85	6	32	61
Corn gluten feed	90	40	36	24	75	80	8.12	5.44	3.2	0.07	0.15	0.27	0.43	7	22	73
Distillers grains with solubles, corn	90	46	4	30	27	92	9.59	6.73	10.6	0.32	0.83	1.1	0.4	11	28	80
Distillers grains with solubles, sorghum	92	43	4	31	47	88	9.13	6.27	10	0.25	0.65	0.5	0.4			68
Grain screenings	90			14	65	65	6.18	3.69	5.5	0.25	0.34					30
Rice bran, full fat	91	23	0	14	70	72	7.10	4.52	19	0.07	1.7	1.8	0.19	12	396	40
Rice mill by-product	91	60	0	7	60	42	3.97	0.00	5.7	0.4	0.31	2.2	0.3			31
Soybean hulls	90	64	28	12	72	77	7.75	5.07	2.6	0.53	0.18	1.4	0.12	18	10	38
Wheat bran	89	46	4	17	72	70	6.83	4.34	4.5	0.13	1.29	1.4	0.24	14		96
Wheat middlings	89	36	2	19	78	79	8.03	5.35	4.6	0.15	1	1.4	0.24	11	128	96
Wheat mill run	90	37	0	17	72	75	7.47	4.89	4.4	0.12	1	1.2	0.22	21		90
Wheat shorts	89	30	0	20	75	80	8.12	5.44	5.4	0.1	0.95	1.1	0.2	13		118
Feed grains																
Corn (maize) grain, whole	88	9	60	10	42	88	9.13	6.27	4.3	0.02	0.3	0.4	0.12	3	8	18
Corn (maize) grain, steam flaked	85	9	40	10	41	93	9.78	6.83	4.1	0.02	0.27	0.4	0.12	3	8	18
Wheat, cracked, rolled or ground	89	12	0	14	77	89	9.22	6.36	2.3	0.05	0.44	0.4	0.14	6	37	40
Sorghum, cracked, rolled or ground	89	16	5	11	45	82	8.39	5.63	3.1	0.04	0.32	0.4	0.14	5	15	18
Sorghum, steam flaked	82	20	38	11	38	90	9.41	6.46	3.1	0.04	0.28	0.4	0.14	5	15	18
High protein meals and seeds																
Cottonseed, whole	91	47	100	23	62	95	9.96	7.01	17.8	0.16	0.62	1.22	0.26	8	12	38
Cottonseed meal, 41%	90	25	23	48	58	77	7.75	5.07	1.8	0.22	1.25	1.7	0.44	17	57	66
Peanut meal, solvent	91	27	23	50	73	77	7.75	5.07	3.6	0.24	0.58	1	0.3	16	29	38
Soybean meal, 48%	91	9	23	54	64	87	9.04	6.18	1.2	0.28	0.71	2.2	0.47	23	41	61
Soybeans, whole	88	15	100	40	72	93	9.78	6.83	18.8	0.27	0.64	2	0.34	15	35	59
Sunflower seeds, high oil	91	24	80	19	75	122	13.10	9.50	42	0.71	0.51	1.06	0.21	20	35	53
Sunflower seed meal with hulls	91		23	26	80	60	6.27	3.87	2.9	0.45	1.02	1.27	0.33	4	20	105
Mung beans	90			23	25	79	8.03	5.35		1.19	0.68					
Feather meal	92	44	23	86	27	69	6.73	4.15	6.5	0.6	0.62	0.2	1.85	14	12	95

Original values from NRC (2000) and adapted from Oklahoma State University (2004).

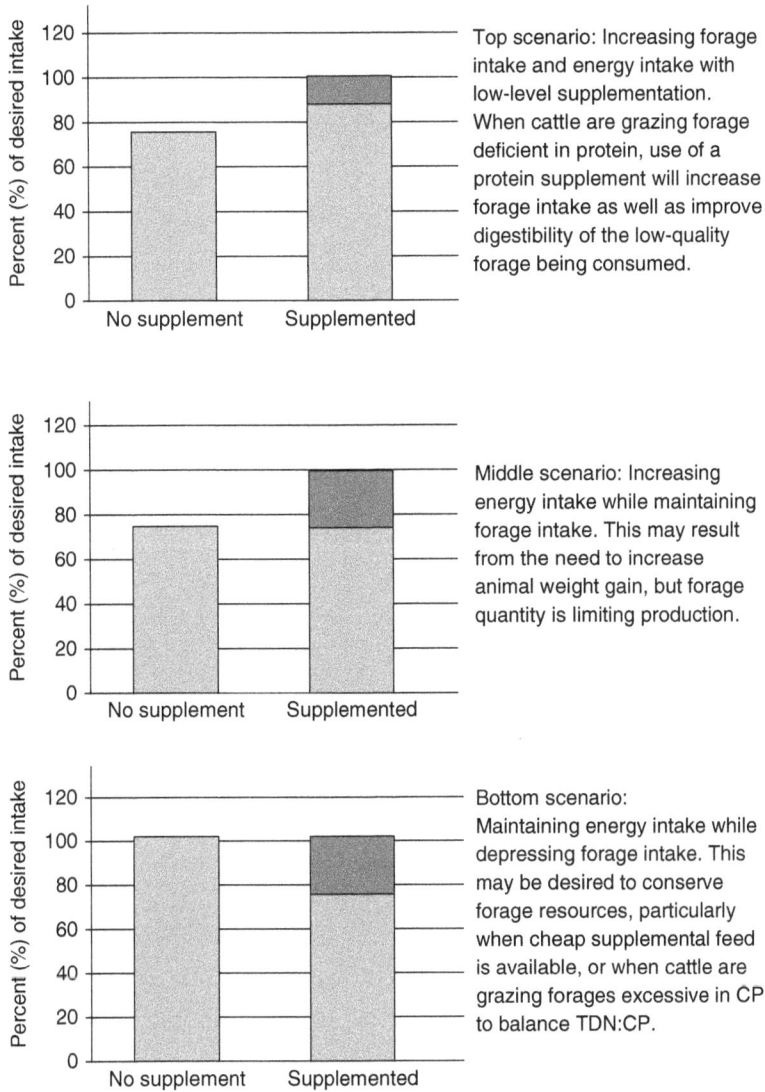

Top scenario: Increasing forage intake and energy intake with low-level supplementation. When cattle are grazing forage deficient in protein, use of a protein supplement will increase forage intake as well as improve digestibility of the low-quality forage being consumed.

Middle scenario: Increasing energy intake while maintaining forage intake. This may result from the need to increase animal weight gain, but forage quantity is limiting production.

Bottom scenario: Maintaining energy intake while depressing forage intake. This may be desired to conserve forage resources, particularly when cheap supplemental feed is available, or when cattle are grazing forages excessive in CP to balance TDN:CP.

Fig. 4.5. Different forage intake considerations associated with feeding supplements. Graphs and concepts adapted from McCollum (1997).

quantity and forage quality. Forage quantity is the amount of forage available, whereas forage quality is the nutritional characteristics of the available forage in regard to protein and digestibility. The forage quantity is what should typically dictate supplementation of energy, while the forage quality is what should dictate supplementation of protein. When forage quality is limiting, supplementation of protein increases the digestibility of the low-quality forage as well as stimulates forage intake,

and this typically occurs with forage that is at or below 7% CP. In many tropical areas it is the low CP of forages that limits their nutritional effectiveness.

If energy is supplemented to cattle grazing low-quality forage, there is little if any increase in forage consumption or digestibility. As the quality (CP %) of forage increases, there is typically a substitution effect where the amount of supplement consumed will be traded off or substituted for a corresponding

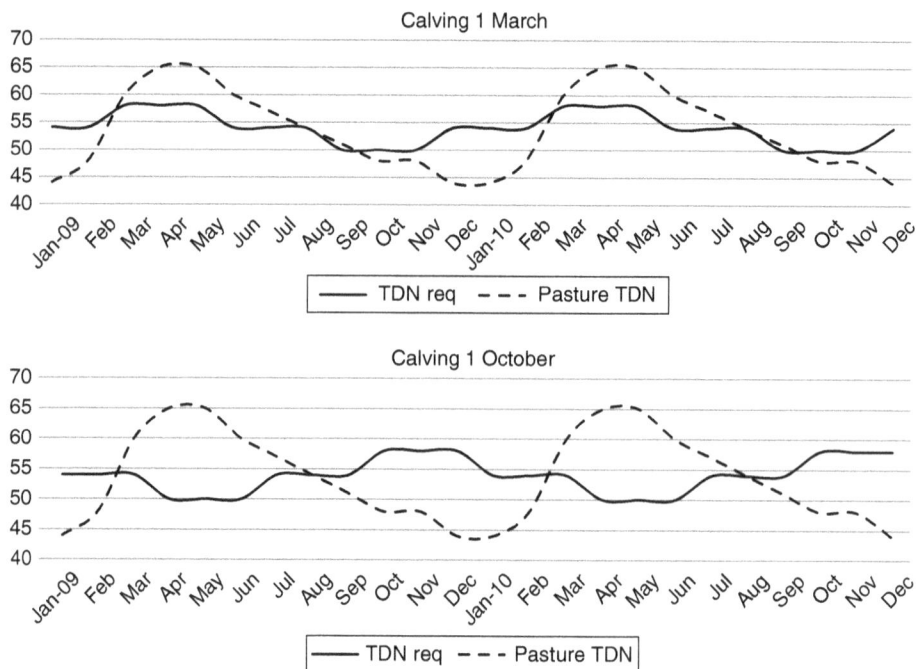

Fig. 4.6. Expected cow total digestible nutrient (TDN) requirements and pasture TDN levels relative to two different calving dates for a single-species, warm season pasture in the northern hemisphere.

amount of forage. For high-quality forages such as those 12% or more CP, there is typically a 1:1 substitution effect with an energy type supplement (0.5 kg of supplement consumed will provide a corresponding 0.5 kg less forage intake, etc.).

Ideally, the ratio of TDN:CP in the diet should be 5:1 (4:1 to 7:1 is acceptable), and this general relationship affects forage intake and digestibility. Forages of low quality have a high TDN:CP ratio (i.e. 50% TDN and 5% CP results in a 10:1 ratio), but highly nutritious actively growing forages may be of such high quality that their TDN:CP is too low for optimal rumen activity and therefore utilization by cattle. Small grain pastures such as wheat may be 75% TDN and 25% CP with a resulting 3:1 ratio. In these cases, an energy supplement would help balance the diet as the forage has more protein than can be utilized by the cattle.

Very rapid changes in feed types and diet composition can cause digestive problems in cattle. This is typically thought of being more severe when transitioning from a low level to a high level as compared to the opposite direction. If cattle are consuming a high roughage diet and they need to be converted to a high grain diet, this should be done incrementally for a smooth transition. Sudden addition of grain to the diet, especially consumed at one time, causes a reduction in rumen pH, and can lead to acidosis as well as bloat (accumulation of gases within the rumen); change in pH of the blood can have impacts on blood flow in the extremities, and can lead to founder (an abnormal growth of the hooves). Rapid switch from a low protein diet to a very high protein diet can produce ammonia toxicity. Cows that have been consuming low-quality roughages or grazing low-quality pasture that are turned into small grain pasture without any type of dietary transition are at risk of ammonia toxicity. A reasonable strategy would be to start to feed cows a protein supplement for a few days before turning them into the high-quality pasture so that the new diet will not be as great a shock to the microbial rumen population.

4.3.2 Economic Considerations

In some cases, the economic aspects of different feeds may be the most important comparison to consider. If the feeds are similar in nutritional value, the price per unit of TDN and CP should be considered. Table 4.11 illustrates the concept of comparing relative energy values of various hays to different grains based on TDN values. For example if a mature grass hay was 47% TDN and maize was 90% TDN, the ratio of 90/47 = 1.9, and this means that a kg (or a lb) of maize would provide the same amount of TDN as 1.9 kg (or 1.9 lb) of that hay. The cost per kg (or lb) of the hay as compared to the cost per kg (or lb) of the grain must be considered to decide which one is better economic value. It may be possible that a portion of the diet supplied by grain (or some other feed) may be more profitable (and this may mean which scenario loses less money in certain cases).

Beef cattle managers need to be familiar with the nutritional content of feeds to best compare their relative values per unit weight of protein and energy. Two different feedstuffs may be hugely different in nutritional content, but two feeds may also be near identical in nutritional content and only differ in price. Table 4.12 shows five example feeds in regard to their cost per kg of TDN and CP.

To calculate the cost per kg (or lb) of TDN or CP, take the cost per kg of feed and divide by the percentage TDN or CP. For example, the milled feed has a market price of $300/1000 kg = $0.30 per kg; dividing this value by the TDN ($0.30/0.75 =

$0.40) gives the cost of $0.40 per kg of TDN, and a cost of $2.14 ($0.30/0.14 = $2.14) per kg of CP. Another way to think about this is that in the 1000 kg of feed there are 750 kg of TDN and 140 kg of CP. So, the cost per kg of TDN was $300/750 kg = $0.40, and the cost per kg of CP was $300/140 kg = $2.14. Protein is always more costly than energy because it makes up a smaller percentage of the total weight of the feed. Simply making feeding management decisions on the price of feed without regard to price per unit of nutrient will lead to unproductive and unprofitable choices (Fig. 4.7). More example calculations with imperial units are provided in appendix Table A8.

4.3.3 Evaluation of Fecal Consistency

The more familiar managers are with the 'normal' behaviors and observations of their cattle, and the more educated they are about the nutritional requirements of animals, the more prepared they are to identify and correct potential nutritional problems. An area of useful observation is in regard to fecal consistency. The fecal consistency changes according to the diets that animals consume, and this can help determine supplementation strategies. In regard to dietary protein, the feces will typically become dark green and more loose

Table 4.11. Relative energy values of grains compared to some hays based on TDN values.

Grain	TDN (%)	Mature grass (47% TDN)	Sorghum–sudan hybrid (56% TDN)	Alfalfa (60% TDN)
Maize (corn)	90	1.9	1.6	1.5
Barley	84	1.8	1.5	1.4
Oats	77	1.6	1.4	1.3
Sorghum	83	1.8	1.5	1.4
Wheat	88	1.9	1.6	1.5

From TCE (2007), originally adapted from Great Plains Beef Cattle Feeding Handbook.

Table 4.12. Feed comparisons on as-fed basis relative to cost per kg of total digestible nutrients (TDN) and crude protein (CP).

Feed	Market price	CP (%)	TDN (%)	Cost ($) per kg CP	Cost ($) per kg TDN
Milled feed	$300 per metric ton	14	75	2.14	0.40
Whole shelled maize (corn)	$240 per metric ton	10	90	2.40	0.27
Cottonseed meal	$400 per metric ton	41	75	0.98	0.53
High-quality grass hay	$70 per 500 kg bale	16	56	0.88	0.25
Low-quality grass hay	$40 per 500 kg bale	7	50	1.14	0.16

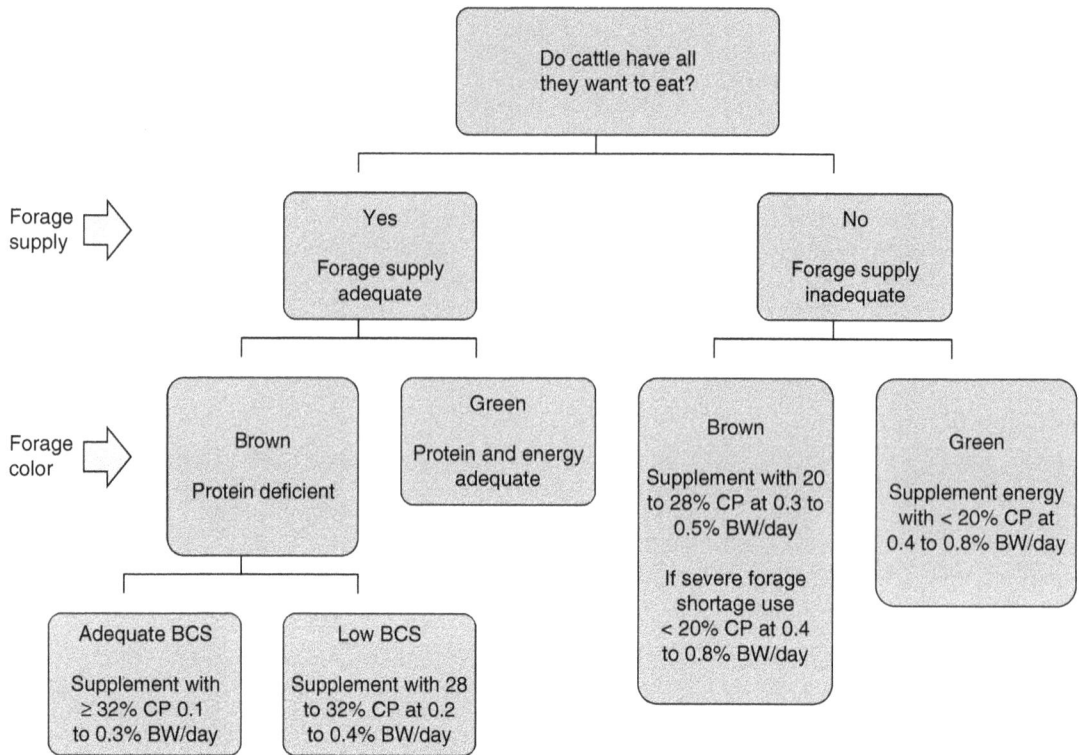

Fig. 4.7. Protein and energy supplementation decision guide. Adapted from Mathis (2004).

(more runny and watery) when animals are consuming diets high in protein. This is characteristically seen when cattle are grazing immature small grain pastures. As the protein in the diet decreases, there is more firmness of the feces, progressing to the point where animals that are consuming very low protein diets will have fecal pads (droppings) that stack very high. Some examples of these fecal appearances are shown in Fig. 4.8. Protein in the diet is not the only factor that affects fecal consistency. The color of the feces can be influenced by the general metabolic status of the animal. Cattle that are experiencing acidosis will often have a steel gray color to their feces. Several other factors can influence appearance of the feces, and more discussion on this topic can be found in Chapter 9.

4.3.4 Nutritional Management of Groups Versus Individuals

It is usually a group of animals that supplemental feed is provided for, not individually fed animals.

When a group of animals is offered feed, the degree of variation among the animals should be considered as it can affect the efficiency of feeding. Although variation in animal weight is discussed in detail in Chapter 6, some discussion is needed here to understand nutrient delivery. Cattle can vary in their weight for a wide variety of reasons (mature skeletal size [frame size], age, stage of production, percent body fat [body condition] and muscle expression. Both the mean (average) weight and the amount of individual variation in weight need to be considered. When the average weight of the group is used to dictate the amount of feed needed, more variation among individual animal weight will result in some animals potentially receiving too much and others receiving too little. Cows of the exact same weight may need different feed amounts if some are lactating and others are not, etc. If the amount fed is appropriate for the group average, but there is wide variation in the animals' nutritional requirements due to the above-mentioned factors, it will probably be effective to sort animals

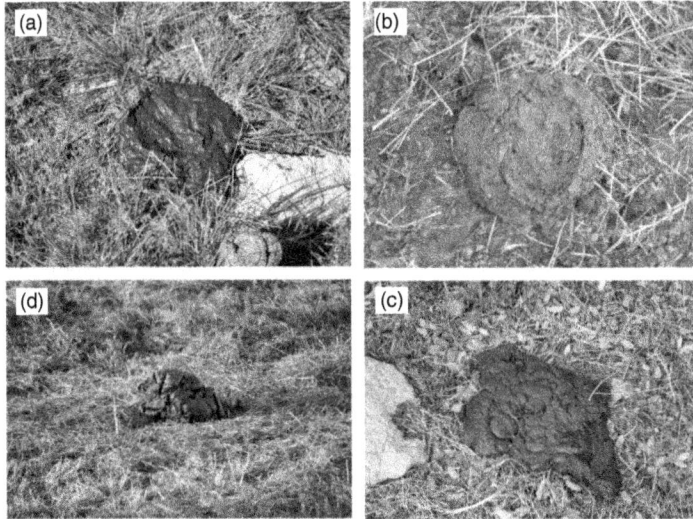

Fig. 4.8. Examples of differences in fecal consistency (clockwise from upper left) for cattle consuming (a) excessive, (b) ideal, (c) marginal, and (d) deficient protein diets.

into distinct groups where the groups may be very different from one another, but there is more consistency within each group. The exact same total amount of feed may be provided overall, but its effectiveness in its delivery will be greatly improved.

4.3.5 Use of Ionophores

Ionophores are a class of feed additives that have been shown to alter the rumen microbial population so that propionate-producing microbes are increased in number and acetate-producing microbes are correspondingly reduced. There are three VFAs produced in the rumen (acetate, propionate and butyrate) that are involved in energy metabolism. In general, more propionate is produced with a starch-based energy source such as grain compared to a cellulose-based source such as roughage. Acetate is produced more from roughage feed sources than grains. Propionate provides more glucose production in the animal, which is generally thought of as a more energy-efficient pathway. Ionophores are fed in very small amounts per animal per day (a few milligrams), and it is important that animals receive a consistent amount of ionophores daily if their use is desired. Use of ionophores has been show to increase feed efficiency in feedlot animals consuming grain-based diets as well as grazing cattle consuming forage-based diets.

Ionophores have also been shown to increase reproductive performance of heifers and cows under pasture conditions (reviews by Geary, 2003; Hess *et al.*, 2005), helping in time to breed back after calving even among those at the same body condition (degree of body fat). There has been some discussion as to whether or not ionophores should be considered antibiotics or not because they do alter the rumen microbes, but they do not appear to completely inhibit bacterial growth of any type (as compared to the typical definition of antibiotic, a product that completely inhibits the growth of a range of bacteria). Where they are available, ionophores are typically inexpensive (a few cents per animal per day), and appear to be cost-effective in most situations for which they are recommended.

4.3.6 Feed Alternatives

Because cattle are ruminants, they have the ability to take advantage of a wide variety of feedstuffs, including non-traditional feeds that are by-products of other industries. Cottonseeds were originally a low-value by-product of cotton lint production. Grazing low-quality grain stubble can be an effective management strategy. Products of human food industries that are no longer desirable for human consumption such as stale bread may be a 'valuable' feed resource for cattle if it is cheap enough and can be fed effectively. Feeding any of these types of feeds to cattle without providing any other additional feeds would be disastrous because they cannot provide all the nutrient requirements on their own. Moreover, some may have unusually high levels of an undesirable nutrient. Some degree of testing is always advisable when a new feed is being considered in the diets of cattle, particularly when there is no experience or historical evidence of its use. These types of opportunities may exist in many geographical areas, and cattle producers should always be on the lookout for regional by-product type feeds that may be equal nutritionally to traditional feeds, but of substantially lower price. There is more discussion on this and related concepts in Sections 13.2 and 13.6.

4.3.7 Growth-Promoting Implants (Hormone Growth Promotants)

There are several growth-promoting products that can be administered to cattle that can impact growth and nutritional considerations, such as hormone-based implants and beta-agonist feed additives. These products are discussed in Section 7.5.

4.4 Feeding Cattle under Intensive or Confined Conditions

In many production or economic situations, feeding cattle a complete mixed diet (which may be called a total mixed ration or TMR) may be warranted, where everything animals consume is provided to them as in a feedlot setting. In some situations there may be conditions that dictate intensive feeding of cattle for a particular stage of production or environmental situation (such as emergency drought management, aftermath of a typhoon, etc.). Tables 4.13 and 4.14 provide some examples for feeding growing cattle.

These diets that are formulated for young, growing cattle are also appropriate for other cattle types such as mature cows or feedlot fattening (finishing) of growing and/or older cattle, but they would be better utilized in combination with other feedstuffs to exactly match requirements. The rate of weight gain associated with any diet fed to cattle depends on their current weight (and stage of production), as well as the level of intake of the diet. Usually with intensive cattle feeding, there is a need to balance increased production potential with associated

Table 4.13. Example diets appropriate for weaned calves.

Example A	Percentage (%)	Example B	Percentage (%)
Sudan hay	55.7	Grass hay	54.7
Wheat	20.0	Wheat	20.0
Maize	18.0	Grain sorghum	19.0
Soybean meal	5.0	Soybean meal	5.5
Dicalcium phosphate	0.5	Dicalcium phosphate	0.5
Limestone	0.5	Limestone	0.5
Salt	0.3	Salt	0.3
Example C	Percentage (%)	Example D	Percentage (%)
Grass hay	52.2	Sudan hay	57.2
Grain sorghum	40.0	Maize	35.0
Soybean meal	7.0	Soybean meal	7.0
Dicalcium phosphate	0.5	Dicalcium phosphate	0.5
Limestone	0.5	Limestone	0.5
Salt	0.3	Salt	0.3
Example E	Percentage (%)	Example F	Percentage (%)
Alfalfa hay	67.7	Alfalfa hay	59.2
Maize	31.5	Grain sorghum	40.0
Dicalcium phosphate	0.5	Dicalcium phosphate	0.5
Salt	0.3	Salt	0.3

Adapted from Lusby and Gill (1982). It is recommended that hays be ground and grains be rolled or ground. Components of these examples may be used as supplements for cattle grazing forage if relative nutritional values of forages are known.

Table 4.14. Example rations for growing cattle.

Example A		Example B		Example C		Example D		Example E	
Ingredient	kg/day	Ingredient	kg/day	Ingredient	kg/day	Ingredient	kg/day	Ingredient	kg/day
Dry hay – 12% CP	5.5	Dry hay – 15% CP	5.2	Dry hay – 15% CP	2.5	Dry hay – 15% CP	0.5	Dry hay – 15% CP	0.5
Grain corn	2.8	Mixed grain	3.5	Corn silage	8.5	Corn silage	7.5	Corn silage	10.0
Supplement	0.5	Supplement	0	Corn screenings	2.5	Haylage	7.5	High moisture corn and cob meal	2.5
Premix	0.03	Premix	0.06	Supplement	0.5	Corn	1.4	Corn gluten feed	2.4
				Premix	0.03	Supplement	0.2	Premix	0.09
						Premix	0.03		
Consideration: Would require about 1100 kg of hay and 560 kg of grain corn per calf.		Consideration: Would require about 1040 kg of hay and 700 kg of mixed grain per calf.		Consideration: Would require about 500 kg of hay, 1700 kg of corn silage and 500 kg of corn screenings per calf.		Consideration: Would require about 100 kg of hay, 1500 kg of corn silage and 1500 kg of haylage per calf.		Consideration: Would require 100 kg of hay, 2000 kg of corn silage, 500 kg of high moisture corn and cob meal and 480 kg of corn gluten per calf.	
Total feed (kg) per calf for feeding period:									
1766		1752		2806		3426		3098	

These examples are taken from Martin (2006) as for Ontario, Canada; diets such as these should result in ADG of 0.9 kg/day over 200 days for cattle growing from 227 to 408 kg of weight with an expected daily DMI of 2.5% of body weight; all are expected to provide approximately 8.6 to 8.7 feed to gain ratio.

increased costs. This same concept is also true with grazing situations, but may not be obvious to managers if they do not know or consider the forage cost and/or consumption. Any ration that is formulated to meet nutritional requirements must be fed at the recommended rate (kg per animal per day) for it to be effective. Feeding 10 kg every day will not produce the same results as feeding 7 kg one day and 13 kg the next, even though the average is still 10 kg per day.

The North American cattle industry relies on confinement feeding of cattle destined for beef for somewhere from 90 to 200 days with a high-concentrate, grain-based diet. In many regions placing cattle into feedlots is desired to increase animal growth as well as fat deposition in order to produce higher quality carcasses. Cattle may also be fed under confined or semi-confined settings during a development stage of their life, such as with replacement heifers or young bulls destined for breeding. In other regions, cattle may be fed forage-based diets such as with cut-and-carry systems. In all of these scenarios, it is critical for managers to know the nutritional requirements of the animals, the nutritional characteristics of the feed

(which is only truly known from feed sample analyses, not just book values), and the intended use and goal for the animals.

4.4.1 Evaluation of Individual Variation in Feed Intake and Efficiency

Feed efficiency can be measured using several different methods (see Archer *et al.*, 1999, for a good general discussion). A widely used industry method has been and continues to be the feed conversion ratio (kg of feed consumed per kg of weight gain), and the inverse of units of weight gain per unit of feed consumed has also been used to evaluate feed efficiency. Feed conversion ratio or its inverse may be negatively correlated with mature size (Koots *et al.*, 1994; Archer *et al.*, 1999), especially when evaluated on an age-constant or time-constant basis. If cattle are evaluated for weight gain per unit of time, this tends to favor those with larger mature size (because they will have different types of growth curves). Consequently, selection for improved (i.e. lower) feed conversion could favor larger animals, and consequently lead to increased mature cow size.

Another method of measuring feed efficiency is residual feed intake (RFI) or net feed efficiency (NFE), which was first proposed by Koch *et al.* (1963). This approach accounts for the size and average daily gain of the cohort of the animals, predicts the expected feed intake relative to these traits, and the difference between predicted intake and actual intake is RFI or NFE; this deviation is relative to the average expected for the cohort, and therefore comparing RFI across contemporary groups is not simplistic. As large-scale individual data collection has become economically feasible due to advances in feeding equipment technology, feed intake of individuals has received more interest and research evaluation. Large differences exist in feed intake and measures of feed efficiency among cattle of all breeds that have been evaluated, and the potential to improve efficiency of production of individual cattle seems to be a realistic goal to pursue.

4.4.2 Transition of Diets

The ability to 'step up' cattle from a lower concentrate diet to a high-concentrate diet should be of concern to anyone providing feed to cattle, but is critical to understand for feedlot managers and associated personnel involved with animal management. Offering cattle too much grain too rapidly alters the rumen digestion processes and can lead to acidosis and potentially death if the pH drop is severe enough, but it can also cause general digestive upset and bloat. Most feedlots will utilize a 3 to 6 diet step-up plan that transitions cattle through a series of formulated rations beginning with a low concentrate, high roughage diet to an eventual high concentrate, low roughage diet for finishing, and the transition avoids potential digestive problems as it allows the rumen microbial population to adjust.

4.4.3 Manure Production

It is appropriate that the last topic discussed in this chapter on nutritional considerations is manure production. Manure is the waste produced from digestion and is composed of both urine (the liquid waste removed from the blood by the kidneys) and feces (the semi-solid waste that comes from the large intestine). The amount of manure, as well as its nutritional composition, will change directly as a function of the diet, according to the digestibility of the particular feedstuffs consumed. Components of cattle manure include water, undigested feedstuffs,

living and dead microbes, plant nutrients and salts. Manure can be used to improve soil fertility but can also be a contaminant for water sources or add excessive nutrients to soil or water sources. Manure management can be a formidable environmental issue for feedlots and geographical areas where cattle are intensively fed; manure management also has potential biosecurity implications for the populations of animals that may come in contact with it. Table 4.15 provides general guidelines that can be followed to predict manure production from different sizes and diets of cattle.

4.5 Summary

Cattle feeding involves science and art. This is true of nutritional management of cattle no matter the circumstance or type of cattle. General knowledge about (i) the nutritional requirements of the animals (and why they change) and (ii) the nutritional aspects of the feed resources (and why they can change) are critical for effective management. In many instances avoidance of potential nutritional disasters may be more of a concern than precise nutritional management. Many feeds that have different nutritional content will produce widely different results in animals when they are fed at the same rate. Many feeds that have the same nutritional content may have widely different costs. The nutritional goals of one operation or in one type of cattle may not be realistic in others due to climatic conditions, labor resources or costs. Managers need to continually observe nutritional inputs and associated production outcomes to determine what is effective for their particular scenarios.

4.6 Study Questions

4.1 If a pasture to be grazed for the next 3 months is 13% CP, explain whether this is adequate for grazing beef cattle. What fecal consistency might be expected in this setting?

4.2 Lucerne (alfalfa) hay (60% TDN, 17% CP) currently costs $190 per tonne, and grass hay (58% TDN, 12% CP) currently costs $100 per tonne. Calculate (a) the price per kilogram of crude protein and (b) the price per kilogram of energy in TDN for these two types of hay.

4.3 For your geographical region, what would be the ideal time(s) of the year to match high nutritional demands of cows with high nutritional values of forages?

Table 4.15. Daily manure expectations from different size classes of cattle.

Item	Rate excreted per 454 kg (1000 lb) live weight[a]				Examples[b]		
	Grazing cow	Grazing calf	Yearling high forage	Yearling high energy	Grazing cow	Grazing steer	Feedlot steer
Manure quantity							
Weight kg (lb)/day	28.57 (63.0)	26.39 (58.2)	26.80 (59.1)	23.22 (51.2)	34.6 (75.6)	23.6 (51.7)	20.5 (44.8)
Volume (cubic m (ft)/day	0.0283 (1.00)	0.026319 (0.93)	0.026885 (0.95)	0.023206 (0.82)	0.0343 (1.2)	0.0237 (0.8)	0.0204 (0.7)
Total solids kg (lb)/day	3.31 (7.30)	3.42 (7.54)	3.07 (6.78)	2.68 (5.91)	4.0 (8.8)	2.7 (5.9)	2.4 (5.2)
Manure organic matter							
Volatile solids kg (lb)/day	2.81 (6.20)	2.91 (6.41)	2.74 (6.04)	2.47 (5.44)	3.406 (7.4)	2.413 (5.3)	2.174 (4.8)
COD kg (lb)/day	2.72 (6.00)	2.72 (6.00)	2.77 (6.11)	2.54 (5.61)	3.296 (7.2)	2.441 (5.3)	2.242 (4.9)
C:N ratio	10	12	11	10			
Manure nutrients							
N kg (lb)/day	0.15 (0.33)	0.14 (0.30)	0.14 (0.31)	0.14 (0.30)	0.18 (0.4)	0.12 (0.3)	0.12 (0.3)
P kg (lb)/day	0.05 (0.12)	0.05 (0.10)	0.05 (0.11)	0.04 (0.094)	0.07 (0.1)	0.04 (0.1)	0.04 (0.1)
K kg (lb)/day	0.12 (0.26)	0.09 (0.20)	0.11 (0.24)	0.10 (0.21)	0.14 (0.3)	0.10 (0.2)	0.08 (0.2)

[a]Values from Hamilton (2004). Assumed weights are: grazing cows 550 kg or 1200 lb, grazing calf 200–340 kg or 450–750 lb, yearling cattle 340–454 kg or 750–1000 lb, grazing and feedlot steer 400 kg or 875 lb. Manure weight is total mass of urine and feces excreted; volatile solids refers to mass percentage of total solids that will ignite when heated to 550°C; COD is chemical oxygen demand and is related to amount of oxygen required to digest manure; carbon to nitrogen ratio (C:N) is related to stability.

[b]The calculation for daily manure production in the examples comes from taking the appropriate rate for the size class in accordance with the animal's weight (for instance daily manure solids from a 550 kg cow takes the rate of 3.31 kg/day/454 kg of weight and multiplies this value by the ratio of the animal's weight of 550 kg as compared to the standard rate, or 3.31 kg is expected from a 454 kg cow, but a 550 kg cow is expected to produce 550/454 = 1.21 times as much, or 4.0 kg, etc.). This same example on an imperial unit basis would be 1200 lb cow takes the rate of 6.2 lb/day/1000 lb of weight and multiplies this value by the ratio of the animal's weight of 1200 lb as compared to the standard rate, or, 6.2 lb is expected from a 1000 lb cow, but a 1200 lb cow is expectec to produce 1200/1000 = 1.2 times as much, or 7.4 lb, etc.

4.4 You are feeding a set of 50 growing calves that average 273 kg in weight under an intensive setting. What are the expected water intakes and manure outputs per day for (a) each calf, and (b) the entire group? Assuming that the calves will eat 2% of their weight per day in dry matter, calculate how much feed would be needed per calf and for the entire group per day when feeding (c) a silage-based diet with 32% dry matter versus (d) a hay-based diet with 89% dry matter.

4.5 If 100 cows were consuming forage that was 48% TDN and 5% CP, discuss whether this forage is adequate for their nutritional requirements, and what potential supplementation considerations might be involved (a) when forage quantity is plentiful versus (b) when forage quantity is limited.

4.6 Do you think it is possible to increase forage intake through supplementation as well as decrease forage intake through supplementation? Explain.

4.7 Are the nutritional values of forage within a pasture expected to be constant for the whole year? Explain.

4.8 A pasture has plenty of standing forage that has been estimated to be 9% CP and 52% TDN. Discuss whether this pasture is adequate for grazing cattle for the next 90 days.

4.9 What does the TDN:CP ratio represent in regard to cattle diets, and why might this be important?

4.10 Cattle are to be fed a hay-based diet for the next 60 days. Whole maize grain (90% TDN) is available for feed as well. If the hay being fed was 56% TDN, what TDN-equivalent weight (kg) of maize would substitute for 2 kg of hay? How many kg of hay would be substituted by feeding 1 kg of maize? How would these numbers change if the hay was 50% TDN?

4.7 References

Annison, E.F. and Armstrong, D.G. (1970) Volatile fatty acid metabolism and energy supply. In: Phillipson, A.T. (ed.) *Physiology of Digestion and Metabolism in the Ruminant.* Oriel Press, Newcastle upon Tyne, pp. 422–437.

Archer, J.A., Richardson, E.C., Herd, R.M. and Arthur, P.F. (1999) Potential for selection to improve efficiency of feed use in beef cattle: A review. *Australian Journal of Agricultural Research* 50, 147–161.

Church, D.C. (ed.) (1988) *The Ruminant Animal, Digestive Physiology and Nutrition.* Prentice Hall, Inc., New York.

Geary, T.W. (2003) Management of young cows for maximum reproductive performance. In: *Proceedings Beef Improvement Federation 35th Annual Research Symposium and Annual Meeting, 28–31 May, Lexington, Kentucky,* pp. 5–8.

Hamilton, D. (2004) Waste management. In: *Beef Cattle Manual,* 4th edn. Oklahoma Cooperative Extension Service, Oklahoma State University System, Stillwater, OK.

Hess, B.W., Lake, S.L., Scholljegerdes, E.J., Weston, T.R., Nayigihugu, V., Molle, J.D.C. and Moss, G.E. (2005) Nutritional controls of beef cow reproduction. *Journal of Animal Science* 83(E. Suppl.), E90–E106.

Koch, R.M., Swiger, L.A., Chambers, D. and Gregory, K.E. (1963) Efficiency of feed use in beef cattle. *Journal of Animal Science* 22, 486–494.

Koots, K.R., Gibson, J.P. and Wilton, J.W. (1994) Analyses of published genetic parameter estimates for beef production traits. 2. Phenotypic and genetic correlations. *Animal Breeding Abstracts* 62, 826–853.

Lusby, K. and Gill, D. (1982) *Formulating Complete Rations,* OSU Cooperative Extension Service, Fact Sheet 3013, Oklahoma State University.

Martin, D. (2006) Typical beef feedlot and background diets. Ontario Ministry of Agriculture and Food, Fact Sheet Agdex# 425/60. Available at: http://www.omafra.gov.on.ca/english/livestock/beef/facts/06-017.htm (accessed 3 September 2013).

McCollum, F.T. (1997) Supplementation strategies for beef cattle. Texas Beef Cattle Management Handbook, Texas AgriLife Extension Service College Station, TX, Pub. no. B-6067.

NRC (2000) *Nutrient Requirements of Beef Cattle: Seventh Revised Edition.* The National Academies Press, Washington, DC.

Oklahoma State University (2004) *Beef Cattle Manual,* 4th edn, E-913. Oklahoma Cooperative Extension Service Division of Agricultural Sciences and Natural Resources.

Owens, F.N. and Goetsch, A.L. (1988) Rumen fermentation. In: Church, D.C. (ed.) *The Ruminant Animal, Digestive Physiology and Nutrition.* Prentice Hall, Inc., New York.

TCE (2007) Substitution of grains for hay. Texas Beef Cattle Management Handbook, Texas Cooperative Extension, Texas A&M AgriLife Extension Service, College Station, TX.

Van Soest, P.J. (1994) *Nutritional Ecology of the Ruminant,* 2nd edn. Cornell University Press Ithaca, NY.

Varga, G.A. and Kolver, E.S. (1997) Microbial and animal limitations to fiber digestion and utilization. Presented at the Conference on New Developments in Forage Science Contributing to Enhanced Fiber Utilization by Ruminants, *Journal of Nutrition* (special edition), pp. 819S–823S.

Yokoyama, M.T. and Johnson, K.A. (1988) Microbiology of the rumen and intestine. In: Church, D.C. (ed.) *The Ruminant Animal, Digestive Physiology and Nutrition.* Prentice Hall, Inc., New York.

5 Grazing and Pasture Management

The concept of cattle and other livestock grazing various types of lands has generated differing opinions about its usefulness and environmental impacts from a wide variety of people. In many cases it is not the concept of whether areas should be grazed or not that is most critical to consider, but how the grazing animals are to be managed on those lands. It is true that many areas can be environmentally sustained or improved through cattle grazing, and it is also true that these same areas may be substantially damaged by livestock grazing. However, the key to effects of grazing (be it positive, neutral or negative) is generally based on the management of the animals. This chapter discusses the primary factors related to what grazing cattle eat and where they go in pastures, the concepts of carrying capacity and stocking rate, the basic aspects of grazing systems, and the general concepts that managers of cattle on grazing lands should consider for optimal management for their particular scenarios. The organization of this chapter follows the broad categories of land, plant, animal and management considerations (see Fig. 5.1), although there are considerable gray areas in making many of these classifications. The understanding of the factors affecting grazing land use and potential interrelationships among those factors epitomizes the concept of systems-based thinking and management.

5.1 Land Characteristics

The land aspects discussed in this section deal with the physical environment (soil, climate, plant species, slope, etc.) where animals graze. In the sense that the plant communities are a function of the physical environment, carrying capacity is a characteristic of the land (although it can be drastically altered by management of grazing animals). The major points of the 'land' characteristics are discussed individually.

5.1.1 Soil

The primary factor that allows for potential plant and animal production is the soil. The term soil can be defined in several ways, but in general refers to the mineral and/or organic material on the surface of the planet that allows for growth of land-based plants. The characteristics of soils are derived from a host of factors dealing with location and climate. The materials from which the soils were derived are usually referred to as parent materials, and often this is related to a type of rock that ends up in the soil. There are many layers (horizons) associated with soil; the uppermost area is referred to as the topsoil with underlying layers the subsoil. Climate (previous as well as current) is related to many soil characteristics; rainfall, wind and temperature patterns all have an effect on soil type. Many soil characteristics influence the type and amount of plants (and other associated living organisms) they can support. Soils can be classified into different categories based on their characteristics, and levels of categories (from broadest to narrowest) have been: (i) orders, (ii) suborders, (iii) great groups, (iv) subgroups, (v) families and (vi) series. The names of soil series are related to their locations as well as their descriptive type. Many countries have developed their own soil classification system. There has been an ongoing effort to have a globally standardized soil classification system since the early 1980s. Information about many aspects of global soil information as well as specific regional information is available through the FAO (2014) website and its Legacy Maps and Harmonized World Soil Database, World Reference Base for Soils Portal (www.fao.org/soils-portal/soil-survey/soil-maps-and-databases/regional-and-national-soil-maps-and-databases/en/) as well as links to many national soils resources.

In the most basic sense, soil can be described based on its particle size and texture. The three particle classifications are sand, silt and clay. Sand

---

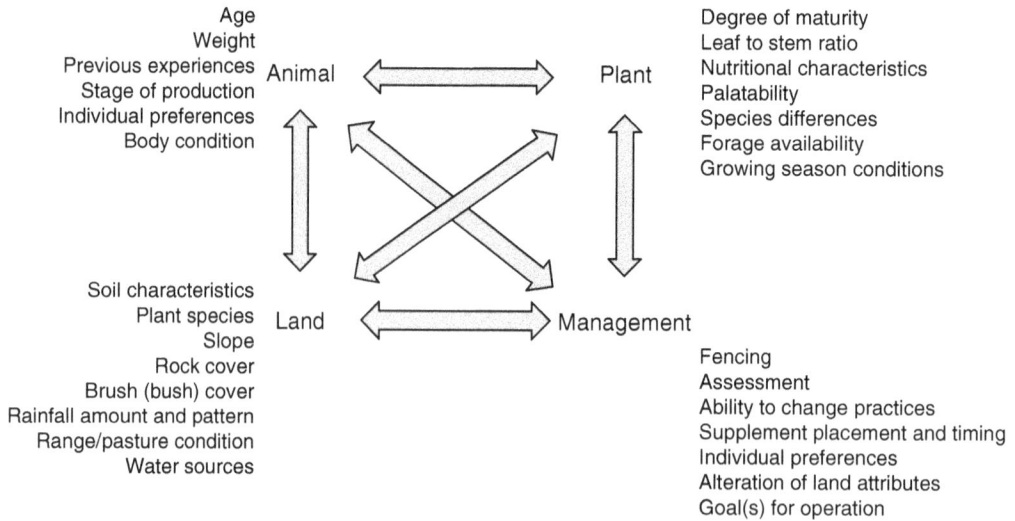

Animal
- Age
- Weight
- Previous experiences
- Stage of production
- Individual preferences
- Body condition

Plant
- Degree of maturity
- Leaf to stem ratio
- Nutritional characteristics
- Palatability
- Species differences
- Forage availability
- Growing season conditions

Land
- Soil characteristics
- Plant species
- Slope
- Rock cover
- Brush (bush) cover
- Rainfall amount and pattern
- Range/pasture condition
- Water sources

Management
- Fencing
- Assessment
- Ability to change practices
- Supplement placement and timing
- Individual preferences
- Alteration of land attributes
- Goal(s) for operation

Fig. 5.1. Primary factors that affect (and interact for) what cattle eat and where they go on grazing lands.

generally has particles from 0.075 to 2 mm in diameter, silt has particles from 0.005 to 0.075 mm in diameter, and clay has particles less than 0.005 mm in diameter. Particles progressively larger than 2 mm would be referred to as gravel, stones, etc., and soils can be classified as mixtures of these three particle types. The soil texture type is strongly related to rainfall penetration and water-holding capacity. Soils that are sandy, silt and clay can all be quite productive relative to many other factors; however the plant types and management will probably be quite different across these soils even if other factors could be constant.

It is a good idea to perform soil testing on areas used for cattle grazing if at all possible. There are several laboratories around the world that can evaluate soil not only for particle texture, but many other characteristics related to soil quality and fertility. If soil sampling is to be conducted to aid in assessment and management, it should be conducted annually within the same season(s), and several sites across the location should also be sampled. Specific guidelines usually accompany soil tests offered by laboratories, and local authorities can also provide guidance. Several factors that impact soil quality can be assessed by soil testing (see Table 5.1); soil testing should also be incorporated, and the results well understood, before major changes in management are carried out (such as fertilizer, herbicide or pesticide application).

Table 5.1. Components of soil testing and associated information.

Type of test	Resulting information
Soil respiration	Biological activity
Infiltration	Ability to take water in from surface
Bulk density	Soil compaction and pore space
Electrical conductivity	Salt concentration
pH	Acidity/alkalinity level
Nitrate	Nitrate level
Aggregate stability	Amount of water-stable aggregates
Slake	Stability of soil fragments in water
Earthworm	Amount of earthworms per unit volume of soil

Taken from USDA-NRCS (2013).

There are many factors that can influence where cattle graze in pastures and how well they utilize available forage resources including water sources, fencing, slope, location and/or density of brush/tree cover, presence of rocks, recent fire, elevation, forage species and distribution, placement of supplemental feed, pasture size, stocking density, and social group aspects. Many of these interact, and it may be difficult to separate many of them. Some of the most obvious ones are discussed here in more detail, but if more in-depth information is sought, some references at the end of the chapter should be consulted.

5.1.2 Slope and Related Topography

The degree of slope is probably related to the type(s) of soil associated with specific land sites. Slope in relation to wind and rainfall influences erosion (loss of soil). Evidence of erosion can include aspects such as gullies, areas where plants grow on higher ground in well-defined spots (may be called pedestals), exposed areas of subsoils, plant damage from wind-blown sand, etc. It should be of utmost priority in any pasture or farming activity to minimize loss of topsoil material. Loss of topsoil severely reduces the production potential of the land, and it is very time-consuming to rebuild. Soil conservation strategies are critical. Different land sites can have the same average percentage slope (2%, 9%, 17%, etc.), but they may be very different in topography due to the presence of ridges, hills, mountains, etc. The general concept used to calculate slope is simply rise over run. For instance, 32 m of increased altitude over 1000 m would be calculated as a 3.2% slope, if elevation changes 39 m over a distance of 400 m, the percent slope is 39 ÷ 400 = 0.0975, or 9.8% slope, etc.

5.1.3 Fire

Fire is a factor that has had a historical significance in many ecosystems. It has also historically contributed to death and destruction through warfare and household accidents. Many people that are not educated about range and pasture management do not realize what a beneficial effect controlled burns can have on many grazing lands. In fact, the overgrowth of shrubs and trees in many areas close to population centers around the world has been due to the suppression of fire. Uncontrolled bush and forest fires can produce massive destruction, but fire in many environments can be one of the most cost-effective ways to promote desirable plants, control invasive undesirable plants, improve carrying capacity and maintain ecological condition over time. The great grassland areas of the world evolved over eons of time where naturally occurring fires were a regular component of the ecosystem. Many producers may not be able to take advantage of controlled fires because they do not leave enough plant residue on lands used for grazing (as with overgrazing), or they are in close proximity to urban and suburban areas. Local authorities must always be included in discussion and consideration of using fire in pasture management programs.

5.1.4 Distance from Water

In general, cattle will not willingly travel further than 3.2 km (2 miles) from water sources and, if utilizing continuous grazing management, these areas around water sources are likely to be overgrazed. Animals will over-utilize these types of areas because they will also be used as leisure areas, particularly if these are the only areas of shade, wind break or other desirable features to guard against environmental elements at certain times of the year and if the area is overstocked.

5.1.5 Brush and Rock Cover

Brush (bush) and rock cover have varying degrees of influence on stocking rate and pasture utilization. At lower prevalence levels these effects reduce forage production and have their influence on reduced stocking rates, but at extreme levels completely eliminate animal use altogether due to both reduced forage production and impaired animal mobility. Extreme levels of rock cover or brush density in small areas can provide 'natural fences' and barriers to land that animals might otherwise graze. It may be cost effective to provide some 'strategic' brush clearing or rock removal when a fairly narrow barrier prevents animal grazing to other areas. This concept is a bit like building a bridge over a ravine or river. In planning pasture utilization, a formal evaluation or documentation of specific areas that have varying brush density may be warranted (see Table 5.2). In many settings, bush or brush is

Table 5.2. Brush (bush) density scores that should be considered in estimating stocking rates and pasture utilization by grazing cattle.

Brush (bush) density score	Description
0	No brush present
1	Very light, few scattered plants
2	Light brush, plants common but mobility of animals or access to areas of pasture not limited
3	Brushy, plant cover thick enough to limit mobility, but cattle can maneuver through it
4	Thick brush, mobility of cattle possible only with pathways
5	Very thick brush, mobility of cattle nearly impossible

Taken from Hohlt *et al.* (2009).

viewed as a nuisance or disadvantage, but it can also be sculpted in regard to its use, and one of the most obvious of these is use of hedgerows to make property lines and natural fences for grazing livestock.

All of the factors mentioned so far in this section can impact how well animals will utilize the land, and how many animals the land will be able to support. This leads to the concept of carrying capacity, which is crucial for grazing land managers to understand.

5.1.6 Carrying Capacity

Recognition of the carrying capacity concept is vital in determining grazing management and its sustainability. This concept may be defined as the number of animals (or animal units) that can be supported by the area without causing long-term ecological damage to the land. Carrying capacity is the inherent ability of the land area that managers need to recognize and utilize in its management. Stocking rate is the number of animals that a manager places on the land area, and if the stocking rate does not correspond to the carrying capacity, there will be a shift in regard to the plant community on that land over time. Both carrying capacity and stocking rate are typically expressed as the number of hectares (or acres) per animal per year. For the discussion in this chapter, this is regarded as a cow producing a calf to a weaning age up to 6–8 months. Furthermore, not all cows are equal in their nutritional demands, and the difference in mature size of the cows is a very important consideration in determining proper stocking rates and associated grazing management.

It is important to also consider the concepts of range condition and forage condition. Range condition refers to the ecological composition of the land in regard to its potential and long-term previous history, whereas forage condition deals with seasonal changes such as nutritional aspects and maturity of the plants on the land. Traditionally, the term 'range' has meant native pastures (plants that are growing there are native to the area and are not the result of having been planted by people), but here the concept of range condition can also be used to consider permanent, improved pastures (pastures where plants have been established or enhanced by people). The type of pasture certainly affects its potential for grazing and should influence its associated management (doing nothing can be an important management decision). An area of land that is planted each year with a seeded crop exclusively for grazing, where almost all of the plant material can be removed by the grazing animals, still needs to have long-term considerations into erosion minimization, nutrient run-off, etc. (Fig. 5.2). Long-term (across years) monitoring of any land resource used for grazing livestock is a must, and a set plan for this monitoring should always be in place.

There are several management practices that can be used to increase carrying capacity. Whether or not increased carrying capacity is desirable depends upon both short-term and long-term goals for economic and environmental sustainability. These practices include cross-fencing and altering grazing distribution through grazing systems, use of improved forage species, irrigation, use of multiple species of grazing animals, use of fertilizer, use of herbicide, use of fire, mechanical removal of invasive plant species, etc. In many areas of the world, costs and/or local policy may prevent use of some of these practices.

5.2 Plant Characteristics

There is a normal expected fluctuation in forage condition associated with every plant species that cattle may eat. These seasonal-based fluctuations are important to understand as they influence when supplementation may be best used, and optimal times for breeding, calving and other production management aspects. Change in forage condition is a short-term management concept, and there may be large differences in forage condition within and/or across years. Change in range condition (or pasture condition) is a long-term concept as it deals with changes across years in regard to the plant species diversity on the land.

Most plants that grow in pastures are nutritious and palatable for grazing livestock (at least for a portion of that plant's life). Other plants may be toxic or poisonous to animals for a portion of the plant's life and/or certain parts of the plants. Local authorities should be consulted for potentially poisonous plants in specific geographical regions. If cattle are given opportunities to choose what they eat, they will eat plants that are both palatable and nutritious. However, in certain situations, animals may not have the opportunity to be as selective, which can lead to them eating less than desirable plants. This is discussed further in Section 5.3.1.

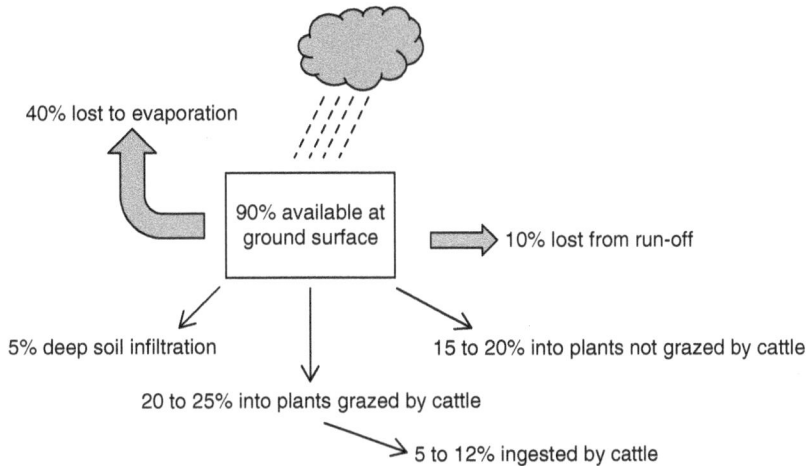

Fig. 5.2. Relative percentages of rainfall use on rangelands. Based on 400 mm annual precipitation. Range condition, soil type, and amounts and species of vegetation and litter present change these percentages. Adapted from McGinty *et al.* (2007).

A general understanding of the classes of forages (cool season vs warm season and grass vs legume or forb, annual vs perennial) is needed for optimal pasture management within and across years.

All species of plants have seasonal growth. The seasonal growth may be different from one year to the next relative to rainfall and other climatic conditions such as temperature. In general, as a young plant grows (or new seasonal growth occurs) it is both highly nutritious and highly digestible. New growth is low in structural carbohydrates and fiber, and much of the mass of the plant is the weight of within-cell wall components (sugars, starches, amino acids and proteins, etc.), but as the plant matures and increases in size, more of its mass is comprised of structural carbohydrates (fiber), it has more stem weight relative to leaves, and the leaves may become more fibrous. As the plant reaches its reproductive phase and produces a seed head, the majority of its nutritional value is represented by the seeds, and the remainder of the plant is low in both protein and digestible energy. If the plant is perennial (meaning that new growth occurs from the same root system each year), the plant may become dormant during a portion of the year, when its growing conditions cannot be met. This occurs in temperate areas with warm season grasses after a severe frost. The plant material above ground becomes dormant (not dead) as it is brown in color and non-growing. Dormancy may

also occur due to drought conditions during the typical growing season. In general, as plants approach maturity, the majority of the plant material decreases in protein, digestible energy and palatability. That does not mean it is not useful for grazing animals, but it does mean that the plant material will not be consumed and/or used the same as it is at other times of the year. It is possible that plants may differ in palatability but not nutritionally, or vice versa, but in general nutritive value and palatability are highly related.

In environments where supplementation can be used to offset nutritional deficiencies, this will also impact pasture utilization. In situations where supplementation is not possible (physically or economically), animals may need to be managed in a fashion so that body condition score is increased enough in the growing season to offset the loss of body condition score in the non-growing season. Pasture management may need to be quite different from one year (or season) to the next simply because of fluctuation in environmental factors that impact plant growth and seasonality. The historical annual average (or median, which is the middle value) rainfall should be used to plan grazing management, but cannot be relied on to be consistent as patterns within a year (not simply total) affect the growing season of forages.

Increases (or decreases) in prevalence of undesirable, invasive plant species are associated with disparity between stocking rate and carrying capacity. As an area is overstocked and overgrazed, there will be over-utilization of nutritious and palatable plants by livestock, and an associated increase in plants less desired for grazing. Overstocking is the term that has been used to represent the level of grazing that removes excessive biomass. This is a short-term concept, and may refer to a single season or time frame such as 7 days, etc. Sometimes it is harder to avoid this than others, as with drought, but planning for undesirable environmental conditions should always be done. Also, the amount of biomass removal that is considered excessive is debatable, and will be further addressed in Section 5.4.3. Overgrazing is the term that has been used to represent the concept of consistently overstocking that leads to land degradation. This is a long-term concept and in the initial stages leads to increases in undesirable plants, and in late stages if severe enough can lead to erosion. Overgrazing comes from a lack of understanding (or care) about matching stocking rate with carrying capacity. This can also be true with undergrazing.

5.2.1 Identification of Grasses and Other Plant Species

To optimally tailor grazing and pasture management for a particular scenario, some knowledge is needed of the exact species of plants that are growing in the pastures. This will aid in several aspects such as (i) knowledge about potentially toxic plants and conditions, (ii) the expected nutritional characteristics of the plant species that are present, (iii) expected seasonal fluctuation in both forage quantity and quality, and (iv) species composition across time to know carrying capacity. It is beyond the scope of this chapter or book to describe grass, forb and browse plant species as these are highly variable across geographical regions (there can be substantial overlap as well), but some sort of plant identification guide, nutritional composition of plant species, and a visual reference guide to estimate forage amount per hectare may be available and should be sought.

Identification of plant species is based on descriptive differences in root system, stems, leaves and seed heads. More pronounced root systems provide more opportunity for plant growth and vigor. Perennial plant species come back from the rootstock year upon year, but annual plants grow from seed origin each year (so the root system for the plant is

established in the same single growing season as plant growth occurs). Many grasses and forbs reproduce and spread by seed production. The plants reach a typical seasonal 'maturity' when the seed head is produced. The nutritional content of the vegetative parts of the plant are highest before the seed head (also called the inflorescence) is produced. The seeds are highly nutritious, but the rest of the plant is not at this time. Understanding and observation of the intricacies of different types of plants is important in their identification, and in communication with others. The evaluation of plant species composition of land areas over time allows for evaluation of ecological condition, but recognition of different species of plants is also critical to evaluation of areas for toxic plants.

5.2.2 Plant Toxicity

As grazing animals have the opportunity to select plants that are nutritious and palatable, they will typically do so even if toxic or poisonous plants are available. Just as numerous species of plants are potentially useful for grazing cattle (and other animals) during various stages of growth, many of these plants may not be toxic throughout their lifetime (only poisonous at specific developmental stages or under certain environmental conditions), or throughout the entire plant (only certain plant components such as berries, etc., may be poisonous), but pose risks in certain situations. Additionally, many plants we think of as not being problem species may have certain environmental conditions that deem them as problems. Plants that are toxic to some animal species are not toxic to others, and some have varying degrees of toxicity across ages or health classes of animals.

The toxic compounds that are most likely to cause problems in grazing cattle are alkaloids, glycosides, nitrates and oxalates (Table 5.3). The species of plants that potentially harbor these compounds vary across geographical regions, and across years within regions due to variation in environmental growing season conditions.

The best guard against cattle becoming threatened from toxic plants is experience with the land resources and regular observation of the animals (Table 5.4). There can be obvious behavioral patterns detected by animals consuming toxic plants, but the timing and rapidity of some instances may not make observation of behavioral signs possible until an animal is severely compromised or found dead.

Table 5.3. Types of toxic compounds found in certain types of plants.

Alkaloids	**Glycosides**
Most powerful toxic substances in plants and most widespread, may be found throughout the plant and typically remains after maturity and/or freeze, nervous symptom disorders are typical signs, animals that do not die can recover quickly; no antidote or treatment exist for most alkaloids; caffeine, nicotine, strychnine and morphine are examples.	Typically have sugars bound to other group(s) and become toxic when sugar is no longer bound due to stressful events to plant; hydrocyanic acid (HCN, or prussic acid) and coumarin are examples, water may increase release of HCN in rumen, cardiac and circulatory disorders are typical; HCN blocks oxygen from blood being absorbed and venous blood is bright red after death with HCN poisoning.
Nitrates	**Oxalates**
Typically more common in many weeds than other plants; nitrogen-rich soil, drought and low light conditions may promote increased nitrate levels, plants treated with some herbicides may increase nitrate levels (and may become more palatable), blood cannot carry oxygen and is dark or brown in color.	May be in soluble or insoluble forms but only soluble oxalates are toxic to livestock; oxalates are normally metabolized by rumen bacteria, but high intakes cannot be metabolized fast enough leading to absorption; oxalates bind calcium in the intestines and the blood lowering serum calcium levels and can cause kidney damage.

Information primarily from Holechek *et al.* (2011), Merck (2012) and Queensland Government (2013). Toxic aspects such as presence of toxins from fungi, molds, etc. that may be associated with plants are discussed in Chapter 9.

Table 5.4. Signs or characteristics that cattle may experience from consuming toxic plants.

Abdominal pain	Collapse	Loss of weight/body condition
Abnormal heartbeat	Coma	Photosensitization
Abnormal urination	Continuous walking/pacing	Pushing or leaning on objects
Abortion	Convulsions	Retained placenta
Anorexia	Depression/weakness	Running into objects or animals
Arched back	Diarrhea	Stiffness
Birth defects	Dilated pupils	Sudden death
Blindness	Excessive salivation	Trembling
Bloat	Incoordination	Unable to eat and/or drink
Blue discoloration of mouth	Irregular breathing (rapid or slow)	Unable to get up
Bright red blood		Unthrifty offspring

None of these signs alone are indicative of poisoning as many are also caused by other conditions or situations.

Many people may think that if plants taste bad to cattle they will not eat them. The problem with this statement is that we really do not know what cattle think tastes bad or good, although we do know that palatability issues impact grazing preferences. Some research has been conducted in the area of food aversion in cattle and other grazing animals and seems to hold great promise in potential pasture and plant utilization. Some people have reported that they have been able to train cattle (or other species) to eat plants that have been previously classified as noxious or invader species, and this is considered in more detail in the next section.

5.3 Animal Characteristics

Grazing cattle will typically consume 2–3% of their body weight per day, and this is a fundamental determinant of forage demand, and therefore stocking rate. For simplicity, we can assume that cows will consume 2.5% of their body weight daily in forage. Of course, many factors will affect forage intake such as forage quality, water content, animal stage of production, soil mineral status and others. Obviously a 450 kg cow will have a different forage demand compared to a 600 kg cow, and therefore stocking rate would need to be modified accordingly. In many cases the estimate of 2.5% of body weight per day (on a dry matter basis) will overestimate the actual feed/forage resource needed and therefore can provide a more conservative stocking rate estimate (it is safer to overestimate feed demand per animal than underestimate it from a budgeting standpoint). Other methods have been proposed and recommended to estimate daily feed intake, such as 10% of the metabolic animal weight

(10% of weight in kg raised to the 0.75 power). All of these methods are flawed in that the intake varies continuously due to many factors, and should merely provide a starting point for initial budgeting purposes. However, close observation and comparison of estimates to actual usage values should be continual by managers and provide for 'adaptive management'. Table 5.5 provides some general guidelines that can be used to estimate grazing demand across some different classes and species of livestock.

5.3.1 Diet Selectivity

Vallentine (2001) reviewed several North American studies evaluating the botanical composition of the diets of grazing animals and classified animals as grazers, intermediate feeders and browsers. Cattle, horses, American bison, elk and mountain bighorn sheep were classified as grazers because grasses (and grass-like species) comprised the majority of their diets. Domestic sheep, burros, desert bighorn sheep, mountain goats and caribou were designated as intermediate feeders because these species relied on grasses, forbs and browse. Domestic goats, mule deer, whitetail deer, pronghorn antelope and moose were classified as browsers because these animals relied the most on browse in their diets. There is variability across seasons in regard to the percentages of these types of plants that all these animal species will eat. There are also differences across these species in their dietary preferences as functions of their adaptation to local environments over time (some of which may be based on palatability and nutritional differences and some of which may be based on availability).

Table 5.5. Animal unit equivalent general comparisons.

Type of animal	Animal unit equivalent
Mature cow (lactating)	1.0
Replacement heifers (18–24 months)	0.8
Replacement heifers (12–18 months)	0.7
Replacement heifers (6–12 months)	0.5
Weaned steers and bulls (6–12 months)	0.6
Young bulls (12–24 months)	1.2
Mature bulls (over 24 months)	1.6
Mature ewes	0.20
Mature nannies/does (goats)	0.17
Mature gelding horse	1.25
Whitetail deer	0.15

There has not been much study on the individual diet selectivity of grazing livestock, but there needs to be more. There is real potential to more effectively utilize pasture resources through understanding and management associated with diet selectivity in cattle and all livestock species. A few studies have documented genetic differences among cattle for diet selectivity and pasture utilization (Howery et al., 1996; De Alba Becerra et al., 1998; Bailey et al., 2001), and more information regarding the potential to select for improved environmental management and/or the potential consequences of correlated changes from selection on other traits such as growth, milk production, etc. is needed to give a more complete picture of environmental impacts.

5.4 Management Considerations

It was mentioned earlier in the chapter that stocking rate is the number of animals that a manager places on the land area, typically expressed as number of hectares per animal or number of animals per hectare. Determining and implementing stocking rate is probably the most important management decision affecting long-term economic and ecological prosperity. If stocking rate and carrying capacity are not well matched for long periods of time, changes in the ecological status of the land will occur. Numerous factors affect calculation of stocking rate/carrying capacity, and many of these have been discussed to some extent in this chapter. A general methodology for calculation of stocking rate is presented in Table 5.6. It cannot be overstated that this type of stocking rate calculation is simply the first step in determining what stocking rate should be (on average), and that adaptive management (continual assessment and modification of impact from practices and environmental influences) is critical. In the strictest sense all approaches to calculating stocking rate rely on imperfect and incomplete data, and the true sense of determining the proper stocking rate can only be determined over time through educated experience (documenting and learning from past experiences) and adaptive management.

Stocking rate and carrying capacity are typically thought of as having fluctuating values over long periods of time, such as across several years, or from one year to the next, because it is understood that rainfall and other climatic influences are not fixed from one year to the next. However, it is also

Table 5.6. General guidelines for estimation of initial annual stocking rates on grazing lands.

Step	Calculation
1.1 Estimate forage allowed for grazing	Total mass (kg) of annual grazable forage multiplied by proper utilization rates:
	• Usually begins with average grazable forage mass per hectare
	• Multiplied by proper utilization rate based on annual rainfall
	• Multiplied by proper adjustment rate due to slope (15% slope = 5% reduction, or 95% utilization, etc.) if needed
	• Multiplied by proper distance to water adjustment if needed (land 1.8 km from water will have 2 × 6.25% = 12.5% less utilization (87.5% utilization rate)

Annual rainfall (mm)	Utilization rate	
Less than 380	40%	Example: 900 hectares with 1600 kg grazable forage per hectare in area with 500 mm rain and average 12% slope. Forage available
380 to 640	45%	= 900 ha × 1600 kg × 0.45 × 0.98
640 to 890	50%	= 635,040 kg usable forage
Over 890	55%	

Amount of slope	Reduction in utilization	
10% or less	No adjustment	
11% to 60%	1% reduction for each percent increase in slope (15% slope = 5%, 22% slope = 12%, etc.)	
61% or more	May not be suitable for grazing, particularly for naïve cattle	

Distance to water		
Less than 1.6 km	No adjustment	
1.6 to 3.2 km	6.25% reduction for each additional 100 m	
Over 3.2 km	Not utilized	

Step	Calculation
2.1 Estimate total forage demand	Multiply animal weight × 0.025 × 365 days
Assume each animal will consume 2.5% of its body weight per day:	Example: 500 kg cow.
	500 × 0.025 = 12.5 kg forage per day
	500 × 0.025 × 365 = 4452.5 kg per cow per year
3.1 Determine number of animals	Divide mass of grazable forage (answer in step 1) by total forage demand (answer in step 2)
This is the total number of animals this scenario will support	From above: 635,040 ÷ 4452.5 = 142.6 or 142 animals
4.1 Calculate stocking rate	Divide number of hectares by number of animals calculated in step 3
This is expressed as number of hectares per animal per year	From above: 900 ha ÷ 142 animals = 6.3 ha/animal

There are numerous approaches to calculating stocking rate, and all have many built-in assumptions. For instance a forage utilization rate of 25% is used in many cases, but that also typically utilizes a forage consumption estimate of 2% of body weight per day, and/or range condition of fair to good, etc. Others assume that all cows have the same forage demand and ignore cow size. The best determination of proper stocking rate involves continual, regular observation and assessment of pasture conditions and resulting adaptive management; this approach only provides an educated starting point.

important to consider these aspects as potentially variable for short-term considerations as well (see Fig. 5.3). Many producers are afraid of 'wasting grass' if they do not understand the components of grazing management systems, and think that if they have too much non-grazed forage, then they were not good managers (when in fact the opposite may be true).

The ability to alter stocking rate across years is obvious for anyone managing grazing animals that have dealt with drought. From a management standpoint, it is not problematic for managers to alter animal numbers from one year to the next as animals in herds are culled due to undesirable performance, particularly in cow herds where non-pregnant females are identified. As a result, it is expected that animal numbers can and should fluctuate across years. This fluctuation in animal numbers may be harder to accept when there is a target level of income needed to pay a fixed level of expenses. It is hard for cow-calf producers, particularly seedstock producers, to reduce cow herd numbers when many years of effort have gone into developing desired genetic type(s) of animals. For this reason alone it is wise to have multiple types of animals in a grazing operation where one or more groups do not have vested long-term ownership implications (such as developing growing animals that will be sold after a growing season). If there needs to be a 20% (for instance) reduction in grazing

pressure and therefore a reduction in animal numbers, the option of selling the non-breeding animals first provides a cushion to alter stocking rate to some extent before the breeding herd(s) become threatened. If 100% of the animals on a ranch are registered purebreds, any necessary reduction in cattle numbers will cut into the genetic base.

Producers with less strict ties to particular groups of animals may feel more comfortable with altering animal numbers (overall and within management groups) to assist in grazing management. Alternation of cattle numbers within groups of seedstock animals causes concern where contemporary group designation and integrity are important because performance records are recorded and submitted to breed societies' databases. As a result, it may be common for seedstock producers to purchase feed or rent additional grazing land before they alter animal numbers in their breeding herds. Carrying capacity is a fluid, continually moving concept, but herd numbers are usually not as fluid. Many managers, whether or not they are seedstock operators, choose not to alter numbers of groups of animals within a season or year until extreme conditions such as drought force them to think about these concepts.

It is critical that managers be prepared to alter stocking rates. Many times in drought situations producers may not reduce animal numbers for a variety of reasons, some of which were discussed above. In many regions, animals with reduced body condition also have reduced price per unit of weight or total value. If pasture conditions continue to deteriorate and supplemental feeding is not adequate, cattle will lose body condition and associated body weight (if this concept is not familiar, refer to the discussion of body condition score in Sections 6.1.2 and 8.3.2). As drought conditions continue and animals are not reduced in number, there will be increased grazing pressure on the land, plus the animals will continue to lose monetary value – this is obviously a losing situation on multiple levels. The severity of the drought cannot be estimated before it happens, so some precautionary actions are usually safer in regard to carrying capacity and feed expense aspects. Table 5.7 provides some guidelines to reduce stocking rate according to varying drought intensities.

The relative condition that the land is in will also dictate adjustments in stocking rates (Table 5.8). Land that is in better condition can withstand higher grazing pressure. This is a bit like the expressions

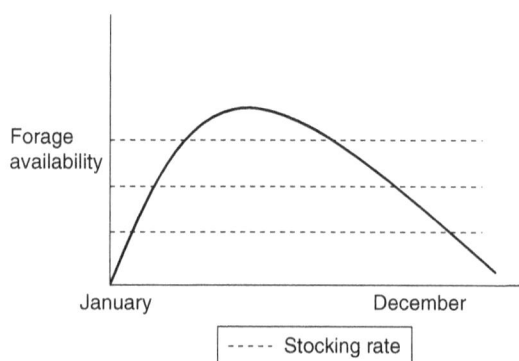

Fig. 5.3. Concept of matching stocking rate with carrying capacity. Forage availability, and therefore carrying capacity (solid line), varies across the year and growing season(s). The scenario above shows stocking rates that are too high (top), adequate (middle) and too low (bottom) as being matched with carrying capacity (on average) for the entire year. None are matched with carrying capacity for the entire year.

Table 5.7. Destocking considerations for grazing lands due to drought conditions.

Level of drought	Reduction of stocking rate
Moderate (55–86% of median growing season rainfall)	25% (20–30%)
Severe (39–55% of median growing season rainfall)	50% (40–60%)
Extreme (less than 39% of median growing season rainfall)	At least 75% (75–100%)

Taken from Ortega-S. and Bryant (2009).

Table 5.8. Suggested stocking rate adjustment due to range condition.

Range condition	Relative to climax	Stocking rate adjustment
Poor	0–25	80%
Fair	26–50	100%
Good	51–75	120%
Excellent	76–100	140%

Values from Ortega-S. and Bryant (2009) and Holechek *et al.* (2011).

'the rich get richer and the poor get poorer', or 'it takes money to make money'. Lands that have been managed better can withstand increased intensity of production (or can rebound quicker from adverse events such as fire, drought or short-term, intensive grazing).

One reason it is critical to evaluate range condition is to prevent its continued reduction. Decreasing range condition begins with reduction of plants desirable for grazing and increases in plant species undesirable for grazing. Undesirable plants may include weeds as well as woody species such as brush/bush. Long-term problems with decreasing range condition can lead to erosion and loss of topsoil. Plant litter and residue is vital for improved soil infiltration of precipitation.

Another factor that can influence how managers view stocking rate (or fail to evaluate stocking rate) is how they evaluate productivity and/or profitability. For instance, if production per animal is the primary focus, there may be reduced performance per animal with increased stocking rates. If performance per unit of land (per hectare or per acre) is the primary focus, this measure of performance is typically increased with higher stocking rate up to a point, and then a resulting decrease. If profit per unit of land is the primary performance measure, again there is typically an increase with increased stocking rate up to a point, and then a resulting decrease as more animals are added. Furthermore, profit per unit of land is usually maximized at a lower stocking rate as compared to that at which production per acre is maximized. This again emphasizes the concept that managers need to optimize production to maximize profit (maximal production may not and usually does not produce maximal profit, especially in regard to an individual trait measured in cattle production such as weaning weight, growth, etc.).

5.4.1 Monocultures vs Mixed Forage Species Pastures

Monocultures of plants can provide high-quality pastures, typically during one season or at least for a limited time of the year in most situations. Many improved pastures are monocultures because the soil had to be prepared and the management of the area can be consistent. Native pastures are the opposite of monocultures. These types of pastures have a wide variety of forage species and can provide nutrients throughout the year because the different plant species grow and mature at different times and intervals; the seasonal supply of nutrients can be improved overall even though individual native species may be much less desirable than individual improved forage species, and there will typically be increased use of the land by other livestock and wildlife species (Fig. 5.4).

5.4.2 Improved vs Native Pastures

The discussion of improved vs native pastures is analogous to the discussion of monocultures vs multi-species pastures, but they are not necessarily the same. Improved pasture is a term which indicates that a pasture has been purposely planted with desired forage, whereas the term native pasture indicates non-planted (or naturally occurring, which may also be a misnomer because native pastures that are mismanaged may give results that are close to monocultures of less desirable plant species). Improved plant species have been utilized the world over to increase carrying capacity as compared to native pastures, because improved species have higher forage production potential and typically withstand higher grazing pressure.

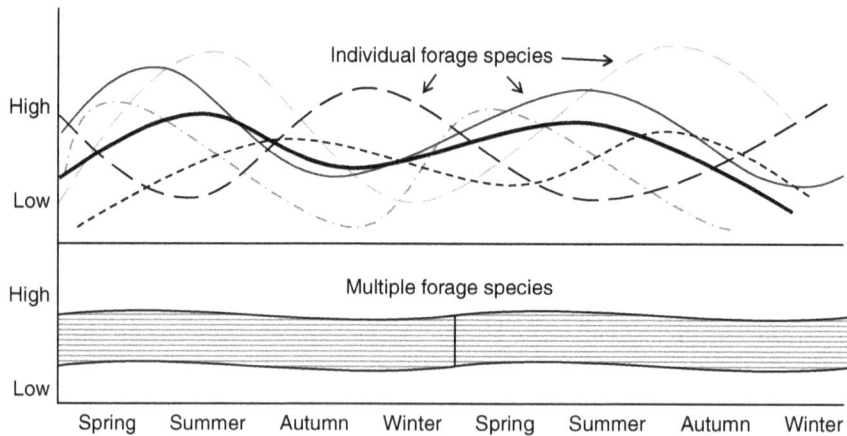

Fig. 5.4. Each individual species of forage has its own seasonal growth pattern (represented by single lines). Having multiple species of plants within pastures helps to smooth out the peaks and valleys of protein and digestible energy across seasons as compared to a monoculture; this is exaggerated even more if warm season and cold season legume and grasses are possible. Much variability can occur across and within regions, years and management scenarios.

Improved species have been developed in many instances to respond to fertilizer. Native species are usually less responsive to artificial fertilizers, but respond well to fire in many instances. The management of these types of pastures needs to be appropriate for their respective types to result in cost-effectiveness. If the added expense to establish improved pastures has been undertaken, but the necessary additional management and costs (such as required fertilizer or water) to allow its increased potential to be expressed is not undertaken, the initial investment was probably a waste of money. Improved pastures provide the potential for increased production per unit of land when the necessary inputs are provided. Planting of forages that are not adapted to local conditions is not sustainable and does not make sense in most cases unless artificial inputs are inexpensive (which is not likely). Hopefully this concept sounds familiar to some of the discussion on animal genetic resources in Sections 3.6.4 and 3.8.

5.4.3 Grazing Systems

Grazing systems can be thought of in three basic categories of (i) continuous grazing, (ii) deferred rotational grazing, and (iii) short-duration grazing, and the general comparison of these types of scenarios are provided in Fig. 5.5. Others may use the terms 'continuous stocking', 'deferred stocking' and 'rotational stocking' for these three concepts

A very intensive version of short-duration grazing may be referred to as mob grazing. There are several grazing systems that fall within deferred and short-duration grazing, some of which are discussed in this chapter. Additional reading on grazing system options can be found in Vallentine (2001), Holechek *et al.* (2011) and several other texts and research articles.

Continuous grazing is as it sounds – animals are continuously on the land area, or the grazing area is not rested. This is the least labor-intensive of all grazing systems, and probably the most risky for managers, as will be discussed further later. Continuous grazing provides for the most grazing selectivity by grazing animals, and the matching of stocking rate with carrying capacity is most critical with this approach. Managers can alter stocking rates, and level of supplemental feeding, but for many both of these approaches can be unattractive. Continuous grazing provides for the lowest ratio of pasture rested time : grazed time because the pasture is never rested. Continuous grazing may refer to year-long occupation of a pasture, or season-long occupation. I have used the term here to mean on a perpetual annual basis; if a pasture is continuously grazed during its growing season, but then destocked during its dormant period of the year, this in fact is a type of rotation. A stocking rate of 10 ha per animal for half of the year during the growing season, and no animals present for the other half of the year, may not be the same outcome and

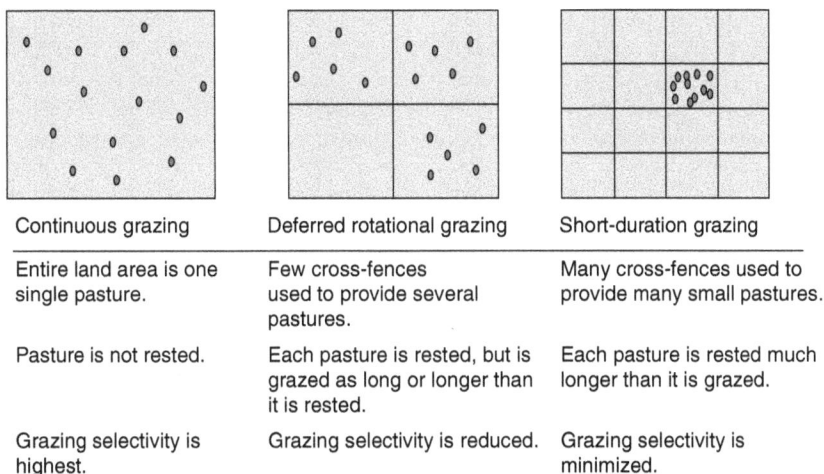

Continuous grazing	Deferred rotational grazing	Short-duration grazing
Entire land area is one single pasture.	Few cross-fences used to provide several pastures.	Many cross-fences used to provide many small pastures.
Pasture is not rested.	Each pasture is rested, but is grazed as long or longer than it is rested.	Each pasture is rested much longer than it is grazed.
Grazing selectivity is highest.	Grazing selectivity is reduced.	Grazing selectivity is minimized.

Fig. 5.5. Three land areas of equal size, but with different numbers of pastures and different intended uses. Dots represent animals.

result in the same pasture conditions as 5 ha per animal per year with the animals continually present.

Deferred rotational grazing typically means that 50% or more of the land is utilized by grazing animals at any one point in time, and a portion of the land is rested. A commonly discussed deferred rotational system is the Merrill four-pasture deferred system. Under this design, there are four pastures, and the number of animals that would be appropriate for the entire area are split into thirds and placed in three pastures. Groups of animals are rotated on 4-month intervals, and this allows each pasture to be rested at different times of year. Over the course of 4 years, each pasture will have been rested in all different seasons. Another deferred rotation is the South African switchback system. This involves two equal pastures, with the number of animals appropriate for the entire area concentrated in one pasture for 5 months or 7 months, and then the animals are switched to the other pasture for an equal amount of time. This sequence allows the pastures to be used at slightly different times of the year each time they are grazed, which helps to guard against animal selectivity and over-utilization of more palatable plant species. Deferred rotation systems are more labor-intensive than continuous grazing because they require animals to be moved regularly; however, the low frequency of animal movement is hardly a labor concern. Stocking density is higher with deferred rotation

systems compared to continuous grazing, and provides for lower grazing selectivity by animals. Research has shown deferred rotation systems can provide effective long-term improvement in range condition and high animal performance.

Short-duration grazing systems provide for the highest pasture rested : grazed time and are the most labor-intensive of these three categories of grazing systems. Under short-duration grazing systems, the majority of the land is rested while a small percentage is being grazed. Under this type of grazing management, the land area is split into many small pastures, and the number of animals for the entire area is concentrated into one pasture (or a very small number of pastures, depending upon pasture size). Under short-duration grazing, stocking density is high, therefore providing for the least selective animal grazing due to competition, animals are moved into fresh pastures often, and there is a long rest/recovery time between each pasture being grazed. One particular example of short-duration grazing is the high-intensity, low-frequency method. Under this scenario, animals graze smaller pastures for something like 15 to 30 days, and pastures are then rested for 90 to 180 days. Another example of short-duration grazing has been called the rapid rotation method. With this strategy, pastures (which may be called paddocks or cells) may be very small and are only grazed for 7 days or less and then are rested for 30 to 60 days. This type of management is the most labor-intensive; however, overall carrying capacity due to improvement in range condition and production per acre can be greatly increased. Short-duration grazing has proved that re-establishment of desirable plants can occur rapidly with proper management. Mob grazing is another term that has been used to describe high-intensity, high-frequency grazing management. A general presentation of these different types of rotational grazing strategies is presented in Table 5.9.

It was stated earlier that continuous grazing may be the most risky type of grazing management. This statement needs further explanation and discussion. If managers are willing to provide continual and regular oversight and monitoring of pastures, rotational grazing provides opportunities to improve economic and environmental resources. It is safer and easier to sustain pasture resources when the pastures receive rest from grazing pressure. When plants are allowed to grow toward maturity, they will possess a more developed root system. This allows longer roots to seek water deep below the soil surface. Plants with more developed root systems are better able to respond to adverse conditions such as severe grazing, fire and drought. Pastures that are continually grazed short will have plants with less developed root systems, and this in turn hinders their forage production potential in desirable environmental conditions and exaggerates undesirable conditions such as drought. If stocking rate is higher than the carrying capacity, rotational grazing provides rest and regrowth.

Regardless of the grazing system, stocking rate that exceeds carrying capacity is undesirable. Short-grazed pastures that are continuously stocked have less canopy cover, will retain less soil surface moisture due to direct sunlight, and the increased temperature and ultraviolet radiation can negatively affect desirable microbes and other beneficial organisms. With all this being said, an intensive, short-duration grazing system that has stocking rate well above the carrying capacity, and where animals are left in pastures too long, will also have undesirable environmental effects. Some producers have tried short-duration grazing and observed undesirable results; very often this may have been the result of keeping with a regular routine for pasture rotation (i.e. rotating strictly every 14 days, etc.) instead of having the continual monitoring of forage conditions and utilizing flexibility with the schedule. Many producers that have been committed to long-term utilization of short-duration grazing and implementation of continual assessment and flexibility in the rotation schedule have observed very desirable results (Fig. 5.6).

There has been much discussion and difference of opinion in the scientific literature about the merits of rotational grazing vs continuous grazing. One thing must be kept in mind: there are no easy answers to complicated questions or issues. And, as has been reiterated throughout this book, there are no one-size-fits-all answers to most important cattle production scenarios. Managers have to find what works best for their own situation(s), and this typically occurs through educated experiences. All grazing system strategies can be made to work and all can be made to fail. One thing that is for certain is that stocking rate and its coordination with carrying capacity is crucial for sustainable health and condition of pastures. If the initial estimate of stocking rate is close to the true carrying capacity, this will be obvious. If it appears to be off, make an

Table 5.9. General considerations for rotational grazing management relative to intensity and frequency of grazing and some associated key concepts.

		Frequency of grazing		
		Low Rest periods may be 90 days to over 365 days	Medium Rest periods may be from 42 days to 90 days	High Rest periods may be 14 to 42 days
Intensity of grazing	High	HILF	HIMF	HIHF
	Medium	MILF	MIMF	MIHF
	Low	LILF	LIMF	LIHF

Adapted from concepts presented by Kothmann (2009). Intensity of grazing is a function of degree of stocking density. Frequency of grazing is a function of how often per year or per season the pasture is grazed (and therefore also a function of the length of time the pasture is rested between grazing episodes). The degree of shading in the table indicates the degree of management oversight needed.

The top photo shows yearling steers in a paddock of 'graze-out' wheat, meaning the wheat pasture was not intended for any use but cattle grazing. The paddock to the left of the steers had animals removed 100 days earlier to allow a wheat crop to be produced. Complete utilization of the forage is expected as seed is planted annually.

The bottom photo shows paddocks being used in a short-duration grazing situation on permanent pasture. The cow herd in the right paddock is about to be moved into the left paddock. The length of time the paddocks are rested to allow plant (re)growth between grazing sessions dictates the sustainability of these types of situations.

Electric fencing is used to separate interior paddocks in both situations. It can be seen in both photos where the cattle have grazed into the adjacent paddocks as far as they could reach without touching the fence.

Fig. 5.6. Two types of grazing scenarios.

adjustment in the right direction, and re-evaluate next year. Another thing that must be remembered is that the carrying capacity (and therefore the stocking rate) is fluid and can change from one year to another with change in climatic factors, mainly amount and distribution of precipitation. As most cattle operations have several types of animals (different ages, sizes, body condition scores, etc.), there can also be a variety in the types of grazing management used within and across time and location. No one probably has 100% continuous grazing 100% of the time, or exclusively high-intensity rotational grazing for all land areas and animals 100% of the time. Adaptive management is the key to sustained productivity and profitability, and the recognition of variability in environmental conditions across time and space is crucial. Implementation of rotational grazing incrementally out of a continuous grazing management plan provides a way for producers to sample this management strategy without

the large-scale risk of an entire system transition too quickly. The results of one trial in one year or growing season can be quite misleading and unrepresentative of the complete or long-term reality. Figure 5.7 shows an example transition for increasing pasture numbers over time.

5.4.4 Establishment of Multiple Pastures

The decision to add cross-fences should be considered from many points of view. It should be obvious that increased numbers of pastures will allow a variety of grazing management scenarios. Just because there are multiple pastures (small pastures may be referred to as paddocks or cells), this alone does not guarantee that improved grazing management will occur; however, there are increased grazing management options with multiple pastures that do not exist with single or very few pastures. In some instances, establishment of multiple pastures

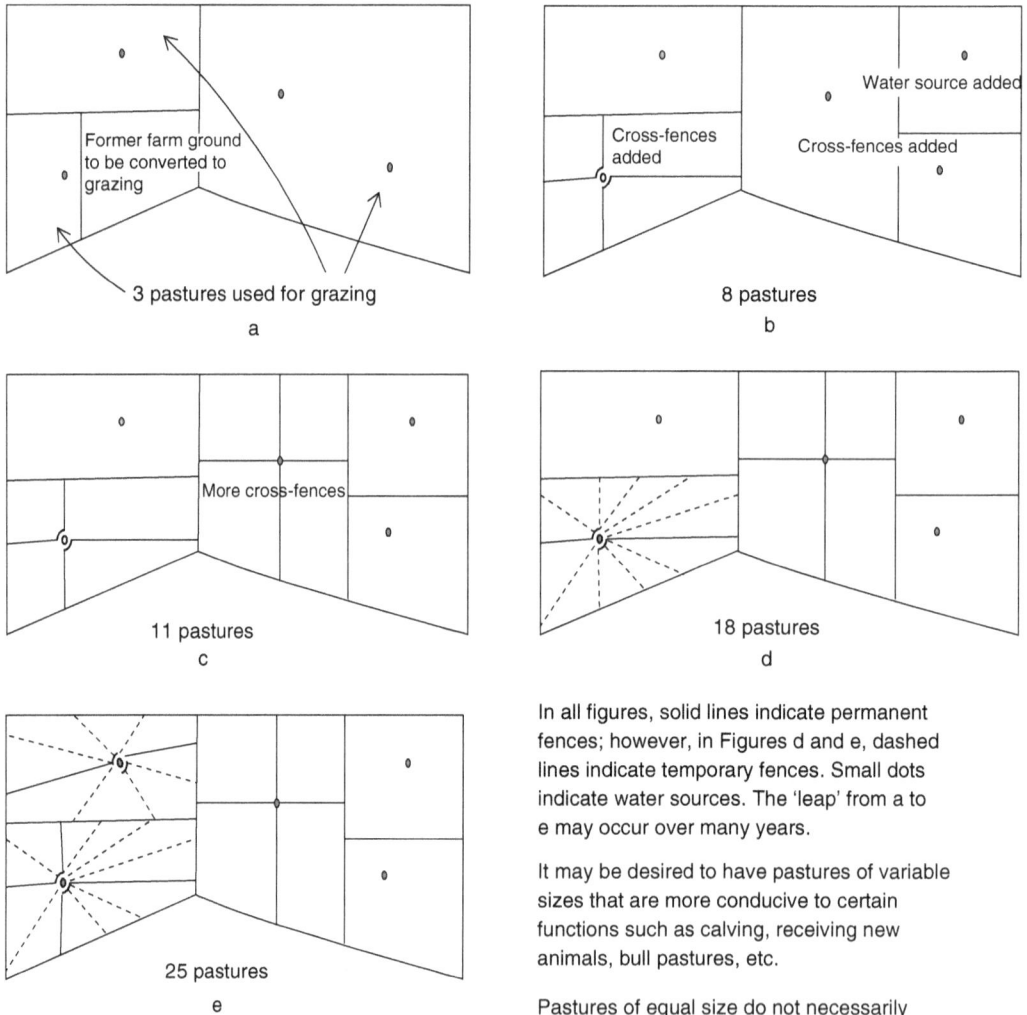

Fig. 5.7. Possible cross-fencing and pasture transition over time.

In the figure:

Figure a: "Former farm ground to be converted to grazing", "3 pastures used for grazing"

Figure b: "Cross-fences added", "Water source added", "Cross-fences added", "8 pastures"

Figure c: "More cross-fences", "11 pastures"

Figure d: "18 pastures"

Figure e: "25 pastures"

In all figures, solid lines indicate permanent fences; however, in Figures d and e, dashed lines indicate temporary fences. Small dots indicate water sources. The 'leap' from a to e may occur over many years.

It may be desired to have pastures of variable sizes that are more conducive to certain functions such as calving, receiving new animals, bull pastures, etc.

Pastures of equal size do not necessarily indicate equal carrying capacity.

can vary in their expenses due to considerations that managers may not think of ahead of time. If four equal-sized pastures have cross-fences that run diagonally (corner to corner) as opposed to at right angles to the perimeter, this can increase the cross-fence lengths substantially.

For smaller producers, a limited number of pastures results in severe management limitations in regard to grazing and general herd management. Multiple pastures are needed for (i) separation of age and/or body condition classes, (ii) separation of bulls from females, (iii) proper pasture management and rotation, and (iv) ability to separate

groups for health/biosecurity reasons. A single water source for multiple pastures can be a convenient and efficient use of resources, but be a disadvantage if groups of cattle need to not be in close proximity to one another. General layouts as to how single water sources might be used in multi-pasture rotational grazing scenarios should also be considered. For instance, a single water source to be used for multiple groups of animals in multiple pastures requires cross-fencing to occur across the water source so that it would be accessible to all pastures simultaneously (Fig. 5.8).

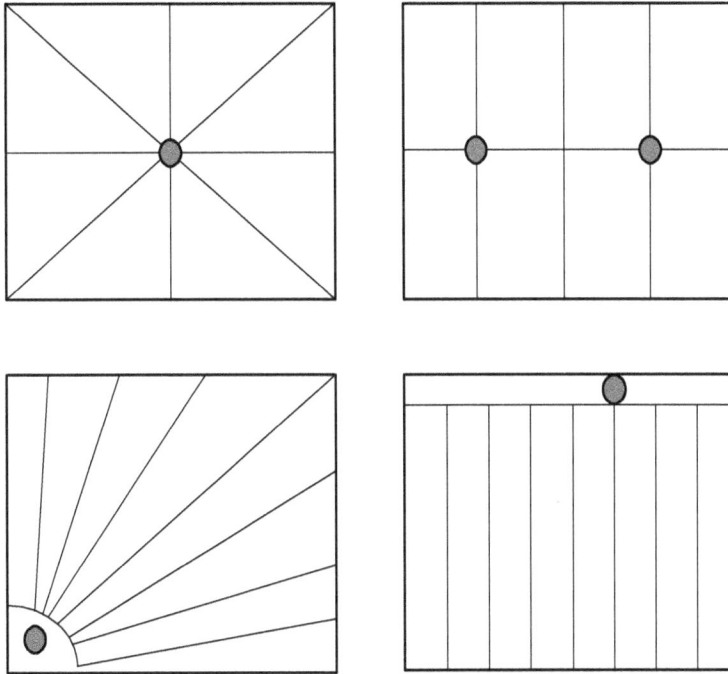

Fig. 5.8. Examples of 8-pasture layout. Contact between groups of animals, rotation of pastures, location and access to water, costs of cross-fencing, etc. are all important considerations related to usefulness and therefore the sustainability of the management system. Circles indicate water sources.

5.4.5 Intra-Pasture vs Inter-Pasture Variation

In most livestock production systems, variation within production units is a potential source of inefficiency. This concept was presented in Chapter 4 when discussing the ability to sort cattle of different nutritional requirements into different management groups. This general concept holds true for pasture management as well. An area used for livestock grazing may have extremes in ecological and topographical characteristics. If these extremes in characteristics are located within a pasture, this will promote less efficient utilization in that pasture because it allows animals the widest degree of choice about where they will decide to graze and therefore may promote increased selectivity and acceleration in differences between ecological regions or increased over-utilization of specific areas with under-utilization of others. Identification of these different types of areas can help decide where to locate cross-fences. If the different ecological conditions are constrained to different pastures (i.e. minimized with pastures), it can make

utilization within those pastures more consistent (Fig. 5.9). This approach will confine and maximize the variation across (inter) pastures and minimize variation within (intra) pastures. This in fact will probably make the utilization time, carrying capacity, etc. quite different between pastures, but will promote less selectivity in grazing and more even distribution within pastures.

5.4.6 Estimating Number of Animal Grazing Days for Fixed Resources

This concept of estimating number of animal grazing days for fixed resources (per pasture, set amount of crop residues, finite hay resources, etc.) is no different from calculating stocking rate, but involves thinking about the process from another angle. The simplest example is to assume that there is a fixed feed resource that will be the sole diet component, such as hay. If there were 20,000 kg of hay, how many cattle would this feed? It depends upon how long the animals are to be fed of course (and the daily intake per animal). Assuming 2.5% body weight per day and 550 kg cows, the daily feed requirement is 13.75 kg per cow. So, this amount of hay would feed approximately 1454 cows (for one day), one cow (for 1454 days), or many other combinations. Dividing the total feed amount (20,000 kg) by the daily amount per animal (13.75 kg in this case) gives the number of animal feeding days (1454.5) for the scenario. As a result, this amount of hay would feed 100 cows for 14.5 days, or 10 cows for 145.5 days, or 50 cows for 29 days, etc.

Feed resource mass ÷ daily feed requirement per animal = number of animal feeding days

20,000 kg ÷ 13.75 kg = 1454.5 animal feeding days

This concept needs to be familiar to managers of grazing livestock, not just for feed purchases, but in regard to pasture management in general because most operations do not have all pastures the same size. Hopefully it is obvious that smaller pastures have fewer animal feeding days than larger pastures. This adds complexity in grazing management and related oversight, but becomes second nature to someone that continually monitors both animal appearance as well as pasture conditions across seasons and across years.

The exact same crop residue across location and/or time will vary in animal feeding days such as pastures will vary in carrying capacity. Consider maize residue following harvest. The seasonal growing conditions, the harvesting conditions,

methodology and equipment used can all impact the amounts and distribution of grain, leaves, etc. (Table 5.10). It was also reported (KZNDAE, 2014) that when grazing maize residue with cattle, sheep or cattle and sheep combinations, the cattle and cattle/sheep combinations produced more weight gain (kg per ha) than sheep alone when comparable animal unit stocking was utilized, illustrating that as with native and permanent range and pastures, different species of animals utilize grazing plants differently.

There can be several features of grazing lands that have 'undesirable' grazing aspects but may have important management features. For instance pastures that are nothing but thick grass may seem like ideal grazing pastures, but at certain times of the year or production calendar these areas may be undesirable because there are no wind breaks, shade or areas for cows to hide while calving. Typically, cattle will regulate their behavior in response to environmental stressors, if they are able to. Cattle do not typically seek out undesirable or toxic plants, and will seek relief from heat and cold stress or predator threat if given the opportunity. Cover from dense brush or woods can allow cows to hide their calves from predatory animals; canyons allow for areas to get away from wind in cold weather; a pasture that has dense, mature grass is not the most desirable from a nutritional standpoint, but it may make a good cover for cows to calve and hide young calves. Also, Provenza (2007) and others have emphasized the need for 'local adaptation' consideration of grazing animals and pasture management, and that grazing animals can be trained to eat a variety of different plants that may be viewed as problematic or invasive and offer real potential for improved management of environmental resources.

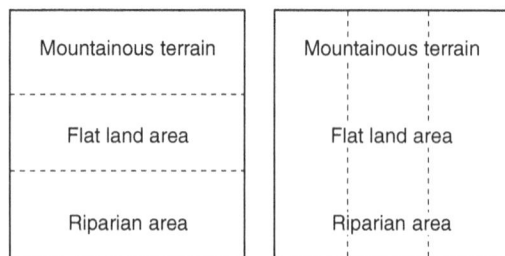

Fig. 5.9. Establishment of cross-fences (broken lines) relative to variation in land characteristics. On the left, cross-fences are used to separate distinctive ecological areas, thereby creating more inter-pasture variation, but less intra-pasture variation. On the right, cross-fences provide for three replicated pastures, and minimize inter-pasture variation, but also do not reduce intra-pasture variation.

Table 5.10. Composition and utilization of maize residue by pregnant beef cows across 3 years.

| Time sampled | Maize crop residue component | | | | |
	Maize grain (kg/ha)	Cob (kg/ha)	Leaf (kg/ha)	Stalk (kg/ha)	Total (kg/ha)
Pre-grazing	182	1334	2876	2286	6713
Post-grazing	60	934	1188	1479	3719
% utilized	67.2	29.9	58.7	35.3	44.6

Data from KZNDAE (2014) at the Dundee Research station (KwaZulu-Natal, South Africa).

There is much more consideration of these types of concepts at the BEHAVE (2014) website.

5.5 Summary

Effective pasture and grazing utilization involves understanding and appreciation of the individual components (land, plant, animal and management) and their potential interactions that can affect various outcomes. The goals of different managers make some pasture management strategies more attractive and feasible than others, but documentation of performance (animal and pasture) is needed for long-term sustainability of natural resources as well as economic evaluation. Cattle and other ruminants take advantage of their ability to harvest and utilize roughages that other species cannot. There are many nuances of grazing management that may result in similar performance levels but different profitability levels (or vice versa), and the exact same management strategy that works in one region may be disastrous in another. Understanding of carrying capacity and matching proper stocking rates will influence ecological conditions, but requires that the personnel making grazing decisions utilize adaptive management for sustainability. This means that management continually changes to keep resources and production conditions constant (which is not how a lot of people think about things).

5.6 Study Questions

5.1 Explain the differences between the concepts of 'carrying capacity' and 'stocking rate'.
5.2 List five factors that could influence where or how cattle utilize pastures for grazing.
5.3 Explain what a four-pasture deferred rotational grazing system means.
5.4 Discuss the concepts of forage condition versus range condition.
5.5 Calculation of stocking rate (Questions **5.5a** to **5.5g**)
Assume that you have 1275 ha of pasture for grazing. The land is in an area that usually receives 750 mm of precipitation per year. The land has an average slope of 5%, and has several water sources, all of which are within 2 km of each other. It has been estimated that there should be 1550 kg of forage produced per hectare annually. Your cows average 550 kg at a body condition score of 5.

5.5a How much forage is there in this operation that you will allow to be used by grazing beef cows?
5.5b How much forage will one of your cows eat per day, and per year?
5.5c How many cattle will this land support on an annual basis?
5.5d What should the annual stocking rate be on this land for this scenario?
5.5e Your neighbor comes by for a visit one day and says 'Why don't you run a cow to every 2 ha on this place? I would!' What would you tell your neighbor?
5.5f Would stocking rate need to be modified if your cows weighed 650 kg at BCS 5? Why?
5.5g Assume that this land is approximately twice as long as it is wide, and you decide to start a short-duration grazing system on one half of the land. Diagram a possible pasture layout for this and state how your cattle would be grouped.
5.6 Assuming 2.5% body weight per day and 500 kg cows, how many cow-days of feeding would be provided by 12,000 kg of hay? How long would this amount of hay last if it was to feed 50 cows?
5.7 What does the abbreviation HIHF refer to, and what does it represent? How would the HIHF concept compare with the LILF concept?

5.7 References

Bailey, D.W., Kress, D.D., Anderson, D.C., Boss, D.L. and Davis, K.C. (2001). Evaluation of F1 crosses from Angus, Charolais, Salers, Piedmontese, Tarentaise and Hereford sires V: Grazing distribution patterns. *Proceedings, Western Section of the American Society of Animal Science* 52, 110–113.

BEHAVE (2014) Behavioral Education for Human, Animal, Vegetation, and Ecosystem Management. Utah State University Extension Service. Available at: http://extension.usu.edu/behave/ (last accessed 11 May 2014).

De Alba Becerra, R., Winder, J., Holechek, J.L. and Cardenas, M. (1998) Diets of 3 cattle breeds on Chihuahuan Desert rangeland. *Journal of Range Management* 51, 270–275.

FAO (2014) Soils Portal: Regional and National Soil Maps and Datasets, Food and Agriculture Organisation of the United Nations. Available at: http://www.fao.org/soils-portal/soil-survey/soil-maps-and-databases/regional-and-national-soil-maps-and-databases/en/ (accessed 18 July 2014)

Hohlt, J.C., Lyons, R.K., Hanselka, C.W. and McKown, D. (2009) Estimating grazeable acreage for cattle. Texas A&M AgriLife Extension Service, publ B-6222.

Holechek, J.L., Pieper, R.D. and Herbel, C.H. (2011) *Range Management Principles and Practices*, 6th edn. Pearson Education, Saddle River, NJ.

Howery, L.D., Provenza, F.D., Banner, R.E. and Scott, C.B. (1996) Differences in home range and habitat use among individuals in a cattle herd. *Applied Animal Behaviour Science* 49, 305–320.

Kothmann, M.M. (2009) Grazing methods: A viewpoint. *Rangelands* (October), 5–10.

KZNDAE (2014) Agricultural publications and production guidelines. Department of Agriculture and Environmental Affairs, Kwazulu-Natal Province, South Africa. Available at: http://www.kzndae.gov.za/en-us/agriculture/agricpublications/productionguidelines.aspx (accessed 11 May 2014).

McGinty, A., Thurow, T.L. and Taylor, Jr, C.A. (2007) Improving rainfall effectiveness on rangeland. Texas A&M AgriLife Extension Service Beef Cattle Management Handbook, College Station, TX, pub. no. L-5029.

Merck (2012) The Merck Veterinary Manual Online. Available at: http://www.merckmanuals.com/vet/ (accessed 11 May 2014).

Ortega-S., J.A. and Bryant, F.C. (2009) Cattle management to enhance wildlife habitat in South Texas. In: *Proceedings of Grazing Management Lectureship*, King Ranch Institute for Ranch Management, Kingsville, TX, 20–23 July, 2009.

Provenza, F.D. (2007) What does it mean to be locally adapted and who cares anyway? *Journal of Animal Science* 86, E271–E284.

Queensland Government (2013) Toxic plant species and symptoms. Available at: http://www.business.qld.gov.au/industry/agriculture/animal-management/disaster-recovery-for-livestock-farms/flood-affected-animals/flood-poison/flood-toxic-plant (accessed 30 October 2013).

USDA-NRCS (2013) Soil Health. Natural Resource Conservation Service of United States Department of Agriculture. Available at: http://www.nrcs.usda.gov/wps/portal/nrcs/main/soils/health/resource/ (accessed 11 May 2014).

Vallentine, J.F. (2001) Grazing Management 2nd ed. Academic Press, San Diego, CA.

6 Live Animal Evaluation

In an age where new technology and information are continually available for cattle producers, some may find it tempting to consider that there may no longer be a need for live animal evaluation. This, however, is not the case. Producers that no longer wish to use live animal evaluation are probably misguided, short-sighted, or perhaps do not enjoy working with livestock. It is true that only utilizing live animal evaluation to make selection, management and marketing decisions is not as effective as incorporation of additional information; conversely, not incorporating live animal evaluation into those same processes also makes them less effective, and could eventually lead to production of animals that despite having desirable breeding values or performance traits, may be physically undesirable to potential buyers, which could negate the progress made in other traits of importance. Some live cattle traits discussed in this chapter are provided in Table 6.1.

As cattle producers have the values of their market animals determined largely by size and weight, live animal measures directly related to weight are of obvious importance. These types of traits can be thought of as both characteristics of the animal and characteristics of the moment. Characteristics of the animal either change slowly over time, or remain relatively constant as the animal grows and matures, whereas characteristics of the moment are

temporary influences that can influence the weight and/or appearance of the animal at a single point in time. Skeletal size, degree of muscularity, amount of body fat and body condition, skeletal conformation, amount of skin along the underline, etc. are characteristics of the animal. Temporary conditions such as the amount of fill in the digestive tract, degree of fatigue, immediate stress from handling and/or transportation, illness, etc. are characteristics of the moment and can influence weight of the animal and behavior. Long-term stress due to nutritional deficiencies, parasitic infestation or infectious illness can cause loss of body condition, change in hair coat and loss of muscle mass, but these take many days or weeks to transition from extremes in appearance. There is no substitute for experience in observing animals as they are handled in a variety of situations, across seasonal conditions, and over time with respect to what is 'normal' vs 'abnormal' in regard to animal appearance. Most if not all of the traits discussed in this chapter follow a continuous distribution, but most are classified into categories to aid in distinctions that need to be made for genetic, management and marketing decisions. Although many live animal traits are related to a variety of production functions, in this chapter the traits are grouped into four categories related to: (i) size and weight, (ii) physical movement and function, (iii) fertility and reproduction, and (iv) adaptation and longevity.

Table 6.1. Live cattle traits that are of economic importance.

• Muscle	• Sheath and underline
• Mature skeletal size (height and long bone diameter)	• Udder and teats
	• Masculinity and femininity
	• Temperament or disposition
• Body condition score	• Teeth and mouth
• Structural (skeletal) soundness	• Weight
	• Evidence of parasite damage
• Hair coat	
• Tick counts	

6.1 Size and Weight Related Traits

There are many factors that affect the weight of cattle. These include (i) degree or percentage of body fat, (ii) amount of muscle expression, (iii) age, (iv) stage of production such as pregnant or not and stage of gestation, (v) skeletal size due to height and thickness of bone, and (vi) fill, or the amount of contents of the gastrointestinal tract such as feed and water. The changes in weight associated with growth and development are discussed in Chapter 7. In this chapter visual evaluation for some of these traits is discussed. Again, traits discussed in this section are related to weight, but they may have their main influence in other categories of traits (such as body condition score is related to weight, but it is a primary driver of heifer and cow reproduction, and may be indicative of health status, etc.).

6.1.1 Muscle Expression and Structure

In most weight-oriented markets where production of cattle for meat is a goal, muscle expression is very important. Many breeds of cattle have distinct differences in muscle expression, and in most breeds variation across family lines can also be observed. Several areas of the body are useful for evaluating degree of muscularity in cattle. These include the size of the forearm, the width of the top line, and the degree of roundness in the rear leg muscles as viewed from both the side and the rear of the animal. Figure 6.1 illustrates areas of the body to evaluate for muscle and body condition (fat).

It is possible that animals which are less muscular in certain areas of the body may have greater degrees of overall muscularity, so it is important to evaluate all areas of the body simultaneously. The shape of the muscle is also important to consider. Cattle that have a mutated form of the myostatin

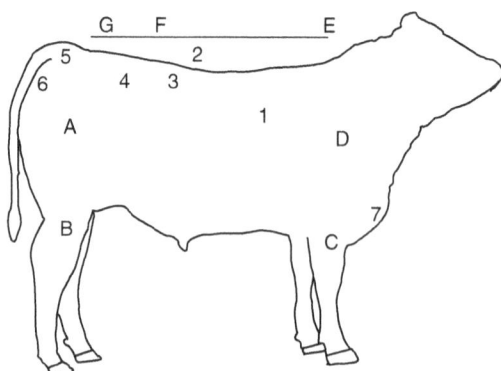

Fig. 6.1. Areas on cattle to evaluate muscle (represented by letters) and body condition (represented by numbers): A = round, B = lower leg above hock, C = forearm, D = shoulder, E–G represent width across the top at shoulder, loin and round, respectively; 1 = over ribs, 2 = over vertebrae, 3 = transverse processes of vertebrae, 4 = hook bone, 5 = around tail head, 6 = pin bones, 7 = brisket area.

gene (Kambadur *et al.*, 1997; McPherron and Lee, 1997) have a pronounced degree of muscle expression termed 'double muscling'. The doubled-muscled condition can occur in high frequencies in some breeds, and several breeds exhibit the condition at close to 100% of the population. Cattle with the double-muscled condition not only show extremely high levels of muscle expression, but the shape of the muscle appears to be shortened and rounded; as a result, the muscles of the rear legs tend to extend more horizontally away from the body when viewed from the side or rear, but tend to not extend down the legs as far as most non-double-muscled cattle. In most non-intensively managed ranching situations, the double-muscled condition is viewed as disadvantageous as there tends to be higher levels of dystocia and more udder problems among double-muscled females, and structurally, it often seems that the set of the rear leg hock joint is extreme (either too straight or too much angle).

Cow-calf producers should pay attention to degree of muscle in the herd bulls they use and the calves produced in their herds. Muscle is dense and adds weight to cattle, so it is very desirable in market animals, but extremely muscular females may have higher maintenance costs, particularly if they are also prone to stay lean. Often cattle that are heavily muscled tend to also be genetically lean, but this relationship probably exists as packaged in different breeds.

Some markets or breed societies may classify cattle in regard to their muscle expression. Cow-calf producers are recommended to evaluate muscle expression in their herds and calf crops. This can be done in a variety of ways, but there has to be some set standard on what is acceptable, and what is not. Producers can use muscle expression as a culling criterion. It is recommended that producers use some sort of scoring system to evaluate their breeding animals and/or calves for muscle expression.

6.1.2 Body Condition Score

Body condition score is one of the most important (probably the most important) live animal evaluation criteria to monitor in cow-calf herds. Low body condition (too little body fat) is the single most important determining factor in regard to beef female reproduction. There has been much research conducted in many genetic cattle types and in many different environments that show adequate body condition is mandatory in the amount of time cows resume estrous cycles after calving. Body condition score is evaluated on a 1 to 9 scale that is described in Table 6.2.

Body condition of cattle is best evaluated on areas of the body where there is little muscle between the skeleton and the skin surface. Evaluation of muscle and of body fat occur on different regions of the body, but these two traits are related because as animals lose body condition because the diet is deficient in energy, the muscles of the body begin to be used as energy sources, and therefore muscle mass will also deteriorate. The most common and effective areas of cattle to evaluate body condition score are over the rib cage, across the top of the backbone (tops of vertebrae), the sides of the vertebrae (transverse processes) through the loin region, around the tail head (fat deposits here called pones), and through the brisket area (front of lower chest). It is recommended that cows and heifers have a body condition score of 5 at the start of the breeding season for highest potential fertility. Figure 6.2 shows the same cow in varying body condition score.

Different scales have been used for body condition score classification in cattle, but these have historically been either a 5-point scale or the 9-point scale mentioned above. Table 6.3 provides an example of how these two body condition scoring scales can relate to one another. Nutritional considerations and the amount of weight change needed across body condition scores in the 9-point scale are discussed in Chapter 8.

6.1.3 Frame Size and Frame Score

Frame size is the concept that each animal has a mature skeletal height and that as animals grow they change in skeletal height correspondingly with their age as they approach maturity. At a given point in time, knowing the animal's age and height can be used to calculate a frame score. Certainly an animal's mature skeletal size affects its weight. The calculations and the concept of frame score

Table 6.2. Descriptions of cattle appearance associated with different body condition scores.[a]

BCS	Description
1	Emaciated condition – bone structure of shoulder, ribs, back, hooks and pins sharp to touch and easily visible from distance. Little evidence of fat deposits or of muscling. Animals extremely weak and in life-threatening state.
2	Poor condition – little evidence of fat deposition but some muscling in hindquarters. The spinous processes feel sharp to touch and are easily seen with space between them.
3	Thin – beginning of fat cover over the loin, back and foreribs. Backbone still highly visible. Processes of the spine can be identified individually by touch and may still be visible. Spaces between the processes are less pronounced.
4	Borderline – foreribs not noticeable; 12th and 13th ribs still noticeable to the eye, particularly in cattle with a wide rib cage and ribs wide apart. The transverse spinous processes can be identified only by palpation (with slight pressure) to feel rounded rather than sharp. Full but straightness of muscling in the hindquarters.
5	Moderate – 12th and 13th ribs not visible to the eye unless animal has been shrunk. The transverse spinous processes can only be felt with firm pressure to feel rounded – not noticeable to the eye. Spaces between the processes not visible and only distinguishable with firm pressure. Areas on each side of the tail head are fairly well filled but not mounded.
6	Good – ribs fully covered, not noticeable to the eye. Hindquarters plump and full. Noticeable sponginess to covering of foreribs and on each side of the tail head. Firm pressure now required to feel transverse processes.
7	Fat – ends of the spinous processes can only be felt with very firm pressure. Spaces between processes can barely be distinguished at all. Abundant fat cover on either side of tail head with some patchiness evident.
8	Too fat – animal taking on a smooth, blocky appearance; bone structure disappearing from sight. Fat cover thick and spongy with patchiness likely. Spinous processes almost impossible to feel. Cows have fat deposited below vulva. Wasty condition.
9	Obese – bone structure not seen and not easily felt. Tail head buried in fat. Animal has extreme patchiness of fat and is blocky in shape. Animal's mobility may actually be impaired by excess amount of fat. Very wasty condition.

[a]Adapted from Richards *et al.* (1986) and Herd and Sprott (2007).

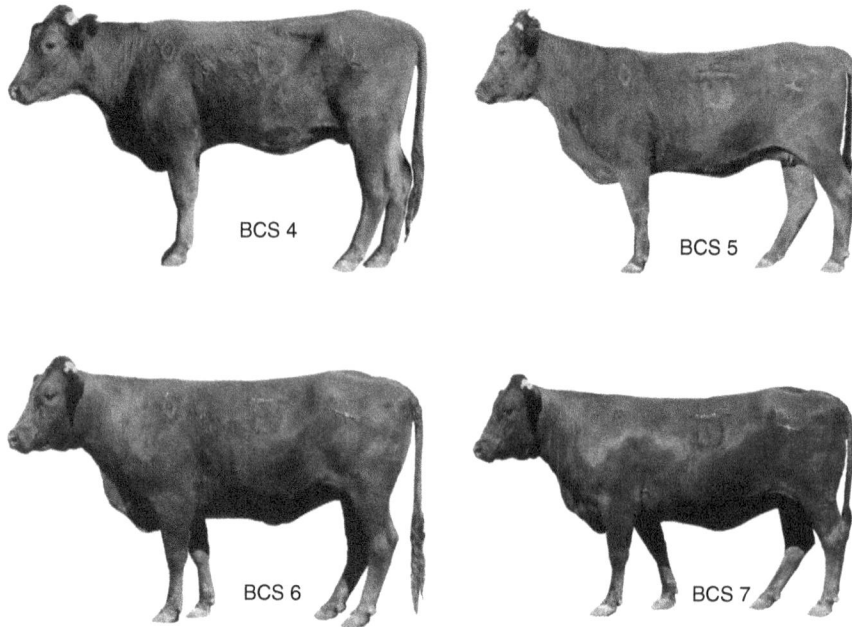

Fig. 6.2. Transition of same cow across body condition scores. Photos courtesy of Clay Mathis.

is discussed in detail in Section 7.4, but it is an important live animal trait that should always factor in live cattle evaluation. The concept of potential mature size is harder to know with certainty in a young, growing animal if the age of the animal is unknown; visually, it can be seen that a calf is relatively young, for instance, but if it is assumed that the animal is 6 months old, and it is in fact 8 months old, or 4 months old, etc., the projected frame size (and mature size) can substantially deviate from what is assumed.

Magnabosco *et al.* (2002) stated that the genetic correlations between traits like skin color, lengths of ears and length of nose were not correlated with weight or other body size measurements. This approach does not however describe the 'type' of weight. It should not be surprising that bigger animals (regardless of how this is evaluated) tend to weigh more, but knowing weight alone does not indicate body shape, muscle pattern, etc. In some cases, the ratio of cattle height to weight has been evaluated. This calculation has different interpretations within a single breed as compared to across breeds. The ratio of height to weight (or weight to height, which is the same concept but provides different calculated values) provides more information than either alone, but again without body com-

position indicators provides an incomplete picture of animals.

6.1.4 Ultrasound Evaluation of Body Composition

Although ultrasound has been used to evaluate livestock body composition for over 50 years, it has recently become more attractive to evaluate breeding animals as well as finishing market animals due to improvements in technology and interpretation software. Typical ultrasound measures of fat thickness over the ribs and/or over the rump give an objective measure of fat thickness at these areas. Longissimus muscle eye area (i.e. rib-eye area) is also possible to evaluate as is percent intramuscular fat (IMF) important in many beef carcass evaluation systems. It is important to use a trained ultrasound technician if these measurements are required.

6.2 Physical Movement and Function Traits

Many of the traits discussed in this section do not have documented values from research trials, at least in recent years. That does not mean they are not

Table 6.3. Relationships of 5-point and 9-point body condition scoring systems.

5-point scale			9-point scale			
Score	Description		Score	Description		Body fat %
1	Poor	The individual short ribs are sharp to touch and no tail head fat can be felt.	1	Emaciated	Little muscle; sunken eyes; tucked up; death imminent.	4
			2	Very poor	Skeleton very clear.	8
			3	Poor	Little muscle over skeleton.	11
2	Backward	The individual short ribs can still be felt, but feel rounded rather than sharp. There is some tissue cover around tail head.	4	Backward store	Light skeletal cover; backbone, hips, shoulders and pins clear.	15
			5	Store	Moderate muscle; backbone and rear ribs clear.	19
3	Moderate	The short ribs can only be felt with very firm thumb pressure. Areas either side of the tail head have some fat cover that can be easily felt.	6	Forward store	Good muscle; low fat; hips evident.	23
4	Prime	The short ribs cannot be felt and fat cover around the tail head is easily seen as slight mounds; folds of fat are beginning to develop over the ribs and thighs.	7	Prime	Good muscle and fat; smooth with no bony protuberances.	26
			8	Fat	Well covered with muscle and fat, but not lumpy appearance.	30
5	Fat	The bone structure is no longer noticeable and the tail head is almost completely buried in fat. Folds of fat are apparent over the ribs and thigh.	9	Overfat	Heavy fat cover with lumpy depots.	34

Scores in the 5-point scale as well as the 9-point scale can be modified further by use of + and – scores in association with the number to approach more continuous ranges. Terms and descriptions used here are based on the subjective scoring systems used in the Australian Beef CRC Project 4.1.3b.

important. Cattle producers that have documented what types of traits in their cattle seem to influence overall functionality, longevity and acceptance in their own herds and those of their customers attest to the importance of these types of traits. Many of these types of traits have been well documented in dairy settings for economic importance, and genetic evaluations for the 'body type' traits is becoming more prevalent, and will probably be of interest in many genomic-based analyses in the near future (provided that databases exist with these kinds of phenotypes).

6.2.1 Structural Correctness

It is important to have a basic knowledge of how the skeleton of cattle is structured as the alignment of the joints associated with the legs is related to the appearance and movement of the animal, and can have direct impacts on animal functionality and longevity in its production environment. Structural soundness should be evaluated on cattle from the front, rear and side views, and should be evaluated on animals moving at a walking pace. It is fine to evaluate the structure of animals when they are standing still, but the animal should be standing at ease in a 'normal' position. However this still image should not be the only evaluation; animals should always be structurally evaluated as they are moving as proper skeletal design and conformation helps ensure adequate animal longevity. The amount of area and distance that animals have to travel each day to find enough forage to eat, in conjunction with their skeletal structure, can dictate longevity of production. Some of these types of traits are beginning to be incorporated into beef breed genetic evaluations (Breedplan, 2013).

Animals as viewed from the side

From the side view both the front and rear legs can be evaluated (see Fig. 6.3). From this point of view the angularity associated with the shoulder, knee and foot can be evaluated for the front legs and angularity though the hip, stifle, hock and foot can be assessed for the rear legs. The ideal angle of the scapula with the humerus is approximately 90°, the ideal angle of the humerus and forearm is 135°, and the ideal angle of the metacarpus bone and the pastern/foot is 140° to 145°. If the angle of the scapula and the humerus is much greater than 90°, animals will appear too straight or too steep in their shoulders. If the knee deviates from the imaginary straight vertical line, the knee can come too far forward (over the front foot) or too far back (behind the front foot).

When evaluating the angularity of the rear legs from the side view, the angle of the femur and the tibia should have an angle of approximately 140°, and the angle of the tibia with the metatarsus should be approximately 160°. If these angles are greater than these recommendations, the animals will be too straight in the rear legs (called post legged), and if the angles are much less than these recommendations, animals will have too much set to the rear legs (called sickle hocked). Figure 6.4 illustrates varying possible sets to the rear leg.

When viewed from the side, an imaginary straight vertical line should descend from the pin bone to the back of the hock. In post legged animals the back of the hock will be too far under the animal when it is standing at rest; in a sickle hocked animal the foot will be very far under the animal if the hock is under the pin bone, or if the animal has its rear leg extended backwards its foot may be behind the pin bone. A sickle hocked animal will be able to stretch out with more distance between its front and rear legs as compared to a post legged animal. Cattle that are sickled hocked will place their feet well under their body when walking, but post legged animals will typically take much shorter strides.

It is generally thought that being too straight in the shoulder or rear legs is more disadvantageous than having too much angularity. In particular, when bulls mount females for breeding there is substantial strain on the hock joints, and bulls that are post legged are thought to be more prone to developing arthritis or other joint problems than bulls that are sickle hocked. It is also believed that longevity may be compromised more in animals that are straight legged as compared to those with too much angularity as there is probably more impact on joints from normal movement (less shock absorbing potential) when these joints have too little angle.

Animals as viewed from the front

From the front view, cattle should have legs in front that follow a vertical line from their forearm through the knee (carpus) to the middle of the toes. When viewing both front legs simultaneously, the imaginary lines should remain parallel the entire length of the legs. Undesirable deviations from this ideal can be due to the knees being too close together, or the knees being too far apart. This may be due to the joining of the bones at the knee joint or it may be because the bones are slightly rotated, or a combination. If the bones are rotated, this will also affect the set of the feet. If the knees are too

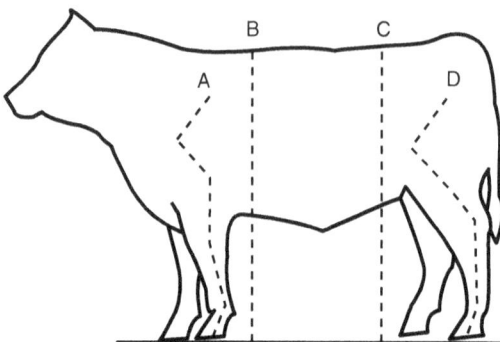

Fig. 6.3. Side view showing areas to evaluate skeletal structure and body depth. Lines A and D trace the general placement of bones in front and rear legs; lines B and C show areas to assess depth of body.

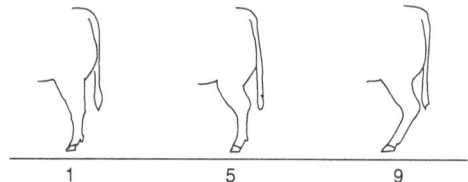

Fig. 6.4. Guidelines for scoring of set to rear legs on a 1 to 9 scale from the side view. Increasing numbers are associated with increasing angularity of rear limbs; 5 is ideal, and increasing deviations away from 5 are more undesirable.

far apart, the toes will be pointed toward the mid-line of the animal (and toward each other), whereas if the knees are too close together the toes will be pointed away from the midline (and away from each other). If the knees are too far apart or too close but the long bones of the legs are not rotated, there will be additional stress at the knee joint and probably uneven distribution of weight across the two toes. It is also possible that the two legs may have different (non-symmetrical) designs, and this is also undesirable.

Animals as viewed from the rear

From the rear view (as it is with the front view and front legs) it is desirable that the rear legs follow an imaginary straight vertical line from the hip through the hock (tarsus) joint and through the foot between the two toes. The hock joints can be too close or too far apart (i.e. the imaginary lines through the center of the legs move toward one another or away from one another at the joint), which can be due to the rotation of the long bones and/or the alignment of the joint itself, and if the bones are rotated off-center this will typically cause the feet to deviate from the ideal design as well, with toes pointing in when the hocks are far apart or toes pointing out when the hocks are too close. Figure 6.5 illustrates evaluation of leg conformation from a rear view.

Feet and toes

Evaluation of the feet is also an important consideration. The hooves (or claws) that cover the toes

provide protection. The two toes should be equal in length and evenly spaced from top to bottom. A healthy animal will have shiny hooves (this does not mean clean hooves, but this is difficult to assess unless the hooves are fairly clean). Unequal length toes will cause cattle to have impeded movement. Appearance of the toes in older cattle may be a reflection of how they are coping in the production environment. Cows may grow long hooves as they age, which may curl toward one another and cause discomfort. Cows that travel over rough ground will usually not have issues with hooves growing too long, but in soft soil this can be an issue for certain animals. Animals with long hooves or other undesirable conformation can catch a toe on saplings, tree branches or other ground structures which could possibly lead to further injury.

Some cattle evaluation guidelines have recommended that the foot make a 54° angle with level ground, and this would correspond to a 144° angle of the metacarpus and the foot. If the angle of the metacarpus and the foot is greater than 145°, the animal will be too steep or too straight in the pastern joint, but if the angle is much less than 140° then the animal has too much set to the pastern joint and is said to be weak in the pastern (see Fig. 6.6). These guidelines for the angle of the foot with the metacarpus bone can also be used as guidelines for the angularity of the metatarsus and foot on the rear leg.

The descriptions of the bones that comprise the front and rear legs are provided in Table 6.4. To properly understand visual assessment of feet and leg structure it is very important to have an understanding of how the bones and joints fit together. Budras *et al.* (2011) provide excellent graphical representations and descriptions of the skeletal structure in cattle.

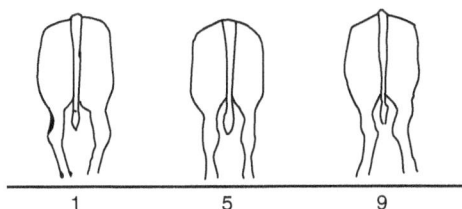

Fig. 6.5. Guidelines for scoring of rear leg conformation on a 1 to 9 scale from the rear view. A score of 5 is ideal, and increasing deviations away from 5 are more undesirable. Smaller numbers are associated with hocks being farther apart (and usually feet closer together), and larger numbers are associated with hocks being closer together (and usually feet being farther apart).

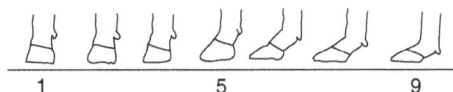

Fig. 6.6. Guidelines for scoring of feet set on a 1 to 9 scale (as viewed from the side of the animal). Score of 5 is ideal; deviations away from this score indicate too little angle (smaller numbers) or too much angle (larger numbers). Feet conformation should be evaluated for front and rear legs, and from all angles (front, back and both sides).

Table 6.4. Description of bones and associated joints of cattle legs.

Front leg (thoracic limb)		Rear leg (pelvic limb)	
Name of bone (and number per limb)	Common name and description of joint	Name of bone (and number per limb)	Common name and description of joint
Scapula (1)	Shoulder blade – the shoulder is attached to the rest of the body by ligaments, tendons and muscles.	Femur (1)	This is the upper thigh bone. It connects into the pelvis as a ball in socket joint.
Humerus (1)	Upper arm – the joint where the scapula and humerus connect is the point of the shoulder.	Patella (1)	This is analogous to a kneecap in humans. The patella is a sesamoid bone.
Radius (1) and ulna (1)	Two bones of lower arm (forearm) – the radius is toward the front of the leg and the ulna lies behind. The portion of the ulna that protrudes backwards may be referred to as the point of the elbow. The elbow joint is where the humerus, radius and ulna come together.	Tibia (1) and fibula (1)	The tibia is the main bone that connects the femur to the hock region; fibula is fused to tibia and is present close to hock. The joint where the femur, patella and tibia meet is the stifle (analogous to the human knee).
Carpal bones (6)	These small bones form the joint referred to as the knee, but this is really more analogous to the wrist in humans.	Tarsal bones (5)	Tarsal bones connect the upper and lower real leg. The calcaneus tarsal bone protrudes toward the back of the leg and results in the back of the hock.
Metacarpal (Mc) bones (3: Mc III, IV, V)	There are two main metacarpal bones that connect the knee to the foot. These together may be called cannon bone; a third, small Mc bone (V) also exists.	Metatarsal bones (2: Mt III and IV); there is also one sesamoid bone associated with Mt III	These form the cannon bone of the rear leg. The hock joint is where the tarsal bones connect the upper leg to the cannon bone. The hock joint is analogous to the ankle in humans.
Digital (6) and sesamoid (6) bones	Bones of front foot.	Digital (6) and sesamoid (6) bones	Bones of rear foot.

Digital and sesamoid bones comprise the bones of the front (manus) and rear (pes) feet. Three long bones (proximal, middle and distal phalanx bones) form the length of the digits, and short sesamoid bones are present at the fetlock and coffin joints. The fetlock, pastern and coffin joints correspond to where the Mc (or Mt) and proximal phalanx meet, where the proximal and middle phalanx meet, and where the middle and distal phalanx meet, respectively. The distal phalanx is referred to as the coffin bone, or the toe. These are presented in order from the top of the leg toward the foot. Descriptions based on Budras et al. (2011).

Vertebrae and topline

Another important consideration in regard to structural soundness is the degree of levelness of the topline. Table 6.5 provides names and descriptions of vertebrae that make up the back and tail. Ideally, animals should be close to level from the shoulders to the hip, although it is common that there is some degree of drop midway between these areas. Animals should slope smoothly and gently behind the shoulders as opposed to a well-defined break or sudden drop in the topline, and animals that drop severely between the shoulders and the hip should not be used as replacements in the breeding herd. Animals that have 'square' hip, meaning that they are close to level from the hook bone to the pin bone are also desirable and tend to be associated with increased muscularity. The set of the tail head is another consideration in regard to structure, but this probably has more to do with eye appeal than it does with production measures, but animals with a more 'square' hip (i.e. level across the top, not sloping from hooks to pins) will usually have a tail head that sets higher on the body. Often cattle with a square hip also appear to have improved muscle expression.

6.2.2 Body Capacity

Body capacity should be assessed by evaluating body depth from the top of the back to the bottom of the chest floor from the side view as well as the roundness of the ribcage from the front or rear view. Body depth can be evaluated as the relative amount of the animal's height that is body torso versus leg length (as with lines B and C in Fig. 6.3). Cattle will typically have approximately 50% of the height from the ground to the topline as leg length and 50% as body depth. Animals with more body capacity typically have a higher percentage of their height represented as body depth. For example maybe 60% of the distance from the ground to the topline is the distance from the chest floor to the topline with 40% of the distance from the ground to the bottom of the chest floor. Another animal may have only 45% of the distance from the ground to the top of the back as body depth, with 55% of the height being the distance from the ground to the bottom of the chest floor. Cattle with more body capacity are desired as they typically can eat more, which is directly related to rate of gain or the ability to maintain body condition. In conjunction with evaluation of body capacity and depth, the amount of balance in body depth behind the elbow along to the rear flank should be evaluated. The ideal beef-type conformation has a relatively similar body depth from front to rear, although a slightly decreased depth at the rear flank is very common. Cattle with more body depth at the rear flank than through the heart girth appear to have more of a dairy-type appearance, whereas

Table 6.5. Description of vertebral column (backbone and tail) in cattle.

Area of body	Neck	Chest	Lower back	Pelvis	Tail
Technical name	Cervical	Thoracic	Lumbar	Sacral	Caudal
Number	7	13	6	5	18 to 20
Comments	Referred to as C1–C7; C1 is the atlas and C2 is the axis and this is where the head is removed at slaughter.	Referred to as T1–T13; these have ribs attached; ribs attached to T1–T8 are also attached to sternum.	Referred to as L1–L6; these have prominent transverse processes protruding perpendicular to topline.	S1–S5 are completely fused to form top side of pelvis (sacrum).	Vertebrae past the pelvis to tip of tail; these have small spinous and tranverse processes.

Vertebrae have a canal through which the spinal cord passes and protrusions on the top (spinous processes) and sides (transverse processes). There is a ligament attached to the back of the skull and the neck vertebrae (nuchal ligament) that runs the length of the vertebral column to the end of the sacrum; the part attached to the skull is called the funiculus, the part attached to the cervical vertebrae is called the lamina, and toward the end of the thoracic vertebrae until the sacrum it is called the supraspinous ligament. This ligament helps support the topline of the animal.

cattle with more pronounced body depth at the heart girth are more of the draft-type conformation. There are tendencies toward body capacity and type being more common in particular breeds than others relative to their ancestry; as there are differences in body capacity and type across breeds, there are substantial differences within most breeds as well.

6.2.3 Femininity and Masculinity

The characteristics associated with femininity and masculinity are vastly more important in breeding animals and herds than in market animals. The concept of expression of desirable secondary sex characteristics is what determines these traits. In summary, bulls should look like bulls, and cows should look like cows. This statement may seem meaningless at first, as there is an obvious difference in the sexes by the appearance and placement of the sex organs. However, if someone was only able to see the animal from the shoulder forward (head, neck and shoulder), it should be obvious whether the animal is a bull, cow (or heifer) or a steer. This appearance should become more apparent as the animal approaches the age for puberty, and approaches maturity. It probably makes very little sense to compare traits of masculinity and femininity across breeds that belong in different genetic groups, but there should be obvious differences among the sexes of the animals within each particular breed. Within one breed, a shorter, wider head may indicate masculinity, but some breeds have longer heads than others, so saying that a bull of one breed is less masculine than the bull of another breed makes no sense when it is only the breed characteristics that are different.

In general, breeding females need to have 'refinement' of the head and neck, which usually means a less massive bone and muscle structure than bulls. Females will have hair on the head and neck that is finer (smaller diameter) than bulls. Bulls of many breeds that have substantial amounts of hair will have coarse, curly hair around the top of the head and neck as compared to castrated males and females. Bulls will also have a pronounced 'crest' to their necks. This should not be confused with the hump that is present in many *Bos indicus* and *Bos indicus* influenced breeds. The crest of the neck is the larger circumference of the neck (due to increased muscle expression in that area) that when viewed from the side appears as a raised area that deviates from the straight line from the head to the shoulder. In general, bulls will have shorter long bone development than steers or heifers of the same frame size, and often appear to have more body depth (higher ratio of their height that is body depth as opposed to leg length) than females of the same breed and frame size.

For many years, probably centuries, cattle breeders have evaluated masculinity and femininity because the desirable expression of these traits has been thought to be related to their fertility, longevity and functionality. The physical appearance of the animal does not guarantee the future (or previous) levels of performance, but it has direct economic importance as most beef breeding animals are valued at least in part on their desirable physical appearance. It is also possible that certain lines of cattle (or combinations) are more desirable for producing bulls versus cows in regard to their expression of femininity or masculinity (as well as their performance). This is not widely documented by breed societies, but several cattle breeders in many different breeds have observed this phenomenon. Breed societies and/or individual breeders could, if they wish, use a scoring standard for masculinity/femininity (such as on a 1 to 9 scale), as is done with other subjectively evaluated traits.

6.3 Fertility and Reproduction Related Physical Traits

The structures of some reproductive organs can be viewed by simple live animal evaluation, such as the vulva in females and the scrotum and sheath in males. Acceptable or normal appearances of these structures do not guarantee reproductive fitness. However, obvious abnormalities may indicate a variety of potential problems in actual performance or in level of acceptability by people purchasing cattle for breeding purposes. The organs of the reproductive tracts in male and female cattle and their functions are discussed in Chapter 8.

6.3.1 The Female Reproductive Tract

The reproductive tracts (organs) are not externally visible in heifers and cows, and largely not visible in bulls. The concept of a reproductive tract scoring system in heifers has been developed and follows a 1 to 5 scale based on size and development of the internal organs. Reproductive tract scoring of heifers is strongly related to their age and level of development, and this system is therefore discussed in Sections 7.1 to 7.3. Likewise, examination of the reproductive organs of bulls is an important component of bull breeding soundness exams, and this concept is discussed in Chapter 8 under reproduction considerations. The visual evaluation aspects of bull scrotum and sheath conformations are discussed here.

6.3.2 Scrotum

The scrotum of bulls used as sires should have a defined neck, should not be extremely close to the body, or too pendulous, as the function of the scrotum is thermoregulation of the testicles and resulting sperm cells. When viewed from the rear, the scrotum should provide evidence of two testicles, and should not appear twisted to either side. When viewed from the side, the shape of the testicles (and therefore the scrotum) may appear perpendicular to the flat ground surface or be slightly curved or rounded, with the rounded bulge toward the front legs and the top and bottom slightly pointed toward the rear legs. Some curvature is typical, but should not be extreme, and should not be in the opposite direction in the two testicles. Bulls with pendulous scrotums may be prone to injury of the testicles.

6.3.3 Sheath and Navel Considerations

In many *Bos indicus* (zebu) and *Bos indicus* influenced breeds, there may be a desire to evaluate and select for conformation of the animal's underline. Several of these breeds may have much more skin than *Bos taurus* breeds. Typically either a 1 to 5 or a 1 to 9 scale is used by breed societies as part

of their herd improvement performance record programs regarding sheath score; these two systems are compared in Table 6.6. The same scoring system is used for both sheath scoring of males and navel scoring of females; however, biologically this is not the same trait across the two sexes, although it is certainly related (for example, see Fig. 6.7).

The goal is usually not to select for underlines that are similar in appearance to *Bos taurus* cattle, but to select against and avoid animals that have too much underline (too loose and pendulous). This is particularly a concern in bulls of these breeds as animals with pendulous sheaths can be more prone to injury, either from other animals stepping on them when penned or confined in close quarters, or from thorny plants such as cactus and/or parasites, particularly when the prepuce (foreskin around the penis) is extended from the opening of the sheath. Some breed societies also record navel scores for females in an attempt to select for optimum amounts of skin on animal underlines, but desirable navel scores in females do not indicate or guarantee desirable sheath scores in male relatives; these traits are related, but not the same.

Zebu and zebu-influenced bulls with pendulous sheaths have been reported to have lower breeding performance in regard to serving capacity tests as well as having an increased likelihood of injury and prolapse of the prepuce (Bertram *et al.*, 1997; Hopkins and Spitzer, 1997). Heritability estimates for sheath score and/or conformation have been moderate (Franke and Burns, 1985; Kriese *et al.*, 1991), and response following selection for favorable sheath conformation has been seen when practiced (McGowan *et al.*, 2002). Figure 6.8 describes the sheath in more detail.

Table 6.6. Description of sheath scores in bulls.

Breedplan score		Description	5-point score
9	Tight	Moderately tight sheath, fairly close to abdominal wall, depth up to about 10 cm with obvious retractor prepuce muscle, moderate-sized preputial opening.	1
7–8	Small	Sheath hangs at 45° angle, depth up to about 15 cm, moderate umbilicus (navel).	2
5–6	Moderate	Sheath hangs at 45° angle, slightly more pendulous than score 2 (on 5-point scale) with depth less than 20 cm and larger umbilicus.	3
3–4	Large	Sheath hangs at up to 90° angle, excessive looseness of umbilical area, with depth just above the hock-to-knee horizontal line.	4
1–2	Very large	Sheath hangs at up to 90° angle, excessive looseness and length of umbilicus, sheath depth at or below the hock-to-knee horizontal line, often with evidence of preputial mucosa (foreskin showing).	5

This is based on the Breedplan scoring system used in Australia (Breedplan, 2013).

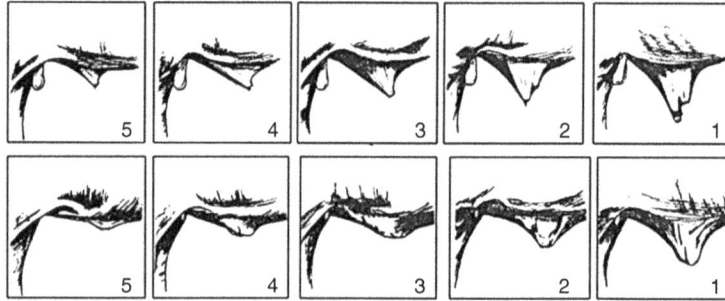

Fig. 6.7. Example of sheath (above) and navel (below) 5-point scoring system. Care needs to be taken when discussing scores across different breed societies or organizations as some may use a 1 to denote the most pendulous conformation (as shown here) where others may use a 5 to denote the same condition. Images taken from and courtesy of the American Brahman Breeders' Association (ABBA) Brahman Herd Improvement Records Program Manual (available at www.brahman.org).

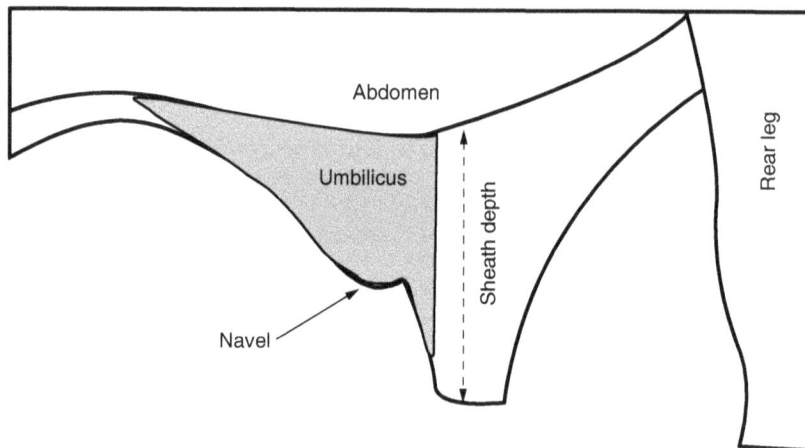

Fig. 6.8. Measurements and description of the sheath. The prepuce consists of two parts: (i) the external hair-covered appendage (sheath) and the inner part (prepuce proper or foreskin) and (ii) the umbilicus (the remnant of the umbilical cord surrounded by the skin through which it passes to the abdominal wall and the external umbilical scar or navel). Figure and descriptions adapted from McGowan *et al.* (2002).

6.4 Adaptation and Longevity Related Traits

If beef cattle are adequately adapted to their production conditions, they will have acceptable (i) survivability, (ii) fertility and (iii) longevity. It is hard to separate these types of traits into distinct concepts as they collectively determine fitness, but this statement means that in order for cattle to thrive, they must first be able to live to an age where they can reproduce, they need to be able to reproduce repeatedly, and they need to reproduce repeatedly for a long time, and the priorities follow that order. All traits discussed to this point may be closely related to longevity and adaptation, particularly when matched with particular environmental and management conditions. However, there are also several additional traits that can be useful to evaluate and can influence the success of cattle in accomplishing the goals of being a productive member of a herd for a long enough time and increasing the chances of being profitable for the manager. Traits that are more directly related to adaptation aspects are discussed here.

6.4.1 Hair Coat Characteristics

The hair coat appearance can change with both seasonal and nutritional influences (which may be almost impossible to separate as season highly influences forage nutritional values). In general a shiny hair coat, regardless of its length, indicates an animal with good nutritional status, and probably good health in general. The length and thickness of the hair is a function of climatic temperatures, and it is typical that cattle in geographical regions that experience seasonal temperature fluctuations will grow new hair coats associated with those seasons. Animals that do not 'slick off' in summer or warmer months may be experiencing stressors (nutritional, parasite or other health issues) or possibly heat stress itself. Recent research in the US Gulf Coast region (Woolfolk *et al.*, 2012) has shown there is substantial genetic variation in the ability of cattle of breeds not considered to be tropically adapted in their ability to shed hair coats, and this should be a consideration in assessing an animal's degree of adaptation to local conditions. Table 6.7 describes an example hair scoring system.

Cattle that begin to lose hair in some areas of the body to the point where the skin is exposed can indicate external parasite problems such as lice or biting flies (these parasites are discussed in Chapter 9), and the relative size and areas associated with these types of hair loss (or skin lesions) can also be scored on a standardized scale, and may also be indicative of an animal's degree of adaptation to stressors or tolerance of parasites. There is a need for increased research efforts to explore whether cattle that are more tolerant of specific stressors are also tolerant to other types, and these types of studies seem to be very limited in the scientific literature. Collection of these types of phenotypes in combination with traditional economically important traits can be done without much expense, but care should be taken so that the scales used are standardized and that evaluators are trained to identify differences. Databases with these kinds of phenotypes will be very important resources for genomic studies. Documentation of these types of traits in imported and local cattle types is also needed and highly encouraged.

6.4.2 Presence of External Parasites

When animals are subjected to the same environmental conditions, there can be individual differences in their tolerance to parasites and differences in parasite loads. This is true with both internal and external parasites, but discussion is restricted here to external parasites as it pertains to live animal, visual evaluation. Table 6.8 provides an example scoring system for the presence of ticks. A similar approach can be taken to flies or other external parasites; however, the scale used for the scoring systems should be related to documented or perceived levels of parasites that correspond to animal performance or stress levels (if known). If the level of parasite load as it relates to performance or stress is unknown, a simple low, medium, high scale can

Table 6.7. Hair coat scoring system developed in Queensland.

Score	Type	Description
1	Extremely short	Hair extremely short and closely applied to the skin. Found in zebu, in some of their crossbreeds, and very rarely in mature Hereford and Shorthorn cows in summer.
2	Very short	Coat sleek, hair short and coarse, lying flat, just able to be lifted by thumb.
3	Fairly short	General appearance of smooth-coated. Hair easily lifted, usually fairly coarse.
4	Fairly long	Coat not completely smooth, somewhat rough, patches of hair being curved outwards, or whole coat showing sufficient length to be ruffled.
5	Long	Hair distinctly long and lying loosely, predominantly coarse.
6	Woolly	Hair erect, giving fur-like appearance. Fingers are partly buried in coat. Fine hair of under-coat gives soft handle.
7	Very woolly	More extreme expression of score 6 with greater length and 'body' and heavy cover extending to neck and rump.

This system was developed by Dr Greig Turner and has been used extensively in research at the CSIRO Belmont Station in Rockhampton, QLD, Australia, with cattle types at that station; breeds mentioned here are specific to those involved in the projects under their tropical environment. This system may be further modified by use of − and + scores within each number score (4−, 5+, etc.) to provide for a 21-point system; this system was used in the Australian Beef CRC project 4.1.3b.

be initiated for use, but the levels of parasite load corresponding to these levels should be standardized and documented.

6.4.3 Pigmentation (and Head Conformation) as Related to Solar Radiation

In cattle with non-pigmented skin (where there is no pigment produced in the hair or the skin from which it emerged, as in the white areas of the body in Hereford), there is increased risk of damage from prolonged exposure to solar radiation, especially where there is no hair cover. The skin of animals where there is this lack of pigment is pink. Not all animals with white or very light-colored hair have underlying non-pigmented skin. Several light-colored breeds may have dark pigmented skin such as Nelore and Chianina for example, and others such as Charolais may have white hair and reddish pigmented skin. It is difficult to say with certainty how much the color pattern alone vs its association with other underlying breed characteristics related to environmental adaptation aspects affect heat tolerance and predisposition to skin damage from environmental exposure. In several breeds where there is variation in color and/or pigmentation distribution, these differences are thought to be related at least in part to adaptation considerations. Cattle breeders that have selected for increased pigmentation around the eyes of white-faced cattle have observed reduced incidence of eye cancers. In some cases the conformation of the skull, such as where a more pronounced brow ridge over the eyes is thought to provide some shading to the eye and surrounding skin, can help reduce exposure to solar radiation. For these types of traits and where recognizable variation exists, standardized scoring systems can be developed and implemented. To study how color pattern alone affects adaptation aspects requires that these different patterns exist within the same breed or population. Figure 6.9 shows variation of pigmentation aspects within a breed.

Table 6.8. Tick resistance scoring system.

Score	Resistance category	Observable ticks
0	Clean	None
1	Very high	≤10
2	High	11 to 30
3	Average	31 to 80
4	Low	81 to 150
5	Very low	>150

Tick resistance is only assessed when variation in tick load is evident, or on occasions when tick burdens on the majority of animals are very high and likely to significantly impact performance (such as an average of at least 20 ticks per side when averaged across 15 animals). Only engorged female ticks between 4.5 and 8 mm are counted. Values and descriptions from Australian Breedplan literature.

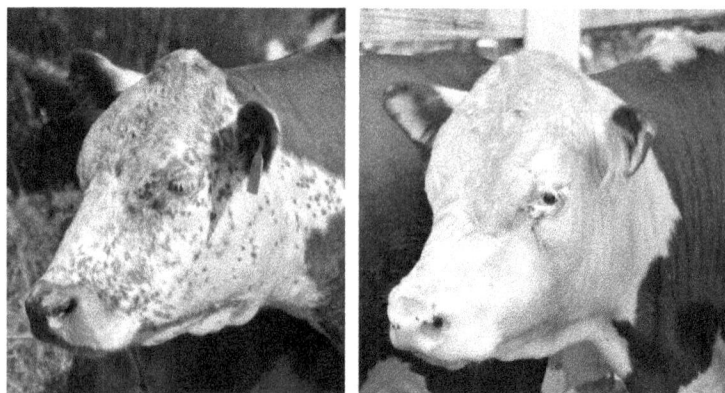

Fig. 6.9. Example of variation in facial pigmentation in polled Hereford cattle. The cow on the left has considerable areas of pigmented hair and skin within the traditional white face Hereford pattern whereas the bull on the left does not. Absence of pigmentation of hair follicles leading to white hair of this type also results in pink (non-pigmented) skin. If desired, variation in color patterns and pigmentation can be scored with a standardized scale as many other traits.

6.4.4 Teeth and Mouth Conformation

Cattle have two types of teeth (front and back). The front or anterior teeth are called incisors, and the back teeth are called molars (some may be premolars). The incisors are easier to evaluate and have been the subject of more general consideration than molars. Cattle have both temporary and permanent teeth of both kinds, and their emergence and sustainability are age related. Young cattle have three pairs of incisors (six total, bottom jaw only), one pair of canines (two total, bottom jaw only), and three pairs of premolars (six total, top and bottom jaws) to give a grand total of 20 temporary teeth. In cattle the canine teeth are very similar in shape to the incisors, and are usually referred to as the fourth pair of incisors (Budras *et al.*, 2011). The temporary teeth are called deciduous or milk teeth. These are lost as a normal aging and maturation process. Mature cattle are expected to have 32 teeth (three pairs of incisors and one pair of canines, bottom jaw only; three pairs of premolars and three pairs of molars, top and bottom jaws). Figure 6.10 shows a guide for evaluating the incisors and corresponding ages.

Cattle and other ruminants have a hard dental pad on the upper jaw as opposed to upper incisors. The incisors of the bottom jaw should make contact with the front of the dental pad as the animal closes its mouth, and typically the tops of the incisors may be slightly in front (toward nose) of the front edge of the dental pad when the mouth is closed. It is possible that cattle may have an unusually deformed conformation of the mouth or jaw. Animals with severe underbites (bottom jaw extends much further toward the nose than the upper jaw) or overbites (bottom jaw extends a much shorter distance toward the nose than the upper jaw) or other problems may have difficulty grazing or eating in general. If supplemental feed is provided to cattle, and an animal loses much of the feed it takes into its mouth while it is chewing, this may indicate problems with mouth conformation. The key to whether or not mouth conformation is a problem is how well the animals with suspect issues maintain their body condition and level of production compared to their peers.

Evaluating the mouths of cows becomes important in older cows as an aid in culling and selection decisions. In general, if the soil contains more sand there is greater potential for accelerated tooth wear, and this can be exacerbated if supplemental

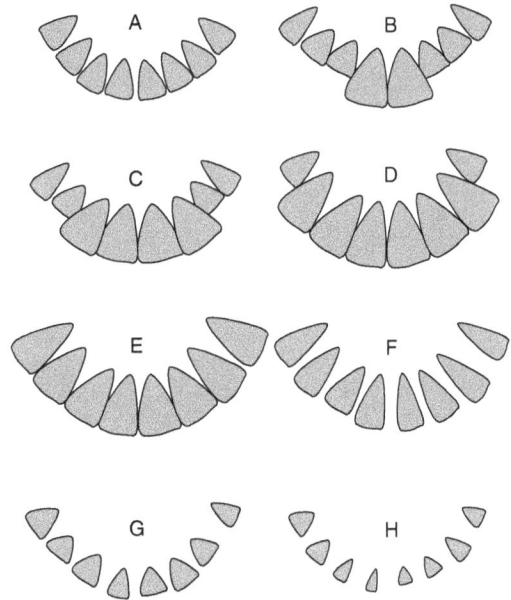

Fig. 6.10. General illustration of incisor development and evaluation in cattle. Expected ages should approximately correspond to: (A) 1 to 18 months, (B) 2 years, (C) 3 years, (D) 4 years, (E) 5 years (F–H) progressing ages and associated tooth wear (ages vary due to both genetic background and environmental conditions). If cows lose all their incisors, they may be called 'smooth-mouthed'.

feeds are provided on bare ground where soil is taken into the mouth, and when forages are grazed too close to the ground as with overgrazing. Several studies (for instance Núñez-Domínguez *et al.*, 1991, and Riley *et al.*, 2001) that have evaluated beef cow longevity have shown that cows with more teeth present are able to stay in production longer, as they are better able to feed themselves through grazing. Most of these studies have only evaluated the condition of the incisors (not the back teeth).

6.4.5 Udder Conformation

Young calves must be able to stand on their own and nurse their mothers very soon after birth to receive the necessary colostrum. Cows with pendulous udders and/or large teats present a challenge for these young calves and can negatively impact calf growth, health and even survival. Several people believe that beef cows must have large udders to produce adequate milk for their calves, because dairy cows have such large udders compared to beef cows; however, there is not an exact relationship between udder size and milk production across beef breeds, and an even lower degree of association within a breed. Evaluation of udder support and teat size needs to be a component of selection programs in all beef herds, but often does not receive much attention (until there is a problem) compared to growth and size type traits. Evaluation of the udder immediately prior to calving will probably be the most useful time for evaluation as the appearance then will be the conformation the newborn calf deals with. This may not be the most convenient time for evaluation and so evaluating the udder during the first 2 months after calving is recommended. Evaluation does not necessarily mean recording a score; however, standardized scoring systems do exist as shown in Fig. 6.11, and are recommended to be followed by several breed society herd improvement programs. In consideration of potential herd sires, where daughters may be kept as replacements, it is wise to evaluate the dams of the sires, if possible while they are nursing calves, to evaluate the udders.

Several studies (Frisch, 1982; Riley *et al.*, 2001, 2004) have shown important relationships between teat and udder conformation with calf survival and growth, and with cow longevity. The conformation of the teats and udder directly influence the time that calves begin suckling and can ingest colostrum, although this general area has been more widely studied in dairy cows than beef cows. Selman *et al.* (1970) provided a classical study on time involved to begin nursing and udder conformation; these authors as well as later studies (Edwards, 1982; Ventorp and Michanek, 1992) found that cows with smaller, less pendulous teats had calves that spent less time searching for their first suckle and that calves from cows with 'low slung' or more pendulous udders cannot be expected to obtain colostrum soon enough by natural suckling.

Simply selecting for increased weaning weight and/or increased milk production could have unintended consequences on udder characteristics. Sapp *et al.* (2004) evaluated udder score data in first-parity Gelbvieh females and found a significant relationship between udder and teat size with calf pre-weaning gain but stressed that female extremes in these traits should be avoided. MacNeil and Mott (2006) studied pre-weaning calf growth and udder conformation in Hereford cattle and concluded that selection purely for increasing maternal pre-weaning gain or milk production may not be the correct thing to do as it could lead to degradation of udder quality and conformation. The degree of genetic relationship reported in these studies indicates that increased calf pre-weaning

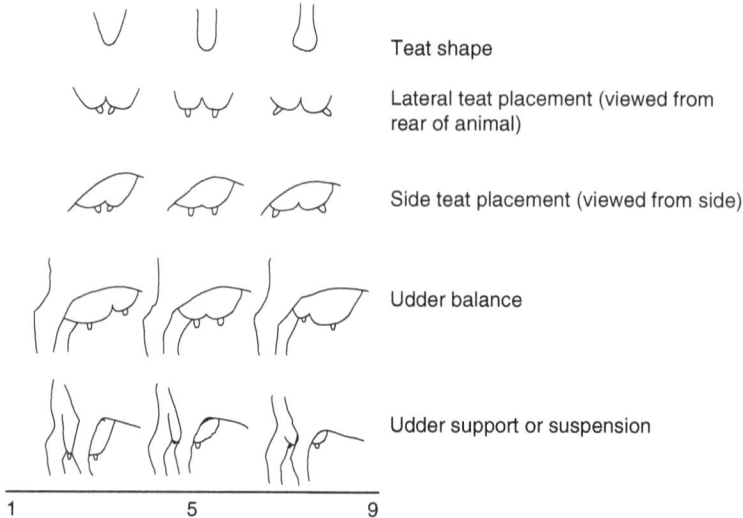

Teat shape

Lateral teat placement (viewed from rear of animal)

Side teat placement (viewed from side)

Udder balance

Udder support or suspension

1   5   9

Fig. 6.11. Guidelines for scoring of teats and udders in beef cows on a 1 to 9 scale (adapted in part from Riley *et al.*, 2001). A score of 5 is the most ideal for teat shape, teat placement and udder balance. Scores of 5 and above are ideal for udder support.

gain can be associated with changes in udder and teat conformation, but one does not dictate the other. As management of herds becomes more extensive (cows are not closely observed or provided intervention), more emphasis on tight udders with small teats will likely increase chances for newborn calf survival. These strategies can also increase the longevity of cows by avoiding undesirable traits of bottle teats and pendulous udders.

6.5 Summary

Many new technologies have been developed to aid in the evaluation and documentation of cattle performance. It may be tempting to think that visual evaluation of livestock is outdated because it is not as 'high tech' as new and improved technologies. There is no substitute for experience in visual evaluation of animals because it can relate to identification of types of animals that are profitable when combined with production data. Visual evaluation alone, or performance data alone, will provide incomplete pictures of cattle performance. The visual appearance of all products is important to consumers. Physical appearance and understanding what is 'normal' or 'abnormal' is crucial as an aid in diagnosis of animal health (Chapter 9). Visual evaluation of muscle expression and body condition and their incorporation into breeding programs in many cattle populations would probably increase overall productivity and profitability in many scenarios. Many visual traits can be effectively scored with a standardized system, and it is probably most convenient to use the same scale for all traits (such as a 9-point scale where a higher number means more of something, etc.). Producers should prioritize which physical traits they really need to track, and be careful not to keep records on too many traits as this is usually not sustainable after the excitement of newness wears off. As with all traits, single-trait selection programs should always be avoided, so incorporation and documentation of live animal, visual evaluation needs to be used in conjunction with performance and financial information for improved beef cattle production systems.

6.6 Study Questions

6.1 List the names of the bones in the front and rear legs of cattle. Also provide the total number of bones of each type in both the front and rear legs.

6.2 List five areas on the bodies of cattle that would be useful to evaluate for body condition score.

6.3 List (in order from head to tail) the names of the different types of vertebrae and the numbers of each type in cattle.

6.4 Describe in general the appearance of a cow with a body condition score of 3 and contrast that description with that of a cow with body condition score of 6.

6.5 List four areas on the bodies of cattle that would be useful to evaluate muscle expression.

6.6 Provide a general description of what you believe to be acceptable teat and udder conformation in beef cows.

6.7 Provide some descriptions of cattle with undesirable feet and leg conformation.

6.8 Describe from the rear view perspective cattle that are quite muscular versus those that are lightly muscled.

6.7 References

Bertram, J.D., Holroyd, R.G., McGowan, M.R. and Doogan, V.J. (1997) Sheath and mating ability measurements and their interrelationships in Santa Gertrudis bulls. *Proceedings of the Association for the Advancement of Animal Breeding and Genetics* 12, 351–354.

Breedplan (2013) Understanding Trial Structural Soundness EBVs. Available at: http://breedplan.une.edu.au/tips/Understanding Trial Structural Soundness EBVs.pdf (accessed 11 May 2014).

Budras, K.D., Grenough, P.R., Habel, R.E. and Mülling, K.W. (2011) *Bovine Anatomy*, 2nd extended edn. Schlütersche Verlagsgesellschaft mbH & Co. KG, Hannover.

Edwards, S.A. (1982) Factors affecting the time to first suckling in dairy calves. *Animal Production* 34, 339–346.

Franke, D.E. and Burns, W.C. (1985) Sheath area in Brahman and grade Brahman calves and its association with preweaning growth traits. *Journal of Animal Science* 61, 398–401.

Frisch, J.E. (1982) The use of teat-size measurements or calf weaning weight as an aid to selection against teat defects in cattle. *Animal Production* 32, 127–133.

Herd, D.B. and Sprott, L.R. (2007) Body condition, nutrition and reproduction of beef cows. Publ. B-1526 Texas Beef Cattle Management Handbook. Texas Cooperative Extension, Texas A&M AgriLife Extension Service, College Station, TX.

Hopkins, F.M. and Spitzer, J.C. (1997) The new Society for Theriogenology breeding soundness evaluation system. *Veterinary Clinics of North America: Food Animal Science* 13, 283–293.

Kambadur, R., Sharma, M., Smith, T.P. and Bass, J.J. (1997) Mutations in myostatin (GDF8) in double-muscled Belgian Blue and Piedmontese cattle. *Genome Research* 7, 910–916.

Kriese, L.A., Bertrand, J.K. and Benyshek, L.L. (1991) Genetic and environmental growth trait parameter estimates for Brahman and Brahman-derivative cattle. *Journal of Animal Science* 69, 2362–2370.

MacNeil, M.D. and Mott, T.B. (2006) Genetic analysis of gain from birth to weaning, milk production, and udder conformation in Line 1 Hereford cattle. *Journal of Animal Science* 84, 1639–1645.

Magnabosco, C.D.U., Ojala, M., de los Reyes, A., Sainz, R.D., Fernandes, A. and Famula, T.R. (2002) Estimates of environmental effects and genetic parameters for body measurements and weight in Brahman cattle raised in Mexico. *Journal of Animal Breeding and Genetics* 119, 221–228.

McGowan, M.R., Bertram, J.D., Fordyce, G., Fitzpatrick, L.A., Miller, R.G., Jayawardhana, G.A., Doogang, V.J., De Faveri, J. and Holroyd, R.G. (2002) Bull selection and use in northern Australia 1. Physical traits. *Animal Reproduction Science* 71, 25–37.

McPherron, A.C. and Lee, S.J. (1997) Double muscling in cattle due to mutations in the myostatin gene. *Proceedings of the National Academy of Sciences USA* 94, 12457–12461.

Núñez-Dominguez, R., Cundiff, L.V., Dickerson, G.E., Gregory, K.E. and Koch, R.M. (1991) Lifetime production of beef heifers calving first at two vs. three years of age. *Journal of Animal Science* 69, 3467–3479.

Richards, M.W., Spitzer, J.C. and Warner, M.B. (1986) Effect of varying levels of postpartum nutrition and body condition at calving on subsequent reproductive performance in beef cattle. *Journal of Animal Science* 62, 300–306.

Riley, D.G., Sanders, J.O., Knutson, R.E. and Lunt, D.K. (2001) Comparison of F_1 *Bos indicus* x Hereford cows in central Texas: II. Udder, mouth, longevity, and lifetime productivity. *Journal of Animal Science* 79, 1439–1449.

Riley, D.G., Chase, Jr, C.C., Olson, T.A., Coleman, S.W. and Hammond, A.C. (2004) Genetic and nongenetic influences on vigor at birth and preweaning mortality of purebred and high percentage Brahman calves. *Journal of Animal Science* 82, 1581–1588.

Sapp, R.L., Rekaya, R. and Bertrand, J.K. (2004) Teat scores in first-parity Gelbvieh cows: Relationship with suspensory score and calf growth traits. *Journal of Animal Science* 82, 2277–2284.

Selman, I.E., McEwan, A.D. and Fisher, E.W. (1970) Studies on natural suckling in cattle during the first eight hours post partum: II. Behavioural studies. *Animal Behaviour* 18, 284–289.

Ventorp, M. and Michanek, P. (1992) The importance of udder and teat conformation for teat seeking by the newborn calf. *Journal of Dairy Science* 75, 262–268.

Woolfolk, M.R., Mayer, J.J., Davis, J.D. and Smith, T. (2012) Determining the relationship between body temperature and hair shedding scores in Angus cows. *Journal of Animal Science* 90(Suppl. 1), 3(Abstr.).

7 Growth and Development

The size of beef cattle affects their value as weight is the primary driver of income when animals are sold, particularly in non-breeding animal markets. The genetic potential for weight and weight gain of progeny also influences the value of many breeding animals, particularly herd sires. The concept of potential cattle growth and development begins when mating decisions are made, but in regard to a particular animal actually begins with its conception. There are mechanisms in place within the first 6 weeks of gestation of a fetus that may influence its productivity as a beef-producing market animal or as a breeding beef cow for many years. The purpose of this chapter is to discuss the main concepts associated with growth and development of cattle that can have impacts on productivity and profitability for beef cattle producers. Considerations in regard to prenatal, pre-weaning and post-weaning aspects are highlighted along with procedures that have been used to evaluate and estimate cattle growth and size traits. Figure 7.1 demonstrates some broad scale ages and stages of production in the lives of beef cattle, beginning with conception, which will be discussed in this chapter.

7.1 Prenatal Growth and Development

Upon fertilization the newly formed, single-cell conceptus (the individual resulting from fertilization/conception) is called a zygote. It begins as one cell, and goes through a series of mitotic cell divisions, doubling the number of cells each time (2 cells, 4 cells, 8 cells, etc.) roughly every 24 hours. During this early phase of development (referred to as the cleavage phase), each cell is identical and has the potential to produce a complete new copy of the individual. The cells are referred to as blastomeres, and are termed totipotent, meaning there has been no cellular differentiation at that point (no cells have yet been programed to develop into any particular type of tissue or function). As a result, early cloning in cattle targeted collection of embyros and either

splitting them, or removing the nuclei from them at the 32-cell stage. The problem with cloning an individual at that point is that it is uncertain what type of animal is being cloned (as opposed to cloning an adult animal). The method of cloning an embryo where the individual nuclei are removed and placed into un-nucleated oocytes is called nuclear transfer or nuclear transplantation. The term used to describe cloning an adult is somatic cell nuclear transfer, which involves taking a differentiated cell, removing its nucleus and placing that nucleus into an un-nucleated oocyte, and then getting the cell to revert back to the totipotent phase and resume development, as occurs immediately after fertilization.

As cell division continues, at some point the number of cells becomes hard to count accurately, and the conceptus is referred to as a morula. In cattle the morula stage is typically from days 4 to 7 or 8 after conception. The structure further develops and a cavity is formed in the center, and this represents the blastocyst stage, which typically lasts from day 7 or 8 to day 12 of gestation. Up to this point the conceptus has been developing within a protein membrane known as the zona pellucida, but between days 9 and 11 of gestation the zona pellucida decreases in strength and the conceptus is said to hatch when it becomes free of the zona pellucida (Fig. 7.2).

The cleavage phase lasts from conception through day 12 to 13 in cattle, and varies among different species. The term embryo is used by many individuals to represent the conceptus early in gestation, but technically the conceptus is officially classified as an embryo after it is free of the zona pellucida, around days 13 through day 45 of pregnancy. The differentiation phase is when the germ layers are formed, the placenta develops, and the organs and specialized tissues begin to form.

After differentiation (from day 46 until birth), the subsequent phase of development is the growth phase, because although organs and tissues continue

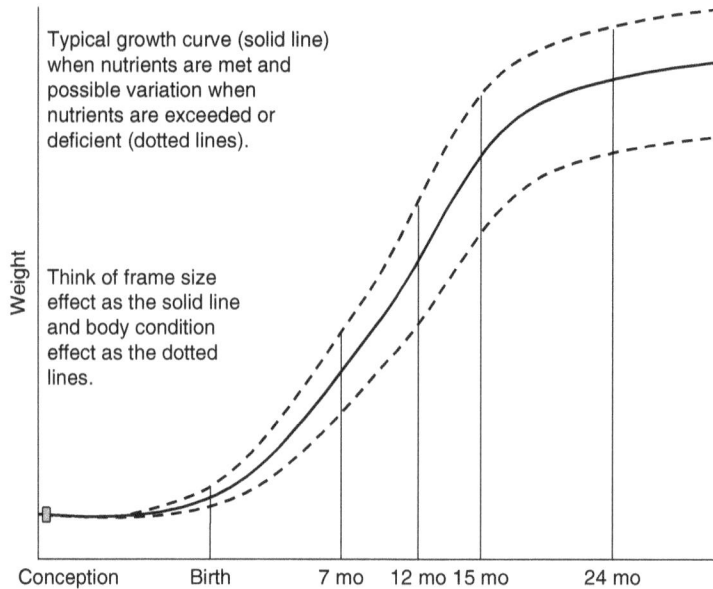

Fig. 7.1. General beef cattle growth curve.

Typical growth curve (solid line) when nutrients are met and possible variation when nutrients are exceeded or deficient (dotted lines).

Think of frame size effect as the solid line and body condition effect as the dotted lines.

Weight

Conception   Birth   7 mo   12 mo 15 mo   24 mo

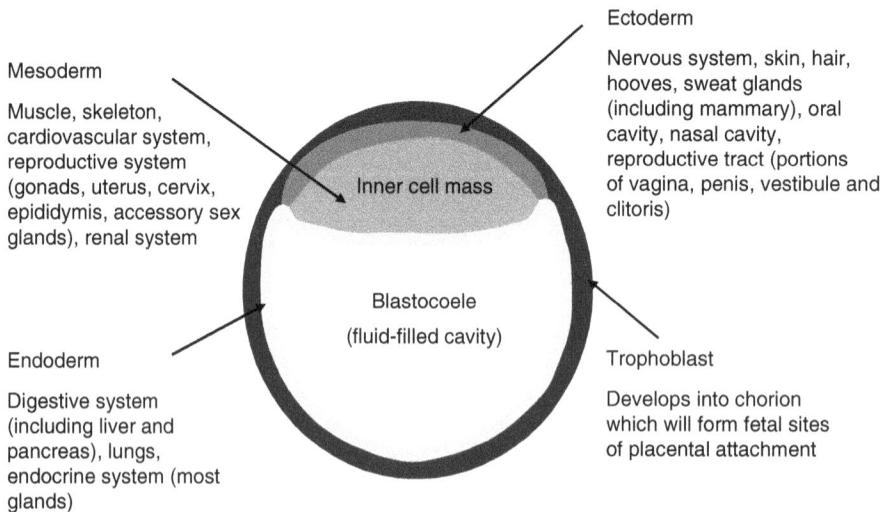

Ectoderm

Nervous system, skin, hair, hooves, sweat glands (including mammary), oral cavity, nasal cavity, reproductive tract (portions of vagina, penis, vestibule and clitoris)

Mesoderm

Muscle, skeleton, cardiovascular system, reproductive system (gonads, uterus, cervix, epididymis, accessory sex glands), renal system

Inner cell mass

Blastocoele
(fluid-filled cavity)

Endoderm

Digestive system (including liver and pancreas), lungs, endocrine system (most glands)

Trophoblast

Develops into chorion which will form fetal sites of placental attachment

Fig. 7.2. Embryonic germ layers that develop from blastocyst. Inner cell mass develops into the embryo. The three primary germ layers develop into the organs and structures of the body. Adapted from Senger (2003) with help from D.W. Forrest.

to develop, the paths of specialized tissues and structures are already in place, and there is primarily continuation of developmental pathways and associated increase in size that occurs in conjunction with the type of development. After differentiation and until birth, the conceptus is referred to as a fetus. Figure 7.3 illustrates the change in length and weight from conception to birth, along with the relative stages of cleavage and differentiation. In measurement of fetuses, crown rump length is a standard measure and refers to the length from the crown of the head to the back of

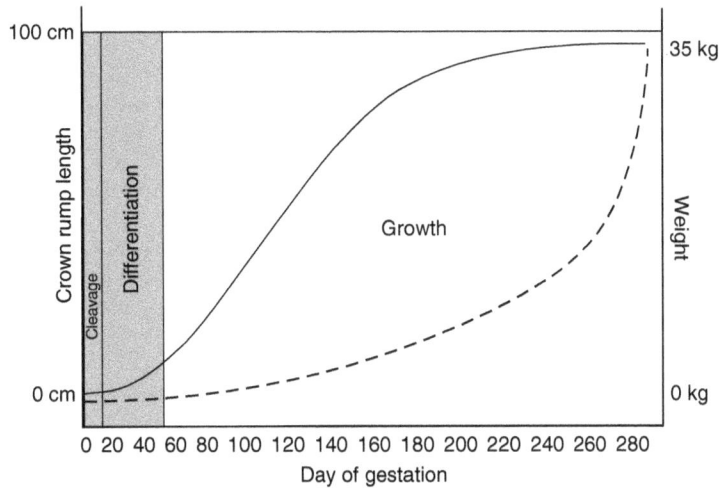

Fig. 7.3. General prenatal growth in cattle. Crown rump length designated with solid line (and left axis) and weight designated with broken line (and right axis). Shaded areas represent cleavage (days 0 to 12) and differentiation (days 13 to 45) phases. The time after the differentiation phase until birth is termed the growth phase, and this marks the transition from embryo to fetus.

the rump (this probably corresponds to the area of pin bones in an older animal).

Figure 7.3 also introduces the concept that growth curves (the increase in size over the time or life of the individual) can vary considerably in shape depending on which particular trait is considered. The change in fetal length happens very quickly early in development and then slows down, but the change in fetal weight is almost the opposite. There is much more occurring with the fetus than just change in weight and length, as specific tissues and organs are known to develop at certain times. Table 7.1 provides some estimates for various tissue, organ or system development in cattle fetuses that have been reported in the literature. As with all measures, there can be variability associated with these developmental processes and their timing, but the variability in timing of these types of traits do not appear to have as much variation as compared to typical production-type traits that are measured later in life.

The term and concept of growth is generally thought of as the change in size in an animal that has not yet reached maturity (it is part of the maturing processes), but there are many component traits associated with growth. This is evident in Fig. 7.3 and there is a large early increase in fetal length, but this slows rapidly at the end of the second trimester.

Conversely, fetal weight increases much more slowly than length early in gestation and increases the most rapidly during the third trimester. Calf fetuses immediately prior to birth may be gaining 0.2–0.5 kg/day in weight, but are not changing much per day in length.

An important aspect that must be considered is that individual variability exists for all components of development, and some of these influences are genetic, environmental, or interactions of the two. Just because a timeline indicates the expected growth rate of a fetus, or a day of gestation that a particular function begins or a particular structure is observable, does not mean that will specifically be the case in all circumstances. It must be remembered that the lines in these types of graphs represent the mean values for a particular function equation. Table 7.2 provides some data on fetal traits in four breeds of cattle. The means (and associated standard errors) are reported for body weight, liver weight and body fat to body weight ratio. There are additional data and traits discussed in the article by Mao *et al.* (2008) from which the data came. These data are interesting because they are from cattle of different biological types and show relationships between size and composition that are variable. This type of variability is also possible within a breed (but may not have the same degree of extremes).

Table 7.1. Some fetal developmental processes in cattle and associated times in gestation.

Process or characteristic	Day of gestation
Interferon tau (bIFNt) produced by conceptus; maternal recognition of pregnancy	15 to 16
Growth and formation of organs	13 to 45
Mammary band present	32 to 39
Development of immune system	45 to 150?
Palate fused	56
Eyelids fused, external genitals differentiated, hooves begin formation, teat development begins	60
Begin calcification of bone matrix	70
Tactile hairs appear on face	76
Differentiation in size between sexes	100
Tooth formation, eruption begins	110
Formation of primary follicles	110 to 130
Hair and color markings appear, hooves become hard, teats formed and defined, descent of testicles complete	150
Extensive bone formation	180
Eyelid separated	196
Hair cover complete	230
Birth	275 to 295

Compiled from multiple sources.

Breeds (and individual animals) may be similar at one point in time, but different in others. In studying the fetal weights at 3 months, the Angus and Belgian Blue fetuses are the most similar, at 6 months the Angus and Holstein are the most similar, but at 9 months the Holstein and Belgian Blue are the most similar. This concept of variability is likely to be true through all phases and stages of growth. Based on discussion in Chapter 3, it is expected that large differences in composition due to muscle mass are associated with many of these weight differences (the general description of weight gain associated with varying body composition is discussed later in the chapter under post-weaning growth).

When discussing fetal growth and particularly birth weight of *Bos indicus–Bos taurus* crosses, there are significant differences in size (and gestation length) depending upon how the cross is made. In particular birth weight is increased when there is more *Bos indicus* influence inherited through the sire line as compared to the dam line, and there is also a much more pronounced difference in birth weight between sexes, with male calves much

heavier than female calves. This is most evident in reciprocal F_1 crosses (*Bos indicus* sire × *Bos taurus* dam as compared to *Bos taurus* sire × *Bos indicus* dam). This has historically been attributed to the genotype of the dam (*Bos indicus* vs *Bos taurus*) being different in regard to nutrient supply and/or maternal–fetal blood flow. In most of these studies, natural service or AI matings have occurred where the genotype of the calf has been confounded with the genotype of the dam. However, even when embryo transfer calves are produced using a single type of recipient (surrogate) dam this same phenomenon has been identified (Amen *et al.*, 2007). Further studies into this phenomenon are needed to characterize the seemingly non-traditional mode(s) of inheritance that may be involved. There may be genetic mechanisms such as genomic imprinting involved, and this broadens the concept of what has been called fetal programming.

7.1.1 Fetal Programming

Fetal programming is the concept that maternal influences during gestation affect the developmental physiology of the fetus and have postnatal growth and health implications for that individual (Barker *et al.*, 1993). This concept was first seriously investigated in humans, but is now of interest in multiple animal species, including food animals. There is very limited information on the effects of maternal gestational influences on food animal production. However, where it has been investigated there appears potential to manipulate some future performance aspects in this manner. Some studies have investigated manipulation of the diets in pregnant cows and compared subsequent performance of the offspring when the only difference in environmental influences occurred during their fetal development inside their dams. The concept of fetal programming stems from the idea that key tissue types and organs are developing at specific stages of gestational development, and that maternal dietary influences may have permanent effects on offspring. Not much research has been done in cattle with production implications considered, but several trials have produced results that warrant consideration of fetal programming in more detail and production systems. There is potential for fetal programming to influence female reproduction, growth rates, carcass yield and quality traits and health.

Table 7.2. Selected cattle fetal traits evaluated at times of gestation.

Breed	3 months	6 months	9 months
		Body weight (kg)	
German Angus	0.22 ± 0.04	8.70 ± 1.41	34.17 ± 4.34
Galloway	0.11 ± 0.07	6.75 ± 2.15	29.60 ± 6.92
Holstein Friesian	0.32 ± 0.03	8.81 ± 1.05	41.00 ± 6.84
Belgian Blue	0.24 ± 0.05	8.39 ± 1.55	46.34 ± 6.28
		Liver weight (g)	
German Angus	8.5 ± 2.4	251.0 ± 35.0	642.2 ± 75.5
Galloway	4.1 ± 2.6	176.0 ± 45.9	481.0 ± 86.1
Holstein Friesian	11.6 ± 0.8	256.8 ± 43.3	832.7 ± 155.0
Belgian Blue	8.6 ± 2.0	251.4 ± 48.2	831.3 ± 132.6
		Body fat:body weight ratio	
German Angus	0.19 ± 0.08	0.84 ± 0.24	1.26 ± 0.37
Galloway	0.15 ± 0.09	0.79 ± 0.21	1.33 ± 0.41
Holstein Friesian	0.21 ± 0.07	0.71 ± 0.15	1.05 ± 0.34
Belgian Blue	0.13 ± 0.03	0.66 ± 0.10	0.95 ± 0.18

Data from Mao *et al.* (2008); 4 to 7 fetuses evaluated for each breed and stage of gestation; means and respective standard errors reported.

Summers and Funston (2011) reviewed several US trials (Stalker *et al.*, 2006, 2007; Larson *et al.*, 2009; Radunz, 2009; Underwood *et al.*, 2010) where dietary treatment in gestating cows resulted in carcass compositional differences in resulting steer progeny. Greenwood and Café (2007) in Australia have also been researching fetal programming on growth and carcass traits. Du *et al.* (2010) discussed fetal muscle and adipocyte (fat cell) development and proposed that nutritional management of the dam during pregnancy and early uterine influences may be the most effective way of manipulating body composition and intramuscular fat (marbling) deposition in market animals. Du *et al.* (2010) stated that the stages in altering marbling (intramuscular fat) through nutritional management were (in priority order): (i) fetal stage, (ii) neonatal stage, (iii) early weaning stage (i.e. 150 to 250 days old) and (iv) post-weaning and older stages. There is still more research needed into various feeding regimes and cattle genetic types (and combinations) to fully assess these recommendations.

Some discussion of the concepts of true growth and other aspects of weight gain is warranted here. In the fetus and in the young animal, increases in weight are growth, even though there can be some variability in the composition of the growth. True growth generally refers to increase in size and mass of structural tissues such as internal organs, muscle and bone; increases in weight associated with increases in fat may not be considered true growth, even though increases in muscle mass may occur concurrently. In regard to fetal development, increases in size are true growth, but as the animal gets older there can be less obvious distinctions. Two terms that are associated with growth need to be defined here as well. Hyperplasia refers to production of new cells and/or an increase in numbers of cells; hypertrophy refers to enlargement of existing cells, or increase in cell size. Another term that is often associated with growth is accretion, which refers to the increase or building up of tissue such as muscle or fat.

7.2 Postnatal and Pre-Weaning Growth and Development

The terms 'postnatal' and 'pre-weaning' are a bit ambiguous in that they can have broad interpretations. They are used here to describe the time in the production process of calves after birth and before weaning when the calf is viewed as the responsibility of the cow (as opposed to post-weaning, where the calf is responsible for its own growth and performance). The term 'neonatal' usually refers to the time immediately following birth and generally is thought of as the few days or weeks following birth.

Growth of calves (primarily considering weight) from birth to weaning (typically at 5 to 7 months of age) is influenced by many factors such as health status and level of nutrition, in concert with the animal's genetic ability for growth. When considering weaning weight (or any other intermediate weight prior to weaning), there are genetic effects related to individual, family and breed differences that influence weight gain as well as heterosis (in the calf as well as the cow). Non-genetic effects that can affect weight gain are sex of the calf, age of the calf, age of the dam, and nutritional influences of diet, ambient climatic conditions and health-associated stresses, many of which can interact with one another and with genetic influences. Management dictates the age at which calves are weaned, and as a result the transition from nursing to weaned calf may not coincide with a major physiological change in the animal.

Early weaning of beef calves is a strategy discussed in more detail in Chapters 8 and 10; it can have significant positive impacts on beef cow reproduction. Some studies have evaluated the nutritional management of early-weaned calves (45–100 days old) as compared to calves weaned at traditional ages (6–8 months old) in regard to weight gain, efficiency of gain and carcass endpoint traits. Many studies have found that early-weaned calves, as compared to contemporaries weaned at 6–8 months of age, have carcass compositional differences later in life, including increased marbling in some studies without increased external fat thickness. It is possible that feeding of animals for optimal nutrition (or conversely, avoiding large deficiencies) early in life is a similar concept to fetal programing; it is expected that there is less potential for future 'programing' from nutritional inputs as animals approach maturity, but this is still not a widely studied area of beef cattle production.

Adjusted birth weight and weaning weight are recommended to be evaluated where the age of the calf and the age of the dam are standardized, as these are known non-genetic influences of weaning weight that if they go undocumented add noise into selection programs. The age of the calf can be adjusted to any age of interest (such as 180 days, 200 days, 205 days, etc.), as can the age of the dam. The Beef Improvement Federation (BIF, 2010) has proposed general age of dam adjustments (Table 7.3) for both birth weight and 7-month-old weaning weight.

The formula to calculate an age-adjusted calf weaning weight is:

$$[(WWT - BWT)/age\ in\ days] \times$$
$$desired\ standardized\ age + BWT \quad (7.1)$$

Example: Suppose that a calf had a weaning weight of 212 kg, it was 214 days old at weaning and its birth weight was 38 kg. Its 200-day adjusted weaning weight would be:

= [(212 − 38)/214] × 200 +38
= [(174/214) × 200] + 38
= (0.813 × 200) + 38
= 200.6 kg

The same procedure can be used for 205-day adjusted weights, 180-day adjusted weights, etc. to any standardized age of calves. In this procedure the pre-weaning average daily gain is calculated by finding the difference between the actual weaning weight and the birth weight and then dividing this weight gain by the age in days. Then, this pre-weaning average daily gain is multiplied by the desired standardized age (the number of days expected to have this level of ADG), and finally, the birth weight is added back because it is part of the animal's current weight. Additionally, if this calf was a male calf from a 3-year-old cow, another 18.1 kg would be added for a standardized age of dam level. If it was a heifer calf from a 4-year-old cow, an additional 8.2 kg would be added, etc.

Many factors can affect the growth rates of calves from birth to weaning, as they do in other developmental stages. We can categorize these effects by multiple means for genetic vs non-genetic factors and direct vs maternal factors. Direct genetic influences refer to genetics for growth possessed by the

Table 7.3. Standard age of dam adjustment factors for beef calf birth weight and weaning weight.

Age of dam (years) at calf birth	Birth weight (kg)	Weaning weight (kg)	
		Male	Female
2	+ 3.6	+ 27.2	+ 24.5
3	+ 2.3	+ 18.1	+ 16.3
4	+ 0.9	+ 9.1	+ 8.2
5 to 10	0	0	0
11 and older	+ 1.4	+ 9.1	+ 8.2

BIF (2010). Adapted from *Bos taurus* breeds; many breed societies have adjustment factors that deviate from these values and are specific to their breeds. It is also possible that geographical regions may deviate in values across age of dam categories. Weaning weight adjustment relative to calf ages of 160 to 240 days.

calves themselves, which can be influenced in turn by additive (i.e. breeding value) and non-additive (i.e. hybrid vigor or heterosis) effects. Maternal influences are thought to be driven primarily by milk production as it affects pre-weaning growth and weaning weight, and maternal influences can be of additive and non-additive genetic origin also. The maternal influence of a cow on her calf's weaning weight is an environmental influence than has a genetic origin (there is genetic influence on milk production in the dam, but the milk production is an environmental influence on the calf). Nutrition, health, etc., are also environmental influences that can impact the direct growth of the calf and/or the maternal influence through milk production. Some of the factors known to impact calf growth and size are shown in Table 7.4.

It is likely that many of the factors in Table 7.4 can interact with one another, and in many cases they may be difficult to differentiate. For instance differences in weaning weights on an operation from one year to another with the same cow herd can be due to combinations of nutrition, stress and health that are interconnected; also, the nutrition of the calf can be due to forages and feed the calf ingests but also milk production from the cow, and these can be influenced by age, breed and heterosis of the cow as well as age, breed and heterosis of the calf. The altered health of the calf could be a function of receiving improper levels of colostrum after birth, but could be observed as reduced nutritional levels in the calf if it eats less vigorously, etc.

In some countries growth-enhancing technologies are available for use in nursing (suckling) calves that alter weight gain and/or efficiency of gain. Growth-promoting implants typically have been shown to increase weaning weights at 6–8 months of age by 7–20 kg (15–45 lb). Feeding of ionophores to calves old enough to have functional rumens also has been shown to increase energetic efficiency of gain by 5–15%. Products should only be used in accordance with label directions and local regulations. Use of growth-promoting implants in young beef heifers is typically discouraged. As most growth-promoting implants are hormone-based, alteration of hormone profiles in developing females could have negative influences on future immediate and lifetime fertility. The ovaries of females are developing continually before puberty occurs (even though this is not obvious), and any practices that jeopardize future fertility or longevity of breeding females should be avoided. Simply putting on more weight in beef heifers from use of these technologies does not have the same potential as putting on weight from increased nutrition. Also, growth-promoting implants should not be used when sub-optimal nutrition is available to the calves. Implants complement proper nutrition levels, they do not substitute for them.

As proper nutrition early in life is desired, overfeeding is not. Just as under-nutrition may lead to permanent effects, so can nutrient excess. If young breeding cattle become too fat early in life they may develop permanent fat deposits in the udder and the scrotum. These could lead to reduced milk production in the females and reduced thermoregulation of the testicles in males.

There has been some research where feeding of grain to calves early in life promotes the proper development of the rumen, but may also alter body composition patterns later in life. Many of these trials have evaluated calves that have been weaned early (as in 60–120 days old). Several studies (Williams et al., 1975; Myers et al., 1999; Fluharty et al., 2000; Schoonmaker et al., 2001) have shown that early weaning of calves and immediately feeding a high-concentrate diet can increase intramuscular fat deposition. Schoonmaker et al. (2001) stated that early weaning and concentrate feeding may allow sufficient marbling deposition to occur early, which is observed at time of harvest. Most of this type of research has been accomplished with British and European types of cattle, and more work needs to be done in this field with tropically adapted breeds.

Table 7.4. Potential factors that can alter calf pre-weaning growth and weaning weight.

Breed/family differences for growth	Nutrition of cow	Disease or illness
Direct heterosis (crossbred calves)	Nutrition of calf	Stress due to handling
Breed/family differences for milk	Age of cow	Geography
Maternal heterosis (crossbred cows)	Age of calf	Parasite load
	Growth implants	

7.3 Post-Weaning Growth and Development

As mentioned above, calves of the same age may be nursing calves or weaned calves, but past weaning the animal is managed for its own growth and development (as opposed to only managing its dam as with very young calves, or managing both the dam and the calf with older nursing calves). The occurrence of puberty is variable across breeds and does represent a significant physiological shift in the life and development of cattle; puberty typically occurs after calves are weaned, but this is a function of the age at which weaning occurs and the genetic background of the calf. Some breeds that have high milk production potential may have at least some calves reaching puberty before typical weaning age (6–8 months); as a result, comparing weaning weights of calves that are pubertal vs ones that are not is not a fair comparison (but is likely not to be observed or documented in many cases). In this discussion it is assumed that puberty is occurring after weaning, such as close to 12 months of age.

Figure 7.4 illustrates weight changes related to growth across multiple seasons. In most economies, the value of cattle used to produce beef is highly related to weight, and weight relative to age in young/growing animals. Many think that smaller animals are 'less productive' because they have less potential or realized growth. This view may or may not be valid and must be evaluated in a broader context. A small animal that is fertile in a harsh environment is adapted to that environment, and it is likely that fertility has been achieved through a lower maintenance requirement (a trade-off of fertility for growth/size). There should be some basis that dictates acceptable and unacceptable sizes of cattle in different areas/societies of the world. In the USA, carcass weight has no price discount at packing plants as long as carcasses weigh between 272 and 408 kg (600 and 900 lb), and at some plants the range may be 45.4 kg (100 lb) or more beyond this range. Some markets may prefer much larger or much smaller carcasses than these ranges. The price discrimination is much harder to accept for underweight carcasses than heavy carcasses because not only does the seller have a reduced price, but there is less weight to base the price on. Producers are therefore scared of producing animals that are too small. There is typically a large discount for calves in North America that are classified as SMALL under the USDA Feeder Calf Grade scheme, but not much difference in price for calves

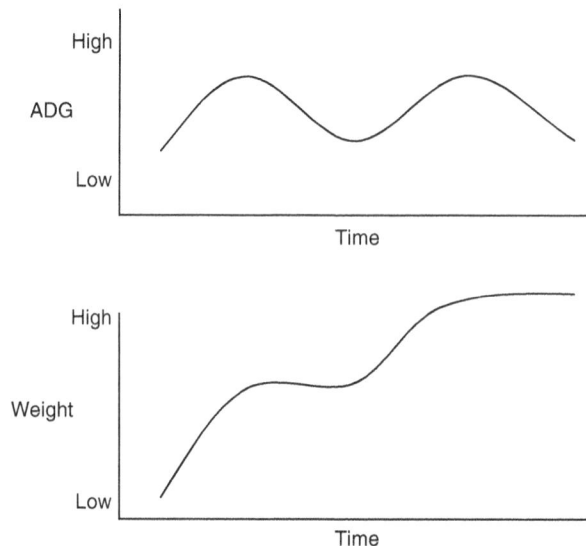

Fig. 7.4. Concepts of average daily weight gain (ADG) and weight in growing cattle on pasture across multiple growing seasons. Cattle in a feedlot experience a constant nutritional environment, but cattle on pasture may not have the same nutritional environment within one season, and certainly do not across multiple forage growing seasons. The terms 'high' and 'low' here represent general relative levels.

classified as MEDIUM vs LARGE. If cow-calf producers see the value of their calf crop determined by how much their calves weigh, then increased calf weight close to sale age (i.e. 6–9 months of age) becomes a priority. If some producers do not have desirable percentages for percent calf crop weaned, they might be able to have more weight of calves to sell by producing larger numbers of smaller calves (if similar types of cows are producing these calves).

In post-weaning heifers that are being developed for breeding purposes, their age obviously affects puberty and the age that they can first successfully reproduce. Nutritional level, which impacts both their weight and their body condition score, is very important. If the birth dates (or approximate dates such as within 1 or 2 weeks) of replacement heifers are known, this gives useful information to consider in combination with other factors used in making breeding decisions. If there is a tight calving season (if all heifers were born within 60 days, 75 days, etc.), then the ages of the heifers in a group will still be relatively known even though birth dates of individuals are not known. If heifers are from a group that can vary in age by 6 months or more, or are of unknown origin/information, then other means of evaluation in regard to their reproductive development and near-term breeding potential are needed. One of these tools is reproductive tract scoring.

The general guidelines to assess reproductive tract scores are provided in Table 7.5. Even when ages of heifers are known, it can be very beneficial to screen them before the breeding season as 'unusual' and underdeveloped animals can be identified. If a herd has several sets of twin calves every calving season, this procedure will identify the freemartin heifers as they have a very underdeveloped reproductive tract

and will be missing functional cervix, uterus and ovaries. If a freemartin heifer is not identified before the breeding season, and she is mounted by a bull, she can be seriously injured and possibly killed because she will not have a normal length vagina, and the bull's penis can rupture the vaginal wall. Use of reproductive tract scores will not guarantee success for those that are deemed adequate in development, but heifers with higher reproductive tract scores at their first breeding season have been shown to have higher fertility rates across many studies and types of cattle.

Producers need to have objective measures of animal size, development, body composition and weight to fully describe their types of animals and to increase the precision of herd management, and marketing strategies. They also need to have methods to separate genetic vs environmental influences (such as proper contemporary group designation, proper consideration of age of dam influences, etc.), and many breed societies offer these services to their members, but there are not very many standardized adjustment programs for non-purebred cattle. Another standardized method of evaluating cattle size that all producers should be familiar with is frame score.

7.4 Frame Score System

A system that has been used in several countries for many years to evaluate mature skeletal size is called frame score. This is based on a system that evaluates an animal's hip height relative to their age and classifies them on a numerical scale. Hip height is measured as the distance from the hip (hook) bones to the ground when the animal is in a relaxed stance. Table 7.6 shows hip heights and ages for corresponding frame scores. Frame scores provide a standardized

Table 7.5. Description of reproductive tract scoring (RTS) system in beef heifers.

RTS	Uterine horns	Ovary length (mm)	Ovary height (mm)	Ovary width (mm)	Ovarian structures	Description
1	Immature, <20 mm diameter, no tone	15	10	8	No palpable follicles	Infantile
2	20–25 mm diameter, no tone	18	12	10	8 mm follicles	Pre-pubertal
3	20–25 mm diameter, slight tone	22	15	10	8–10 mm follicles	Peri-pubertal
4	30 mm diameter, good tone	30	16	12	>10 mm follicles, CL possible	Cycling
5	>30 mm diameter	>32	20	15	CL present	Cycling

System developed by LeFever and Odde (1986). Estrous response has been shown to increase among heifers that were more reproductively developed at the beginning of estrus synchronization and/or breeding season in many reports. The term tone refers to the feel/firmness of the uterine horns when evaluated by rectal palpation.

Table 7.6. Cattle frame score based on age (months) and hip height (cm).[a]

Males[b]

Frame score[c]	3.0	4.0	5.0	6.0	7.0	8.0
Age (months)						
5	95	100	106	111	116	121
6	99	104	109	114	119	124
7	102	107	112	117	122	127
8	105	110	115	120	125	130
9	107	113	118	123	128	133
10	110	115	120	125	130	135
11	112	117	122	128	133	138
12	114	119	124	130	135	140
13	116	121	126	132	137	142
14	118	123	128	133	138	143
15	120	125	130	135	140	145
16	121	126	131	136	141	146
17	122	127	132	137	142	147
18	123	128	133	138	143	148
19	124	129	134	139	144	149
20	125	130	135	140	145	150
21	125	130	135	140	145	150
Maturity	133	137	142	147	152	157
Finished steer weight (kg)[e]	454	499	544	590	635	680
Mature bull weight (kg)[f,g]	712	785	857	930	998	1070
Expected carcass weight[h]	272	299	327	354	381	408

Females

Frame score[c]	3.0	4.0	5.0	6.0	7.0	8.0
Age (months)						
5	94	100	105	110	116	121
6	97	102	107	113	118	123
7	100	105	110	115	120	125
8	102	107	112	117	122	128
9	104	109	114	119	124	130
10	106	111	116	121	126	131
11	107	113	118	123	128	133
12	109	114	119	124	130	135
13	111	116	121	126	131	136
14	112	117	122	127	132	137
15	113	118	123	128	133	138
16	114	119	124	129	134	139
17	115	120	125	130	135	140
18	116	121	126	131	136	140
19	116	121	126	131	136	141
20	117	122	126	132	136	141
21	117	122	127	132	137	141
Maturity[d]	122	127	132	137	142	146
Finished heifer weight (kg)[e]	408	454	499	544	590	635
Mature cow weight[g] (kg)	454	499	544	590	635	680
Expected carcass weight[h]	245	272	299	327	354	381

[a]Adapted from Beef Improvement Federation (BIF, 2010) and Hammack (2009).
[b]Steers continue to bone growth for longer than bulls and will be 1.0–2.5 cm taller at 18–21 months.
[c]USDA Feeder Calf Frame grade of medium is frame score of 4.0 to 5.5.
[d]If calved first at 2 years of age, add 2.5 cm if calved first at 3 years.
[e]At 9–10 mm of fat cover over last two ribs.
[f]At 12 months of age, bulls weigh 50–60% of mature weight under non-extreme conditions.
[g]At body condition score of 5 (moderate body fatness) where 1 = extremely thin and 9 = obese; cow weights vary 7–8% per condition score and as much as ± 10% for extremes in muscle thickness.
[h]Assuming 60% dressing percent and average muscle of 11.6 cm² of longissimus muscle area per 45.4 kg of carcass weight and 9–10 mm of fat thickness at 12th to 13th rib interface.
This table is provided in imperial units in Table A9.

method to calculate expected mature skeletal size, and the idea is that as cattle grow, their height increases but their height relative to their age (and therefore their frame score) should remain constant across ages. This also assumes that all animals have the same growth curve, which we know is not true; this does not mean that utilization of frame scores is not meaningful, it just means that its interpretation should not be oversimplified. This is discussed more in Section 7.5 There is a predicted/expected weight associated with cattle of different frame scores, but there is also considerable variation in weight among animals of the same frame score due to body condition and degree of muscularity, all of which are related to weight. Many selection programs that have increased weight of calves over time have increased their frame score, and have also resulted in corresponding increases in mature cow size; however, it is also possible to increase mature weight without changing frame score. Skeletal size is a trait that is high in heritability, and therefore is quite easy to change through selection.

Under this approach the same function (or curve) is associated with all male and all female frame scores (they are shifted up or down the y-axis), but a different curve is used for males vs females. Frame score is a widely used concept and can be quite useful, but like many other concepts it is not perfect. Also, only using a single component of size (weight, frame score, etc.) provides an incomplete picture of the real situation. Selection programs can change weight without changing frame score (and vice versa), but the type of animal will also change. Animals of the same frame score can have widely different weights, and animals of the same weight can have widely different frame scores (see Fig. 7.5). Think about what factors would be involved for these types of scenarios (the reader may want to review aspects discussed in Chapter 6 regarding live animal evaluation if this does not make sense). Gilbert *et al.* (1993) evaluated a variety of body size measurements in Angus and Hereford following 20 years of selection for post-weaning gain. These authors stated that measurements of height were not useful in explaining total body size and that after 20 years of selection cattle had almost twice the rate of gain than previously, but essentially had the same hip height.

The discussion of frame size is important because it is one evaluation tool among several to evaluate both market and breeding cattle. Selection for increased weight may or may not increase frame size; it depends what traits are simultaneously considered in selection decisions. The heritability of skeletal size is higher than the heritability of weight, particularly at many stages of development in growing cattle. As a result, selection should not emphasize weight alone (higher weights alone over time could produce cattle with less desirable body composition and/or conformation) because cattle could weigh more, but be less muscular relative to their body size. If someone wanted to increase weight, they might not alter frame score, or might even reduce it if more muscular but shorter stature cattle were favored. It must also be noted that although animals of the same breed name in different countries come from the same ancestor

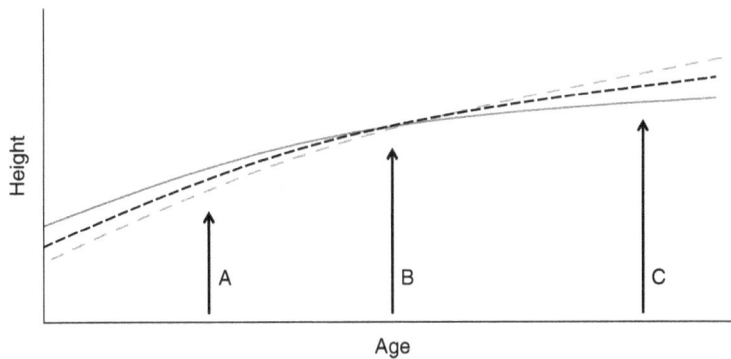

Fig. 7.5. Different growth curves at ages of evaluation. Three animals represented by three different lines that are potentially evaluated at times (A, B, C). If these animals are only evaluated at a single point in time, different calculations and rankings of frame scores are expected. If animals are only evaluated at time B, they will have the same frame score calculation, and the same expected future sizes (including mature size).

base, they can be quite different in frame size and many other traits over time (as can family lines within the breed within a country).

It is very likely that early influences in the life of an animal may have lasting or permanent impacts on its performance for a variety of traits later in life. For instance, we know that many things can affect the future health and growth of people when they were exposed to undesirable effects as a fetus. This is why pregnant women are urged to refrain from smoking and consuming alcohol, and to maintain a proper diet while they are pregnant. It should not surprise us then that similar negative effects (or positive effects) may be possible in calves due to the management of their dams during pregnancy. There is more research needed in beef cattle to evaluate the impacts of fetal programming on subsequent animal productivity.

As previously mentioned, some scientists have reported increased marbling amounts in carcasses from calves that were fed grain early in life with no increased subcutaneous fat as opposed to contemporaries not fed grain early. It is human nature in many instances to over-use practices, substances, etc. out of good intentions. If a little grain helps, then a lot of grain will help more, for instance. However, in the growth and development of calves, severely restricting nutrients or offering excessive nutrients are both undesirable. Just as malnutrition can have permanent effects for reduced size, overly excessive energy can make animals too fat, and deposition of fat in the udder of heifers may reduce future milk production, and deposition of fat in the scrotum of young bulls may lower fertility after puberty; also, producing obese young nursing calves may influence efficiency of gain after weaning. Optimal management and production where extremes are avoided are usually safe strategies.

7.5 Growth Curves

Weight is the primary objective measure of overall body size of cattle and is instrumental in determining animal value in most markets. However, weight alone does not explain animal skeletal size or body composition. The amount of fat (body condition) and the degree of muscularity greatly influence weight. Figure 7.1 shows a generic growth curve where weight increases from conception through 24 months of age, which follows a sigmoid-shaped pattern; there is little measurable weight change for

many days into gestation, and then an increasing weight gain at an increasing rate (showing exponential growth), then a portion of the curve that is close to linear, and finally closer to maturity there is increased weight at a decreasing rate with the curve eventually reaching a plateau. In older animals, this weight may begin to decrease. Changes in body size such as length or height also follow a sigmoid-type curve as does weight, however the ages when animals reach a set percentage of their mature weight do not correspond to the ages when they reach the same percentage of their mature height. For instance when growing conditions are favorable, cattle may reach 80% of their mature height by 7–8 months of age, but typically have only reached 35–45% of their mature weight. Moreover, by 12 months cattle may be 90% of their mature height, but only 50–60% of their mature weight. Under more stressful conditions, these percentages may be substantially lower. This concept is the same aspect that was discussed in regard to fetal growth.

The historically most used curve to estimate weights over time is referred to as a Brody curve, named after the scientist that first developed and reported the formula (Brody, 1945). The formula for a Brody growth curve is:

$$\text{Weight}_t = A^* \, (1 - Be^{-kt}) \qquad (7.2)$$

In this formula, Weight_t is the estimated weight at time t, A is the asymptotic weight (mature weight, or the weight that the curve approaches and eventually plateaus at), B is a constant that is specific to an individual curve (or type of animal), e is the base of the natural logarithm, which is raised to the power $-kt$ where k is the maturing rate parameter and t is the time point relative to the weight being estimated. The Brody function has been reported to fit some cattle populations well and others not so well. Several growth curve functions have been proposed and evaluated in different cattle populations. Table 7.7 provides four of these functions which were described by Goonewardene *et al.* (1981).

7.5.1 Composition of Weight Gain in Growing Cattle

As calves grow from birth to maturity, the composition of the weight gained changes, and therefore the efficiency of the weight that is gained also changes. When animals are young, a high percentage of their weight gain is muscle and protein, and this growth is much more efficient (amount of weight gained per unit of feed consumed) than later in life (closer to maturity), when both the rate of weight gain slows and there is more body fat being deposited. Young calves consuming a high-concentrate diet may have a feed conversion ratio (amount of feed consumed divided by the amount of weight gained) of 4:1 (4 kg of feed required for each kg of weight gain), but cattle that are close to reaching a fat-constant end-point (such as 10 mm of fat cover over the ribs, 28% body fat, etc.) will have a much less efficient gain and feed conversion ratio may be 12:1 or so.

Table 7.8 shows expected composition of weight gain in regard to protein and fat for a medium frame size steer across various weights and various rates of gain. It can be seen that as animals grow (get older and larger), the percent protein of the weight gain goes down, and correspondingly, the percent fat of the weight gained increases. It should also be observed that at a constant weight (stage of production or age

in growing animals), as the energy level in the diet increases, there will be increased weight gained, and that increased weight gained will also have lower protein and more fat as compared to the weight gained at a slower rate of gain.

This concept of body composition is why in growing and developing animals it is important to know what the nutritional requirements are and what the animals are consuming because managers can maintain a constant plane of nutrition in order to get efficient weight gain and avoid having animals overly fed or malnourished during development, which could have long-term production implications. This same concept should be considered for the developing calves *in utero*. Table 7.9 shows reported protein percentages among different body components of cattle across different ages.

7.5.2 Compensatory Growth

The concept of compensatory growth or weight gain is when there is a period of faster or more efficient growth following a period of stress that has restricted growth (NRC, 2000). Feed intake and quality of diet can both cause periods of restricted growth and may result from climatic and/or seasonal conditions, physical restriction of intake by a manager, or health problems. In North America, cattle buyers will pay a slightly increased price for calves that appear to have been held back a little from optimum conditions (a little thinner than expected), because they expect compensatory weight gain (increased weight gain per day at a more efficient rate) in these cattle. Animals that appear to have been severely restricted, however,

Table 7.7. Different growth functions proposed to describe cattle weight over time.[a]

Function	Formula	Inflection point
Richards	$Y_t = A(1 - Be^{-kt})^m$	
Brody	$Y_t = A(1 - Be^{-kt})$	–
Von Bertalanffy	$Y_t = A(1 - Be^{-kt})^3$	$0.296A$
Logistic	$Y_t = A(1 - Be^{-kt})^{-1}$	$0.5A$

Y_t = weight at time t in days; B = constant of integration; e = base of natural logarithm; A = asymptote; k = maturing rate; m = inflection parameter.
[a]Taken from Goonewardene *et al.* (1981).

Table 7.8. Composition of weight gain in medium frame size steer across weights and rates of average daily gain.

ADG (kg)	Weight (kg)						
	200	250	300	350	400	450	500
	Protein in weight gain (%)						
0.6	20.4	19.5	18.8	18.0	17.3	16.6	16.0
0.8	18.7	17.6	16.5	15.5	14.6	13.6	12.7
1.0	17.0	15.6	14.2	13.0	11.7	10.5	9.3
1.3	14.4	12.5	10.7	9.0	7.3	5.7	4.2
	Fat in weight gain (%)						
0.6	5.9	9.7	13.2	16.6	19.9	23.1	26.2
0.8	13.6	18.7	23.6	28.2	32.8	37.1	41.4
1.0	21.4	27.9	34.1	40.1	45.6	51.5	56.9
1.3	22.3	29.0	35.4	41.5	47.4	53.2	58.7

Adapted from NRC (2000).

Table 7.9. Distribution of protein in empty body of cattle.

Age (months)	EBW (kg)	Total protein (%)	Distribution of protein in tissues (%)				
			Blood	Organs	Hair and hide	Skeleton	Fat and lean
3	98	19.7	5.7	8.7	17.1	23.2	45.5
8	171	18.3	5.0	9.0	15.6	18.7	51.8
11	281	17.3	4.6	8.2	17.4	16.6	53.3
18	454	16.6	4.5	6.7	14.4	17.2	57.8
21	475	16.1	5.7	7.7	18.5	16.5	51.6
48	813	12.3	5.9	6.6	17.2	19.8	50.5

From Aberle *et al.* (2001).

receive large price discounts because they are deemed unthrifty or unhealthy. Severe and prolonged nutrient restriction of animals can have permanent environmental effects on their growth and final weight, and can impact mature size. It is realistic to expect that animals undergoing compensatory growth may have up to 20% reduced maintenance energy, and the length of this compensatory period may last 60–90 days (NRC, 2000). The concept of body condition score was introduced in Chapter 6, and Table 7.10 provides general compositional differences in cattle associated with varying body condition.

As mentioned, each animal has an expected growth curve relative to its genetic and environmental influences. Positive and negative influences will result in deviations from the expected growth curve. Growing animals that receive excess amounts of energy will increase their rate of weight gain, and the composition of the weight gain. Consequently, these animals may undergo reductions in weight gain when more stressful (less favorable) conditions arise. For instance, cattle buyers in North America will bid a lower price for calves that have a little extra body condition compared to a higher price for calves that are a bit thin for the same underlying reason. Calves that are being weaned, and/or are entering a feedlot that are viewed as 'fat' are expected to experience reduced weight gains and more stress than calves that are 'thin' because the thin calves have already been stressed or else they would be in a higher body condition (Fig. 7.6).

7.5.3 Use of Ultrasound to Evaluate Composition and Growth

As discussed in Chapter 6, a valuable tool to assess body composition and potential for marbling development is ultrasound. Ultrasound is a technology that

Table 7.10. Empty body composition (%) of cattle across different body condition scores.

BCS	Fat	Protein	Ash	Water
1	3.8	19.4	7.5	69.4
2	7.5	18.8	7.0	66.7
3	11.3	18.1	6.6	64.0
4	15.1	17.0	6.2	61.4
5	18.8	16.8	5.7	58.7
6	22.6	16.1	5.3	56.0
7	26.4	15.4	4.8	53.4
8	30.2	14.8	4.4	50.7
9	33.9	14.1	4.0	48.1

Adapted from NRC (2000).

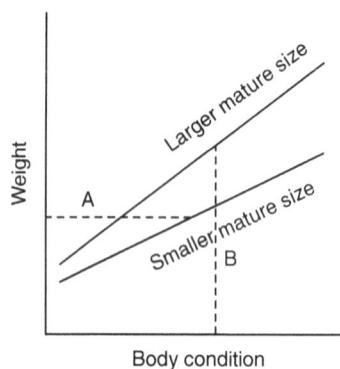

Fig. 7.6. Comparisons of weight and body condition across frame size. Line A indicates two animals of the same weight can be in different developmental stages, ages, body condition, etc. Line B indicates two animals of the same body condition may be of quite different weights.

uses high-frequency sound waves to produce visual images. Several research trials have evaluated the use of ultrasound to help predict carcass attributes and that might be useful to sort animals into outcome or management groups based on growth and development predictions (for instance Brethour, 2000; Rouse *et al.*, 2000; Bruns *et al.*, 2004; Wall *et al.*, 2004; Albrecht *et al.*, 2006). It is widely accepted that use of ultrasound closer to the time of harvest provides a higher degree of association (referred to as a correlation) between the live animal trait and the carcass trait. However, for ultrasound to be useful as an animal sorting tool to improve management, it must be reliable well before time of harvest. There is growing evidence that a single ultrasound measurement may not be the best use of this technology, but that ultrasound in combination with other information can be used for more informative prediction tools (Dean, 2006).

7.6 Mature Size and the Concept of Maturity

The implications for manipulating calf size and growth are different in terminal breeding programs vs programs designed to produce replacement animals. As previously discussed, measures of size and growth throughout an animal's life are related (but may not always have the same relationships). The ability to predict future size, including mature size, from early measures on calves is of interest to all cattle producers, but the measurements of mature size and weight of breeding animals maintained in herds are typically not practiced. Mature cow size has implications in regard to nutritional and pasture management of breeding herds. If there is a change in mature cow size but it is not recognized, input costs and output measures may also change, but how these changes impact overall productivity and profitability is also likely to go unrecognized and is hard to quantify in many production systems. Change in weight of mature animals is not the same concept as weight change due to growth in young animals, but there is change in muscle and fat when there is substantial change in weight.

Mature cow size may not occur until animals are 4–8 years old, depending upon the genetic background, but the change in size is more gradual past 2–3 years of age as compared to earlier in life. In an industry publication a few years ago a producer was quoted as saying that if 2-year-old first calf heifers (females that just produced their first calf as a 2-year-old) were still growing when they were expected to breed back while nursing calves, the results could be disastrous. But in fact all 2-year-old first calf heifers are still growing, and recognition of this fact is important for balance of current productivity and improvement of beef cow longevity. Table 7.11 illustrates potential weights and respective percentages of mature weight that may be expected in immature cattle.

Cunningham (2013) reviewed studies that classified cows into different mature size categories and evaluated associated productivity measures. The results as well as the classification schemes from these studies are quite variable (i.e. a small cow classification from one study may be a medium classification in another, etc.). Buttram and Willham (1989) stated that frame size and breed is probably completely confounded in many studies and considerations. Weight without consideration of body composition or skeletal size gives an incomplete picture of animals as well as their carcasses; similarly, any one of these measures without reference to others also provides for an incomplete characterization.

Use of the BIF frame score system provides a standardized reference point for predicting mature skeletal size; however, it is based on European *Bos taurus* cattle types and may have a slightly different interpretation in straight *Bos indicus* cattle, and does not guarantee weight or body shape. Menchaca *et al.* (1996) evaluated hip heights relative to various ages in Brahman cattle in Florida, USA, and found that weaning hip heights in 8-month-old cattle yielded frame scores that were one value below the frame scores calculated on hip height at 5 years of age. Horimoto *et al.* (2007) evaluated a new type of frame score calculation in Nelore cattle that was based on weight and hip height at 18 months of age as compared to hip height and age, with phenotypic and genetic correlations evaluated among the traits. These authors also expressed a concern that the BIF frame score system may not mean the same thing in zebu cattle and British and European types, and therefore proposed a new type of frame score calculation that incorporated weight and hip height at 18 months of age as compared to hip height and age (at 18 months, etc.), as is the basis of the BIF approach. These authors found higher phenotypic correlations of muscle score (0.42) and scrotal circumference (0.27) with the new frame calculation as compared to these traits with BIF frame (0.15 and 0.11, respectively). However, the genetic correlations of the traits and frame were very close to zero, a slight exception being muscle score and BIF frame (−0.12). The genetic correlations (discussed

Table 7.11. Example weights expressed as percentage of mature cow weight (kg) that may correspond to important stages of lifetime production.

Birth weight			Weaning weight				Yearling weight			2-year weight	Mature weight
6%	7%	8%	35%	40%	45%	50%	55%	60%		75%	100%
24	28	32	140	160	180	200	220	240		300	400
27	31.5	36	157.5	180	202.5	225	247.5	270		337.5	450
30	35	40	175	200	225	250	275	300		375	500
33	38.5	44	192.5	220	247.5	275	302.5	330		412.5	550
36	42	48	210	240	270	300	330	360		450	600
39	45.5	52	227.5	260	292.5	325	357.5	390		487.5	650
42	49	56	245	280	315	350	385	420		525	700
45	52.5	60	262.5	300	337.5	375	412.5	450		562.5	750
48	56	64	280	320	360	400	440	480		600	800
51	59.5	68	297.5	340	382.5	425	467.5	510		637.5	850

The percentage weight that is expected and realistic will vary across different body conformation and breed types.

in Chapter 3) indicate how certain traits could respond in a selection program when the other trait is changed.

It should not be of concern that the same calculated frame score can yield differences in growth rates and/or mature size, or that different ages when hip height (or some other measure) is taken may be more indicative of mature skeletal size in some breeds than others. A mature 500 kg Hereford cow can (and probably will) have different physical appearance than a mature 500 kg Charolais, or a mature 500 kg Brahman cow, or a mature 500 kg Jersey cow, etc. It is expected that breed differences (and family differences) exist, and these differences can be effectively used in selection and crossbreeding programs. Use of calculated frame scores provides a reference point for evaluation of animal size, and if producers actively monitor the mature size of their cows and associated input costs it is more likely they will find the proper balance of production and input costs for increased profit (this concept is discussed at length in Chapter 12).

The concept of beef cattle maturity is truly a long-term concept as it takes cows and bulls (or oxen for that matter) many years to stop growing and developing. With respect to beef cow longevity, heifer development programs that ensure that adequate growth and fertility expectations (but that do not expect or utilize extreme levels of growth or production) are met will typically allow for increased length of productive life. For many cattle, 75–90% of the mature skeletal height may be attained by 12 months of age. This may lead people to believe that the cattle are close to finished in growth, but this is a misconception. Several studies in many different breeds have shown cow weight to increase up until 6, 7 or 8 years of age. In Chapter 6 the importance of proper skeletal structural soundness was discussed as it related to longevity of production, but Fig. 7.7 illustrates that many growth plates in the feet and legs do not ossify until several years of age, and these ages can also vary within and across breeds.

Use of dentition as an age measure can be useful as a benchmark type of trait (discussion and examples in Chapter 6), but again there is expected to be variation in this type of developmental trait, as with others. Andrews (1974) described ages and associated variation in emergence of incisors in Friesian cattle of known age classes in Great Britain. The general ages as related to incisor eruption and presence have been used in beef carcass

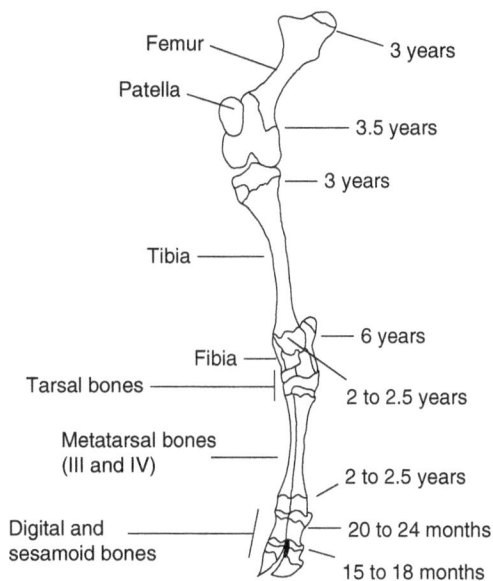

Fig. 7.7. Bones of cattle rear leg and ages that growth plates fuse (adapted from Figure 14 of Budras *et al.*, 2011).

evaluation systems as age of animal relates to several carcass quality traits. There are also physiological changes regarding bone ossification (turning of cartilage into bone) that can be evaluated on the vertebrae and other bones in the carcass. The heads of cattle are matched to their carcasses in most larger-scale processing facilities as the daily harvest sequence number stays with both the head and the carcass until grading. The relationships between the skeletal maturity of beef carcasses and the presence of incisors was evaluated by Lawrence *et al.* (2001) for US carcasses and reinforced the fact that these traits were still highly related in cattle as compared to standards and relationships reported many years earlier.

7.7 Summary

Beef cattle production systems can utilize the efficiency of young animal growth in the development of muscle, typically measured as animal weight. There are numerous genetic and environmental influences that can contribute to variation in rate and extent of an animal's growth and size at many stages in its life. Many factors affect animal size and weight, and comparisons of only one measure of size or only weight, or only body composition, gives an incomplete picture of the actual value of the animal. The maturing process in cattle takes several years, and to state that cows or bulls are mature when they first begin to reproduce is a gross understatement and non-realistic assumption. There are many potential influences on growth, development and future performance of cattle that may have important programming influences established during fetal development that need further research. There is biological variation in all cattle that have been measured, and it is very important to document known management effects that could contribute to additional variation in animal size or growth that if not documented could be mistaken for genetic differences. Connection of young animal performance with productivity later in life, as with cow and bull performance and longevity, is important to evaluate total production systems and to realize what optimal size and growth levels are needed.

7.8   Study Questions

7.1 Discuss the general considerations associated with weight gain in growing cattle that can be associated with varying energy levels in the diet.

7.2 Consider the birth and weaning information below on four calves of the same breed and born in the same herd and season. BWT = birth weight, WWT = weaning weight, WAGE = weaning age, AOD = age of dam.

ID	Sex	BWT (kg)	WWT (kg)	WAGE (days)	AOD (years)
862	Male	37	242	189	4
871	Male	41	267	220	12
884	Female	34	249	205	2
895	Female	38	253	215	6

7.2a Based on the above information, calculate (a) the adjusted birth weights and (b) the 200-day adjusted weaning weights of these four calves.

7.3 Can animals of the same frame size have different weights? Provide an example.

7.4 Can animals of the same weight have different frame size (or other physical differences)? Provide examples.

7.5 For the following scenarios, calculate the frame score (BIF system) for the following cattle:

Heifer that is 7 months old and has 102 cm hip height. Bull calf that is 6 months old and has 114 cm hip height.
Heifer 15 months old and has 125 cm hip height.
Four-year-old bull that has 150 cm hip height.

7.6 Is it likely that individual variability exists for the ages at which cattle attain a certain percentage of their mature weight? Provide some example values and ranges for consideration.

7.9   References

Aberle, E.D., Forrest, J.C., Gerrard, D.E. and Mills, E.W. (2001) *Principles of Meat Science*, 4th edn. Kendall Hunt Publishing Company, Dubuque, IA.

Albrecht, E., Teuscher, F., Ender, K. and Wegner, J. (2006) Growth- and breed-related changes of marbling characteristics in cattle. *Journal of Animal Science* 84, 1067–1075.

Amen, T.S., Herring, A.D., Sanders, J.O. and Gill, C.A. (2007) Evaluation of reciprocal differences in *Bos indicus* x *Bos taurus* backcross calves produced through embryo transfer: I. Birth and weaning traits. *Journal of Animal Science* 85, 365–372.

Andrews, A.H. (1974) A comparison of two different survey methods for the study of intra-oral development of the anterior teeth in cattle. *Veterinary Record* 94, 130–138.

Barker, D.J.P., Martyn, C.N., Osmond, C., Hales, C.N. and Fall, C.H.D. (1993) Growth *in utero* and serum cholesterol concentration in adult life. *BMJ* 307, 1524–1527.

BIF (2010) Chapter 3, Animal Evaluation. In: *Guidelines For Uniform Beef Improvement Programs, Ninth Edition*. Beef Improvement Federation, Raleigh, NC. Available at: http://www.beefimprovement.org/content/uploads/2013/07/Master-Edition-of-BIF-Guidelines-Updated-12-17-2010.pdf (accessed 12 May 2014).

Brethour, J.R. (2000) Using serial ultrasound measures to generate models of marbling and backfat thickness changes in feedlot cattle. *Journal of Animal Science* 78, 2055–2061.

Brody, S. (1945) *Bioenergetics and Growth*. Reinhold Publishing Corporation, New York.

Bruns, K.W., Pritchard, R.H. and Boggs, D.L. (2004) The relationships among body weight, body composition, and intramuscular fat content in steers. *Journal of Animal Science* 82, 1315–1322.

Buttram, S.T. and Willham, R.W. (1989) Size and management effects on reproduction in first-, second-, and third-parity beef cows. *Journal of Animal Science* 67, 2191–2196.

Cunningham, S.F. (2013) Early life measures of size as related to weights and productivity in beef cows and carcass traits in steers. PhD dissertation, Texas A&M University, College Station, TX.

Dean, D.T. (2006) Evaluation of ultrasound and other sources of information to predict beef carcass traits and final carcass value. PhD dissertation, Texas A&M University, College Station, TX.

Du, M., Tong, J., Zhao, J., Underwood, K.R., Zhu, M., Ford, S.P. and Nathanielsz, P.W. (2010) Fetal programming of skeletal muscle development in ruminant animals. *Journal of Animal Science* 88, E51–E60.

Fluharty, F.L., Loerch, S.C., Turner, T.B., Moeller, S.J. and Lowe, G.D. (2000) Effects of weaning age and diet on growth and carcass characteristics in steers. *Journal of Animal Science* 78, 1759–1767.

Gilbert, R.P., Bailey, D.R.C. and Shannon, N.H. (1993) Linear body measurements of cattle before and after 20 years of selection for postweaning gain when fed two different diets. *Journal of Animal Science* 71, 1712–1720.

Goonewardene, L.A., Berg, R.T. and Hardin, R.T. (1981) A growth study of beef cattle. *Canadian Journal of Animal Science* 61, 1041–1048.

Greenwood, P.L. and Café, L.M. (2007) Prenatal and pre-weaning growth and nutrition of cattle: Longterm consequences for beef production. *Animal* 1, 1283–1296.

Hammack, S.P. (2009) Texas Adapted Genetic Strategies for Beef Cattle X: Frame Score, Frame Size, and Weight. Texas A&M AgriLife Extension Publication E-192.

Horimoto, A.R.V.R., Ferraz, J.B.S., Balieiro, J.C.C. and Eler, J.P. (2007) Phenotypic and genetic correlations for body structure scores (frame) with productive traits and index for CEIP classification in Nellore beef cattle. *Genetics and Molecular Research* 6, 188–196.

Larson, D.M., Martin, J.L., Adams, D.C. and Funston, R.N. (2009) Winter grazing system and supplementation during late gestation influence performance of beef cows and steer progeny. *Journal of Animal Science* 87, 1147–1155.

Lawrence, T.E., Whatley, J.D., Montgomery, T.H. and Perino, L.J. (2001) A comparison of the USDA ossification-based maturity system to a system based on dentition. *Journal of Animal Science* 79, 1683–1690.

LeFever, D.G. and Odde, K.G. (1986) Predicting reproductive performance in beef heifers by reproductive tract evaluation before breeding. CSU Beef Program Report, Fort Collins, CO, pp. 13–15.

Mao, W.H., Albrecht, E., Teuscher1, F., Yang, Q., Zhao, R.Q. and Wegner, J. (2008) Growth- and breed-related changes of fetal development in cattle. *Asian-Australasian Journal of Animal Science* 21, 640–647.

Menchaca, M.A., Chase, Jr, C.C., Olson, T.A. and Hammond, A.C. (1996) Evaluation of growth curves

of Brahman cattle of various frame sizes. *Journal of Animal Science* 74, 2140–2151.

Myers, S.E., Faulkner, D.B., Ireland, F.A., Berger, L.L. and Parrett, D.F. (1999) Production systems comparing early weaning to normal weaning with or without creep feeding for beef steers. *Journal of Animal Science* 77, 300–310.

NRC (2000) *Nutrient Requirements of Beef Cattle: Seventh Revised Edition.* The National Academies Press, Washington, DC.

Radunz, A.E. (2009) Effects of prepartum dam energy source on progeny growth, glucose tolerance, and carcass composition in beef and sheep. PhD dissertation, The Ohio State University, Columbus, Ohio.

Rouse, G.S., Greiner, D., Wilson, C., Hays, J.R., Tait and A. Hassen. (2000) The use of real time ultrasound to predict live feedlot cattle carcass value. Beef Research Report, Iowa State University. A.S. Leaflet R1731.

Schoonmaker, J.P., Fluharty, F.L., Loerch, S.C., Turner, T.B., Moeller, S.J. and Wulf, D.M. (2001) Effect of weaning status and implant regimen on growth, performance, and carcass characteristics of steers. *Journal of Animal Science* 79, 1074–1084.

Stalker, L.A., Adams, D.C., Klopfenstein, T.J., Feuz, D.M. and Funston, R.N. (2006) Effects of pre- and postpartum nutrition on reproduction in spring calving cows and calf feedlot performance. *Journal of Animal Science* 84, 2582–2589.

Stalker, L.A., Ciminski, L.A., Adams, D.C., Klopfenstein, T.J. and Clark, R.T. (2007) Effects of weaning date and prepartum protein supplementation on cow performance and calf growth. *Rangeland Ecology Management* 60, 578–587.

Summers, A.F. and Funston, R.N. (2011) Fetal programming: Implications for beef cattle production. In: *Proceedings, Applied Reproductive Strategies in Beef Cattle*, 31 August – 1 September, Joplin, MO, USA, pp. 283–294.

Underwood, K.R., Tong, J.F., Price, P.L., Roberts, A.J., Grings, E.E., Hess, B.W., Means, W.J. and Du, M. (2010) Nutrition during mid to late gestation affects growth, adipose tissue deposition and tenderness in cross-bred beef steers. *Meat Science* 86, 588–593.

Wall, P.B., Rouse, G.H., Wilson, D.E., Tait, R.G., Jr and Busby, W.D. (2004) Use of ultrasound to predict body composition changes in steers at 100 and 65 days before slaughter. *Journal of Animal Science* 82, 1621–1629.

Williams, D.B., Vetter, R.L., Burroughs, W. and Topel, D.G. (1975) Effects of ration protein level and diethylstilbestrol implants on early-weaned beef bulls. *Journal of Animal Science* 41, 1525–1531.

8 Reproduction

In many instances, reproduction is the limiting or deciding factor in controlling profitability for cow-calf operations and heavily influences total output potential in beef production systems. The potential value of a calf that is never born is not realized. Reproduction (fertility) is the successful culmination of a multitude of factors, each one being a potentially limiting factor. It has been said that there may not be much difference between a cow that almost gets pregnant and a cow that barely gets pregnant, but there is a huge difference in the measurement of that trait (as opposed to the continuous distribution seen with weight traits and many of the component traits of reproduction such as hormone levels) because either she is pregnant or she is not pregnant (only two possible outcomes). This is true in regard to pregnancy, calving and weaning a calf. Typically the calf is thought of as the responsibility of the cow (and as a trait of the cow) until it is weaned. There are genetic differences, nutritional influences and health considerations that affect numerous aspects of cattle fertility, as discussed in Chapters 3, 4 and 9, respectively, and the potential for reproductive success of bulls and cows probably begins shortly after they themselves were conceived; many of these aspects were discussed in Chapter 7. The goal of this chapter is to discuss the basic components of beef cattle reproduction that need to be understood and that will aid in management and increased productivity and profitability for beef cow-calf producers.

8.1 Female Reproduction Fundamentals

The goal of reproduction in a beef cow is that she obtains puberty at an early enough age, conceives early in the breeding season, has a calf without problems, breeds back in a timely manner, and stays on a regular pattern of reproduction for many years. This is easier said than done in many settings. This section on female reproduction gives an overview of the female reproductive organs and their functions,

the estrous cycle, hormones involved in reproduction, estrus (i.e. heat), heifer development, artificial breeding and parturition.

8.1.1 Female Reproductive Tract and Function

Most structures of the female reproductive tract can be thought of as a series of tubes with concentric layers of serosa, muscularis, submucosa and mucosa from outermost to innermost, respectively. The open area inside the mucosa (the interior of the structure) is called the lumen. As discussed by Senger (2005), the serosa provides a covering to the structure, the muscularis is a double layer of smooth muscle (each running perpendicular to the other where the outer layer is in the same direction as the tube and the inner layer runs around the tube) which allows contractions and movement of gametes, embryo and fetus; the submucosa contains nerves and blood vessels, and the mucosa lines the interior and is comprised of a layer of secretory epithelium. Figure 8.1 shows reproductive organs in female cattle.

The ovary is the organ that produces the reproductive cells (egg or ovum, ova plural) in female mammals. Within the ovary are many follicles that can give rise to ova over the lifetime of the female. During each typical estrous cycle there will be a dominant follicle that will give rise to the egg destined to be ovulated. Ovulation is the process where the egg is released from the ovary. The follicle itself has defined structures. It is surrounded by a basement membrane, and on either side of this membrane are theca cells (outside membrane) and granulosa cells (inside membrane). The interior of the follicle has a fluid-filled cavity called the antrum; the oocyte is in the layer of granulosa cells until ovulation. The surface of the ovary is covered in a layer of cells referred to as the germinal epithelium. Following ovulation, the area corresponding to the follicle prior to ovulation forms the corpus luteum (CL). The CL can be felt as a small bump on the ovary

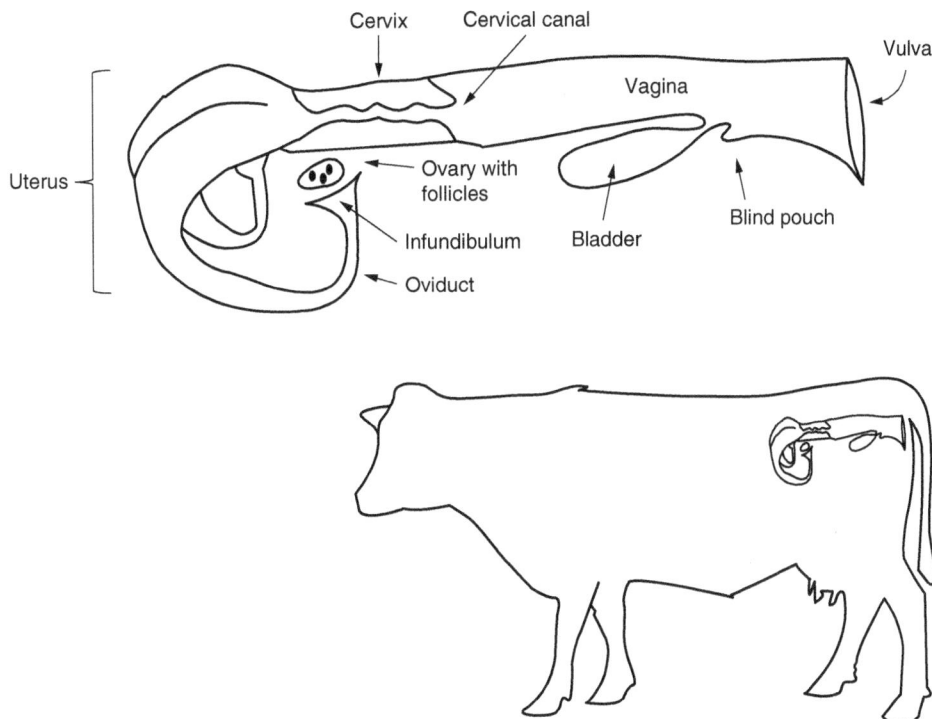

Fig. 8.1. Female reproductive tract. Semen from a bull is deposited in the vagina, but semen from AI is deposited in the uterus, just past the end of the cervix. Fertilization (union of sperm and egg) occurs in the oviduct. The lower diagram shows the relative position of the tract in the animal. The entire tract is supported by the broad ligament (not shown).

and will remain present for a few days during each estrous cycle, and if pregnancy occurs, it will remain for the entire gestation period. A former CL that is no longer functional is termed a corpus albicans. The organs and structures of the female reproductive tract are summarized in Table 8.1. Knowledge of the structures of the reproductive organs can be helpful in assessing reproductive health and aid in artificial insemination and pregnancy determination by rectal palpation.

8.1.2 Estrous Cycle

The estrous cycle is the time between when a female can become pregnant during the estrus (or heat) period and the subsequent estrus period. In cattle the estrous cycle is an average of 21 days in length. The length of the estrus period itself is approximately 12–18 hours, and ovulation typically occurs 24–30 hours after the onset of estrus (after the beginning of heat period), so this typically corresponds to about 12 hours after the end

of heat; this delayed timing is the basis of the long-standing recommendation (AM-PM rule) that if a cow is in heat in the morning, she should be bred in the evening of that day, and if she is in heat in the evening, she should be bred the following morning, so that the semen deposited in the uterus can travel to the egg at the correct time.

The cattle estrous cycle has defined phases and stages. There are two phases, known as the follicular phase (dominated by the follicle) and the luteal phase (dominated by the CL). There are four stages to the estrous cycle, known as prestrus, estrus, metestrus and diestrus. The follicular phase and the proestrus and estrus stages are characterized by high estrogen (estradiol, E_2) levels, and the luteal phase and the metestrus and diestrus stages are characterized by high progesterone (P_4) levels (Fig. 8.2).

If fertilization (the union of a sperm cell with an oocyte) occurs, the CL will remain and produce the hormone progesterone, which is the primary hormone responsible for maintaining pregnancy.

Table 8.1. Summary of female reproductive structures.

Organ or structure	Function
Ovary	Produces the gametes (oocytes, eggs) and a variety of hormones that influence reproduction. Oocytes develop in follicles in preparation for ovulation. Corpus luteum (CL) forms on the ovary following ovulation.
Oviduct (fallopian tube)	Transports the oocyte from ovary to uterus. Fertilization occurs in the oviduct. Provides the proper environment for fertilization and pre-attachment before uterus. Three components of the oviduct are the infundibulum, ampulla and isthmus (in order progressing from ovary to uterus).
Uterus	Provides sperm transport to oviduct, proper environment for pre-attachment embryo development, provides environment for attachment development of placenta, and holding fetus during gestation. The three layers of the uterus are the perimetrium (serosa layer), myometrium (muscularis layer) and the endometrium (submucosa and muscosa layers).
Cervix	Secretes mucus during estrus and provides transport of sperm from vagina to uterus, and provides a barrier between uterus and vagina during pregnancy with a thick mucosal seal.
Vagina	Provides receptacle for sperm during copulation, produces mucus for lubrication during copulation.
Vulva	Entry into vagina, protection of vagina from foreign substances.

Descriptions here follow those of Senger (2005).

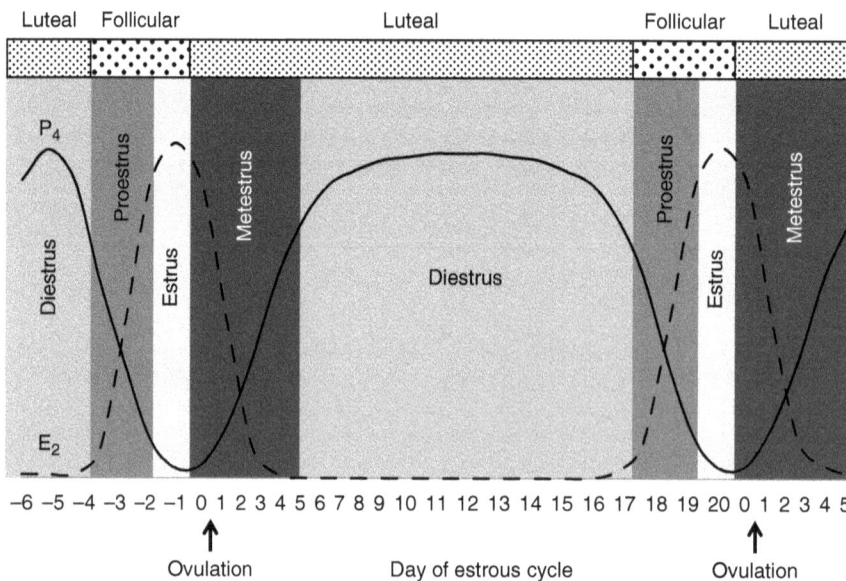

Fig. 8.2. Phases and stages of the cattle estrous cycle. The follicular phase and the proestrus and estrus stages are characterized by high estrogen (E_2) levels whereas the luteal phase and the metestrus and diestrus stages are characterized by high progesterone (P_4) levels. Adapted from Senger (2005).

In early pregnancy administering another hormone, prostaglandin $F_{2\alpha}$, will cause the CL to regress (go away) and this will result in abortion. After the first trimester of gestation, the placenta produces enough progesterone that an injection of prostaglandin $F_{2\alpha}$ will typically not cause abortion. If fertilization does not occur, the CL will regress after approximately 11 days, and the next estrous cycle will initiate.

8.1.3 Hormones

The following discussion relies heavily on information from Senger (2005), an excellent general reproductive physiology reference for domesticated mammals. The relationships involved between the nervous system, the endocrine system and the reproductive organs control reproduction, and hormones are crucial for regulating these activities. Hormones are products secreted from glands or nerves that enter the bloodstream and influence tissues and organs that have the proper receptor cells. Hormones can be classified by their source, their mode of action or their biochemical description and are released in specific quantities as well as specific patterns associated with various reproductive activities. Protein hormones act through cell membrane receptors and influence action in the cytoplasm. Steroid hormones act on receptors in the cell nucleus and stimulate production of new proteins. Glycoproteins are proteins that have carbohydrate components. A peptide is a very short protein or hormone (such as only 20 or so amino acids). Hormones produced in the hypothalamus (such as GnRH) cause release of other hormones from the anterior pituitary gland (such as FSH, luteinizing hormone [LH] and prolactin). Hormones produced in the gonads influence development of secondary sexual characteristics associated with masculinity and femininity in intact (non-castrated or sterilized) animals and provide feedback to the hypothalamus anterior pituitary and reproductive organ tissues. In general, reproductive hormones originate from the hypothalamus, pituitary, gonads, uterus and placenta.

Table 8.2 provides a summary of reproductive hormones and their functions. Reproduction texts such as Senger (2005) provide more detailed description of hormonal function than offered here. In general, the estrous cycles and pregnancy are influenced by hormone levels. The anterior pituitary of the brain produces gonadotropin hormones that control ovulation. The hormone GnRH (gonadotropin releasing hormone) influences the release of LH and follicle stimulating hormone (FSH); FSH stimulates follicle development and LH influences ovulation. Estradiol and progesterone are produced in the ovary. These hormones are released in specific patterns in surges or pulses (may be referred to as pulsatile secretion), and the amount of hormone produced as well as the pattern regulate the major processes associated with reproduction.

8.1.4 Replacement Heifer Development Programs

Heifers must reach puberty at an early enough age that their future fertility (and that of the herd overall) is not impeded. Breeds that are known for high levels of milk production have an earlier age of puberty than breeds with lower milk production potential. Within the same level of milk production, breeds that are larger in mature size tend to reach puberty later than those of smaller mature size. Increased plane of nutrition will decrease age of puberty as well because rate of weight gain and percent body fat also influence puberty. All these influences are also found in age of puberty of males. Heifers of most *Bos taurus* breeds reach puberty around 10–13 months of age when nutrition is not limiting. Target calving age of approximately 24 months is realistic in many environments and with *Bos taurus* breeds and crosses and most *Bos indicus*–*Bos taurus* crosses if they are F_1s or less than 50% *Bos indicus*. In many dual-purpose breeds puberty can be as early as 6 months, and in many *Bos indicus* breeds or severely nutrition-limited settings puberty may be as late as 18 months.

It is important to have a good idea of what the mature size and weight of heifers will be because it is recommended that 60% of the mature weight at BCS 5 is the target breeding weight for heifers and 85–90% of their mature weight is the recommended target calving weight. Both of these targets ensure that heifers will be in the proper BCS of 5 to 6. Historically, a target breeding weight of 65% had been a common recommendation for *Bos taurus* heifers with intended age of first calving at 24 months of age; however, recent reports have shown that a target breeding weight of 65% is not necessary, and in several cases may be 50–60% (Lynch *et al.*, 1997; Freetly *et al.*, 2001; Funston and Deutscher, 2004; Martin *et al.*, 2008). The target calving weight also includes the weight of the calf inside the heifer before she calves. In some circumstances, these target weights may be in reference to heifers calving first at 2 years of age, 2.5 years of age or 3 years of age, so the ages they obtain these target weights may vary drastically across breed types and production environments.

It is a disadvantage to get replacement heifers too fat early in life. A BCS of 6 is more than adequate for reproductive success. Heifers that get too fat (BCS 7, 8, 9) could have increased risk of reduced milk production as fat deposits in the udder early

Table 8.2. Summary of reproduction hormones.

Hormone	Biochemical classification	Source	Male target tissue	Female target tissue	Male primary action	Female primary action
Gonadotropin releasing hormone (GnRH)	Neuropeptide	Hypothalamus	Anterior lobe of pituitary	Anterior lobe of pituitary	Release of FSH and LH	Release of FSH and LH
Luteinizing hormone	Glycoprotein	Anterior lobe of pituitary	Testes (Leydig cells)	Ovary (theca interna and luteal cells)	Stimulates testosterone production	Stimulates ovulation, formation of CL and progesterone secretion
Follicle stimulating hormone (FSH)	Glycoprotein	Anterior lobe of pituitary	Testis (Sertoli cells)	Ovary (granulosa cells)	Sertoli cell function	Follicle development and estradiol synthesis
Prolactin	Protein	Anterior lobe of pituitary	Testis and brain	Mammary cells (CL in some rodent species)	Can induce maternal behavior in males and females	Lactation, maternal behavior (CL function in some species)
Oxytocin (OT)	Neuropeptide	Synthesized in hypothalamus, stored in posterior lobe of pituitary, synthesized by CL	Smooth muscle of epididymal tail, ductus deferens and ampullae	Myometrium and endometrium of uterus, myoepithelial cells of mammary gland	$PGF_{2\alpha}$ synthesis and pre-ejaculatory movement of spermatozoa	Uterine motility, promotion of uterine $PGF_{2\alpha}$ synthesis, milk release
Estradiol (E2)	Steroid	Granulosa cells of follicle, placenta, Sertoli cells in testis	Brain, inhibits long bone growth	Hypothalamus, entire reproductive tract, mammary gland	Sexual behavior	Sexual behavior, GnRH, elevated secretory activity of entire reproductive tract, increased uterine motility
Progesterone (P4)	Steroid	CL and placenta	–	Uterine endometrium, mammary gland, hypothalamus	–	Endometrial secretion, inhibits GnRH release, inhibits reproductive behavior, maintains pregnancy

Hormone	Chemical classification	Source	Target tissue	Target tissue	Action	Action
Testosterone (T)	Steroid	Interstitial Leydig cells, theca interna cells in females	Accessory sex glands, tunica dartos of scrotum, seminiferous epithelium, skeletal muscle	Brain, skeletal muscle, granulosa cells	Anabolic growth, promotes spermatogenesis, promotes secretion of accessory sex glands	Substrate for E2 synthesis, abnormal masculinization (hair, voice, behavior, etc.)
Inhibin	Glycoprotein	Granulosa cells in females, Sertoli cells in males	Anterior pituitary	Anterior pituitary	Inhibits FSH secretion	Inhibits FSH secretion
Activin	Glycoprotein	Granulosa cells in females, Sertoli cells in males	Anterior pituitary	Anterior pituitary	Inhibits FSH secretion	Inhibits FSH secretion
Prostaglandin $F_{2\alpha}$ ($PGF_{2\alpha}$)	Prostaglandin (C-20 fatty acid)	Uterine endometrium, vesicular glands	Epididymis	CL, uterine myometrium, ovulatory follicles	Affects metabolic activity of spermatozoa, causes epididymal contractions	Luteolysis, promotes uterine tone and contraction, ovulation
Prostaglandin E_2 (PGE_2)	Prostaglandin (C-20 fatty acid)	Ovary, uterus, embryonic membranes	–	CL, oviduct	–	Ovulation, stimulates CL secretion of P_4 by ovary
Placental lactogen	Protein	Placenta	–	Mammary gland	–	Mammary stimulation

Taken from Senger (2005), Table 5-2, pp. 122–123. These are the hormones of major importance in cattle in most circumstances. Other hormones may be of importance in other species, and source and/or action may be different in some species.

in life can have permanent effects on milk production. There is also growing evidence that certain genes involved with growth, development and fertility may be differentially expressed later in life due to these early environmental influences through fetal and early-life programming. A steady plane of growth where heifers gain 0.4–0.8 kg/day is more than adequate for reproductive development and does not endanger any reduced future productivity as cows. Reproductive tract scoring (Chapter 7) of heifers may be considered.

8.1.5 Artificial Insemination

Artificial insemination (AI) is a reproductive process that can greatly enhance breeding programs and reproductive management. Semen is commonly collected from highly prized bulls and is available for purchase. Just because semen is available from a bull, this does not automatically mean this sire is superior or useful for particular scenarios. In general, bulls are kept at facilities specialized in semen collection and processing. Semen that is processed and stored properly can be kept almost indefinitely. Specific guidelines about techniques to use with associated AI products are available from many AI and semen companies. AI is both a science and an art, meaning that practice and experience are needed to ensure success, but anyone can successfully learn to artificially inseminate cattle.

It must be remembered that the time of breeding of a cow in heat needs to be altered when using AI as compared to when she would be bred by a bull. A bull will breed a cow when she is in standing heat, and the semen is deposited into the vagina. However, because a small dosage of semen is used with AI, it is directly placed into the uterus. It takes the sperm cells several hours to travel from the vagina, through the cervix and into the uterus with natural service mating. Consequently, the AM-PM rule is recommended for breeding cows with AI where if the cow was in standing heat in the morning, she should be bred by AI later in the day (the evening), and if she was in heat in the evening, then she should be bred by AI the following morning. For many producers, the labor and time associated with estrus detection is the limiting step in the use of AI. Use of proven, calving ease bulls through AI of heifers can be desirable, and in some cases sexed semen (as discussed in Chapter 13) may be considered.

8.1.6 Estrus Signs and Detection

It is important for producers to understand and recognize reproductive behavior in female cattle as it can provide insight into reproductive success (or failure) and is needed to determine timing for AI use in many instances. Although it is typical for heifers and cows to exhibit classical standing heat signs, there can be widely variable expressions of this and most other reproductive behaviors; this variability in estrus timing has led to increased desire for synchronization and timed-AI recommendations.

8.1.7 Synchronization of Estrus

Several hormonal products are commercially available to aid in altering the estrous cycle of cattle and allow for the synchronization of estrus. The products can vary across countries, and the dosages of hormones in the products can also vary. Synchronization of estrus has been shown to be effective through a variety of strategies and products in many research trials. The largest advantage to estrus synchronization is that a concentrated number of heifers and/or cows can be bred in a short time frame (with AI or with natural service by bulls). There may be select groups of cattle that are candidates for estrus synchronization and/or AI, such as first-calf heifers, so that proven low birth weight sires can be used and the animals will be more concentrated at calving time. Simply because several females are bred and become pregnant the same day, this does not mean that they will all calve the same day. Results have been documented where a set of females were all bred on the same day to the same AI bull, but that calves were born over a 21-day range, due to individual differences in gestation length (Burns *et al.*, 2010).

Several research projects have evaluated the use of fixed-time AI, where as opposed to observing when cows are in standing heat, the animals are bred by AI at a predetermined (i.e. fixed) time. Using a fixed-time AI estrus synchronization protocol, the day of AI can be scheduled (for instance Saturday afternoon, etc.), and the resulting protocol requirements scheduled on the corresponding previous days. Many estrus synchronization programs for fixed-time AI use a vaginal implant that is a controlled internal drug release (CIDR) device in conjunction with hormone injections, and these require gathering animals and bringing them through facilities two or three times for protocol requirement before the time to AI. Specific protocols should be developed with local veterinarians and/or artificial breeding company personnel for estrus synchronization and fixed-time AI in various scenarios (season of year, age of females, breed, lactation status, manager goals, etc.).

8.1.8 Pregnancy Detection

The act of pregnancy determination by rectal palpation (touching with fingers) is easy to learn, but requires considerable practice and a fundamental knowledge of the female reproductive tract. Rectal palpation is a cost-effective practice that can be learned by anyone. Technicians and veterinarians that have considerable experience with rectal palpation are highly accurate at 45 days of gestation, and possibly at 35 days or so. The usefulness of pregnancy determination depends upon when it is used in the management calendar and what is done with the results.

Although rectal palpation is the most commonly used technique to determine pregnancy in cattle, other techniques do exist. Ultrasound can be used for pregnancy determination as well as examination of the reproductive tract. Advantages of ultrasound for pregnancy determination over rectal palpation include the ability to visualize the fetus and other reproductive structures as desired, but the main disadvantage of ultrasound compared to traditional rectal palpation would be increased cost. However, unlike other techniques, ultrasound makes it possible to observe the sex of the fetus. There is also at least one commercial blood-based pregnancy determination test available in cattle in some countries. The cost of the test is comparable to the cost of rectal palpation, but is based on detection of a fetal-produced protein and requires that at least 2 ml of blood be drawn and shipped to a company. This type of test can detect pregnancy 28 days after fertilization in females not lactating or that are 90 days post-calving.

8.1.9 Parturition

A few days before a heifer or cow is ready to give birth, her udder will become filled with milk, swollen and prominent. Very near the time of birth she will also show an increasingly swollen vulva. Her behavior will also change. Many cows will seek solitude and shelter for the act of calving. As parturition approaches, cows will typically become more restless – they may stand and lie down several times repeatedly. They may also stand with the appearance that they are about to urinate, with their rear legs slightly underneath them and their tail raised somewhat, and appear to be straining. The birth process itself is called parturition. The time immediately before parturition is termed labor. Careful observation may be warranted with cows and particularly heifers when they are in labor. Most heifers and cows do not need help to give birth, but if possible it is good insurance to monitor them closely around this time in case some assistance is needed. When a cow or heifer is ready to deliver the calf she will often lie down flat on her side as she is pushing the calf out. Some cows may not fully lie down on their sides and some may even remain standing, but that is not very common. More detailed discussion of cow behavior associated with calving and calf rearing is provided in Chapter 10.

8.1.10 Dystocia

Dystocia is the term that refers to a difficult birth process or problematic delivery. There are several reasons for dystocia, but it happens most frequently in first-calf heifers, particularly when giving birth to large calves. Dystocia can also occur because the calf is in the improper presentation or position during labor. Figure 8.3 provides some examples of possible calf positions during labor.

The normal presentation of the calf is for it to have its front two legs extended through the cervix and vagina (referred to as the birth canal at this point) with its head appearing next. The two front feet can be seen extending from the vagina before birth. The time from the first appearance of the two front feet until delivery may be a few minutes to a couple of hours. If the front feet are visible for over 2 hours or if the cow has been actively pushing for 2 hours and no apparent progress has been made, assistance may be needed. However, just because two feet are visible does not mean that the calf should be pulled at that point unless it is known that these are the front feet and that the head is in the proper position. It is possible that the calf may have its head turned back. If one foot is visible, the calf should not be pulled by just one leg. Calves can be successfully assisted during birth for several of these abnormal presentations and even pulled backwards (with the back feet). If dystocia results from the calf being too large for the pelvic

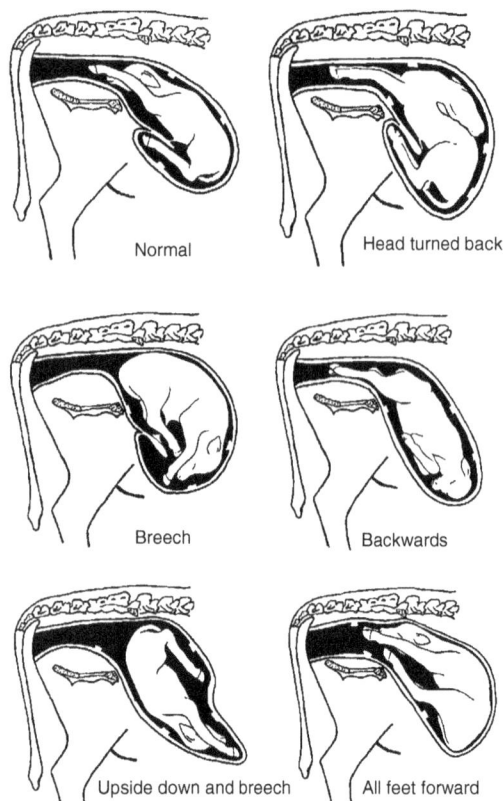

Fig. 8.3. Possible positions (presentations) of calf during labor. Graphics courtesy of Texas A&M AgriLife Extension Service.

opening that presents problems as well. In extreme cases of dystocia, surgical intervention may be required to save the calf and/or cow. It is much more common for heifers to experience dystocia than older cows due to the size of the calf and this is why it is important to breed heifers to bulls with low birth weight for the first calf.

8.2 Bull Reproduction Fundamentals

Very often the failure of cows to get pregnant is assumed to be the fault of the cow. However, the fertility of bulls, which includes both mating ability and libido, is also very important. Reproductive success/failure per cow or per heifer is a binary trait (pregnant or not pregnant, calved or did not calve, etc.). However, reproductive potential and success in bulls follow a more continuous distribution. Bulls can be evaluated for several component traits related to

their fertility such as sperm numbers, number of cows mounted and/or serviced, live/dead sperm ratios, etc. that show a continuous (or nearly continuous) distribution; plus they can be evaluated as to number of calves sired following a breeding season, which is also quantitative. Fertility in bulls is a function of both mating ability (the physical ability to deliver sperm cells and cause fertilization) and libido (the willingness and eagerness to identify females receptive to breeding). This section focuses on the male reproductive tract and the evaluation of traits in bulls that are directly related to their ability to successfully impregnate females.

8.2.1 Male Reproductive Tract and Function

The discussion of bull reproduction begins with an understanding of the male organs and their functioning and role in successful delivery of sperm cells and subsequent fertilization of oocytes. The structures are illustrated in Fig. 8.4 and briefly described in Table 8.3.

8.2.2 Evaluation of Scrotal Circumference

Scrotal circumference has been a widely measured trait in breeding bulls. It is a function of both testes size as well as overall body size. Numerous research trials have shown that increased scrotal circumference in bulls is positively related to their own sperm producing capacity as well as sperm production capacity in their sons, and negatively related to age of puberty in sons and daughters. All measures of bull fertility should include measurement of scrotal circumference, which is easily measured. Scrotal circumference is included in many points of discussion throughout the rest of this section.

8.2.3 Young Bull Growth and Development

Puberty in bulls can be defined in many ways, as can puberty in heifers. Some definitions of bull puberty have included age when reproductive behavioral traits are expressed, age at first ejaculation, age when sperm cells first appear in the ejaculate, or age when the semen contains a set number of spermatozoa. As a result, it is important to know the criteria used when comparing ages of puberty reported in different studies to determine whether it is biological effects or trial criteria that are different. The connotation

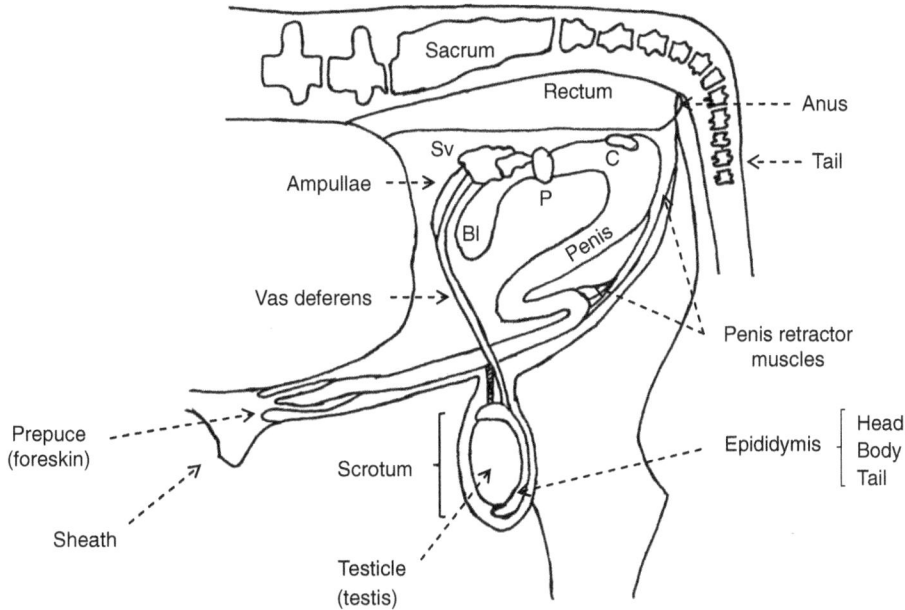

Fig. 8.4. Diagram of the male reproductive organs. The urethra is the canal inside the penis that transports urine from the bladder and semen. Bl = bladder, C = Cowper's gland, P = prostate, SV = seminal vesicles.

Table 8.3. Summary of structures of male reproductive tract.

Organ or structure	Function
Spermatic cord	Suspends the testes from the body cavity into the scrotum and transports sperm and blood supply. It houses the ductus deferens for sperm transport, cremaster muscle that raises and lowers testes and vascular network known as pampiniform plexus.
Scrotum	Two-lobed sac that houses testes outside of body for thermal regulation. It consists of four layers of skin, tunica dartos, scrotal fascia and parietal vaginal tunic (outer to inner, respectively).
Testis (testicle)	Produces the gametes (spermatocytes, sperm cells) and a variety of hormones that influence reproduction. Testis has two distinct types of area: interstitial (solid structure that has seminiferous tubules, Leydig cells, capillaries, lymph vessels and connective tissue) and tubular (seminiferous epithelium, Sertoli cells, developing sperm cells).
Excurrent duct system	Consists of efferent ducts (transport sperm cells out of testicle proper), epididymal duct (epididymis, allows for maturation and storage of spermatozoa), and ductus deferens (vas deferens, transports spermatozoa away from epididymis, and ampullea, connection to urethra).
Accessory sex glands	These include vesicular glands (seminal vesicles), prostate and bulbourethral glands (Cowper's glands) that produce the seminal plasma (the non-cellular fluid of semen that transports sperm) excreted into the urethra at ejaculation.
Penis	Reproductive organ that deposits sperm into the vagina. The urethra is the tube through which semen passes out of the penis. Bulls have a fibroelastic penis housed inside the body in an S-shaped (sigmoid flexure) configuration. Retractor penis muscles keep the penis inside the body until copulation.
Sheath	Skin that provides external protection and passage for penis. The inner layer of skin inside the cavity of the sheath is the prepuce or foreskin.

Descriptions here follow those of Senger (2005).

of bull puberty here is the age when a bull can first make a female pregnant under natural mating conditions.

Several research trials have been conducted that show semen quality and scrotal circumference increase dramatically for 4–6 months following puberty (Fig. 8.5). However, it should not be assumed that a bull is fully fertile simply because he has reached puberty. Puberty is the ability to impregnate some heifers or cows, not the ability to be a fully useful herd sire. It is also recognized that puberty in bulls occurs close to when the bull has 26–28 cm scrotal circumference, and the age that this happens can vary widely across breed types. Some indication of the bulls' potential to impregnate females is needed prior to the breeding season, and is typically accomplished through a breeding soundness evaluation or examination, which considers scrotal circumference, but also includes more detailed information.

8.2.4 Breeding Soundness Evaluation or Examination

Ideally, bulls should be evaluated in regard to some form of breeding soundness examination 60 days before the start of the breeding season. The length of time to produce new, functional sperm cells (spermatogenesis) in bulls is 54 days. Therefore if bulls are to be re-evaluated before the start of the breeding season, problems resulting from temporary illness and fever or injury and inflammation may have subsided. The breeding soundness exam is comprised of three distinct parts: (i) general physical exam, (ii) examination of the reproductive tract and

organs, and (iii) evaluation of semen. The general physical exam is to assess the overall health and physical capabilities of the bull. Aspects such as clearness and alertness of eyes, feet and leg soundness, and body condition score are scrutinized. During the examination of the reproductive tract, the testes are palpated for obvious problems, scrotal circumference is recorded, the penis and prepuce (foreskin) and sheath are evaluated, and the accessory sex glands such as the prostate are palpated rectally. The semen evaluation includes the color of the ejaculate, the total sperm cell number, the motility of sperm cells, the live to dead sperm cell ratio, and the percent of sperm cells with abnormalities.

Although only semen evaluation traits are given a score, failure in any category can result in failure of the overall breeding soundness exam (see Tables 8.4 and 8.5). Serving capacity (libido) is not evaluated as a component of breeding soundness exams under the Society of Theriogenology (USA), but is included under the Australian Association of Cattle Veterinarians (McGowan *et al.*, 1995) evaluation. The Western Canadian Association of Bovine Practitioners (Barth, 2000) has a place on the form for libido assessment but states the producer of the bull should evaluate this.

8.2.5 Activity vs Fertility – Age and Learning Aspects

Just because a bull has the adequate reproductive capacity from sperm production, this does not indicate fertility perfectly. Young bulls show much more breeding activity than older bulls when evaluated in serving capacity tests. A serving capacity test is where females are confined in pens or small paddocks and bulls are turned in with the females and recorded for number of mounts, number of cows serviced, etc. This type of evaluation provides supplemental information that can be used in combination with other fertility indicators, but as with the other measures of fertility, this trait alone does not adequately predict fertility. In general, studies have shown that young bulls tend to exhibit more mounting activity than older bulls, but this activity is not indicative of more services or fertility; it is also common that although younger bulls may show greater activity as compared to older and experienced bulls, it is the older bulls that have higher percentages of females become pregnant. There have also been some trials illustrating that young bulls (as well as older bulls) can perform to higher levels of breeding capability from increased female:bull ratios (for

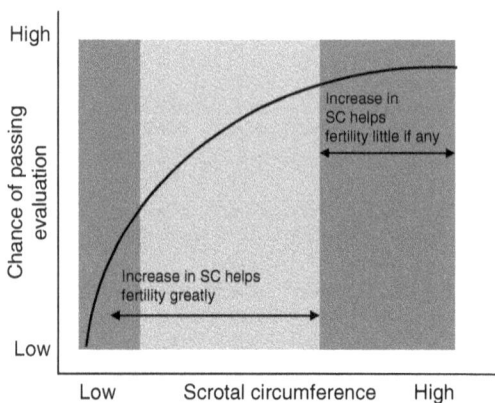

Fig. 8.5. General relationship of scrotal circumference (SC) and breeding soundness evaluation score.

Table 8.4. Considerations for evaluation of bull semen.

Collection technique	Semen density and volume	Gross motility	Individual motility	Sperm abnormalities
• Electroejaculation • Artificial vagina (AV) • Rectal massage Electroejaculation and rectal massage produce less volume than AV; use of AV requires more labor and specialized facilities. It is usually not critical to obtain a full ejaculate for adequate semen evaluation.	Very good – creamy, grainy consistency with 750 to 1000 million sperm cells per ml Good – milk-like with 400 to 750 million sperm per ml Fair – skim milk-like with 250 to 400 million sperm per ml Poor – translucent with < 250 million sperm cells per ml	Very good – very rapid, dark swirls Good – slower swirls and eddies Fair – no swirls but prominent individual sperm cell motion Poor – little or no individual cell motion	Very good – 80 to 100% Good – 60 to 79% Fair – 40 to 59% Poor – < 40%	Primary – include those that occur in the testicle during spermatogenesis Secondary – include those that occur in the epididymis after spermatogenesis Abnormalities may include problems with the head, acrosome and tail. Usually 30% abnormalities is the maximum allowed.

Taken largely from Barth (2000). Gross motility, individual motility and sperm abnormalities must be assessed with a microscope.

Table 8.5. Example components for assessment of bull breeding soundness evaluation.

Breed of bull: _____	Age of bull: _____		Name/ID of bull: _____	
History:				

Sex drive (libido)	Physical components		Semen components	
Many times this may be entirely for the producer to evaluate, but could be part of formal evaluation.	Body condition score	Scrotal circumference	Collection method	Abnormalities (%) for head, mid-piece, principal piece, detached heads, proximal droplets, acrosome and total
	Eyes	Scrotum	Penis response Volume, density, gross motility, individual motility, live:dead cells	
	Accessory sex glands			
	Feet	Scrotal shape		
	Inguinal rings	Testicles		
	Legs	Penis	Epididymides	
	Mouth	Prepuce		
	Pelvic area	Spermatic cord	Presence of blood cells	

Overall assessment: Satisfactory, Decision deferred, Questionable, Unsatisfactory
Comments:

Combined from the Australian Association of Cattle Veterinarians (McGowan *et al.*, 1995), the Society of Theriogenology (USA), and the Western Canadian Association of Bovine Practitioners (Barth, 2000).

instance Hawkins *et al.*, 1988) and frequent ejaculation (such as in Almquist *et al.*, 1976) compared to average values used for planning purposes, but this ability is quite variable across individuals.

8.2.6 Social Hierarchy

There is considerable evidence that multiple-sire breeding groups will produce higher conception rates than single-sire mating groups. There is known to be competition among bulls, and it is also likely that some bulls will sire more calves than others. Mixing of bulls that are to be used together during the breeding season should be done well in advance of turning them out with the cows for the breeding season in order for them to establish a social order. There are no problems with mixing bulls of different breeds, different sources or different ages for use during the breeding season as long as it is done at least 30 days prior to the beginning of the breeding season.

8.2.7 Fertility Associated Antigen

A test to evaluate an additional component of bull fertility is available in some countries, which evaluates semen for certain heparin-binding proteins and is referred to as fertility associated antigen (FAA). Research has been conducted on bulls that had passed a breeding soundness exam, but if the bull's semen contained this particular protein, they had higher pregnancy rates than the bulls without this protein in their semen. This test is based on placing a few drops of semen into a small plastic container, and if the bull has the protein in its semen a line shows up on a small piece of paper inside the container (much the same way that over-the-counter pregnancy tests work for women). The cost of the test is close to the cost of breeding soundness exams, and may provide valuable information to complement such exams, but should not be used as a substitute.

8.3 Reproductive Management of a Herd

Management of a group of individuals is not the same concept as management of an individual. Understanding the biological principles of reproduction (and the inherent potential for variability) is a cornerstone of effective management of a group of cows, heifers, bulls, etc. This section focuses on some specific considerations associated with the reproductive management of groups of cattle.

8.3.1 Controlled Breeding and Calving Seasons

Farm and ranch managers make decisions, plan activities and pay bills on a regular repeating time schedule, and a huge challenge in cow-calf production is to make the breeding females reproduce annually and in a timely manner as with a set schedule. Consequently, a target annual calving interval of 365 days is a major goal for beef cow reproduction. Proper reproductive management of beef cows revolves around having a short and controlled calving season for groups of cows. It is typical that beef cows will not return to estrous activity until approximately 30 days or more after calving (termed post-partum interval, or post-partum anestrus). Gestation length (Table 8.6) in beef cattle is approximately 283 days among many *Bos taurus* breeds, 285 days among *Bos indicus–Bos taurus crosses*, and approximately 290 days in *Bos indicus* (zebu). In order to maintain a 365-day calving interval, there will only be $365 - 285 - 30 = 50$ days (or less) in which cows can breed back and remain on an annual calf production cycle. In planning and evaluation of herd-level reproductive management, it can be helpful to simply think about a timeline and

begin to consider when and why various components occur (such as calving season, breeding season, when calves will be weaned, etc.). Figure 8.6 provides an example timeline for management considerations.

The length of time for the calving season is also important. Many production records for contemporary groups are based on individuals that are born within a 90-day period (individuals within this age range can be fairly compared to one another for performance traits (as discussed in Chapters 3 and 7). If all individuals within a herd (there could be multiple herds within a ranch/operation) are born within this 90-day window, this herd can be classified as a single contemporary group, provided they are all managed the same. The length of the calving season will usually be 10–15 days longer than the length of the breeding season because of individual variability in gestation length among calves.

It is desirable to have groups of calves that are uniform for a variety of reasons. Producers want to give vaccinations, castrate bull calves, wean, etc. on a single day. The more uniform the ages of the calves, the more uniform the management can be when all calves are treated the same. This will aid in replacement heifer development if someone is producing their own replacements and the group they are in is uniform. There should also be economic advantages if marketing of groups of uniform-sized calves is possible. There is a balancing act between calving/breeding seasons that are short enough for increased uniformity vs too short for the production environment limitations.

8.3.2 Body Condition Score

The single most widespread limiting factor of reproduction among breeding age beef females is low body condition (percent body fat). Body condition score (BCS) is a visual assessment of body fat that is represented on a 1 to 9 scale. On this scale, 1 is extremely emaciated, very weak and in a life-threatening state, whereas 9 is extremely obese. Under normal production conditions, BCS values close to these extremes will be avoided. Continual regular monitoring of cattle is needed so that intervention can be taken when animals appear to be losing too much body condition compared to others in the herd. This could be due to illness from infectious disease, injury, parasite load, old age/tooth loss, or competition among animals when supplemental feed is offered.

Table 8.7 shows the expected fat cover over the last two ribs and the percent empty body fat across BCS as well as the required weight change across

Table 8.6. Expected calving dates based on 285-day gestation.

	1	2	3	4	5	6	7	8	9	10	11	12	13	14	15	16	17	18	19	20	21	22	23	24	25	26	27	28	29	30	31	
Jan	1	2	3	4	5	6	7	8	9	10	11	12	13	14	15	16	17	18	19	20	21	22	23	24	25	26	27	28	29	30	31	
Oct	13	14	15	16	17	18	19	20	21	22	23	24	25	26	27	28	29	30	31	1	2	3	4	5	6	7	8	9	10	11	12	Nov
Feb	1	2	3	4	5	6	7	8	9	10	11	12	13	14	15	16	17	18	19	20	21	22	23	24	25	26	27	28				
Nov	13	14	15	16	17	18	19	20	21	22	23	24	25	26	27	28	29	30	1	2	3	4	5	6	7	8	9	10				Dec
Mar	1	2	3	4	5	6	7	8	9	10	11	12	13	14	15	16	17	18	19	20	21	22	23	24	25	26	27	28	29	30	31	
Dec	11	12	13	14	15	16	17	18	19	20	21	22	23	24	25	26	27	28	29	30	31	1	2	3	4	5	6	7	8	9	10	Jan
Apr	1	2	3	4	5	6	7	8	9	10	11	12	13	14	15	16	17	18	19	20	21	22	23	24	25	26	27	28	29	30		
Jan	11	12	13	14	15	16	17	18	19	20	21	22	23	24	25	26	27	28	29	30	31	1	2	3	4	5	6	7	8	9		Feb
May	1	2	3	4	5	6	7	8	9	10	11	12	13	14	15	16	17	18	19	20	21	22	23	24	25	26	27	28	29	30	31	
Feb	10	11	12	13	14	15	16	17	18	19	20	21	22	23	24	25	26	27	28	1	2	3	4	5	6	7	8	9	10	11	12	Mar
Jun	1	2	3	4	5	6	7	8	9	10	11	12	13	14	15	16	17	18	19	20	21	22	23	24	25	26	27	28	29	30		
Mar	13	14	15	16	17	18	19	20	21	22	23	24	25	26	27	28	29	30	31	1	2	3	4	5	6	7	8	9	10	11		Apr
Jul	1	2	3	4	5	6	7	8	9	10	11	12	13	14	15	16	17	18	19	20	21	22	23	24	25	26	27	28	29	30	31	
Apr	12	13	14	15	16	17	18	19	20	21	22	23	24	25	26	27	28	29	30	1	2	3	4	5	6	7	8	9	10	11	12	May
Aug	1	2	3	4	5	6	7	8	9	10	11	12	13	14	15	16	17	18	19	20	21	22	23	24	25	26	27	28	29	30	31	
May	13	14	15	16	17	18	19	20	21	22	23	24	25	26	27	28	29	30	31	1	2	3	4	5	6	7	8	9	10	11	12	Jun
Sep	1	2	3	4	5	6	7	8	9	10	11	12	13	14	15	16	17	18	19	20	21	22	23	24	25	26	27	28	29	30		
Jun	13	14	15	16	17	18	19	20	21	22	23	24	25	26	27	28	29	30	1	2	3	4	5	6	7	8	9	10	11	12		Jul
Oct	1	2	3	4	5	6	7	8	9	10	11	12	13	14	15	16	17	18	19	20	21	22	23	24	25	26	27	28	29	30	31	
Jul	13	14	15	16	17	18	19	20	21	22	23	24	25	26	27	28	29	30	31	1	2	3	4	5	6	7	8	9	10	11	12	Aug
Nov	1	2	3	4	5	6	7	8	9	10	11	12	13	14	15	16	17	18	19	20	21	22	23	24	25	26	27	28	29	30		
Aug	13	14	15	16	17	18	19	20	21	22	23	24	25	26	27	28	29	30	31	1	2	3	4	5	6	7	8	9	10	11		Sep
Dec	1	2	3	4	5	6	7	8	9	10	11	12	13	14	15	16	17	18	19	20	21	22	23	24	25	26	27	28	29	30	31	
Sep	12	13	14	15	16	17	18	19	20	21	22	23	24	25	26	27	28	29	30	1	2	3	4	5	6	7	8	9	10	11	12	Oct

Date of conception is on the first line and expected calving date is on the line below. For example, a cow bred on 1 January is expected to calve on 13 October, a cow bred on 31 January is expected to calve on 12 November, etc.

Concept of a beef cow reproducing on a 365-day interval in various stages of production across time.

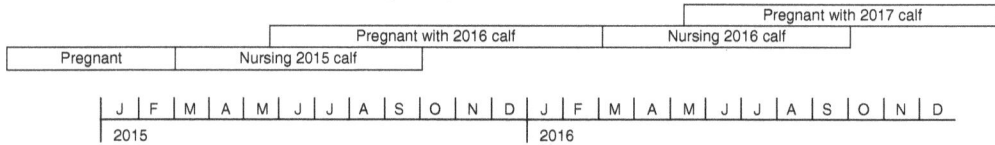

Fig. 8.6. Stages of production in beef cows. The nutritional demands of beef cows are dictated by the primary stage of production (early pregnancy, late pregnancy, early lactation, etc.) they are in, but the management difficulty can be compounded because very often in the lives of cows they are in more than a single stage of production, such as lactation and pregnancy.

Table 8.7. Expected percentage of body weight change relative to body condition score change in mature cows.

Body condition score	Empty body fat %	Carcass fat %	Carcass fat cover (mm)	Ratio of weight relative to BCS 5	Percent weight gain needed to increase to next BCS
1	0	0.7	0	0.740	5.8
2	4	5.0	0	0.798	6.2
3	8	9.3	1	0.860	6.7
4	12	13.7	3	0.927	7.3
5	16	18.0	5	1.000	8.0
6	20	22.3	7	1.080	8.7
7	24	26.7	10	1.167	9.1
8	28	31.0	14	1.258	10.2
9	32	35.3	17	1.360	

Weight at BCS 5 is the reference point. A cow in BCS of 4 will weigh 92.7% of what she weighs in BCS of 5, and this cow at BCS of 6 will weigh 108% of what she weighs at BCS of 5, etc. The last column indicates the percentage (%) of the weight when at BCS of 5 required to increase to the next BCS score. From Herd and Sprott (2007).

BCS in reference to mature cows. The target BCS at breeding time is a 5 because many research trials have shown markedly reduced fertility when cows were in BCS of 4 or 3 at breeding compared to those in BCS 5. Furthermore, it is widely recommended to have cows at BCS of 6 at calving time because many cows lose body fat because of lactation stress after calving; this strategy will help ensure that cows are at BCS 5 for breeding time, when body condition is critical. If supplemental feeding is used to increase BCS, this should be done during the last 90–120 days of gestation. It will be much easier for cows to gain weight before calving as opposed to during lactation. If the desire is to increase BCS, then an energy supplement will be needed.

The single most important time to evaluate cows is 100 days prior to calving because if cows need to increase BCS, they may need to gain 40–50 kg to change body composition, and this allows enough time to accomplish this. It should

be noted that cows will also need to gain approximately 30–35 kg during the last trimester to account for the growing calf inside them. If a cow in late gestation is maintaining her weight, she is losing body condition because the calf continues to grow. As a result a cow in late gestation that needs to increase 1 full BCS will probably need to gain 70–85 kg of weight. This is extremely hard to accomplish in less than 100 days in most circumstances. If BCS is only assessed at or near calving time, it is too late to make any necessary changes (or becomes much more expensive). There is substantial research also showing that cows gaining weight before calving have a rebreeding advantage over cows that lost weight before calving, even if these two types of cows were in the same BCS at calving. Whatever the circumstances, cows pregnant in late gestation need to gain at least 0.4 kg per day until calving to ensure there is not increased post-partum anestrous.

8.3.3 Early Weaning of Calves

In many geographical areas is it common to wean calves when they are around 7 months old. There is nothing magical about this calf age for weaning, but many breed societies and associations adjust weaning weight to a standardized calf age of 200–205 days. Some producers weigh calves around this age to submit weights to breed societies, but keep calves on the cows until 9 or 10 months of age. Many commercial producers also wean calves when they are around 6–8 months old on average. Some producers may let cows wean the calves on their own; under extensive conditions a cow will usually stop letting her calf nurse when she is getting close to having another calf, but not always. Other commercial producers may wean calves early, such as at 45, 60, 75 or 90 days of age. Early weaning has illustrated known advantages for over 40 years (Laster *et al.*, 1973) and been recommended as one of the most important management tools for first-calf heifers to breed back (Geary, 2003). Funston (2006) reviewed several trials where fertility was substantially increased with early weaning of calves, particularly in young cows (26–38% improvement). Whether or not this is a good economic decision depends on what can be done with the calves as increased nutrition and health management of early weaned, young calves present serious management considerations (Guyer, 1983; Amaral-Phillips *et al.*, 2000).

8.3.4 Use of Strategic Nutritional Supplementation

Although low BCS is probably the most common way that improper nutrition limits reproduction, many other nutritional considerations are important. When grazing situations are sub-optimal and/or when deficiencies are known to be common (as with minerals in many geographical areas) specific nutrients may need to be supplemented (Table 8.8).

There has been interest in certain nutritional supplement programs for beef cows such as fat and use of ionophores in regard to beef female reproduction. Most seed lipid is linoleic acid, but most forage lipid is linolenic acid. Linoleic acid is a substrate for $PGF_{2\alpha}$, which causes regression of CL at the end of the non-pregnant estrous cycle and is also responsible for uterine involution (level related to rate) following parturition. Linolenic acid seems to inhibit $PGF_{2\alpha}$ synthesis. As a result it

Table 8.8. Examples of reported reproductive problems in beef cattle from nutrient deficiencies or excesses.

Nutrient consumption	Reproductive impact
Excessive energy	Low conception rate, abortion, dystocia, retained placenta, reduced libido
Inadequate energy	Delayed puberty, suppressed estrus, reduced libido, impaired sperm production
Excessive protein	Low conception rate
Inadequate protein	Suppressed estrus, low conception rate, fetal loss and reabsorption, premature birth, weak offspring
Vitamin A deficiency	Impaired sperm production, anestrus, low conception rate, abortion, weak offspring, retained placenta
Phosphorus deficiency	Anestrus, irregular estrus
Selenium deficiency	Retained placenta
Copper deficiency	Reduced reproduction, impaired immune system, impaired ovarian function
Zinc deficiency	Reduced sperm production

From Funston (2006).

is thought that feeding of grain as an energy source might be more effective than forage as energy source, and that fat supplementation could alter some metabolic pathways leading to hormone secretion (some of this was reviewed by Hess *et al.*, 2005). Likewise, feeding of ionophores alters the rumen microbial population so that more proprionate is produced leading to more glucose production, and glucose is the fuel source of the central nervous system. Feeding of ionophores has been shown to have beneficial effects for return to estrus following calving in several trials (for instance see the review by Geary, 2003). There is no nutritional shortcut that will overcome low body condition in breeding females; however some of these nutritional strategies can improve reproductive performance in females that are borderline in body condition. As there is variation in individual cows' ability to rebreed relative to body condition, documentation of individual animal performance along with documented protocols used across seasons and years can help explain differences in overall reproduction for both environmental and genetic influences.

8.4 Recommended Reproductive Performance Measures

Reproductive measures can and should be monitored at levels of: (i) the herd, (ii) the individual cow and heifer and (iii) the individual bull. All producers have the ability to measure herd-level reproductive measures if they have adequate inventory information, and these herd-level measures provide benchmark values to monitor overall performance over time. Many cow-level and bull-level reproductive measures require individual animal identification, and represent a higher level order of data collection and decision making. The fact that it is possible to record and measure too many traits (i.e. it is not cost-effective to measure 50 different traits per animal, etc.) is also important for managers to recognize; the traits discussed here will aid in monitoring reproductive management and help identify non-productive as well as highly productive animals. Many of these recommendations follow those of the Beef Improvement Federation (BIF, 2010).

8.4.1 Herd-Level Measures

Percent pregnant

Percent pregnant is the number of cows that are determined pregnant divided by the number of cows exposed to bulls multiplied by 100.

Percent calf crop born

Percent calf crop born is the number of cows that are calved divided by the number of cows exposed to bulls multiplied by 100.

Percent calf crop weaned

Percent calf crop weaned is the number of cows that weaned a calf divided by the number of cows exposed to bulls multiplied by 100.

It is important to realize that all three of these herd-level reproductive measures are relative to the number of cows that were exposed to bulls because the basis of monitoring reproductive performance is relative to the total number of cows that had the potential to become pregnant, calve and wean a calf. Some people confuse percent calf survival with percent calf crop weaned; percent calf survival is the number of calves that are weaned divided by all live calves that were born multiplied by 100.

Calving distribution

Calving distribution describes the pattern of births over the course of the calving season. It is important for cows and heifers to calve as early as possible in each calving season so that they will have the maximum possible time to rebreed before the breeding season ends. Evaluation of the percentage of the herd calving in 21-day increments (first 21 days, second 21 days, etc.) can be a useful measure of herd reproductive efficiency across years.

8.4.2 Cow-Level Measures

Individual cow identification

Individual, unique identification (ID) is needed on beef cows for precise reproductive monitoring and improvement. A hide brand accompanied by an ear tattoo is probably the most guaranteed way to have a permanent identification for each cow. A hide brand along with a tissue or DNA sample also provides a secure, permanent ID record. A detailed discussion of individual animal identification methods and strategies can be found in Chapter 13 in regard to supply chain management. If a cow does not have an individual and unique ID, it is almost impossible to track her fertility and productivity across time, and this adds to imprecision in the production system.

Pregnancy status

This can simply be recorded as a 1 or 0 (for yes or no, respectively). If recorded as 1 and 0 in a spreadsheet, pregnancy rate for the herd can automatically be calculated from the individual values. Some producers may record pregnancy status directly on the animal somehow. For instance, a small notch can be taken from an ear each time she is pregnant (or each time she is not), a year brand can be put on her back each time she is pregnant, etc. This is better than not tracking any data; however, all of the information is lost if the animal becomes lost, and there is no way to tie this reproductive measure to other productivity traits, or to family information.

Calving status

Calving status is simply recording whether or not a cow gave birth as 1 or 0 (or yes or no).

Calving dates

This may be hard to obtain in extensive conditions, but it is important to monitor cows and heifers regularly during the calving season in case any assistance/intervention is needed. Recording the date when the cow is first observed with a calf is as good as knowing the exact calving date, particularly if cows are observed at 2–3 day intervals. It is also important to recognize that if there is no individual identification on the cow, there is no way to track her calving dates (or most other reproductive measures) across years. Calving status and calving date should be recorded at the same time. If calving dates are recorded, the difference between these dates also produces the value of calving interval.

Weaned calf status

Recording the weaning status for each cow may not seem important to all producers, but if someone wants to identify cows that are far below or above herd average across several years, there is no other way to do this unless cows are individually identified and their records are updated each year.

Weaning date

Weaning date is typically a single date for a group of calves weaned on the same day. If the weaning date is recorded as well as the calving date, the weaning age of each calf can be calculated automatically.

Body condition score at calf weaning

As weaning is one practice that will be performed in each herd each year, this provides a useful time to assess cow BCS. In herds on a 365-day calving cycle when calves are weaned around 6–7 months of age, this typically provides for 4–5 months before the upcoming calving season begins. If cows need to gain 40–80 kg of weight to increase their BCS (not counting an additional 30 kg of fetal calf weight), with 120–150 days to work with, these types of weight increases can be realized.

Longevity and culling criteria

The main reason for cows being culled from beef herds is due to some problem with reproduction. It is important to note the age and the reason that females leave the breeding herd (when and why they were culled).

8.4.3 Databases for Tracking Cow Reproductive Performance

A simple spreadsheet can be used to record and evaluate cow herd productivity (although several sophisticated software programs are available and may be appropriate). Simply having a spreadsheet where the data can be sorted by year (or breeding season, etc.) as well as by individual animal provides a convenient way to access both herd-level and cow-level productivity. It is important that each animal that was in the breeding herd is in the database, even if the cow did not become pregnant. This is the basis of whole-herd reporting, and many breed societies have begun to transition to this method of data tracking. If a cow is not in the database because she failed to become pregnant, this misrepresents the breeding group. When all cows in the breeding herd are in the database with a 0 or 1 recorded for failure and success, respectively, the average value for the group is the percent calf crop born, weaned, etc. Figure 8.7 shows the importance of tracking females when they may be switched across breeding season groups.

8.4.4 Bull-level Measures of Reproductive Efficiency

It is not as common to track reproductive performance on individual bulls, particularly when used in multiple-sire pastures for non-registered commercial herds. However, the same measures that can be calculated on a per herd basis can be determined on a per bull basis if the information is available. For instance in single-sire mating groups, if a bull was placed with 30 cows, the percentage calf crop born and weaned for that set of cows represents those measures on the bull as well. Combining breeding groups after the breeding season is over often happens, but this information can still be tracked by breeding group (and therefore bull), but may not explain whether observed differences were caused by the bull or some other factor unique to the breeding group.

In multi-sire breeding groups, documentation of parentage through DNA analyses can be performed by various companies for $20–30 per calf based on hair, blood or ear notch samples. It may be worthwhile knowing how evenly the bulls are siring calves. We may assume that if four bulls are placed with 100 cows, and if 100 calves are produced, they are close to equally split (close to 25 for each

calving breeding

Potential advantages:

- More cows per bull per year
- Calve home-raised heifers at 30 months if desired
- Cows failing to breed back once don't have to be held over an entire year
- Spreads risks over time

Potential disadvantages:

- Temptation to let less productive cows slide into new season groups repeatedly
- Some groups may not be 'best matched' to pasture resources

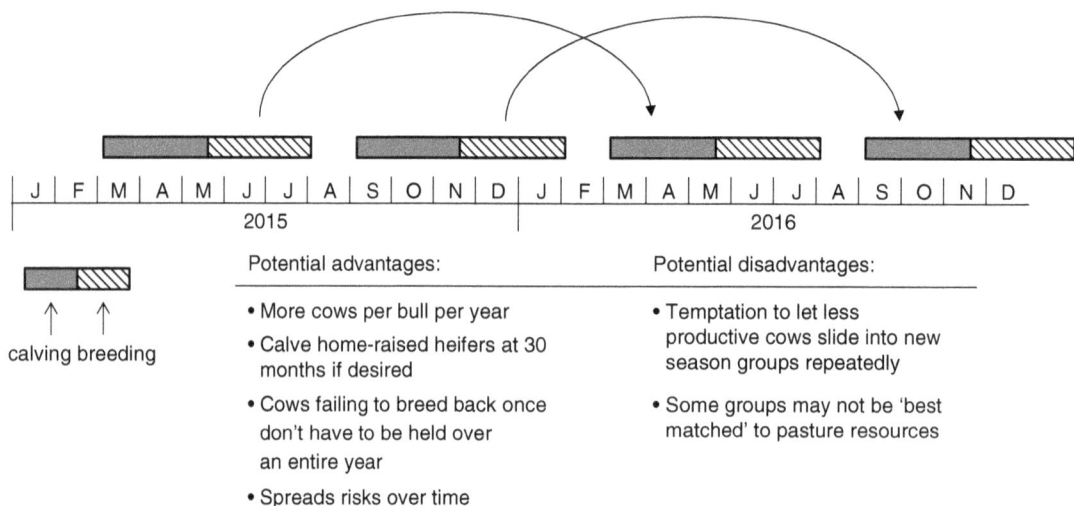

Fig. 8.7. Concept of split calving seasons (spring and fall) as opposed to one calving season.

sire), but this may deviate drastically from expectation. Use of DNA testing for parentage offers seedstock producers a way of taking advantage of potential increases in fertility through multi-sire breeding groups yet still know the parentage for registration purposes.

8.4.5 Cow Culling Strategies

If genetic progress for reproductive ability is desired, there needs to be a consistent culling policy in place and accepted before it needs to be enforced. This policy may need to be modified in extraordinary circumstances such as with exceptional drought or other problem scenarios, but a contingency plan in place ahead of time (or at least the concept of it) is also recommended. The culling strategy may be different in commercial herds than in purebred or seedstock herds, but both need to have a consistent plan within the operation. All herds must be concerned about fertility of both males and females. Seedstock herds that do not have a justifiable culling strategy based on acceptable fertility are at risk of long-term survival, particularly if feed resources become more limited (and expensive). A reasonable strategy for all types of herds in environments that are not extremely harsh would be to cull females from the herd when they fail to wean a calf for the second time. In very good environments, this may need to be after a single failure to wean a calf.

In harsh environments, perhaps this should be after the third failure to wean a calf. Many seedstock producers have strictly adhered to culling strategies where any female that fails to wean a calf is removed from the herd no matter what the reason, and over the long term this has resulted in highly productive and fertile herds whose animals are highly sought for improvement in other herds. Many assorted management factors can relate to fertility and calf survival across environments (Fig. 8.8).

In commercial herds, it is typically less expensive to keep a non-pregnant (open) female for an additional year than to purchase her replacement. The problem with this decision is that it is not known whether she will remain open for the subsequent, second year. If it was known she would remain open for two consecutive years, it would be better to cull her when she was discovered to be open (so there is risk involved). In many instances with young cows, they may fail to breed back as a 2-, 3- or 4-year-old; however, there are many of these young cows that may produce a calf annually for many years following a skip. The culling of older cows (such as those 10 years or more) is probably warranted if they are discovered open the first time at an advanced age, as the chance increases that she will decrease in productivity more each year as she reaches advanced age. The economic and other associated implications from various cow longevity scenarios are discussed in detail in Chapters 12 and 13.

Many natural features of pastures can be effective for providing protection for females at calving and for newborn calves.

Pastures that have forests, woods, brush or bush cover adjacent to open areas provide cover.

Tall, dormant grass is low in nutritional quality, but can provide protection from wind and predators in calving pastures in areas without trees.

Canyons and other low-lying areas can provide protection from wind in areas without trees or other features.

Matching of features in calving pastures with associated environmental problems (heat stress, cold stress, predators, etc.) is instrumental for newborn calf survival.

Fig. 8.8. Natural features useful for providing protection.

8.5 Summary

Fertility is important for all beef cattle production systems. In cow herds, level of reproduction is the primary driver of profit potential. Fundamental understanding of the reproductive tract organs and their functions will aid in identification of abnormalities and potential reproductive problems. Understanding of the hormone influences and interactions involved in fertility can provide for necessary management intervention and manipulation. There are several management strategies that can be used in concert with production conditions to improve herd reproductive efficiency. No strategies to improve reproduction will be able to compensate for inadequate nutrition. Recording of important performance measures in herds will set benchmark values to monitor reproductive success within and across years; however, herd-level measures alone will not allow for identification of under-performing females or bulls, and increased reproductive management efficiency can be obtained with individual animal identification in many scenarios. There are also numerous reproductive technologies than may be of varying benefit under different circumstances, but these need to be evaluated relative to cost vs benefit on a case-by-case basis.

8.6 Study Questions

8.1 A short, controlled calving season produces more uniformity. List five potential advantages of this increased uniformity.

8.2 Many cow-calf producers attempt to get as many cows as possible to calve early in the calving season (i.e. the first 21 days). Why would this be desirable?

8.3 A cow weighs 630 kg when in body condition score (BCS) of 5. Approximately what would she weigh at (a) BCS of 4 and (b) BCS of 6?

8.4 Some bulls are capable of successfully breeding up to 50 cows per breeding season. Explain whether or not you would take a recently purchased 18-month-old bull and put him with 50 cows.

8.5 Reproductive management scenario
Use the following information to answer Questions 8.5a to 8.5e below about a cow-calf operation. For this scenario the target age of heifers calving is 24 months. Show your calculations where appropriate.

Average mature weight of cows in BCS of 5: 520 kg
Target calving season of mature cows: 1 February to 1 April annually

Target weaning date: 1 May annually
Average weaning weight of heifers (from mature cows): 250 kg

8.5a What are the target breeding weight and calving weight for replacement heifers produced from this herd?

8.5b If you have mature cows that need to change BCS from a 5 at weaning time to a 6 at the start of the upcoming calving season, determine the weight gain and the average daily gain needed, assuming that the cows must gain an extra 34 kg just due to the pregnancy.

8.5c Determine the proper breeding season for (a) mature cows and (b) heifers under this scenario.

8.5d Determine the proper ADG needed for replacement heifers (a) from weaning until start of their first breeding season and (b) from the start of the breeding season until start of calving.

8.5e How would the answers and the considerations to Questions 8.5a to 8.5d above change if you calved heifers at 36 months of age and used 65% expected mature weight for a target breeding weight?

8.6 In regard to a mature cow herd, there were 180 cows exposed to bulls during the previous breeding season. There were 168 cows that calved, and 159 cows that weaned calves with an average weaning weight of 265 kg.

8.6a Calculate (a) the percent calf crop born, and (b) the percent calf crop weaned.

8.6b Calculate the kilograms of calf weaned per cow exposed to breeding.

8.6c State your recommended number of bulls needed to cover these 180 cows (single number or range is fine) and what your recommendation is based upon.

8.6d If you felt confident that bulls could cover 40 cows each, how many bulls would you use to breed these cows?

8.7 If it cost $50 per bull for breeding soundness evaluation, and the bull is expected to breed 25 cows in the upcoming breeding season, what is the cost of this evaluation on a per cow basis?

8.7 References

Almquist, J.O., Branas, R.J. and Barber, K.A. (1976) Postpubertal changes in semen production of Charolais bulls ejaculated at high frequency and the relation between testicular measurement and sperm output. *Journal of Animal Science* 42, 670–676.

Amaral-Phillips, D.M., Scharko, P.B., Johns, J.T. and Franklin, S. (2000) Feeding and managing baby calves from birth to 3 months of age. University of Kentucky Cooperative Extension Service Publication ASC-161. Available at: http://www2.ca.uky.edu/agc/pubs/asc/asc161/asc161.pdf (accessed 17 December 2013).

Barth, A.D. (2000) *Bull Breeding Soundness Evaluation*, 2nd edn. Western Canadian Association of Bovine Practioners Lacombe, Alberta.

BIF (2010) Chapter 3, Breeding herd evaluation. In: *Guidelines For Uniform Beef Improvement Programs, Ninth Edition*. Beef Improvement Federation, Raleigh, NC. Available at: http://www.beefimprovement.org/content/uploads/2013/07/Master-Edition-of-BIF-Guidelines-Updated-12-17-2010.pdf (accessed 12 May 2014).

Burns, B.M., Herring, A.D., Laing, A., Fordyce, G., Bertram, J., Grant, T. and Hiendleder, S. (2010) Unrecognized variation in gestation length and birth weight of Droughtmaster calves produced through fixed-time AI. In: Lucy, M.C., Pate, J.L., Smith, M.F. and Spencer, T.E. (eds) *Reproduction in Domestic Ruminants*. Ruminant Reproduction 2010: 8th International Ruminant Reproduction Symposium, Anchorage, AK, 3–7 September 2010, pp. 585.

Freetly, H.C., Ferrell, C.L. and Jenkins, T.G. (2001) Production performance of beef cows raised on three different nutritionally controlled heifer development programs. *Journal of Animal Science* 79, 819–826.

Funston, R. (2006) Nutrition and reproduction interactions. In: *Proceedings, Applied Reproductive Strategies in Beef Cattle*, 3–4 October, Rapid City, South Dakota, pp. 215–230.

Funston, R.N. and Deutscher, G.H. (2004) Comparison of target breeding weight and breeding date for replacement beef heifers and effects on subsequent reproduction and calf performance. *Journal of Animal Science* 82, 3094–3099.

Geary, T.W. (2003) Management of young cows for maximum reproductive performance. In: *Proceedings Beef Improvement Federation 35th Annual Research Symposium and Annual Meeting*, 28–31 May, Lexington, Kentucky, pp. 5–8.

Guyer. P.Q. (1983) Management of early weaned calves. University of Nebraska Lincoln Cooperative Extension, Institute of Agriculture and Natural Resources. Publication G83–655.

Hawkins, D.E., Carpenter, B.B., Sprott, L.R., Beverly, J.R., Hawkins, H.E., Parish, N.R. and Forrest, D.W. (1988) Proportion of early conceiving heifers is increased by high serving capacity bulls. *Journal of Animal Science* 66(Suppl. 1), 246(Abstr.)

Herd, D.B. and Sprott, L.R. (2007) Body condition, nutrition and reproduction of beef cows. Publ. B-1526 Texas Beef Cattle Management Handbook Texas Cooperative Extension, Texas A&M AgriLife Extension Service, College Station, TX.

Hess, B.W., Lake, S.L., Scholljegerdes, E.J., Weston, T.R., Nayigihugu, V., Molle, J.D.C. and Moss, G.E. (2005) Nutritional controls of beef cow reproduction. *Journal of Animal Science* 83(E. Suppl.), E90–E106.

Laster, D.B., Glimp, H.A. and Gregory, K.E. (1973) Effects of early weaning on postpartum reproduction of cows. *Journal of Animal Science* 36, 734–740.

Lynch, J.M., Lamb, G.C., Miller, B.L., Brandt, R.T., Cochran, R.C. and Minton, J.E. (1997) Influence of timing of gain on growth and reproductive performance of beef replacement heifers. *Journal of Animal Science* 75, 1715–1722.

Martin, J.L., Creighton, K.W., Musgrave, J.A., Klopfenstein, T.J., Clark, R.T., Adams, D.C. and Funston, R.N. (2008) Effect of prebreeding body weight or progestin exposure before breeding on beef heifer performance through the second breeding season. *Journal of Animal Science* 86, 451–459.

McGowan, M., Galloway, D., Taylor, E., Entwistle, K. and Johnston, P. (1995) *The Veterinary Examination of Bulls*. Australian Association of Cattle Veterinarians, Indooroopilly, Queensland.

Senger, P.L. (2005) *Pathways to Pregnancy and Parturition*, 2nd revised edn. Current Conceptions, Inc., Redmon, OR.

9 Health

There are many health threats to cattle that can place severe constraints on beef cattle production and profitability. These vary across geographical regions – in some areas, environmental threats from parasites, infectious diseases or harsh climatic conditions (and/or combinations thereof) are quite severe, but in other areas pose little threat. Some infectious diseases have vaccines available to offer increased protection, but some do not, or are not even fully understood in some cases in regard to their cause(s). Many diseases (or general health problems) are region-specific, and it is always recommended that local information and expertise be consulted where available. This chapter does not discuss all disease and health threats to beef cattle, but aims to provide the reader with a basic understanding of the appearance of healthy (and unhealthy) cattle, basic aspects of animal immune system function, some potential cattle disease threats, and fundamental biosecurity and herd health management considerations that can influence beef cattle production.

9.1 Healthy vs Non-Healthy Cattle

The concept of an animal being classified as 'healthy' has historically been that of an animal that is not obviously ill. Health is a complex consideration that involves nutrition, environmental stresses, pathogen exposure, animal age and genetic background and many management factors. The line of distinction between an animal that is diseased vs one that is not is usually quite clear in regard to health; however, the relative health status of an animal that is viewed as a 'hard doer' or appears stressed vs one that is not stressed is a more difficult concept. The discussion of cattle health considerations in this chapter begins with expected signs of cattle that appear healthy, or without obvious physical illness or stresses.

9.1.1 Vital Signs of Healthy Cattle

It is important to recognize what the observable signs are in healthy cattle to know when something might be wrong, and it is also important not to make a hasty decision based on only a brief or limited observation (Table 9.1). Healthy cattle will have alert, moist eyes, and beads of perspiration on the nose. Their heads are alert, and they are attentive to their surroundings and humans when approached. When standing, their feet are below the shoulder and hips. Feces of healthy cattle have a high water content compared to many animal species and should make a flattened pad when they hit the ground and are not watery or in formed balls when they exit the body. The normal body temperature of cattle (typically measured rectally) is reported to be 101.5°F (38.5°C), but varies among individuals (as it does in people), with time of day and ambient conditions, and level of activity or stress. Cattle that are content do not typically vocalize, but cattle that are unrestrained and not moving and constantly making sounds may be in pain or distressed. Hair coats of cattle change in appearance with climatic and nutritional seasonal changes, but a shiny (high luster) hair coat is an indication of good overall health.

Effective health management programs always begin with proper nutrition. Nutritional deficiencies (and sometimes excesses) contribute to an animal's increased stress levels. Proper dietary levels of energy and protein are needed to mount effective immune responses. Proper dietary levels of vitamins and minerals are needed to maintain proper skin and membrane function. The skin is the largest physical barrier to potentially infectious pathogens entering the body. Several physical aspects of live animal evaluation discussed in Chapter 6, such as body condition, hair coat, appearance of the skin, etc., as well as behavioral aspects discussed in Chapter 10, are important considerations in effective diagnosis of health-related problems. General physical appearance in regard to several of these traits can relate directly to the level of adaptation of animals in regard to the physical environment.

Table 9.1. Expected indicators of healthy cattle and potential indications for concern.

Trait	Typical value	Potential concern
Rectal temperature	38.5°C (101.5°F)	Many antimicrobial products are administered when over 40°C
Breaths per minute	30	Rapid, shallow breaths, long pauses between breaths, irregular pattern
Heartbeats per minute	60 to 70	Slow and rapid heart rates can result from a variety of causes
Nose	Beads of perspiration, shiny and even skin	Dry, very warm to touch, cracked or scaly surface
Eyes	Moist, smooth and shiny, brown iris	Dull and non-reflective, unusual color or presence of growths
Demeanor[a]	Alert, but at ease; attentive but not combative	Actions that substantially deviate from 'normal' behavior
Feces	Greenish color, not extremely dry or watery; makes a plop sound and forms patties/pads	Color may be dark, light, grayish; consistency too firm or too loose; appearance of blood
Movement[a]	Even and smooth actions of walking, moving neck, standing, no trembling	Limping, shaky in movements such as walking, holding head at unusual angle, wobbly or jerky
Hair appearance	Shiny, and evenly distributed	Dull without luster, patchy areas of thin or lost hair, unusual color

[a]What is 'normal' behavior as well as other traits can vary drastically across individuals and circumstances (such as with a newborn calf), and abnormal is a marked deviation from what is typical. History of observation provides useful information. Documentation of those signs that are suspected to be abnormal will aid veterinarians in diagnoses.

In many cases, allowing cattle to have access to areas that can alleviate or reduce environmental stresses can impact health and productivity. This could be as simple as having pastures with trees, where cattle can rest in the shade during the hottest parts of the day, or having areas where trees or other vegetation can provide windbreaks in the winter. If a non-tropically adapted breed is utilized in an area where there is substantial heat and solar radiation, pastures without shade and/or water sources large enough for cattle to enter could potentially result in reduced health and productivity as compared to pastures where animals can attempt to self-regulate from these stresses. When animals are more stressed in general, the potential for infection increases. Placing animals into production environments to which they are not adapted also promotes stress. Table 9.2 provides some general guidelines for considerations of heat and cold stress conditions in cattle.

9.2 Infectious Diseases

Many diseases affect cattle, and several are caused by infectious causative organisms (called pathogens).

Some infectious diseases of cattle also affect humans (these are called zoonoses when they can infect animals as well as humans). Zoonotic pathogens can be acquired during close contact with infected animals through inhalation, ingestion or direct contact, and sources of organisms can include body fluids, secretions and excretions, and lesions (Roth and Spickler, 2012). Some organisms can be spread by the ingestion of contaminated food or water to large numbers of people. Sources of zoonotic pathogens in foodborne disease from cattle can include undercooked meat or other edible tissues and unpasteurized milk and dairy products, and insect vectors may be important in transmitting some organisms (Roth and Spickler, 2012). Diseases affecting cattle that also affect other species are listed in Table 9.3. Some of the most serious diseases that affect only cattle are given in Table 9.4. Several useful resources related to cattle health and animal health in general are available on the internet, including the World Organization for Animal Health (OIE, www.oie.int) and the Merck Veterinary Manual Online with searchable databases; both of these were heavily used as references for this chapter.

Table 9.2. Guidelines for environmental conditions of heat and cold stress in cattle.

Temperature (°C) for heat stress

RH[a]	27.8	28.3	28.9	29.4	30.0	30.6	31.1	31.7	32.2	32.8	33.3	33.9	34.4	35.0	35.6	36.1	36.7	37.2	37.8
90	32.8	35.0	36.7	38.9	40.6	42.8	45.0	47.2	50.0	52.2	55.0	57.8	60.6	63.9	66.7	70.0	73.3	76.7	80.0
85	32.2	33.9	35.6	37.2	38.9	41.1	43.3	45.0	47.2	50.0	52.2	54.4	57.2	60.0	62.8	65.6	68.3	71.7	75.0
80	31.7	32.8	34.4	36.1	37.8	39.4	41.1	43.3	45.0	47.2	49.4	51.7	53.9	56.7	58.9	61.7	64.4	67.2	70.0
75	31.1	32.2	33.3	35.0	36.1	37.8	39.4	41.1	42.8	45.0	46.7	48.9	51.1	53.3	55.6	57.8	60.6	62.8	65.6
70	30.0	31.1	32.2	33.9	35.0	36.7	37.8	39.4	41.1	42.8	44.4	46.7	48.3	50.6	52.2	54.4	56.7	58.9	61.7
65	29.4	30.6	31.7	32.8	33.9	35.0	36.7	37.8	39.4	40.6	42.2	43.9	45.6	47.8	49.4	51.7	53.3	55.6	57.8
60	28.9	30.0	31.1	31.7	32.8	33.9	35.0	36.1	37.8	38.9	40.6	41.7	43.3	45.0	46.7	48.3	50.6	52.2	53.9
55	28.9	29.4	30.0	31.1	31.7	32.8	33.9	35.0	36.1	37.2	38.3	40.0	41.1	42.8	44.4	45.6	47.2	48.9	51.1
50	28.3	28.9	29.4	30.0	31.1	31.7	32.8	33.9	35.0	36.1	37.2	38.3	39.4	40.6	42.2	43.3	45.0	46.1	47.8
45	27.8	28.3	28.9	29.4	30.6	31.1	31.7	32.8	33.3	34.4	35.6	36.7	37.8	38.9	40.0	41.1	42.8	43.9	45.6
40	27.2	27.8	28.3	28.9	29.4	30.6	31.1	31.7	32.8	33.3	34.4	35.0	36.1	37.2	38.3	39.4	40.6	41.7	42.8
35	27.2	27.8	28.3	28.9	29.4	30.0	30.6	31.1	31.7	32.2	33.3	33.9	35.0	35.6	36.7	37.8	38.9	40.0	41.1
30	26.7	27.2	27.8	28.3	28.9	29.4	30.0	30.6	31.1	31.7	32.2	33.3	33.9	34.4	35.6	36.1	37.2	38.3	38.9
25	26.7	27.2	27.8	27.8	28.3	28.9	29.4	30.0	30.6	31.1	31.7	32.2	32.8	33.9	34.4	35.0	36.1	36.7	37.8

Temperature (°C) for cold stress

WS[a]	−23.3	−22.2	−21.1	−20.0	−18.9	−17.8	−16.7	−15.6	−14.4	−13.3	−12.2	−11.1	−10.0	−8.9	−7.8	−6.7	−5.6	−4.4	−3.3
42	−50.6	−48.9	−47.2	−45.6	−43.9	−42.2	−40.6	−38.9	−37.2	−35.6	−33.9	−32.2	−30.6	−28.9	−27.2	−25.6	−24.4	−22.8	−21.1
38	−49.4	−47.8	−46.1	−44.4	−42.8	−41.1	−39.4	−37.8	−36.1	−34.4	−33.3	−31.7	−30.0	−28.3	−26.7	−25.0	−23.3	−21.7	−20.0
35	−47.8	−46.1	−45.0	−43.3	−41.7	−40.0	−38.3	−36.7	−35.0	−33.3	−32.2	−30.6	−28.9	−27.2	−25.6	−23.9	−22.2	−20.6	−19.4
32	−46.7	−45.0	−43.3	−41.7	−40.0	−38.3	−37.2	−35.6	−33.9	−32.2	−30.6	−29.4	−27.8	−26.1	−24.4	−22.8	−21.1	−20.0	−18.3
28	−44.4	−43.3	−41.7	−40.0	−38.3	−37.2	−35.6	−33.9	−32.2	−31.1	−29.4	−27.8	−26.1	−25.0	−23.3	−21.7	−20.0	−18.9	−17.2
25	−42.8	−41.1	−39.4	−38.3	−36.7	−35.0	−33.9	−32.2	−30.6	−29.4	−27.8	−26.1	−25.0	−23.3	−21.7	−20.6	−18.9	−17.2	−15.6
22	−40.6	−38.9	−37.8	−36.1	−34.4	−33.3	−31.7	−30.6	−28.9	−27.2	−26.1	−24.4	−22.8	−21.7	−20.0	−18.9	−17.2	−15.6	−14.4
18	−37.8	−36.1	−35.0	−33.3	−32.2	−30.6	−29.4	−27.8	−26.7	−25.0	−23.9	−22.2	−21.1	−19.4	−18.3	−16.7	−15.6	−13.9	−12.8
15	−34.4	−33.3	−32.2	−30.6	−29.4	−27.8	−26.7	−25.6	−23.9	−22.8	−21.1	−20.0	−18.9	−17.2	−16.1	−14.4	−13.3	−12.2	−10.6
12	−31.1	−29.4	−28.3	−27.2	−26.1	−24.4	−23.3	−22.2	−20.6	−19.4	−18.3	−17.2	−15.6	−14.4	−13.3	−12.2	−10.6	−9.4	−8.3
8	−26.1	−25.0	−23.9	−22.8	−21.7	−20.6	−19.4	−17.8	−16.7	−15.6	−14.4	−13.3	−12.2	−11.1	−10.0	−8.9	−7.8	−6.1	−5.0
5	−20.0	−18.9	−17.8	−16.7	−15.6	−14.4	−13.9	−12.8	−11.7	−10.6	−9.4	−8.3	−7.2	−6.1	−5.6	−4.4	−3.3	−2.2	−1.1

[a]RH = relative humidity (%); WS = wind speed in km/h.
Shaded areas represent potential danger conditions with temperatures in table indicating heat index and wind chill values (what temperature it feels like to the animals). Ambient temperatures between 0 and 26°C generally present no or minimal environmental stresses to Bos taurus breeds. The range with no to minimal stress for Bos indicus breeds is probably 10–35°C.
This table is presented in °F Table A10.

Table 9.3. Infectious diseases that affect multiple animal species including cattle.

Disease name	Causative organism	Geographical areas	Transmission	Body areas, functions affected	Main clinical signs	Preventative measures
Anthrax (also known as charbon, woolsorters' disease, ragpickers' disease, malignant carbuncle, malignant pustule)	Bacteria (*Bacillus anthracis*)	Worldwide	Ingestion of bacterial spores, possibly by biting flies	Many soft tissues affected from endotoxins	Fever, depression, loss of appetite, weakness, prostration, death (infected animals usually found dead)	Vaccination, animal movement
Aujeszky's disease (pseudorabies, mad itch)	Virus (alphaherpesvirus)	Natural host is pigs, areas where pigs are common	Directly from infected animals	Central nervous system, respiratory system	Nervousness, restlessness and unusual behavior	Vaccination, animal movement, not major problem in cattle
Bluetongue	Virus (bluetongue virus)	Tropical and subtropical areas (may be seasonal related to insect survival)	Not contagious from animals, only from insects (certain species of midges)	Found in RBC and semen; may cause abortion and malformed fetuses	No apparent symptoms in infected cattle, cattle may shed virus for extended time	Vaccination, animal movement
Brucellosis (also known as Bang's disease, contagious abortion, and, undulant fever in humans)	*Brucella* spp. (*B. abortus and B. melitensis* in cattle)	Worldwide, several areas have become certified-free through eradication programs	Direct from infected animals through milk, uterine discharges, and aborted fetus and/or placenta	Uterus and udder, possibly accessory sex organs in bulls	Abortion with first infection, general infertility may follow	Vaccination of heifers, testing, and animal movement
Foot and mouth disease (FMD)	Virus (FMD virus with seven serotypes)	Worldwide potential, several countries FMD-free	Direct contact between animals, exposure of excretions and secretions from infected animals	Through pharynx, spread by lymphatic system to site of replication (muzzle, feet, teats); high fever	Sores (vesicles) on feet, in and around mouth and udders of females; cannot be clinically distinguished from other vesicular diseases	Most contagious disease of mammals (100% morbidity possible); animal and human movement, vaccination

Continued

Table 9.3. Continued.

Disease name	Causative organism	Geographical areas	Transmission	Body areas, functions affected	Main clinical signs	Preventative measures
Heartwater (cowdriosis)	Rickettsia (*Ehrlichia ruminantium*)	Sub-Saharan Africa and neighboring islands, Caribbean	Ticks (*Amplyomma*), not contagious from animals	Vascular tissues	Sudden high fever, depression, diarrhea, nervous signs later, death may occur without noticeable signs	Recovered animals remain carriers, some vaccines in development
Hemorrhagic septicemia	Bacteria (certain serotypes of *Pasteurella multocida*)	Asia and Africa, southern Europe, Middle East	Associated with high humidity and temperatures as with monsoon seasons	Ingestion or inhalation of pathogen; endotoxin effects throughout body	Septicemia, fever, respiratory distress, nasal discharge, frothing from mouth, high morbidity and mortality	Vaccination, sanitation, animal movement
Leptospirosis	Bacteria (*Leptospira* spp.)	Worldwide	Direct contact among animals	Liver, kidneys, lungs, genital tract, CNS	Abortion, stillbirth, premature birth, weak offspring	Vaccination, reduce animal contact from those infected
Screwworm (New World and Old World forms)	Larvae of flies (*Cochliomyia* – NW or *Chrysomya* – OW)	Historically, subtropical and tropical areas of North and South America, Africa, India and South-East Asia	Screwworm flies	Female flies lay eggs at edges of wounds or at body orifices, larvae burrow head-first into tissue	Larvae (maggots) in wounds with characteristic foul odor	Control of flies with dipping, spraying or injectibles, animal movement, release of sterilized male flies
Paratuberculosis (Johne's disease)	Bacteria (*Mycobacterium paratuberculosis*)	Worldwide	Ingestion from contaminated environment, may be passed to offspring in milk and found in semen	Infection of intestine wall	Infected animals may not show signs until later in life, eventually weight loss is observed from reduced gut function	Vaccination does not prevent disease and may interfere with tuberculosis diagnosis and control, termination of infected animals

Disease	Cause	Distribution	Transmission	Affected system	Symptoms	Control/Prevention
Q fever (Query fever, Coxiellosis)	Bacteria (*Coxiella burnetii*)	Worldwide except New Zealand	Direct contact from excretions and secretions of infected animals, primarily birth and placental fluids	Certain white blood cells; shed in urine, feces and milk as well as placental fluids	Sporadic abortions, dead or weak offspring, infertility; infected animals typically subclinical carriers for many years	Some vaccines available; prevent contact with infected animals
Rabies	Virus (*Lyssavirus*)	Worldwide	Bites containing saliva from infected animals	Central nervous system (CNS)	Marked behavioral changes; always fatal unless identified soon after exposure	Vaccination, control of animal bite threats
Rift Valley fever	Virus (genus *Phlebovirus*)	Africa	Mosquitoes	Liver damage	Fever, abortion, high mortality in newborn offspring	Vaccination, animals from areas without disease most at risk, control of mosquitoes
Rinderpest	Virus (genus *Morbillivirus*)	Historically, Europe, Africa and Asia	Secretions and excretions of infected animals, ingested or inhaled	Lymphoid tissue, GI tract mucosa, upper respiratory tract	Fever, shallow erosions of gums, tongue, cheeks and hard palate, eye and nasal discharge, diarrhea, highly fatal	Global eradication program launched in 1992, recently announced to have been eradicated
Surra (*Trypanosoma evansi* infection, also called El Debab, El Gafar, Tabourit, MBori, Mal de Caderas, or Murrina)	Protozoa (*Trypanosoma evansi*)	Africa, Asia, Central and South America	Spread by several species of biting flies (bats may also be involved in Brazil)	Causes immunodeficiencies that can lead to other diseases	Recurrent fever, progressive anemia, loss of weight	Animals stressed are more susceptible, animal movement and testing

Continued

Table 9.3. Continued.

Disease name	Causative organism	Geographical areas	Transmission	Body areas, functions affected	Main clinical signs	Preventative measures
Tuberculosis (TB)	Bacteria (*Mycobacterium bovis* for bovine TB)	Worldwide, but incidence highly variable across countries	Aerosol exposure primarily, but also ingestion of contaminated materials, exposure from wildlife species possible	Lesions occur in lungs and lymph nodes, and may be found in liver, spleen and other organs	Infection may be subclinical, signs may include weakness, anorexia, weight loss, dyspnea, enlarged lymph nodes, cough	Primarily animal testing and movement, vaccines exist but may interfere with TB tests, control of contact with wildlife species
Vesicular stomatitis	Virus (vesiculoviruses)	North, Central and South America	Transmission is unclear, virus has been isolated from flies and mosquitoes	Muzzle and mouth interior, feet and teats	Clinically indistinguishable from FMD	Vaccines are available but efficacy is unknown, animal movement

Information for this table was taken from the OIE and Merck Veterinary Manual Online websites. This is not necessarily meant to be an exhaustive list, but includes major diseases known to affect large geographical regions. Many of these diseases are zoonoses (affecting people as well as livestock).

Table 9.4. Infectious diseases that affect primarily or only cattle.[a]

Disease name	Causative organism	Geographical areas	Transmission	Body areas, functions affected	Main clinical signs	Preventative measures
Anaplasmosis	Rickettsia (*Anaplasma* spp.)	Tropical and subtropical regions, some more temperate regions	Primarily ticks, but can be spread through syringe needles	Red blood cells	Anemia, jaundice in later stages	Control of biting insects, some vaccines available
Babesiosis (also called tick fever)	Protozoa (*Babesia* spp.)	Tropical and subtropical areas	Ticks, predominantly *Boophilus* spp.	Red blood cells	High fever	Tick control
Campylobacteriosis (formerly referred to as vibriosis)	Bacteria (*Campylobacter fetus* subsp.)	Worldwide	Primarily through mating (natural or AI)	Reproductive tract	Infertility, early embryonic loss, abortion	Vaccination, clean instruments between animals
Bovine spongiform encephalopathy (BSE, or mad cow disease)	Prion	Europe, North America, some Asian countries	Feeding ruminant-derived meat and bonemeal	Brain and central nervous system	Loss of neurological function in older animals	No feeding of ruminant-derived meat and bonemeal
Bovine viral diarrhea (BVD, or pestivirus)	Virus, (BVD virus, BVDV); many different strains	Worldwide	Direct contact between animals; will be passed to progeny from persistently infected cows	Many tissues may be affected; virus can be found throughout body in all secretions	Highly variable from subclinical to diarrhea, high fever, nasal discharge, abortion, stillbirth	Vaccination, screening of breeding animals for PI
Contagious bovine pleuropneumonia	*Mycoplasma mycoides*	Africa and Middle East currently	Inhalation and ingestion from infected animals; highly contagious	Lungs, possibly kidneys and other organs	Variable from subclinical to fever, weight loss and respiratory signs; animals may be chronic carriers	Vaccination, control of animal movement and contact

Continued

Table 9.4. Continued.

Disease name	Causative organism	Geographical areas	Transmission	Body areas, functions affected	Main clinical signs	Preventative measures
Bovine leukosis	Virus (bovine leukemia virus, or BLV, a retrovirus)	Europe, North and South America	Transfer of infected cells between animals, through contaminated needles, gloves, etc., possibly through blood-sucking insects	White blood cells initially; virus found in tumor cells and cells in body fluids	Highly variable; lymph nodes may be enlarged, some animals develop lymph node tumors; affected organ(s) dictate clinical signs	Eliminating the movement of blood between animals, especially through tools, instruments, gloves, needles, etc.
Infectious bovine rhinotracheitis (IBR, red nose disease)	Virus (bovine herpesvirus-1)	Worldwide, but has now been eradicated from several European countries	Direct animal contact	Replicates in mucous membranes of upper respiratory tract and tonsils, then conjunctiva and ganglia; animals remain infected for life	Variable from subclinical to nasal discharge, redness of the nose, fever, depression, loss of appetite, abortion	Vaccination, animal movement
Lumpy skin disease (LSD, knopvelsiekte), a type of pox disease	Virus (multiple strains of capripoxvirus)	Africa, Middle East	Primarily insects	Painful nodules over entire skin of body and mucosa of the GI, respiratory and genital tracts	High, persistent fever, enlargement of superficial lymph nodes, large decrease in milk production, large nodules may develop and produce necrotic plugs penetrating the hide	Vaccination, limit exposure to insects, *Bos taurus* more susceptible

Disease	Cause	Areas	Transmission	Location	Signs	Control
Malignant catarrhal fever (MCF)	Virus; herpesvirus from natural hosts of wildebeest (AlHV-1) or sheep (OHV-1)	Areas with wildebeest and/or sheep	Exposure to infected wildebeest (which show no symptoms)	Cattle are highly susceptible to AlHV-1 but relatively resistant to OHV-1	High fever, loss of appetite, nasal discharge, loss of vision, may produce skin lesions	Isolation from wildebeest
Theileriosis (East Coast fever in Africa caused by *Theileria parva*)	Protozoa (*Theileria* spp.)	Europe, Africa, Asia	Carried by ticks	Blood, mainly WBC	Enlarged lymph nodes, fever, increasing respiratory rate, dyspnea, possibly diarrhea	Tick control and vaccination
Trichomoniasis	Protozoa (*Tritrichomonas fetus*)	Worldwide	Spread by sexual contact	Bulls (particularly older bulls) are primary long-term carriers; most cows clear infection; organism lives in prepuce cavities	Infected bulls show no signs; infected cows may have inflamed vulva, abortion, infertility, uterine infection	Vaccination, use of AI, use of young bulls, thorough screening of bulls
Trypanosomosis (tsetse-transmitted; tsetse-fly disease)	Protozoa (*Trypanosoma* spp.)	Africa	Spread by tsetse fly and possibly by other biting flies	Blood plasma, lymph and various tissues	Intermittent fever, enlarged lymph nodes, anemia, edema, abortion, infertility, loss of appetite and weight, eventually death	Control of flies, use of adapted, resistant cattle types

[a]Several of these diseases may also exist in other species, but these diseases in cattle have different causative pathogens than the disease of the same name in other species. Information for this table was taken from the OIE and the Merck Veterinary Manual Online websites. This is not necessarily meant to be an exhaustive list, but includes the main diseases known to affect large geographical areas. In some cases, related species (bison, yak, banteng, etc.) may also be affected. Some specific geographical areas of the world have other diseases not listed here that can be very important considerations for cattle production.

9.3 Immune System and Function

A properly functioning immune system is crucial for animals to remain healthy. There are two broad classifications of immunity: (i) innate, or non-specific, and (ii) adaptive, acquired or specific (Fig. 9.1). When cattle are exposed to a potentially infectious disease-causing pathogen such as a virus or bacteria, or other organisms such as parasites, there is a multifactor inflammatory response in the area of the body that is affected, as well as generalized responses that allow the body to gear up to fight infection. With innate immune responses there are rapid responses, but there is no specialized response if the pathogen has been seen in the body before. With adaptive immunity, there are memory cells that produce antibodies so that the body can mount a rapid response if exposed to that pathogen in the future, but these types of responses are still slower than most innate responses. These components of the immune system are discussed below.

9.3.1 Immune System Components

The discussion for this section is based on several immunology references (Mosmann and Fowell, 2002; Yewdell and Bennink, 2002; Coico et al., 2003; Actor, 2007; Playfair and Bancroft, 2008). Areas of the discussion below are targeted toward bovine respiratory disease (BRD). There can be particular nuances that vary due to the type of infectious pathogen, but this discussion should serve as a general basis for vaccination and immune response.

Immune function can be classified as innate (nonspecific) and acquired or adaptive. Furthermore, adaptive immunity can be both humoral (production of immunoglobulin antibodies) mediated through B lymphocytes (B cells) and cell-mediated (CMI) responses through T lymphocytes (T cells). Although serum antibody production (specifically IgG) has been the primary immune measure for vaccine comparisons and represents 80% of total serum antibodies,

this alone does not explain antibody immune response to vaccination or pathogen exposure; IgA comprises 13% of serum antibodies.

There is growing evidence about and interest in the study of cell-mediated responses to pathogens. The bovine major histocompatibility complex (MHC) is involved in immune response as it is a region of the genome that codes for cell surface glycoproteins that play critical roles in interactions among immune system cells. There are T cells that will kill virus-infected cells (these are referred to as cytotoxic T cells or CD8 cells) and their action is dependent upon the MHC Class I genes. There are also T cells that assist B cells in antibody production (these are referred to as T helper cells or CD4 cells), and actions of these are dependent upon the MHC Class II genes. T helper cells are classified as Th1, which aid in antibody production, and Th2, which are involved in allergic and parasite reactions. In addition to cytotoxic and helper T cells, memory T cells also exist. Figure 9.2 summarizes white blood cell types, and Fig. 9.3 illustrates antibody production.

In addition to the above immune system components, cytokines are also involved. Cytokines are low molecular weight secreted proteins that are thought to regulate the intensity and duration of both innate and acquired immune responses and are secreted from a variety of cell types. It is known that at least six cytokines may be involved in various bovine immune functions pertaining to BRD: interleukin 1 (IL-1) is thought to be involved with innate response to bacterial infection and Th2 activity, interleukin 2 (IL-2) is involved in CMI lymphocyte proliferation, interleukin 4 (IL-4) is secreted by Th2 cells and involved with B cell differentiation, interleukin 6 (IL-6) is thought to be involved with innate response to bacteria and B cell differentiation and thus humoral response, interferon gamma (IFN-γ) is secreted by several types of T cells and involved in both innate and Th1 activity, and tumor necrosis factor alpha (TNF-α) is thought to be involved in activation of macrophages, neutrophils and NK

Physical and mechanical	Innate (native or non-specific) immunity	Adaptive (acquired or specific) immunity
Skin, beneficial microbes, mucous membranes, secretions,coughing, diarrhea	Chemical (cytokines) and physical (complement and white blood cell) aspects	Cell-mediated and humoral (antibody) aspects from B and T lymphocytes

Fig. 9.1. Types of defences against infection. Mechanisms can interact within and between these categories.

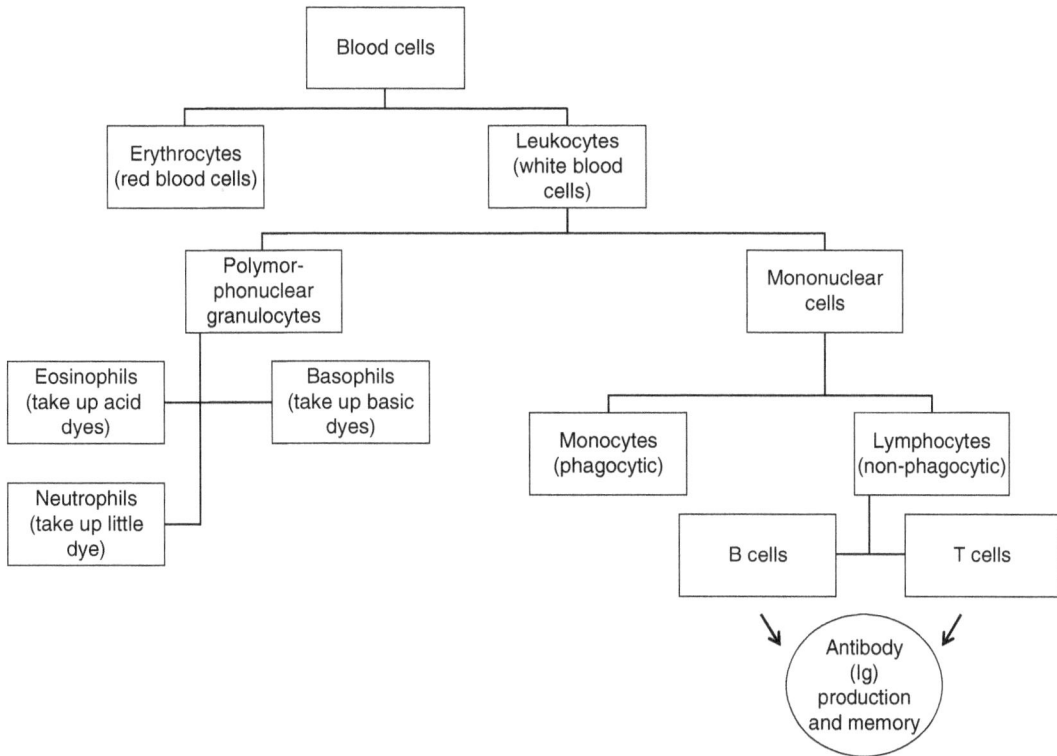

Fig. 9.2. Types of blood cells. Many of these types of cells interact through multiple and complex mechanisms.

T cells. Study of these multiple responses in regard to vaccination and pathogen challenge will provide for a more complete picture of immune response to disease. There is generally much more information needed in most cattle diseases about host–pathogen interactions and particular cascades of immune response.

9.3.2 Immune Response to BRD Vaccines and BVDV

The immune system is complex and multifaceted. However, most vaccines are structured to provide high serum-neutralizing antibody titers post-vaccination, and this has been the primary basis for evaluating vaccine immune response (Endsley *et al.*, 2003; Ridpath *et al.*, 2003). It is widely recognized that young cattle with maternally derived antibodies do not show increased antibody levels following vaccination until the maternally derived antibodies have subsided. This is viewed by many in the industry as the vaccination being ineffective. However,

there are a growing number of studies that have shown administering a modified live BRD vaccine or exposure to BVDV in young calves that have circulating colostrum antibodies will have increased protection even though they themselves show no antibody response (Ellis *et al.*, 1996; Cortese *et al.*, 1998; Endsley *et al.*, 2003; Ridpath *et al.*, 2003; Zimmerman *et al.*, 2006). Figure 9.4 shows why varying immune responses may occur.

9.3.3 Genetic Aspects of Livestock Health

Many components of animal immune function have been reported to have significant genetic influence. Antibody response in pigs has been shown to have heritability of 0.16 to 0.27 (Wilkie and Mallard, 1999). Heritability of antibody response following vaccination in dairy cows has also been reported to be moderate to high (Wagter *et al.*, 2000). Heritability estimates of neutrophil function in dairy cows have shown a wide range from 0 to 0.88 (Detilleux *et al.*,

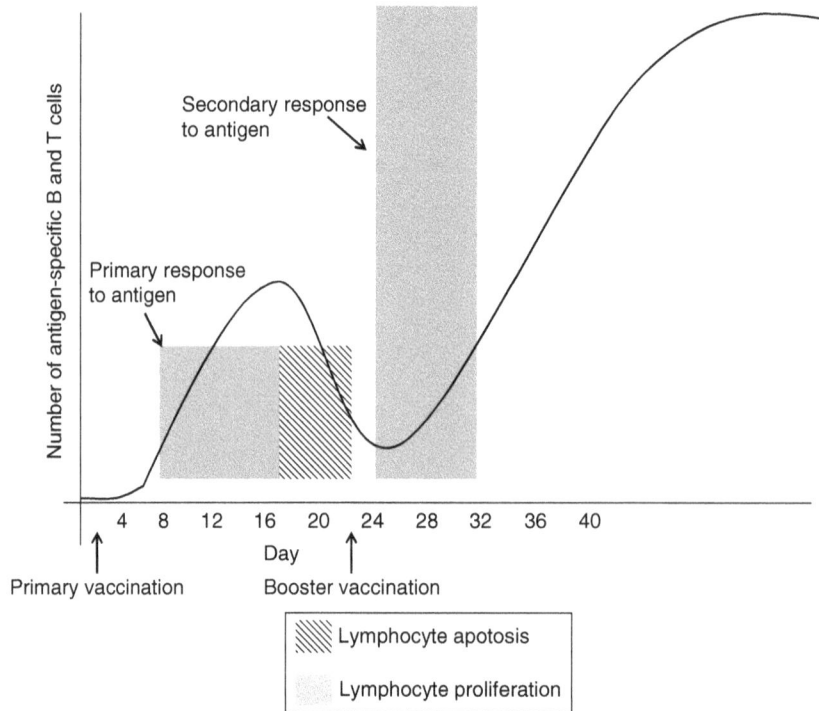

Fig. 9.3. Response to vaccination. The whole process from vaccination to achieving homeostasis takes at least 3 weeks for the development of a primary response, which can then be boosted to get a true anamnestic secondary response (information and graph taken from Chase *et al.*, 2008).

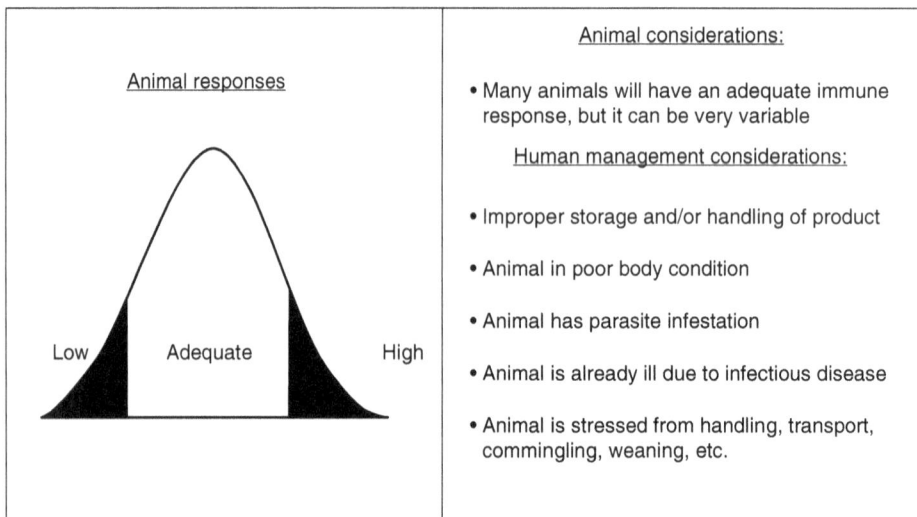

Fig. 9.4. Reasons vaccines may not work. It is assumed that an effective vaccine is available, and the reason that the vaccination does not work is not due to the vaccination itself. However, it is important to consult with local veterinarians and health authorities for reliable information on useful vaccine products.

1994). Wilkie and Mallard (1999) selected for high and low immune response in pigs (antibody and cell-mediated immunity) with respective values combined into estimated breeding values. They selected these lines for a particular pathogen; however, the high-response line produced better immune responses than the low line when exposed to a new pathogen, indicating the possibility to select for increased overall health.

Genetic aspects of BRD and many other cattle diseases have not been well studied. Heritability of BRD incidence in calves pre-weaning was reported to be 0.10 to 0.20 by Muggli-Cockett et al. (1992). Snowder et al. (2006) re-evaluated these data and estimated heritability to be 0.08 in the same data, but stated that characterization of accurate phenotypes is very important in regard to study of animal health. Imprecise phenotypes will result in reduced heritability estimates in any trait, even when substantial genetic variation exists. Larson (2005) pondered whether or not cattle health may have been compromised through intense selection for growth and stated that large datasets with parentage and health information were needed to study genetic impacts of health.

9.3.4 Identification of Sick Cattle and Subclinical Illness

Animals are thought to be ill when their behavior and/or other signs deviate from expected normal levels. Cattle in feedlots are pulled from pens when clinical symptoms such as runny eyes and noses, coughing, depressed demeanor or lack of alertness are noticed by feedlot personnel, and are typically treated with a BRD antimicrobial when exhibiting elevated body temperature (i.e. 40°C or greater) or severe clinical symptoms. However, several have reported a low correspondence between cattle that are pulled for BRD treatment based on clinical symptoms and cattle that have lung lesions at harvest (Wittum et al., 1996; Buhman et al., 2000; Groschke et al., 2006; Thompson et al., 2006). Cattle with respiratory disease are expected to have damaged lungs, with the extent relative to the level of illness severity. Basing overall health solely on visual symptoms is probably an inefficient assessment method, particularly with pathogens that can produce subclinical illness.

9.3.5 Effects of Stress on Immune Response

Many cow-calf producers vaccinate calves at weaning, and many stocker operators and feedlots vaccinate calves upon arrival. However these occasions, when animals are stressed, are probably poor times to vaccinate for BRD or other diseases. Additionally, there are a few reports that have linked temperament to health or immune response. Oliphant et al. (2006) found calm bull calves had a more favorable serum antibody response to clostridial vaccine than did temperamental contemporaries. Fell et al. (1999) also reported poor temperament to negatively impact cattle immune response. Studies in humans regarding temperament and immune function warrant attention in livestock industries. Feng et al. (1991) stated stress could not only affect antibody production, but could influence antibody isotype distributions, Burns et al. (2003) reported that stress may impact primary immune responses through B cell and T cell activity, and Burns et al. (2002) found stress to influence rate of antibody deterioration. More detailed animal stress effects in general are discussed in Chapter 10. Figure 9.5 shows antibody production from three vaccine groups when exposed to the same pathogen.

9.3.6 Colostrum and Passive Immunity

The first milk produced after calving is called colostrum. This milk is very high in protein and fat and contains a high concentration of immunoglobulins (antibodies) from the dam. It is critical for calves to receive colostrum in the first few hours after birth because there are porous openings in their small intestines that can absorb these maternal antibodies. However, pathogens can also be absorbed through these openings, so they begin to close 6 hours after birth, and are completely closed within 24 hours. If maternal antibodies are not absorbed during this time, the calf will not receive protective immunity from the dam as antibodies are not passed across the placental barrier in cattle. Any stressor that inhibits the calf's ability to receive colostrum shortly after its birth can greatly compromise its health and survival chances. Stress to calves and dams from dystocia, poor health or body condition of the dam, physical hindrances of the dam such as threats from predatory animals, etc. are potential threats to adequate colostrum delivery. As discussed in Chapter 7, the degree to which adequate colostrum supply and absorption affect early life programming is generally unknown and merits more research.

The next section discusses the broad categories of diseases and illnesses that can affect productivity and profitability of beef cow operations. The management associated with the different categories of

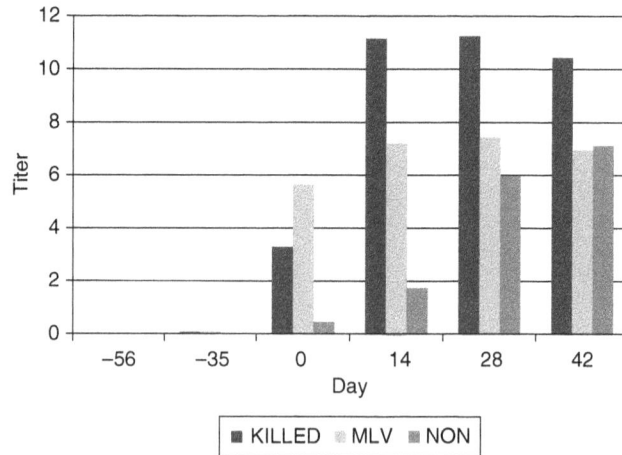

Fig. 9.5. Antibody titer (log base 2) response to BVD type 1b virus exposure across three vaccine treatments in *Bos indicus–Bos taurus* crossbred yearling steers; Primary killed vaccine on day − 56; Killed booster and single MLV on day − 35; BVD virus challenge administered on Day 0. Data from Downey *et al.* (2012).

diseases is vastly different in many cases. It is well beyond the scope of this chapter to attempt to discuss most of the potential diseases that could affect beef cattle production, and in many cases they are diseases specific to localized individual regions.

9.4 Categories of Production-Related Diseases

There are many ways to classify and categorize diseases (by area of the body affected, by type of causative organism, etc.). Here, diseases are categorized according to how cattle managers typically think about management of operations (genetics, reproduction, nutrition, etc.), with a very broad, introductory coverage provided. The goal of this discussion is for the reader to understand the different categories of disease in order to make associated management decisions, and to discuss the most widely recognized health threats to cattle that can have global implications.

9.4.1 Genetic and Inherited Diseases

There are many diseases in cattle that are inherited as genetic recessive traits, which means that in order to have one of these diseases, the animals must have two copies of the allele that causes the disease (one being inherited from each parent). There are probably

many genetic diseases in cattle that go unrecognized because the frequency of the undesirable allele is very low in the population, and very few animals are carriers (heterozygotes) of the defective allele. In turn, when there are very few carriers the chance that these carriers mate is also very low, and when they do mate the chance of producing a recessive offspring is only 25% (if the reason for this is not clear, refer to the discussion on inheritance in Chapter 3). All mammalian species have numerous undesirable genetic recessive conditions, and many of these can be lethal (i.e. offspring do not survive). Genetic diseases may affect many body functions, from absence of hair to deformed skeletal system, to a single non-functioning enzyme. As a result, these have not been grouped into production-oriented categories as has been done with reproductive, digestive, etc. classifications, but there are genetic diseases that can fall within each of those categories. What makes a disease genetic is that the affliction is based upon the transmission of undesirable alleles from parent to offspring. Some diseases can be transmitted from parent to offspring by the passing of a pathogen from parent to offspring (such as with BVD or some venereal diseases), but these are not genetic diseases because they are not based on inheriting a faulty allele.

Typical management considerations with combatting genetic recessive diseases have been to identify heterozygous animals and prevent them from breeding (the ultimate goal is to keep the frequency of the recessive allele from artificially increasing, and to reduce its frequency as much as possible). Genetic tests for many genetic recessive diseases have been developed based on DNA evaluation, and these tests continue to be developed and have their associated costs decline as molecular biology technologies improve.

9.4.2 Reproductive Diseases

As reduced fertility is the single biggest obstacle to profitability for cow-calf producers in most regions of the world, minimizing exposure of cattle to infectious pathogens known to cause direct reproductive problems such as abortions is of paramount importance. There are probably several diseases that may not be obvious reproductive problems because their effects are early-gestation embryo loss or abortions that go unrecognized, and their symptoms are reduced fertility through increased calving intervals. These types of diseases are hard to recognize in herds where bulls are left with cows for long periods or year-round because increased time between calves can be caused by so many factors. Table 9.5 provides a list of reproductive diseases that should be considered by most cattle producers.

9.4.3 Metabolic and Digestive Diseases

Many metabolic and digestive health problems can be avoided with proper feeding and pasture management. There are many plants that can be toxic to cattle, but animals will usually avoid these plants if they have access to other palatable forages; many of these toxic plants may also be ingested in small quantities without detrimental results if a wide variety of non-toxic plants are abundant. Moreover, there are environmental conditions that may produce temporary toxicity issues in several forages that cattle commonly consume. Many toxic plants may not be problematic unless they are consumed in substantial quantities. Leaving cattle in pens where there are high concentrations of weeds could lead to toxicity issues (Fig. 9.6). There are certain types of forages that are known risks under specific environmental conditions. Some plants have toxic compounds that can lead to abortion. It is also possible that certain toxic compounds can disrupt the fetal development and produce deformed offspring (which when first observed may lead one to

believe that a genetic disease is the cause). Most of these conditions are highly specific to localized regions, and risk of potential toxic and poisonous plants should be thoroughly investigated with local authorities before bringing new animals or new operations to any region. Toxic compounds that may be problematic in grazing cattle are discussed in more detail in Section 5.2.2. Table 9.6 lists some general classes of compounds that can be poisonous to cattle. Table 9.7 lists some diseases that can be caused by molds and fungi possibly found on feedstuffs that cattle producers need to be familiar with.

As discussed in Chapter 4, the microbial population in the rumen is influenced by the type of diet consumed. Rapid changes in diet, particularly in transitioning from a forage-based diet to a grain-based diet, must be done over several days. Consuming large quantities of grain causes a decrease in rumen pH, and if this is extreme enough it can cause acidosis. The best way to prevent acidosis in transitioning cattle to a grain-based diet such as maize is to gradually start feeding grain at 2–5 kg per animal per day for a minimum of 5–7 days, and then begin to incrementally increase the grain amount and feed that level for another 5–7 days, etc. Often cattle with acidosis will have grayish colored feces. If cattle become deprived of feed, they can overeat when feed is reintroduced and this can cause a problematic cycle where microbial balance is not maintained and lowered ruminal pH occurs repeatedly, greatly increasing the risk of sub-acute ruminal acidosis (Merck, 2012). With subclinical acidosis, lameness and overgrowth of the hooves (which can be referred to as founder) can occur, but the exact relationship between acidosis and foot problems is not known.

Bloat is another digestive problem that can be caused by changes in diet. Bloat results when there is a physical impediment to animals expelling gas from the rumen out of the mouth. As the rumen is situated on the left side of cattle, animals that become bloated will have a distended left side that extends further away from the midline of the body and upwards as it increases in severity. Bloat can be caused by animals consuming too much grain at one time, feeding of moldy hay or grain, grazing certain legume pastures, or grazing small grain pastures (see Fig. 9.7). Bloat in cattle grazing lush growing forage may be related to specific climatic conditions. There are several management considerations that can help reduce the incidence of bloat. These include avoiding dramatic swings in nutritional values of

Table 9.5. Diseases that can impact cattle reproduction.

Disease name	Causative organism	Transmission	Body areas, functions affected	Main clinical signs	Preventative measures
Campylobacteriosis (formerly referred to as vibriosis)	Bacteria (*Campylobacter fetus* subsp.)	Primarily through mating (natural or AI)	Reproductive tract infection	Infertility, early embryonic loss, abortion	Vaccination
Bovine viral diarrhea (BVD, or pestivirus)	Virus (BVD virus); possibly hundreds of different strains	Direct contact between animals; will be passed to progeny from persistently infected cows	Virus found throughout body	Highly variable from subclinical to diarrhea, high fever, nasal discharge, abortion, stillbirth	Vaccination, screening of breeding animals for PI (ear notch IHC test); respiratory disease in calves
Infectious bovine rhinotracheitis (IBR, red nose disease)	Virus (bovine herpesvirus-1)	Direct animal contact	Replicates in mucous membranes of upper respiratory tract and tonsils, then conjunctive membrane and ganglia; animals remain infected for life	Variable from subclinical to nasal discharge, redness of the nose, fever, depression, loss of appetite, abortion	Vaccination, animal movement; may cause respiratory disease in calves
Trichomoniasis	Protozoa (*Tritrichomonas fetus*)	Spread by sexual contact	Bulls (particularly older bulls) are primary long-term carriers; organism lives in prepuce cavities; most cows clear infection	Infected bulls show no signs; infected cows may have inflamed vulva, abortion, infertility, uterine infection	Vaccination?, use of AI, outside use of young/virgin bulls, thorough screening of bulls
Leptospirosis	Bacteria (*Leptospira* spp.)	Direct contact among animals	Kidneys, udder, reproductive tract	Abortion, stillbirth, premature birth, weak offspring	Vaccination, clean water sources, limit exposure to feral pigs
Neosporosis	Protozoa (*Neospora caninum*)	Ingestion of oocytes in feces of infected canines, passed from infected dams to offspring	? Found in several organs of infected individuals	Abortion, calves born underweight, weak or paralysed, or may become paralysed within 4 weeks of birth	Limit exposure to canine feces, particularly stored feed and water
Brucellosis	Bacteria (*Brucella abortus*)	Ingestion of bacteria from infected animals, aborted fetus, fluids or milk	Udder and uterine infection, also repro tract of bulls	Abortion, long rebreeding time	Vaccinate females only, between 6 and 12 months old by vet

Fig. 9.6. Weeds in a temporary pen for holding cattle. Pens that are occasionally used to process or house cattle may contain plants that are not consumed by cattle on pasture when desirable plants are plentiful. Unplanned housing of animals (especially calves) in pens such as these can pose health threats if toxic plants are present, especially if the animals are held there for several hours.

the diet, providing animals with a choice of feed-stuffs to consume and certain feed additives, depending upon geographical availability. If cattle become severely bloated it can cause intense pressure on their internal organs, including the heart and lungs, and can cause death. Cattle that are bloated have to be relieved by extending a flexible tube down the esophagus into the rumen to expel the gas, and if the animal's life is in immediate danger, the rumen may need to be punctured with a large-gauge needle (or sharp knife as a last resort) high on the abdominal wall. Animals that are not having difficulty breathing can be fed mineral oil over hay, and this will reduce the gas pressure and promote movement through the GI tract.

Rapid transition from a very low protein diet (as when grazing dormant forage or old, low-quality hay) to a very high-protein diet (as with a lush, growing pasture) should also be avoided. This is likely to be a problem when cattle are moved from a very low-quality pasture or diet and turned into a new, high-quality pasture that is very high in protein. Growing small grains and legumes can be 20% crude protein or more. Cattle that are not transitioned to become accustomed to the high-protein diet can suffer from ammonia toxicity (by-product of protein metabolism); if cattle are provided a high-protein

Table 9.6. Classes of substances and compounds that can be toxic or poisonous to cattle.

Class of substance	Example(s)
Inorganic elements	
Metals	Beryllium, cadmium, lead, mercury, silver, thallium, tin
Metalloids	Arsenic
Minerals	Copper, chromium, cobalt, iron, manganese, selenium, zinc
Non-metals	Fluorine, chlorine
Organic	
Phosphorus containing	Organophosphates, aluminium phosphide
Nitrogen containing	Cyanide
Carbohydrate containing	Methanol, ethanol, ethylene glycol
Pharmaceutical	Numerous compounds used for medicinal purposes
Biological	
Animal origin	Numerous compounds produced from reptile, amphibian and fish species; stinging insects, scorpions, spiders
Microbial origin	Numerous from many species of bacteria and protozoa
Plant and fungi origin	Numerous compounds in many species

Table 9.7. Mycotoxicoses (diseases from fungi or molds) that affect cattle.

Disease name	Toxins	Regions reported	Contaminated toxic feedstuff(s)	Signs and lesions
Aflatoxicosis	Aflatoxins	Widespread (warmer climatic zones)	Moldy peanuts, soybeans, cottonseed, rice, sorghum, maize (corn), other cereals	Major effects in all species are slow growth and hepatotoxicosis, unthriftiness, weakness, anorexia, and sudden deaths can occur. May be seen with BRD.
Diplodiosis	Unknown	South Africa	Moldy maize (corn)	Nervous system disorders, cold and insensitive limbs. Recovery usual on removal of source.
Ergotism	Ergot alkaloids	Widespread	Seed heads of many grasses, grains	Lameness, peripheral gangrene (limbs, tail, ears), late gestation, suppression of lactation initiation.
Paspalum staggers	Paspalinine and paspalitrems, tremorgens	Widespread	Seed heads of paspalum grasses	Acute tremors and ataxia, trembling of the large muscle groups, movements are jerky and uncoordinated
Estrogenism	Zearalenone	Widespread	Moldy maize (corn) and pelleted cereal feeds, standing corn, corn silage, other grains	Weight loss, vaginal discharge, nymphomania, uterine hypertrophy, early abortion in heifers, multiple returns to service, testicular atrophy in males.
Facial eczema (pithomycotoxicosis)	Sporidesmins	Widespread	Toxic spores on pasture litter	Photosensitization and jaundice appear about 10–14 days after toxin intake, animals frantically seek shade, exposure to sun rapidly produces erythema and edema in non-pigmented skin.
Fescue foot	Ergovaline	USA, Australia, New Zealand, Italy	Tall fescue grass (*Festuca arundinacea*)	Lameness, weight loss, hyperthermia, dry gangrene of extremities, agalactia, thickened fetal membranes.
Mycotoxic lupinosis (as distinct from alkaloid poisoning)	Phomopsins	Australia, South Africa, New Zealand, Europe	Moldy seed, pods, stubble and haulm of several *Lupinus* spp. affected by Phomopsis stem blight	Lassitude, reduced appetite, stupor, icterus, marked liver injury. Usually fatal.
Perennial ryegrass staggers	Lolitrems	Australia, New Zealand, Europe, USA	Endophyte-infected ryegrass pastures	Tremors, incoordination, collapse, convulsive spasms.
Pulmonary edema, emphysema	4-Ipomeanol	USA	Moldy sweet potatoes	Pulmonary edema, leading to interstitial pneumonia and emphysema.
Slobbers	Slaframine (and swainsonine)	USA	Blackpatch-diseased legumes (notably red clover) eaten as forage or hay	Salivation, bloat, diarrhea, sometimes death. Recovery usual when removed from clover.

Sweet clover poisoning	Dicumarol	North America	Sweet clover (*Melilotus* spp.)	Stiffness and lameness, due to bleeding into the muscles and joints, hematomas, epistaxis or GI bleeding may be seen, death may occur suddenly.
Tremorgen ataxia syndrome	Penitrems, verruculogen, paxilline, fumitremorgens, aflatrems, roquefortine	USA, South Africa, probably worldwide	Moldy feed	Tremors, polypnea, ataxia, collapse, convulsive spasms.
Trichothecene toxicosis Fusariotoxicosis	Non-macrocyclic trichothecenes (deoxynivalenol, T-2 toxin, diacetoxyscirpenol, many other trichothecenes)	Widespread (except for deoxynivalenol, more likely in temperate to colder climates)	Cereal crops, moldy roughage	Loss of appetite and milk production, diarrhea, staggers, skin irritation, immunosuppression; recovery on removal of contaminated feed.
Stachybotryotoxicosis	Macrocyclic trichothecenes (satratoxin, roridin, verrucarin)	Former USSR, south-east Europe	Moldy roughage, other contaminated feed	Stomatitis and ulceration, anorexia, leukopenia, extensive hemorrhages in many organs, inflammation and necrosis in the gut, immunosuppression.
Myrotheciotoxicosis, dendrodochiotoxicosis	Macrocyclic trichothecenes (verrucarins, roridins, etc.)	South-east Europe, former USSR	Moldy rye stubble, straw	Acute – diarrhea, respiratory distress, hemorrhagic gastroenteritis, immunosuppression, death. Chronic – ulceration of GI tract, unthriftiness, gradual recovery.
	Macrocyclic trichothecenes (baccharinoids)	Brazil	Plants of *Baccharis* spp. that contain the toxins	Epithelial necrosis of GI tract.

Information taken from the Merck Veterinary Manual Online (20⁻2).

Fig. 9.7. Steer that is bloated. Accumulation of rumen fermentation gases that cannot be expelled through eructation can lead to death if the condition becomes severe.

supplement for 3–7 days before turning into the new pasture this will greatly reduce the potential threat.

9.4.4 Diarrhea or Scours and Newborn Calf Health

There are numerous pathogens that can cause diarrhea or scours in cattle, especially young animals. Diarrhea is not a disease, but is a condition that, if it becomes severe enough and persists long enough, can threaten the life of the animal and is associated with management and health of newborn calves. Bacteria such as *E. coli* and *Salmonella* are known to cause diarrhea in very young calves, but so are several other organisms. The most severe threat from diarrhea itself is dehydration, which affects many processes and functions of the body. Naylor (2009) provided guidelines for fluid loss considerations associated with diarrhea in young calves and stated that 50 kg calves with severe diarrhea and 10% dehydration required 11.5 liters of fluid per

24 hours. Table 9.8 provides some guidelines for calves that can be followed to reduce the risk of infectious diarrhea in young, intensively housed calves.

9.4.5 Clostridial Diseases

There are several diseases around the world caused by *Clostridum* spp. bacteria. These seem to affect growing cattle more than mature animals, and are therefore important to consider in nursing calves to calves post-weaning. Many of these bacteria can be region-specific, but many are common across wide regions of the world. *Clostridum* spp. bacteria can affect multiple areas of the body. In many regions vaccines are available for the major threat species. In many instances, severe animal illness or death occur rapidly with these diseases. The vaccines for clostridial diseases are highly effective (Merck, 2012). The number of clostridial antigens in the vaccine is denoted by its description (such as 4-way, 7-way, 8-way, etc.), and recommended species to vaccinate against depend on local prevalence. Some more common clostridial diseases in North America and several other regions, along with the causative bacteria, are: blackleg (*Clostridium chauvoei*), malignant edema (*C. septicum*), bacillary hemoglobinuria (*C. novyi*, type D [hemolyticum]), infectious hepatitis (*C. novyi*, type B), tetanus (*C. tetani*) and enterotoxemia (*C. perfringens*, types B, C and D).

9.4.6 Respiratory Diseases

Bovine respiratory disease complex (BRD) affects cattle on pasture and in feedlots in many areas of the world. Although these can affect calves of all ages, in many instances it is cattle shortly after weaning that are most at risk from these diseases. In this disease complex there are four viral diseases (bovine viral diarrhea or BVD; infectious bovine rhinotracheitis, referred to as IBR; bovine respiratory syncytial virus, referred to as BRSV; and parainfluenza, PI3) that contribute to BRD, and may lead to pneumonia if these diseases are severe enough, particularly if secondary bacterial infection is involved. There are several commercially available vaccines (as killed or modified live products) to guard against the viral diseases, and many are available that also protect against pneumonia-causing bacteria. The term 'shipping fever' is an old term that was used in North America for many years to describe the sickness that many calves got after being transported

Table 9.8. Factors associated with risk of diarrhea or Cryptosporidial infection in hand-reared calves.

Factor	Increased risk	Decreased risk
Calving area	Born in loose housing	
Many cows in maternity pen	Yes	
Floor of maternity pen	Soil	Concrete
Daily removal of soiled bedding	No	Yes
Calving ease	Assistance required	No assistance required
Cleanliness of cows	Dirty	Clean
Feeding		
Method	Through nipple	From bucket
Frequency per day	One	Two
Concentrate	None	Yes
Building		
Overall condition	Damp	Dry
Bedding	Damp	Dry
Ammonia smell	Present	Absent
Calf restraint method	Tied with collar	Loose (not restrained)
Quarantine facilities for sick animals	Yes	No
Density (space per calf)	< 1.6 m^2 for tie stall	> 1.6 m^2 for tie stall
	< 1.0 m^2 for free stall	> 1.0 m^2 for free stall
Calf pens against wall	No	Yes
Vaccination against enteric pathogens	No	Vaccination against enteric disease caused by *E. coli*
Calf rearer	Adult male	Women and children
Herd size	Large	Small

Taken from Naylor (2009).

(i.e. shipped) to a feedlot, and this disease was BRD. Several management practices have been developed to better prepare calves for weaning and subsequent transportation to a feedlot or another operation that includes vaccination against BRD pathogens and reduction of stress.

Although many producers may think of BRD as a feedlot disease, two of the BRD-contributing viral diseases (BVD and IBR) have reproductive consequences as well and should be considered in cow herd-level management. Both of the diseases can cause abortions, and the source of BVD problems at a feedlot or a ranch may be due to the presence of persistently infected (PI) animals (this should not be confused with the abbreviation for parainfluenza, PI3). Cows that are pregnant and become infected with the BVD virus from early to mid-gestation have a high chance of producing a PI calf (which may be normal in appearance and show no signs of illness). These PI animals will shed BVD virus their entire lives and be a continual threat and source of infection for other cattle. If a heifer is PI, every calf she will produce will also be PI. Only calves that were exposed to the BVD virus during gestation can become PI.

9.5 Parasites

Parasitic infections are a cause of lost productivity in livestock worldwide (Ballweber, 2006) and very often are not obvious, but in many cases application of parasitic control has improved productivity and/or efficiency of production. As a result (and as is probably the case with multiple other pathogens), subclinical illness may be more of a threat than outright diseased conditions in many situations. The species of parasites vary widely with geographic region, and discussion of specific parasites problematic to a particular region should be investigated before management decisions are made for operations in that region. Parasites can be broadly classified as internal (parasite adult stage occurs inside the body) vs external (parasite adult stage is outside of the body), and some of the most prominent parasites that can affect cattle in large geographical areas or globally are discussed here. More in-depth descriptions are found in many veterinary and animal health references, and as with all animal health issues, local authorities should always be consulted for specific regional concerns and recommendations.

9.5.1 Internal Parasites

Many internal parasites can severely limit production and endanger cattle health in many areas of the world. Most of these types of parasites are known as helminths and are generally referred to as stomach worms. Livestock producers and veterinarians may rely heavily on anthelmintic drugs to reduce infections and disease when in fact they overlook essential sanitation and nutritional considerations (Craig, 2006). Treatment with anthelmintics has been proven to be cost-effective when used properly; however, used improperly they may interfere with the host animal's ability to mount an effective immune response (Craig, 2006). A low degree of parasite occurrence in animals may keep their immune system active and guard against that species of parasite when environmental conditions propagate the number of parasites.

Several species of worms can be parasites of cattle. These parasites have particular areas in the body where they live and reproduce. On a broad scale these can be classified as nematodes (roundworms that affect the abomasum or intestine), cestodes (tapeworms in small intestine) and trematodes (flukes). The nematodes have direct life cycles and live solely in cattle (they do not require other species as intermediate hosts), whereas tapeworms and flukes have indirect life cycles involving other animal species. In regard to nematodes, the eggs are shed from adults in the body and the eggs are passed in the feces. After eggs develop into larvae, the larvae are ingested on vegetation to infect new animals (or re-infect animals). Figure 9.8 depicts the general life cycle of nematode parasites of cattle, but there may be exceptions with particular species. Liver flukes require snails as intermediate hosts. The larvae from snails climb on vegetation and become ingested by grazing livestock. Control of the snails may be the most important control mechanism, which may be done through artificial control or through pasture use (or avoidance) at certain seasons or for certain lengths of time. Table 9.9 provides a general discussion and comparison among gastrointestinal helminth parasites of cattle.

There are also other types of internal parasites that may be of concern to cattle producers. Nematodes are not solely found in the gastrointestinal tract and may infect other parts of the body, as happens with lungworms (*Dictyocaulus viviparus*). There are similarities in the life cycle of lungworms with nematodes of the intestines, in that eggs are passed

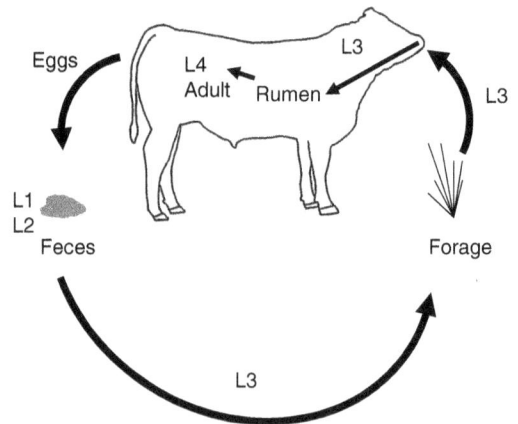

Fig. 9.8. General life cycle of gastrointestinal nematodes of cattle. Parasites exist in egg, larval and adult stages. L1, L2, L3 and L4 represent consecutive larval stages. L3 is when most genera become infective and move from feces to vegetation. Descriptions adapted from Craig (2009).

in the feces, infective third-stage larvae are ingested, become active in the abomasum, penetrate the intestine and then migrate to mesenteric lymph nodes near the intestines; fourth-stage larvae reach the lungs 7–14 days after ingestion (Panuska, 2006). There are also protozoan parasites that can infect cattle. These may affect the GI tract (as with coccidiosis) or other areas of the body such as with trichomoniasis. Table 9.9 summarizes GI protozoan parasites that may be of concern in cattle. The most common concern of these diseases is coccidiosis, which is commonly thought of when bloody diarrhea is seen in younger cattle. Other protozoan diseases are more of a concern for humans than cattle because even though cattle may harbor them, they do not produce obvious symptoms. As opposed to larvae of helminths, oocysts of protozoa do not require vegetation and can therefore be acquired in both drylot and pasture settings (Craig, 2009). Table 9.10 summarizes characteristics of some helminth parasites in cattle.

It appears that most internal parasites are primarily threats to young/naïve animals, and that animals will develop tolerance and/or active immunity to these parasites following initial infection. This probably happens much more than most producers realize, because the presence of these parasites is not obvious unless an animal becomes severely ill from infestation and develops pronounced clinical signs. But even then,

Table 9.9. Protozoan parasitic diseases of the gastrointestinal tract.

Parasite/disease	Organism	General	Clinical symptoms
Giardiasis	*Giardia* spp.	Disease in humans; not known to cause disease in cattle although it may be present in most or all animals in an area	None
Coccidiosis	*Eimeria* spp.	Infection may be almost universal, but naïve and stressed animals are susceptible to disease; coccidiostats can be used, but continual use may lead to resistance; feeds with ionophores help reduce incidence	Diarrhea (may or may not have blood in it); bloody diarrhea has been thought of as classical symptom; may have excessive mucus in feces
Cryptosporidiosis	*Cryptosporidium* spp.	Disease in humans; *Cryptosporidium parvum* infections cause diarrhea in neonatal calves; *Cryptosporidium andersoni* may cause infection of abomasum in older cattle	Diarrhea in young calves (*C. parvum*); none for *C. andersoni*

From Craig (2009a).

it is not obvious what the exact cause of the illness is. On the other hand, infection with external parasites is obvious (and concerning) to producers. Continual use of artificial compounds to treat parasites is probably not helping cow herds cope with parasitic loads in a sustainable way, and is likely propagating resistant strains or even entire species of these parasites. Misuse of these products is worse than their lack of use. Effective and sustainable parasite control programs must utilize information about the life cycle of the organism in conjunction with pasture management and selection of cattle that are suitable for the environment. Rotational grazing in combination with use of adapted cattle can substantially reduce the threat of parasitic load and/or economic threat.

9.5.2 External Parasites

The presence of external parasites on cattle becomes obvious to the manager as animals lose hair, develop lesions, or by the sheer number of parasites seen. This is not the case with internal parasites. In areas of the world that have screw worm flies (Old World screw worm is *Chrysomyia bezziana*; New World screw worm is *Cochliomyia hominivorax*), tsetse flies and cattle fever ticks (*Boophilus* spp.), these parasites can cause major economic losses. Many external parasites of cattle such as the many fly species, mites and lice are more economic nuisances than serious health threats to animals. Nonetheless, they can cause economic losses directly due to reduced

productivity, and may be vectors of other parasites or bacterial pathogens such as *Salmonella* and *E. coli*, certain viruses, and some potentially devastating diseases carried by ticks. Additionally, threat of some external parasites such as biting flies can provide opportunities for additional parasites or pathogens to invade the body. The desire to remove external parasites through use of chemical compounds can lead to misuse of these products, which can contribute to increased resistance of the insects. Table 9.11 summarizes some of the primary cattle external parasites.

9.6 General Health Management and Biosecurity Considerations for Herds

The concept of biosecurity is very broad and includes the overall health-related management practices and their potential interactions, so by definition it is a systems-based approach for improved health and/or reduced threat of illness. Different types of cattle operations (cow-calf, yearling grazier, heifer developer, feedlot, etc.) will have very different biosecurity concerns and management. Many people may confuse the procedures of biosecurity with the disease threats, and although related these are different considerations. For instance specific diseases may be unique to geographical areas, but if they are caused by the same type of pathogen, the effective management strategies of the herds are likely to be similar. The broad categories of biosecurity include vaccination, use of instruments and

Table 9.10. Helminth parasites of the gastrointestinal tract.

Name	Type of organism	General	Area of affliction
Ostertagia	Nematode	Abomasal damage caused by maturing worms appearing from abomasal glands, and there are two types of disease (Type I and II); can be seen 10–14 days after ingestions of infective larvae (Type I) or months after pasture exposure (Type II); typically associated with overstocking pasture; lack of weight gain and dark green diarrhea (Type I) or emaciation and brown to green diarrhea (Type II) may be seen	Abomasum
Trichostrongylus	Nematode; stomach hairworm	Often associated with *Osteragia* presence; may not be problematic individually	Abomasum
Haemonchus	Nematode	Some species restricted to tropical environments, others seen in temperate and tropical settings; warm season parasite that undergoes hypobiosis during winter or prolonged drought; blood sucking and attached to stomach lining; increased resistance has been reported	Abomasum
Cooperia	Nematode	Often found in association with abomasal helminths; not as susceptible to many compounds as some other types of heminths; can produce decreased weight gains, diarrhea and emaciation	Small intestine
Nematodirus	Nematode; wireworms	Found primarily in dry or frigid climates; may not be as susceptible as other genera to certain drugs; threat to cattle is probably not high	Small intestine
Bunostomum	Nematode; hookworms	Cattle can pick up larvae through ingestion or penetration through skin; typically more associated with moist tropics; can produce anemia and black, thick feces in calves; sanitation and dry bedding is probably best control	Small intestine
Strongyloides	Nematode	First helminth usually seen in calves; usually does not cause disease, but may cause loose feces; sanitation is key to control	Small intestine
Toxocara	Nematode	Trans-mammary acquisition is the only proven method of transmission; disease is rare; calves usually undergo cure by 3 to 5 months of age	Small intestine
Oesophagostomum	Nematode	Larvae found in gut mucosa from large intestine to anus; transmission can be from ingestion or skin penetration of infective larvae; larvae in nodules can exist for up to 12 months, and are not exposed to anthelmintics in that state	Small intestine, large intestine
Trichuris	Nematode; whipworms	Usually not pathogenic in cattle, even with large numbers	Small intestine

Continued

Table 9.10. Continued.

Name	Type of organism	General	Area of affliction
Tapeworms	Cestode	Can grow up to 2 m in length, but presence has not been proven to be economically important in cattle; has indirect life cycle and requires mites or psocids as intermediate hosts	Small intestine
Rumen flukes	Trematode; many species	Has indirect life cycle with snails as intermediate host; infective cysts are ingested; immature flukes burrow into intestinal wall and migrate to rumen/reticulum where they remain as adults; disease is not common but diarrhea and emaciation can occur during immature infestation	Rumen and reticulum
Liver flukes	Trematode; *Fasciola hepatica, Fasciola gigantica, Fascioloides magna*	Liver flukes are only transmitted on specific pasture lands and are of no significance elsewhere unless infected cattle are moved; deer and wapiti are primary host, and cattle sharing pastures with these animals are most at risk; larval flukes enter liver and may remain there or in bile ducts; anemia, hypoalbuminemia and edema under the jaw (bottle jaw) may be symptoms; eggs are never seen in feces with *Fascioloides* or during acute and subacute infections of *Fasciola*; general reduced performance, weight gain and liver condemnation at harvest are likely	Liver

Taken from Craig (2009b).

tools, commingling of animals, handling of animals, and pasture management. These concepts are briefly discussed individually.

9.6.1 Intermingling or Commingling of Animals

Facilities for isolation and treatment of sick animals and for isolation of newly introduced animals are both important biosecurity considerations. The ability to separate sick animals can reduce the threat of infection spreading to other animals. Keeping naïve or high-risk cattle such as newborn calves away from infected animals, as well as the facilities that recently housed them, is important. New animals that are introduced to the herd can harbor pathogens that threaten the existing herd, and new animals can also be at risk from pathogens harbored by the existing herd. As a result, it is often recommended to isolate new animals for 14–21 days before intermingling with the existing group(s) of cattle.

Figure 9.9 provides some examples of animal housing and associated biosecurity considerations. There is a very different mindset needed by producers if groups of animals are to be put with other groups either at the ranch level, or as a component of marketing as compared to individual contemporary groups of cattle staying together throughout the supply and production chain.

9.6.2 Recommended Biosecurity Measures for Cow Herds

It is very important for producers to have a well thought out and designed herd health management plan. Components of this plan should include considerations for (i) vaccinations for infectious diseases, (ii) parasites and, (iii) animal movement and exposure. It is critical for producers to be informed of health risks for their particular geographical area(s), and they should always consult with local veterinarians or other animal health authorities about recommended

Table 9.11. Potential external parasites of cattle.

Type of parasite	General description	Health concerns of cattle
Mosquitoes, black flies, biting midges and sand flies	Suborder Nametocera; adults are small and delicate with long, segmented antennae; considered micropredators due to feeding on blood of hosts in small amounts, and are not generally host species specific.	Mosquitoes can transmit Rift Valley fever in Africa; some species of sand flies can transmit vesicular stomatitis virus in North America; general nuisance if numbers are large and may reduce weight gain; this group is not severe threat in most cases.
Horse flies, deer flies, snipe flies	Large and painful biting flies with strong, cutting mouth parts to take blood meals; prefer heavily shaded and moist areas; pasture management best control tool.	Painful bites that annoy animals and disrupt normal activity; bites may continue to bleed after these flies leave, attracting other potential parasites.
Muscid flies – house flies, stable flies, horn flies, face flies, buffalo flies	Have similar body shape but can vary in size; house flies do not take blood meals but feed on almost any available organic matter; stable flies take blood meals and prefer legs and underside and deposit eggs in decaying vegetative matter; horn flies prefer necks and backs to feed and deposit eggs in feces; face flies prefer moist areas (around eyes, nostrils, mouth).	House flies can carry *Salmonella*, *E. coli*; stable flies and horn flies are viewed as general nuisance; face flies can spread bacteria (*Moraxella bovis*) involved with pinkeye; severe fly bites as with buffalo fly may provide wounds for screw worm infestation.
Bottle flies, blow flies, screw worm flies	Bottle flies and blow flies are attracted to decaying animal tissue as well as infected or necrotic tissues of living animals; screw worm flies are attracted to fresh, open cuts and wounds; eggs are deposited in tissue and maggots feed on surrounding area; dehorning and castration considerations and timely treatment of wounds is important.	Infestation of these parasites provide animal welfare concerns, can greatly reduce performance, and many lead to death if untreated or heavy infestation present; eradication of screw worm fly from much of North America can been accomplished.
Bot flies	Biting flies where the larvae migrate through part of animal's body before completing life cycle; cattle grubs (larvae of heel fly, *Hypoderma lineatum*) is most common bot fly threat in many areas of world.	Animal welfare considerations, general depressed performance, potential for other parasites from exit wounds, condemnation of meat from infested animals.
Lice	Two basic types of lice (biting and sucking); sucking lice take blood meals (referred to as anoplurans); biting lice (mallophagans) feed on skin; all lice are wingless; species may be localized to specific areas of body; entire life cycle is spent on host.	Loss of blood to point of anemia with anoplurans; itching, irritation and loss of hair; general depressed performance.
Mites	Burrow into skin or hair follicle; individual scabies mites may produce tunnels up to 1 cm long; some species specific to certain body areas; entire life cycle usually on host.	Intense itching, loss of hair, depressed performance, bleeding or pustules may attract other insects.
Ticks	Two primary classifications of soft ticks (Argasidae) and hard ticks (Ixodidae); soft ticks feed multiple times over life cycle and females may have multiple reproductive cycles, but hard ticks feed only once and females have one reproductive cycle.	Many ticks produce itching and irritation and can disrupt behavior and weight gain; however, there are several diseases of cattle (and many other mammals including humans) where ticks are the vector; cattle diseases carried by ticks include babesiosis (tick fever), anaplasmosis, theileriosis (East Coast fever) and cowdriosis (heartwater).

The majority of this information was condensed from the review of Cortinas and Jones (2006), but also relied on the Merck Veterinary Manual Online (2012).

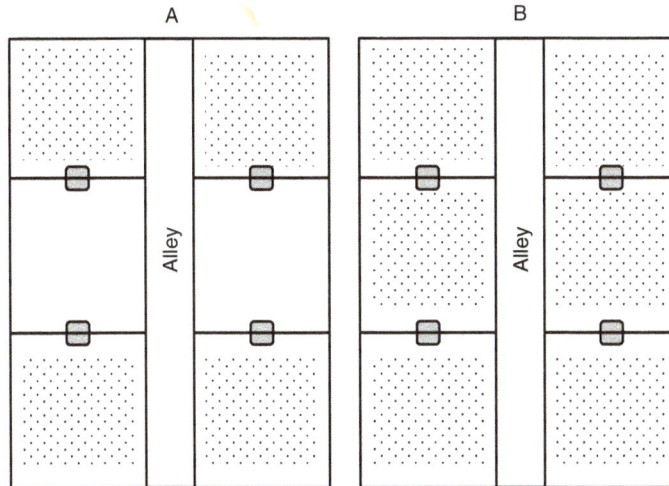

Fig. 9.9. Animal housing and biosecurity considerations. Under scenario A, groups of animals (represented by shading) do not share fence-line contact or water sources; under scenario B animals can have nose-to-nose contact across fences and share water sources (with multiple groups for center pens). This same concept applies to pastures as well as confinement situations.

practices, and always read and follow label directions for all animal health products.

Cleere *et al.* (2007) summarized that transmission of infectious diseases could occur by (i) aerosol – pathogens carried in the air on moisture droplets from sneezing or coughing, (ii) direct contact – pathogens passed from open wound, saliva, blood or mucous secretions from nose-to-nose touching, rubbing, licking and biting, (iii) oral – animals consume disease-causing pathogens in contaminated feed and water or lick or chew contaminated objects, (iv) mechanical – contaminated objects, such as needles, equipment, tools, trailers, trucks, bandages, etc. or clothing, (v) reproductive – pathogens spread by sexual acts or parturition, (vi) vector – living organism such as insect, animal or human serves as intermediate host, and (vii) fomite – soil, water and feed that contains pathogen. It is wise to have defined health plans in place for the different types of animals in an operation. For instance, an operation that only buys and sells replacement heifers will have some similar, but many different, considerations as compared to an operation that has cows, bulls and nursing calves, and grass finishes its own steers. Each of these different types of animal should be considered, but if the different types of animals are housed together versus separately, this also influences biosecurity considerations and decisions. Producers are encouraged to consider health-related management specific to different classes and ages of cattle.

9.6.3 Facilities

Many infectious agents can be killed through exposure to sunlight and dry conditions. Animals that are in confinement are more susceptible to spread of infectious disease than animals on pasture. Cattle in confinement may also be more exposed to certain pathogens due to the nature of the facility (such as concentration of some bird species, rodents or insects more common in urban areas).

The overall health program and associated biosecurity considerations should be thought of for the functional group of animals that are being managed together; this is typically referring to a herd-level way of thinking as opposed to an individual-animal way of thinking. For instance an operation may have a single cow herd, or multiple herds (groups) of animals. If the different groups of animals for that operation do not intermingle, they are essentially separate units, but if heifers, cows, bulls, etc. are moved among the different herds, or if the different herds are simply separated by fences (but never actually housed together), they should be thought of as a single unit for health management.

Even though it may appear that animals across fences do not mingle, this may not be true because several cattle may jump fences if they are less than 2 m (6 feet) tall, and the nose-to-nose contact across fences, shared water sources, etc. provide real potential for transfer of infectious pathogens even if the animals are physically separated.

The level of health risk associated with bringing new animals into cattle herds is directly related to the amount and accuracy of background information associated with the new animals, and this level of risk is also influenced directly by the infectious, transmissible threats in the particular geographical area(s).

9.6.4 Beef Quality Assurance Programs

In many instances, beef quality assurance (BQA) programs or best management practices (BMP) have been demonstrated to be effective in improving animal health and well-being as well as consumer satisfaction. Recent reports (Garcia *et al.*, 2008; Moore *et al.*, 2012) show that major improvements have been made by cow-calf producers since 1991 in regard to injection site lesions, changes in vaccination programs, and changes in preconditioning for improved BQA programs. It is hard to overstate the importance of reading and adhering to label instructions and directions for animal health products. Not only is optimal efficiency of product use important, but overdosing of animals as compared to the label directions can endanger the lives of the animals, and potentially the safety of consumers that eat meat products from these animals. Dosage amounts on labels have been determined through numerous research trials for both efficacy of use and withdrawal times. It is the responsibility of the person to adhere to the withdrawal time to ensure food safety for all consumers. The withdrawal time is the number of days post-treatment that the animal must be withheld from the food chain. Animals metabolize pharmaceutical compounds at rates relative to the molecular structure of the compound, and at the amount administered to the animal. As a result, overdosing of medications may also prolong the withdrawal times, and disregard for these types of considerations jeopardize consumer safety and trust. There is likely to be increased public interest in animal health and well-being in the future as consumers and policy makers scrutinize animal antibiotic use and overall sustainability issues.

9.7 Summary

This chapter provides a broad overview of cattle health considerations and is meant to provide enough background information that students and producers can ponder important questions about the necessary health considerations for their specific scenarios. Knowledge about infectious diseases within geographical areas and the ability to minimize their threats are critical. Understanding of general concepts such as sanitation, naïve and/or non-adapted animals vs immunized and/or adapted animals, types of disease threats, reduction of stress, proper nutrition, along with animal movement, is needed by all cattle producers, not only to minimize health threats to their individual operations, but on a broader scale to control diseases on a regional or even national scale. As control measures and understanding of cattle health improve there will probably be more of a focus on subclinical illness as a thief of production efficiency. Improved cattle health in general allows for increased production efficiency and the ability to increase breeding herd size and total beef production for improved food security.

9.8 Study Questions

9.1 You have a calf that is ill, and the local veterinarian has given you an antibiotic for the animal. The dosage on the label says 1.4 ml per 45 kg of animal weight. This calf weighs 205 kg. How much antibiotic does this animal need? If today's date is 14 July, and the drug has a withdrawal time of 21 days, what is the first acceptable date that this animal could be processed for human consumption?
9.2 Based on a scenario of your choice, how many pastures (or pens) would you need to have an optimal animal biosecurity program? Discuss why.
9.3 If neosporosis is a potential threat in your region, what management considerations might be appropriate to reduce its chance of occurrence in a beef cow herd?
9.4 If annual vaccinations are available and recommended for your region, think about and discuss when in your typical management program these vaccinations could be incorporated.
9.5 List five signs or symptoms of cattle that may indicate they have some type of health problem.
9.6 If vaccination was not an option for a potential infectious disease in some geographical region, discuss what health management strategies could be useful to decrease the chance of infection.

9.7 List five internal parasites of cattle.

9.8 Discuss why zoonotic diseases require more attention than diseases that affect only cattle.

9.9 List five external parasites of cattle.

9.10 Identify five mycotoxicoses diseases (from fungi or molds) that affect cattle, along with the name of the toxin and a likely contaminated feed source for each.

9.9 References

Actor, J.K. (2007) *Elsevier's Integrated Immunology and Microbiology.* Mosby Elsevier, Philadelphia, PA.

Ballweber, L.R. (2006) Diagnostic methods for parasitic infections in livestock. *Veterinary Clinics of North America: Food Animal Practice* 22, 695–705.

Buhman, M.J., Perino, L.J., Gaylean, M.L., Wittum, T.E., Montgomery, T.H. and Swingle, R.S. (2000) Association between changes in eating and drinking behaviors and respiratory tract disease in newly arrived calves at a feedlot. *American Journal of Veterinary Research* 61, 1163–1168.

Burns, V.E., Carroll, D., Ring, C., Harrison, L.K. and Drayson, M. (2002) Stress, coping, and hepatitis B antibody status. *Psychosomatic Medicine* 64, 287–293.

Burns, V.E., Carroll, D., Ring, C. and Drayson, M. (2003) Antibody response and psychosocial stress in humans: relationships and mechanisms. *Vaccine* 21, 2523–2534.

Chase, C.C.L., Young, A., Zimmerman, A., Gilkerson, K., Barling, K. and Scholz, D. (2008) The induction of a T helper 1 immune response with an inactivated viral vaccine. In: *Proceedings 4th U.S. BVDV Symposium,* 25–27 January, Phoenix, AZ (poster 2).

Cleere, J., Gill, R. and Dement, A. (2007) Biosecurity for beef cattle operations, Publication L-5506, Texas Beef Cattle Management Handbook Texas Cooperative Extension, Texas A&M AgriLife Extension Service, College Station, TX.

Coico, R., Sunshine, G. and Benjamini, E. (2003) *Immunology,* 5th edn. John Wiley & Sons, Inc., Hoboken, NJ.

Cortese, V.S., West, K.H. and Hassard, L.E. (1998) Clinical and immunologic responses of vaccinated and unvaccinated calves to infection with a virulent type-II isolate of bovine diarrhea virus. *Journal of the American Veterinary Medical Association* 213, 1312–1319.

Cortinas, R. and Jones, C.J. (2006) Ectoparasites of cattle and small ruminants. *Veterinary Clinics of North America: Food Animal Practice* 22, 673–694.

Craig, T.M. (2006) Anthelmintic resistance and alternative control methods. *Veterinary Clinics of North America: Food Animal Practice* 22, 567–582.

Craig, T.M. (2009a) Gastrointestinal protozoal infections in ruminants. In: Anderson, D.E. and Rings, M. (eds) *Current Veterinary Therapy: Food Animal Practice,* 5th edn. Saunders Publishing, Philadelphia, PA, pp. 91–95.

Craig, T.M. (2009b) Helminth parasites of the ruminant. In: Anderson, D.E. and Rings, M. (eds) *Current Veterinary Therapy: Food Animal Practice,* 5th edn. Saunders Publishing, Philadelphia, PA, pp. 78–90.

Detilleux, J.C., Koehler, K.J., Freeman, A.E., Kehrli, Jr, M.E. and Kelley, D.H. (1994) Immunological parameters of periparturient Holstein cattle: genetic variation. *Journal of Dairy Science* 77, 2640–2650.

Downey, E., Fang, X., Runyan, C.A., Sawyer, J.E., Hairgrove, T.B., Ridpath, J.F., Gill, C.A. and Herring, A.D. (2012) Sire and vaccine treatment effects on immune response to BVDV 1b challenge. *Journal of Animal Science* 90(Suppl. 3), 403(Abstr.).

Ellis, J.A., Hassard, L.E. and Cortese, V.S. (1996) Effects of perinatal vaccination on humoral and cellular immune responses in cows and young calves. *Journal of the American Veterinary Medical Association* 208, 393–400.

Endsley, J.J., Roth, J.A., Ridpath, J. and Neill, J. (2003) Maternal antibody blocks humoral but not T cell responses to BVDV. *Biologicals* 31, 123–125.

Fell, L.R., Golditz, I.G., Walker, K.H. and Watson, D.L. (1999) Associations between temperament, performance, and immune function in cattle entering a commercial feedlot. *Australian Journal of Experimental Agriculture* 39, 795–802.

Feng, N., Pagniano, R. and Tovar, C.A. (1991) The effect of restraint stress on the kinetics, magnitude and isotype of the humoral immune response to influenza virus infection. *Brain, Behavior, and Immunity* 5, 370–382.

Garcia, L.G., Nicholson, K.L., Hoffman, T.W., Lawrence, T.E., Hale, D.S., Griffin, D.B., Savell, J.W., VanOverbeke, D.L., Morgan, J.B., Belk, K.E., Field, T.G., Scanga, J.A., Tatum, J.D. and Smith, G.C. (2008) National Beef Quality Audit-2005: Survey of targeted cattle and carcass characteristics related to quality, quantity, and value of fed steers and heifers. *Journal of Animal Science* 86, 3533–3543.

Groschke, D.W., Herring, A.D., Sawyer, J.E. and Paschal, J.C. (2006) Evaluation of feeder calf grade, background information and health status on feedlot weight gain and carcass traits of calves fed in South Texas. *Journal of Animal Science* 84(Suppl. 2), 38(Abstr.).

Larson, B.L. (2005) A new look at reducing infectious disease in feedlot cattle. *Proc. 2005 Plains Nutrition Council.* pp 9–18.

Merck (2012) The Merck Veterinary Manual Online. Available: http://www.merckmanuals.com/vet/ (accessed 11 May 2014).

Moore, M.C., Gray, G.D., Hale, D.S., Kerth, C.R., Griffin, D.B., Savell, J.W., Raines, C.R., Belk, K.E., Woerner, D.R., Tatum, J.D., Igo, J.L., VanOverbeke, D.L., Mafi, G.G., Lawrence, T.E., Delmore, Jr, R.J. Christensen, L.M., Shackelford, S.D., King, D.A., Wheeler, T.L., Meadows, L.R. and O'Connor, M.E. (2012) National Beef Quality Audit–2011: In-plant survey of targeted carcass characteristics related to quality, quantity, value, and marketing of fed steers and heifers. *Journal of Animal Science* 90, 5143–5151.

Mosmann, T.R. and Fowler, D.J. (2002) The Th1/Th2 paradigm in infections. *Immunology of Infectious Diseases.* Edited by S.H.E. Kaufmann, A. Sher, and R. Ahmed. ASM Press, Washington, DC.

Muggli-Cockett, N.E., Cundiff, L.V. and Gregory, K.E. (1992) Genetic analysis of bovine respiratory disease in beef calves during the first year of life. *Journal of Animal Science* 70, 2013–2019.

Naylor, J.M. (2009) Neonatal calf diarrhea, Chapter 21. In: Anderson, D.E. and Rings, M. (ed.) *Current Veterinary Therapy: Food Animal Practice,* 5th edition. Saunders Publishing Philadelphia, Pennsylvania USA pp. 70–77.

Oliphant, R.A., Welsh, Jr., T.H., Randel, R.D., Laurenz, J.C. and Carroll, J.A. (2006) Calm cattle have better responses to weaning vaccinations. *Beef Cattle Research in Texas,* pp 113–117.

Panuska, C. (2006) Lungworms of ruminants. *Veterinary Clinics of North America: Food Animal Practice* (Ruminant parasitology issue) 22, 583–594.

Playfair, J. and Bancroft, G. (2008) *Infection and Immunity.* Oxford University Press, Inc., New York.

Ridpath, J.E., Neill, J.D. and Endsley, J. (2003) Effect of passive immunity on the development of a protective immune response against bovine diarrhea virus in calves. *American Journal of Veterinary Research* 64, 65–69.

Roth, J.A. and Spickler, A.R. (2012) Transmission of zoonoses between animals and humans. March 2012 Merck (2012) The Merck Veterinary Manual Online. Available: http://www.merckmanuals.com/vet/ (accessed 11 May 2014).

Snowder, G.D., Van Vleck, L.D., Cundiff, L.V. and Bennett, G.L. (2006) Bovine respiratory disease in feedlot cattle: Environmental, genetic, and economic factors. *Journal of Animal Science* 84, 1999–2008.

Thompson, P.N., Stone, A. and Schultheiss, W.A. (2006) Use of treatment records and lung lesion scoring to estimate the effect of respiratory disease on growth during early and late finishing periods in South African feedlot cattle. *Journal of Animal Science* 84, 488–498.

Wagter, L.C., Mallard, B.A., Wilkie, B.N., Leslie, K.E., Boettcher, P.J. and Dekkers, J.C.M. (2000) A quantitative approach to classifying Holstein cows based on antibody responsiveness and its relationship to peripartum mastitis occurrence. *Journal of Dairy Science* 83, 488–498.

Wilkie, B. and Mallard, B. (1999) Selection for high immune response: and alternative approach to animal health maintenance. *Vet. Immunol. Immunopathol.* 72, 231–235.

Wittum, T.E., Woollen, N.E., Perino, L.J. and Littledike, E.T. (1996) Relationships among treatment for respiratory tract disease, pulmonary lesions evident at slaughter, and rate of weight gain in feedlot cattle. *Journal of the American Veterinary Medical Association* 209, 814–818.

Yewdell, J.W. and Bennink, J.R. (2002) Overview of the viral pathogens. In: Kaufmann, S.H.E. Sher, A. and Ahmed R. (eds) *Immunology of Infectious Diseases.* ASM Press, Washington, DC.

Zimmerman, A.D., Boots, R.E., Valli, J.L. and Chase, C.C.L. (2006) Evaluation of protection against virulent bovine diarrhea virus type 2 in calves that had maternal antibodies and were vaccinated with a modified-live virus vaccine. *Journal of the American Veterinary Medical Association* 228, 1757–1761.

10 Behavior, Temperament, Handling and Welfare

Cattle have certain behavioral characteristics that are crucial parts of their daily lives. Animals have inherent differences in personality and therefore their responses to various stimuli differ. Knowledge of various normal and abnormal behaviors of cattle is crucial to understanding expected animal reactions to handling and transportation as well as general management for overall improved production efficiency. There are widely documented genetic variants for behavior in cattle. Producers that understand fundamental principles of cattle behavior are better positioned to identify and rectify potentially stressful situations, identify sick animals, and improve facility, pasture and transportation management, all of which can reduce risk of injury and illness for improved animal well-being and animal performance. In general, improvements in animal welfare within the context of a production system will typically lead to improvements in production and/or product quality. This chapter discusses the principles of behavior and associated management of cattle.

10.1 Behavior

The term behavior can indicate many considerations of how animals act. Some behavioral aspects deal with animals in their 'undisturbed' state, such as when people are not around. Other aspects deal with how animals respond to different types of stress. There are some cattle behavioral actions that are considered typical and some actions that are considered atypical or abnormal. Some discussion is provided on what is expected to be observed with cattle behavior in general and how that knowledge can relate to various management considerations.

10.1.1 Abnormal Behaviors

Under extensive conditions such as cattle on pasture when there is adequate forage availability, cattle will typically spend approximately 8 hours per day grazing, 8 hours resting and 8 hours ruminating and chewing their cud. Cattle do not vocalize much when they are not stressed. Cows with young calves will make various sounds to communicate with their calves. Animals that vocalize often are generally not happy about something. This may be due to being separated from calves/dams, as at weaning, but it may also be because they are hungry or in pain. Cows that are accustomed to receiving supplemental feed at a set time of day may become quite vocal in anticipation of feeding. Animals usually vocalize considerably when they have procedures such as branding and dehorning performed on them, but simply being restrained without physical pain may also cause cattle to vocalize if they are not used to being handled, or have been mistreated in the past. Producers typically think about cattle being content or not in distress if they are not vocalizing, but this may not always be true. In extreme pain or illness animals may be non-vocal, but this generally is accompanied by lack of movement and other signs of distress.

Cattle that are not distressed are typically calm and quiet. Animals that are agitated may make repeated low sounds, swish their tails repeatedly, pace along a fence, turn their heads abruptly toward sound or movement, urinate and/or defecate small amounts frequently, have their eyes wide open and their heads held high; these behaviors can occur individually, or in combination. These types of behaviors are commonly seen in cattle that have been separated from a group and are isolated such as in a pen, particularly if they are not frequently handled. Animals may show wide variation in these types of traits. When handling cattle, it is important to realize that animals will remember experiences (good and bad); this is why they can be trained. Therefore, the more calmly and quietly animals can be handled through pens and working facilities, the easier it will be to bring them through the same facilities in the future; this is referred to as low-stress handling. On the other hand, animal handlers must also be aware of the signs of upset cattle, to avoid situations where they or other people could be injured, and to be thoughtful of what might need to be done if an animal does something unexpectedly. The best advice about handling cattle is usually to be quiet and remain aware of how animals are responding to their situation, and to have an escape plan if the need arises.

It is more common to see certain 'unusual' behaviors in cattle that are in confinement conditions. Animals may vocalize, pace, chew and/or push on fencing, and fight with one another when in confinement due to boredom, especially when fed high-energy diets, because they spend less time per day eating, ruminating and chewing their cud. Knowledge about cattle grazing behavior has led to improved feedlot management practices. Some US feedlots have altered their feeding schedules so that cattle are fed at times of the day when they are typically more active, such as early morning and around dusk. Research has shown that altering feeding patterns of cattle can change animal activity patterns and reduce dust from feedlots (Wiggers et al., 1998; Mitlöhner et al., 1999).

If cattle have the opportunity, they will modify behavior to cope with environmental stresses. Animals will seek shade or may submerse themselves in water sources in high temperatures. In cold conditions, cattle will seek shelter from rain or wind. Shelter from wind may be artificially provided by barns or constructed walls but can also be provided by natural conditions such as trees, brush, tall grass, etc. Cows with young calves will attempt to find appropriate shelter for their calves. These various types of behaviors of cattle are discussed below.

10.1.2 Maternal Behavior

Often when a cow is ready to calve she will venture away from the rest of the herd. It is important if there are undesirable climatic or other environmental conditions for cows to be able to find shelter during calving so that newborn calves are not subjected to extreme temperatures or predators. For instance some predatory birds may flock around a cow that is calving in an open area and kill the calf as it is delivered, but if the cow can go into a wooded area so that the calf can be better protected, most will do so. If cows are confined to low, wet areas while calving without access to high ground,

calves could drown immediately after birth. Reports from a few research projects that documented maternal behavior in cattle are provided.

Selman et al. (1970) published a detailed record of behavior for beef cows, dairy cows and dairy heifers in the UK for the 8 hours following parturition (30 females total). Particular interest was paid to position at parturition, licking and grooming behavior, maternal orientation, maternal ejection and resting behavior. In 20 (eight beef cows, seven dairy cows and five dairy heifers) of the 30 total calvings, the calves were delivered to their hips while the dams were recumbent (lying down but still upright). At this stage, the dams rose to their feet and dropped their calves, with little further effort, on rising or shortly afterwards. One cow (a beef cow) remained standing throughout the act of parturition, and remaining dams (one beef cow, three dairy cows and five dairy heifers) completed parturition while lying down. Most dams initiated licking of their calves immediately after delivery and typically was accompanied by distinct and varying sounds. Beef cows spent more time (48.3 minutes) continuously licking their calves compared to dairy cows (32.9 minutes) and dairy heifers (11.0 minutes).

Selman et al. (1970) also described that distinct sounds were usually discernible in dams associated with specific behaviors. During the initial phase of calf licking, the dams frequently became very excited and loud drawn-out bellows were emitted and delivered with the mouth open and the head extended in the direction of the calf. Later, when the dams had quietened down, a series of soft, pharyngeal grunts would be made with the mouth closed. This series of sounds was frequently repeated during calf licking and while nursing or eating contaminated bedding. A similar sound to this nursing sound, but louder and only repeated once or twice, was made when dams appeared to be worried, for example when a calf wandered too far from its resting dam. Twenty-one dams (70%), including 9 of 10 beef cows, did not lie down until their calves had either suckled or at least carried out prolonged teat-seeking activities. All of the beef cows ate contaminated bedding and placentas. The authors also stated that there appeared to be more variability in various behaviors among the dairy heifers as opposed to the beef and dairy cows.

Licking or maternal grooming of offspring serves several functions (Selman et al., 1970). Licking at the time of parturition serves an obvious hygienic function, in addition to stimulating the young to stand and eventually suckle. In addition, Gubernik (1980) has proposed that a mother may 'label' her offspring by maternal grooming at this time, thus providing a mechanism for offspring discrimination. Later, as the young mature, the hygienic significance of grooming is gradually replaced by social functions relating to the establishment and maintenance of a social bond or attachment between mother and young (Schloeth, 1958).

Price *et al.* (1981) evaluated maternal behavior in cows giving birth to twins vs single calves. Twinning did not affect the probability that a dam would nurse her offspring during morning and evening observation periods. However, twins suckled longer than single calves (6.4 minutes per hour per calf vs 4.4 minutes per hour per calf), presumably because of a sub-optimal milk supply. Twin calves received less licking time from their dams (1.0 minutes per hour) than did single calves (2.2 minutes per hour). Twin calves were more persistent in their attempts to suckle other cows (referred to as alien dams), and dams with twins were successfully suckled by other (alien) calves more frequently than dams of single calves (47 vs 13%).

It is desired by many managers that maternal behavior be aggressive enough to provide for protection of the newborn calf against potential predators, but not be so aggressive toward people that someone cannot get close enough for observation or some type of intervention if deemed necessary. Some studies have classified cows (as on a 1 to 5 scale where 1 = completely docile and 5 = extremely aggressive and combative) as equally acceptable below some threshold value, but it is possible that having cows and heifers that are all extremely docile and unresponsive may also be undesirable, particularly immediately after calving has occurred. A cow that is seemingly unconcerned about her calf is probably not as desirable as a cow that is attentive and concerned, but not combative toward handlers.

Sandelin *et al.* (2005) evaluated maternal behavior scores across several breeds and ages of *Bos taurus* females on a 4-point scale (where 1 = very aggressive – cow was willing and fought the handler to protect calf; 2 = very attentive – cow remained in close proximity with mild aggression, but did not fight the handler to protect calf; 3 = indifferent – cow remained in proximity, showed no aggression toward handler, but remained in sight of calf; 4 = apathetic – cow showed no emotion toward calf

in presence of handler, grazed away or moved out of proximity). In general, cows in thin body condition had higher scores that those in average or above average body condition, and cows 5–10 years of age had lower scores than young cows. However, the most striking values reported were those of calf survival. Cows with scores of 1, 2, 3 and 4 had calf survivability values of 93%, 86%, 77% and 60%, respectively, which indicates that the more attentive cows at birth enhanced survivability of their offspring, whether caused by protection from predation or other factors that might have been involved (Sandelin *et al.*, 2005) or potential ancillary relationships with other influential traits.

10.1.3 Grazing and Feeding Behavior

Detailed description and consideration of grazing behavior is covered in Chapter 5. Numerous intrinsic characteristics of animals and management influence both grazing and feeding behavior. Bailey (2004) presented a thorough review of influences of grazing behavior in arid and semi-arid regions. Much of the work studying confinement feeding behavior has revolved around associated animal health, where both decreased feed intake and water intake is associated with bovine respiratory disease and other health problems. Recent developments, where feed intake and feeding behavior can be monitored through automated systems in feedlot settings, has prompted more research into characterizing feeding traits of typical healthy cattle. Table 10.1 provides data on some eating behavioral traits in grazing beef heifers.

10.1.4 Bull and Cow Sexual Behavior

Both breed and family differences have been widely documented in reproductive activity in bulls. Discussion of bull breeding aggressiveness and overall sexual activity as well as cow sexual behavior related to estrus are discussed in more detail in Section 8.1.6. Social dominance and social group interactions are thought to potentially influence bull sexual activity more than cow or heifer sexual activity as cows will only stand to be mounted by a bull or other animal when they are in active estrus (standing heat). If there is a group of only females (no males in the group), there may be less pronounced mounting behavior of females (as compared to having a bull in the group); female-only breeding groups are common when the females are

Table 10.1. Eating behavioral traits in grazing heifers of differing expected mature size.

Year	Small frame early maturing Angus, expected mature weight 387 kg	Conventional Angus, expected mature weight 413 kg	Hereford and Red Poll, expected mature weight 468 kg	Charolais and Chianina, expected mature weight 589 kg
	Bite size in mg DM per bite			
1986	479	551	453	458
1987	532	676	561	687
1988	525	575	533	588
	Bite rate, number of bites per minute			
1986	30.5	28.3	28.7	31.5
1987	34.6	30.5	33.6	34.2
1988	46.8	46.1	48.4	46.9
	Grazing time, minutes per day			
1986	391	442	452	453
1987	395	453	460	485
1988	415	400	443	475
	Predicted daily intake from grazing behavioral traits, g DM per day[a]			
1986	5,712	6,892	5,876	6,535
1987	7,271	9,340	8,671	11,395
1988	10,197	10,603	11,428	13,099
	Daily intake expressed as percentage of expected mature weight, %[a]			
1986	1.48	1.67	1.26	1.11
1987	1.88	2.26	1.85	1.93
1988	2.63	2.57	2.44	2.22

From Erlinger *et al.* (1990). Bite here refers to new mouthful of forage during grazing bout, not associated with chewing of cud. Forage characteristics for 1987 and 1988 measurements, respectively: plant height, 10.6–15.8 cm, 9.1–9.5 cm; forage mass, 4323–6313 kg/ha, 4179–6286 kg/ha; N%, 1.4–1.7, 1.4–1.8; ADF%, 33.6–38.9, 35.3–35.5.
[a] Not reported in source but calculated from reported values.

being bred by artificial insemination and/or an embryo transfer program. Standing to be mounted and attempting to mount may be two quite different behaviors, and not solely in response to sexual stimuli. This is discussed further in the following section.

10.1.5 Social Behavior

Social behavior can be thought of as how an animal interacts with other animals. As a result many maternal behaviors are social behavior, but that is a specific, unique type of social interaction. Here social behavior refers to those animal–animal interactions that are not specific to maternal-offspring care, but the social aspects related to being in a group of animals. The degree of expression of many of these types of behaviors can be modified in the presence of new observers or new settings for the animals. Some observational studies of feral cattle are considered here, and it is expected that the same behaviors are present in most if not all groups of cattle to varying extent.

Reinhardt *et al.* (1986) studied a herd of semi-wild Scottish Highland cattle in Germany for various behaviors and found that social standing (dominant or subordinate) depended on age and sex, with older animals generally being dominant to young animals and males dominant to females. Young bulls began to dominate adult cows when they reached about 2 years of age. These sex and age influences on dominance were similar to observations by Hall (1989). Mock fighting was observed as a friendly contact behavior by both sexes and all ages. Licking was seen as a social service primarily performed by subordinate animals and received by dominant animals. Reinhardt *et al.* (1986) also stated that mounting was a playful behavior shown by calves of both sexes and by bulls (but not by cows) and that mounting was not used as a dominance demonstration.

Hall (1989) studied behavior in the Chillingham cattle in northern England as a model for feral cattle. He classified behaviors observed in these cattle into maintenance and social in nature, and these are provided in Table 10.2. Social activities that

Table 10.2. Classifications of behavior in cattle with little human contact.

Maintenance	Social	
	Self-expression	Interaction with others
Grazing	Calling	Sniffing (with and without flehmen)/
Standing	Lowing	receiving sniff
Walking	Digging with nose	Approaching/response to approach
Rubbing on objects	Pawing ground	Displacing/being displaced
Lying (doing nothing else)	Rubbing neck on ground	Make/receive lateral threat
Standing and ruminating	Beating vegetation with horns	Guard/be guarded
Lying & ruminating		Rub/be rubbed
Sleeping		Attempt to mount/receive mount
Licking own body		Lick/be licked
Drinking		Suckle
Eating mineral supplement		
Running		
Scratching with own limb		
Defecating		
Urinating		

From Hall (1989).

could be performed alone were deemed 'self-expression' and those requiring other animals were deemed 'interaction with others.'

Sato *et al.* (1991) evaluated the effects of social licking in a small sample of Holstein in Japan. Social licking decreased on rainy days and tended to increase in a dirty barn and when feed was restricted. Solicitation or recruitment of social licking occurred not only from dominant animals of pairs but also from subordinates. Of the licking interactions, 31% occurred following solicitation, and these accounted for 39% of the total time spent licking. Following solicitation, 78% of social licking was oriented to the head and the neck areas inaccessible to self-licking animals. Unsolicited licking, however, was oriented not only to the head and the neck but also to the back and the rump regions, and the two latter regions receiving most licking. Factors investigated were dominance–subordinance, kinship, familiarity and sex of calf, but only familiarity had a significant effect on licking. These authors also stated that exchanges of social licking increased with length of cohabitation and suggested that social licking could serve cleaning, tension reducing and bonding functions.

10.1.6 Grooming Behavior

It is likely that grooming behavior is both a maintenance behavior and a social behavior. Hall (1989)

stated that cattle of European descent have very little grooming behavior compared to wild bovid species, and that this lack of behavior could at least in part be related to their susceptibility to ticks, but evidence of this has not been found in the scientific literature. Also at the time of publication, no reports could be found in the literature in regard to grooming behaviors in tropically adapted cattle. If genetic variation existed for this type of behavior in cattle, and it was shown to have beneficial effects, selection programs could target grooming behavior if desired.

10.1.7 Buller Steers

In some cases of confined feeding, there may be animals that are continuously mounted by other animals in the pen. This phenomenon has been studied in several North American trials with feedlot steers. The animal that is being mounted by others is referred to as the 'buller'. If this happens repeatedly, the animal being ridden may suffer physical exhaustion and/or injury. There is no apparent consistent cause of this phenomenon, but many believe it may be a pheromone-related issue. Taylor *et al.* (1997) stated that 1–3% of feedlot steers may experience bulling and that the stress from its incidence might promote release of corticosteroid hormones that could interfere with sexual activity-related hormone levels. Studies that have evaluated different hormone-based growth promoting implants have not seen consistent trends. There is evidence that if the buller animal is removed from the pen for a short time (3–5 days) and then reinstated, its susceptibility may subside (for example, Stookey, 2001). Blackshaw *et al.* (1997) provided a review of buller steer syndrome and stated the social dominance–submissiveness relationships associated with this behavior were unclear.

10.2 Temperament

There are many behaviors inherent to cattle, such as how they move their legs to walk, how they use their tongue and mouth during grazing and eating, how females interact with their newborn calves, etc., and many of these types of behaviors have been discussed in the preceding section. However, there are also aspects of behavior referred to as temperament or disposition that describe inherent differences in how cattle respond to certain handling or stressful situations, and these aspects of behavior might also be thought of as personality. All cattle have a 'fight or flight' response (as do other species of animals and people) in particular situations, but the degree to which animals feel comfortable/uncomfortable (i.e. the size of their flight zone) varies according to an animal's previous experiences, their inherent genetic predispositions to stimuli, and even potential interactions of experience and genetic background. The first principle to understanding

temperament and its potential management implications in cattle is that of the flight zone.

10.2.1 Flight Zones of Animals

Understanding how cattle perceive their surroundings and people has been observed and emphasized by several people for many years. Many of these concepts should be thought of as components in systems-based considerations and have important production as well as animal welfare implications. The size of the flight zone is different in animals of varying temperament and previous experiences. The 'newness' of a situation provides a stress to the animal, and even some situations that are 'undesirable' at first may become familiar to animals. For instance, Grandin (2005, 2008) has stated that cattle used to being around people on horseback may perceive an entirely different situation the first time the same people are walking around them on foot. The same could be true when animals are used to seeing vehicles close to them in a pasture, but a person on a horse or on foot may be perceived differently. Temple Grandin's recommendations on cattle perceptions, flight zone considerations (Fig. 10.1) and point of balance provide useful insights into animal handling and movement and heavily influence the discussion of handling, facilities and transportation in this chapter. The reader should consult her web page (www.grandin.com) or the book *Livestock Handling and Transport* for more information and more detailed discussion about many of these concepts (Grandin, 2000).

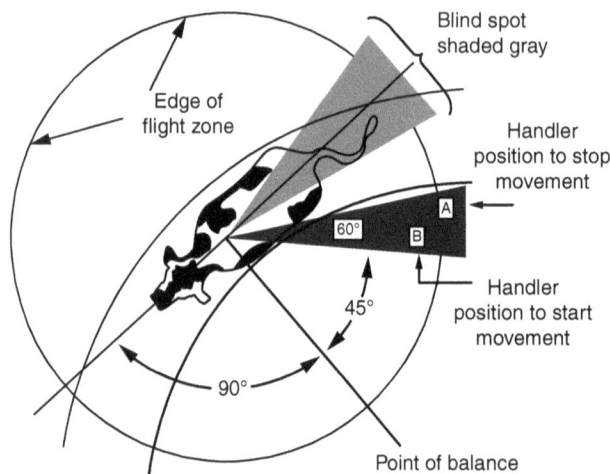

Fig. 10.1 Concept of flight zone.

10.2.2 Methods used to Measure Temperament

People that are used to being around cattle or other livestock recognize that not all animals have the same reaction to the same situation. Most if not all cattle can be trained to perform a variety of tasks such as pulling a plow or wagon, being led by halter, standing to be milked, eating from a specific feeding spot, etc. The best time to characterize inherent genetic differences for temperament among individual animals is probably before they have had extensive contact with people and before they have been trained to perform any task (which may be almost impossible except in very young animals, as the time calves spend with their dams can certainly influence learned behaviors). Two primary ways that have been used to evaluate temperament have been through use of subjective scores (such as a 1 to 5, 1 to 6 or 1 to 9 scale, etc.) or through use of an objective measure such as exit velocity (also called escape velocity).

Exit velocity is typically measured over a standardized distance (such as 1.8 m) with two sets of lasers that trigger a timer that displays the amount of time taken for each animal to cover that distance (see Fig. 10.2). This is commonly measured as the animals leave a restraint (such as squeeze chute/cattle crush, but could be as they are allowed to pass through a gate), and the amount of time to travel this distance is converted to a speed such as meters per second, hence the name exit velocity. Because this trait is automatically recorded, it is objective. The placement of the restraint relative to facility structures such as leaving a covered area, a long open alley vs a sharp turn following release from the restraint, etc. must be considered in making comparisons of exit velocity across animals, time and locations.

Our research group has evaluated cattle temperament in calves approximately 4–6 weeks after weaning (average of 8 months of age) for component traits of temperament including aggressiveness, nervousness, flightiness, gregariousness and overall temperament on a 1 to 9 scale (Fig. 10.3). In this scoring system, these component traits are individually scored and the overall temperament score is based on the evaluator's overall assessment (not an average of the component values). When these cattle were taken to new facilities as yearlings, the initial exit velocity of these animals was not correlated with their weaning temperament scores; however, for subsequent evaluation times, the correlations between the weaning temperament scores and the exit velocity measurements increased (Fang et al., 2012), probably indicating that as the animals became more accustomed to the facilities their inherent temperament differences may have become more evident.

Temperament evaluation by subjective scoring has most commonly been done while animals are

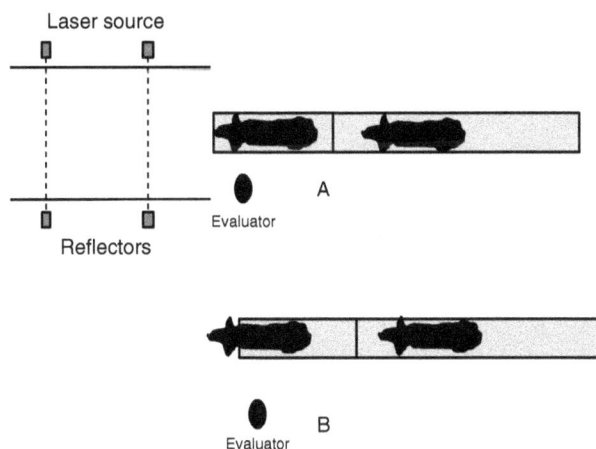

Fig. 10.2. Example of exit velocity and chute or crush scoring. Both have been used to assess cattle temperament. Animals are restrained in a chute and scored by an evaluator in close proximity to the chute. Animals may have their heads restrained (B) or not (A). The position of the evaluator should be standardized for each group of animals. Laser lights connected to a timer are used to collect flight time, or exit velocity when animals are released from restraint.

restrained (chute score), when animals are not restrained but part of a group (pen score, see Fig. 10.4), or when animals are not restrained but are separated from a group (alley score). A summary of the temperament evaluation systems that have been used in cattle are summarized in Table 10.3.

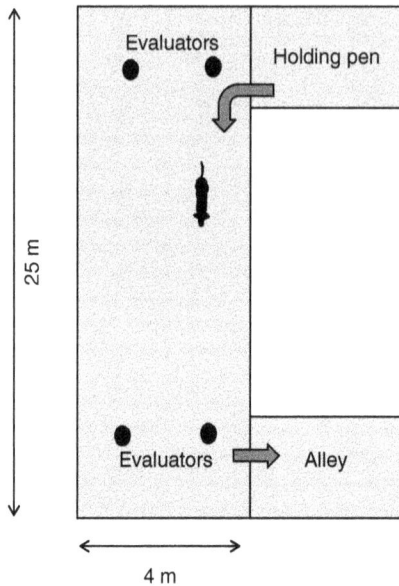

Fig. 10.3. Example of temperament scoring of individual cattle without restraint. Cattle are brought individually into an evaluation area from a holding pen with 12 to 18 animals and are evaluated by multiple people for approximately 1 minute before being released into an alley.

Some temperament recording systems have been developed that automatically monitor the movements of animals while restrained. For instance a squeeze chute or weigh box that is part of a working chute can have vibration recorders attached and these can objectively measure how much the animal is moving while in the chute. This (or other scoring systems) may or may not be measuring the same traits in all cattle as some animals that are nervous may show little movement or even balk and/or lie down while being restrained, and these animals would show little movement as an objective measurement, but balking is an undesirable trait, as is hyperactivity and excitability or aggressiveness.

10.2.3 Training

It is widely recognized that cattle may be trained to alter their behavior in regard to different tasks or activities. Many producers may not recognize that animals have to be trained or taught to perform many 'routine' activities in beef cattle production settings (such as how to drink water out of a trough or to eat feed out of a bunk, as when naïve calves go from a ranch to a feedlot). Cattle can also be trained for things such as how to utilize pastures and how to proceed through working facilities in an orderly fashion. Failure to understand and incorporate principles of cattle behavior may result in them being trained to act undesirably. Providing supplemental feed, placement of salt and mineral licks, addition of watering sources, etc. can be very useful in influencing how animals utilize large pastures. Likewise, handling cattle calmly and quietly

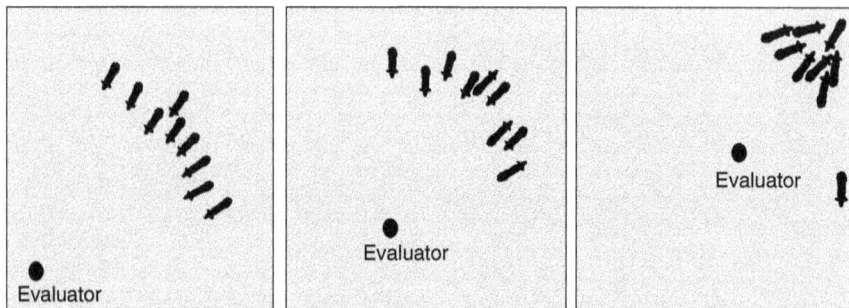

Fig. 10.4. Example of pen scoring of cattle behavior. When the evaluator is outside of the flight zone, the animals turn to face the evaluator. As the evaluator approaches animals, they will turn and move away from the evaluator when the flight zone is entered. As the evaluator moves closer to animals they will attempt to get away. Each animal may have a quite different site flight zone.

Table 10.3. Systems for assessing cattle temperament (disposition).

Chute score	Pen score	Individual alley score	Exit velocity/time (flight speed/time)
Subjective	Subjective	Subjective	Objective
1 to 5 or 1 to 6 scale	1 to 5 or 1 to 6 scale	1 to 9 scale	Time to travel fixed distance (i.e. 1.8 m) or distance traveled per second for the fixed distance.
Animal is evaluated while restrained in squeeze chute (or some other confined space).	Animals in a group are approached by the evaluator and scored individually on how they respond.	Animal is placed individually in a small alley or pen and scored by 1 to 4 evaluators.	After being released from a squeeze chute, the time and/or speed is recorded for each animal.
For these subjective scoring systems, lower values indicate very calm or non-reactive animals and high scores indicate very aggressive, wild or agitated animals.			Animals that are very wild or agitated are expected to leave very quickly when released.

as they are moved from pasture to pasture or though alleys and chutes in working facilities will make these same tasks much easier in the future as animals are not as stressed by this manner of handling, and they will be less averse to similar situations in the future.

Proper cattle handling and transportation guidelines are now viewed as important aspects of beef quality assurance (BQA) and best management practices (BMP) programs. It is important for cattle producers to think about the impacts of how they handle their cattle; this not only impacts their own future interactions with these same animals, but also how other people in the industry will interact with their animals. Many consumers and their suppliers have growing concerns about food animal welfare and the processes involved in food production and are interested in knowing that animals have not been mistreated. Previous handling and treatment of animals affects their perception of humans, working pens and transportation, and can impact production responses such as health and meat quality. Successfully training livestock to act (and react) in production settings can be quite rewarding for animal handlers and managers.

10.2.4 Consequences of Poor Temperament

Several production traits have been evaluated as to how they might be influenced by animal temperament. Several scientists have reported that animals with undesirable temperament may have reduced weight gain and reduced meat tenderness. There have also been reports that cattle with poor temperaments

may not have as desirable immune response to vaccination as contemporaries with calm temperaments. Temperament is probably another trait where there is an optimum intermediate level. Animal handlers want animals that are easy to handle and move. Animals that are extremely wild or excitable may be very hard to control, but animals that are extremely used to people may be hard to move due to their lack or concern about human contact. It is recognized by many people who collect birth weights on calves that a cow overly protective of her calf is dangerous to be around, but a cow that runs away from the calf and caretaker is probably much less desirable in regard to calf survival. An alert and attentive cow is a happy intermediate, but that might not be the case in regions where potential predators threaten calf survival.

10.2.5 Genetic Components of Temperament

It has been reported for several years that there are underlying genetic differences in temperament (may also be termed disposition) in cattle in several different breeds (Shrode and Hammack, 1971; Stricklin et al., 1980; Hearnshaw and Morris, 1984; Le Neindre et al., 1995) with heritability estimates ranging from approximately 0.20 to approximately 0.50 across studies. This is a similar level of genetic influence to that reported for many growth and size traits in cattle. Furthermore, differences in temperament where females have less desirable scores than male contemporaries have also been reported in several cattle populations (Shrode and Hammack, 1971; Stricklin et al., 1980; Voisinet et al., 1997;

Gauly *et al.*, 2001), and Voisinet *et al.* (1997) speculated that sex differences in temperament might be more exaggerated in some breeds than others (another potential type of genotype by environment interaction). Not all studies that have evaluated breed or family influences of behavior have found significant effects.

There have also been several instances where no significant heterosis was observed in crossbred cattle, or where there has been unfavorable heterosis, including instances where *Bos indicus* crossbred cattle have had less desirable temperament scores than contemporary purebred animals (Bonsma 1975a,b, as cited by Grandin and Deesing, 2005; Riley *et al.*, 2004) which would indicate heterosis in an undesirable direction. Kabuga and Appiah (1992) reported N'Dama-Holstein crossbreds to have greater flight distance as well as higher temperament scores (evaluated with 5-point scale) as compared to the purebred N'Dama and purebred Holstein contemporaries. Morris *et al.* (1994) found temperament scores in Angus–Hereford F_1 cows to rank higher than the Angus–Hereford mid-parent value for temperament at calving, but not at other times (when the F_1 cows ranked below mid-parent values). Riley *et al.* (2004) reported substantially unfavorable heterosis in both Brahman–Romosinuano (21%) and Brahman–Angus crosses (10%). This is curious as most production traits in cattle are favorably influenced by heterosis. These types of reports make the study of underlying genetic mechanisms associated with cattle temperament more intriguing, and potentially more complicated as what is 'good' or 'bad' temperament can vary due to the production situation, the animal genetic background, and the animal caretaker perceptions and abilities. There has also been some documentation of genetic markers and identification of QTL for cattle temperament and behavior traits.

10.3   Cattle Handling

Understanding the concept of flight zone and general principles of cattle behavior, as well as that there are breed and individual variations in the behaviors animals exhibit can lead to improved animal handling efficiency.

10.3.1   Animal Handling Strategies

Understanding how cattle normally move in response to human contact has been widely studied and should be a basis for movement of animals as well as design and use of animal handling facilities. Curved, sweeping turns as opposed to 90° corners greatly aid in the flow of cattle through alleys that they are not accustomed to. Minimal noise from animal handlers or other sources, without abrupt movements, helps to keep animals calm. As cattle are curious by nature about new objects and facilities, allowing them time to get accustomed to their surroundings helps them remain calm. Allowing a few seconds for cattle to smell a fence post at a gate opening may make the person impatient, but forcing animals through alleys, gates and other facilities quickly can contribute to higher stress levels in the animals and possibly lead to obvious undesirable consequences such as running into fences, turning and attacking the handlers, etc. and possibly unobserved consequences of stress such as impaired immune system function or reduced carcass quality, such as dark-colored beef at harvest. Rough handling of animals is more likely to cause bruising from hitting structures, and allowing animals to enter a squeeze chute at speed, when they may hit the head gate hard, may cause damage to lymph nodes located along the neck. Low-stress handling of cattle reduces stress for both animals and personnel, and is now a recommended component of many BQA programs. Understanding the natural curiosity of cattle (Fig. 10.5) to new surroundings and situations is a consideration of low-stress handling.

Pajor *et al.* (2000) used aversion learning techniques to determine which handling practices cattle are most averse to. Cows were repeatedly walked down a race with treatments applied when they reached the end, with the time and force required for the cows to walk down the race measured. In one experiment cows were assigned to four treatments corresponding to hitting and shouting, brushing, control (nothing done), and food being provided. Cows in the hitting and shouting treatment took more time and required more force to walk through the race than cows in other groups, and cows that were given food took less time to move through the race than cows that were brushed. In another experiment cows were assigned to five treatments of electric prod, shouting, hitting, tail twisting, and control (no action). Cows in the shouting and electric prod treatments took more time and required more force to walk through the race than control cows. Pajor *et al.* (2000) stated that use of the aversion race successfully distinguished between handling treatments

Fig. 10.5. Cattle are curious by nature and use multiple senses to explore situations and novel items. Positive (and negative) experiences are remembered by cattle and influence their movement through pastures, yards and restraint facilities.

that differed greatly in aversiveness but lacked the sensitivity to distinguish between treatments that were similar in nature.

10.3.2 Facilities and Associated Considerations

Producers do not need to have facilities that are elaborate, but they need facilities that are functional. Animal handling and the facilities need to complement one another, and properly handling cattle will always improve the use of the facilities. Perfectly designed pens, alleys and chutes do not give freedom for rough animal handling; however, well-designed facilities can greatly improve handling and processing of unruly animals or animals that are naïve to the location and handlers. All cattle operations need to have facilities to gather and restrain cattle so that fundamental husbandry practices can be performed. These include as a minimum a set of 'gathering' pens to hold animals, and typically involve a squeeze chute (also referred to as a cattle crush or a cage) or at least a narrow chute with a head gate or some other method of restraining animals.

In special cases, where all animals are very docile or halter broken, many of these considerations can be modified and the necessary minimal facilities will change. In other cases, where animals will pass through a facility structure only a few times (once or twice, as with cattle in a feedlot), as compared to animals that will repeatedly use facilities many times over the course of their lives (as with cows that experience sorting, vaccination, artificial insemination, etc. over the course of many years) also provide different considerations for facility design (Fig. 10.6) and use because things that make some cattle balk at first such as shadows on the ground, change in lighting, activity on the other side of the pen, etc. will have different effects when they are new to animals than when the animals become accustomed to them. The same overall principles of sound, low-stress cattle handling is desired in all situations.

A wide variety of materials may be used to construct various cattle handling facilities. An important overall consideration is that no matter the building material, the same principles are employed so that injury to animals and handlers is minimized. Solid-sided alleys provide less potential distractions from the outside to cattle in them, but if a person

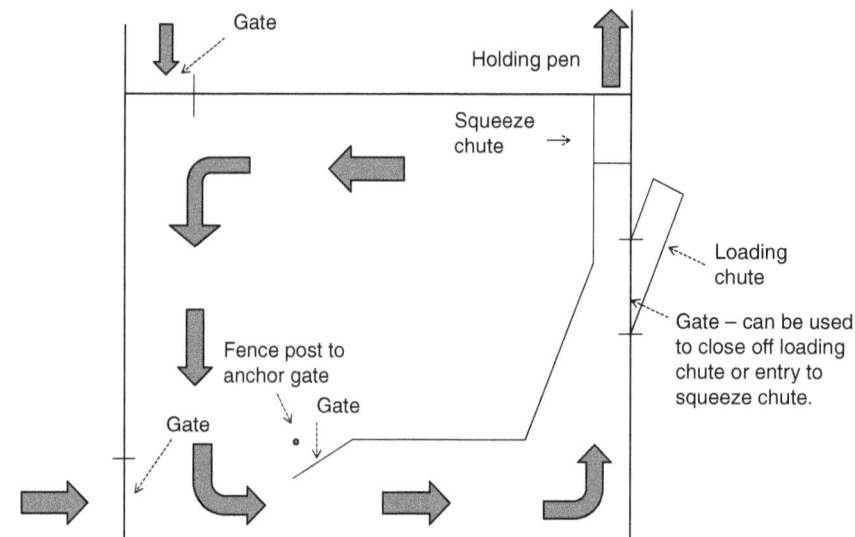

Fig. 10.6. Minimum considerations for cattle working facility (not drawn to scale). Entry into this pen could be from a small pasture or a holding pen. It is recommended that cattle leaving the squeeze chute flow into an additional pen in case animals slip past the head gate operator or have to be released for emergency situations. This type of set-up can be made with portable equipment or as a permanent fixture. Arrows show intended animal flow.

is in the alley with the cattle and needs to get out quickly, completely solid sides with no escape opening or steps can be problematic if an emergency develops. It is possible that one or more animals may become aggressive or combative in close quarters (especially when isolated from other animals), even though they appear calm in large pens or on pasture. No matter what the situation is, facilities must be built in a manner such that they are sturdy enough to hold the most aggressive and largest animal possible, and must have gates that are easy to close and will latch securely if problem animals act in an undesirable fashion. Figure 10.7 shows a potential cattle sorting pen.

Convenience for both cattle and personnel is an important consideration in the use of facilities and equipment (Fig. 10.8). There may be desirable features of handling facilities that provide relief for both. Covered working structures provide protection for both animals and people. A solid roof provides protection from sun and heat in warm weather and dry areas in wet conditions (which could be desirable on both warm and cold days). In areas with mild and dry conditions, this type of structure may not be considered that useful, but it can also provide protection from ultraviolet radiation when vaccinations are being administered, and extend

the life of equipment and structures underneath it. In many situations it may be highly desirable to have access to water and electrical power at the working facility. Water storage units and solar or petroleum-fuel generators could allow some producers to utilize several management techniques or technologies. The investment in useful, functional cattle working facilities should be taken seriously.

10.3.3 Portable Pens and Facilities

Several manufacturers offer portable cattle scales, squeeze chutes, loading chutes and pens. If investing in portable equipment, the operator needs to make sure that these items are constructed well enough to withstand their intended uses and longevity. If cattle are located in multiple physical locations, and there are not enough cattle or frequency of use to justify investment in permanent facilities, portable equipment may be a wise investment. Due to the portable nature of these items, they are likely to use less materials than recommended permanent facilities (such as solid-sided pens) to minimize their weight. However, even portable panels and pens can be adapted for improved cattle handling, such as by having panels of plywood to attach to the outsides of panels to create solid-sided effects. It is critical

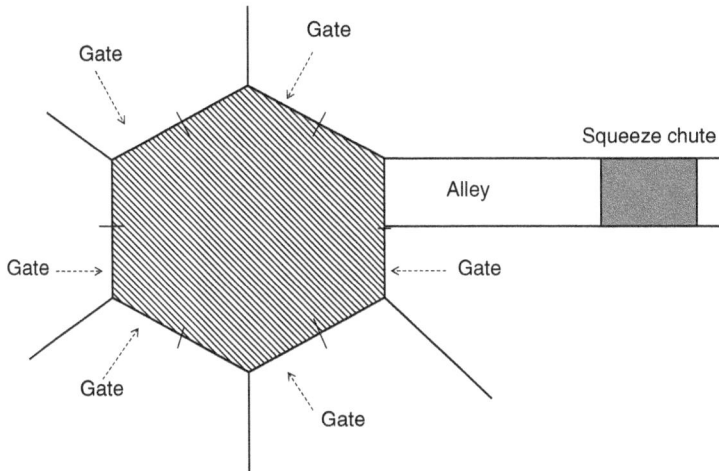

Fig. 10.7. Cattle sorting area after animals leave a working chute. The ability to sort animals into management groups is critical for improved management efficiency. In this example there is a common area that opens into six different holding pens. If heavily used by many animals and/or frequently used, the floor of the sorting area would be recommended to be concrete with a scarified surface.

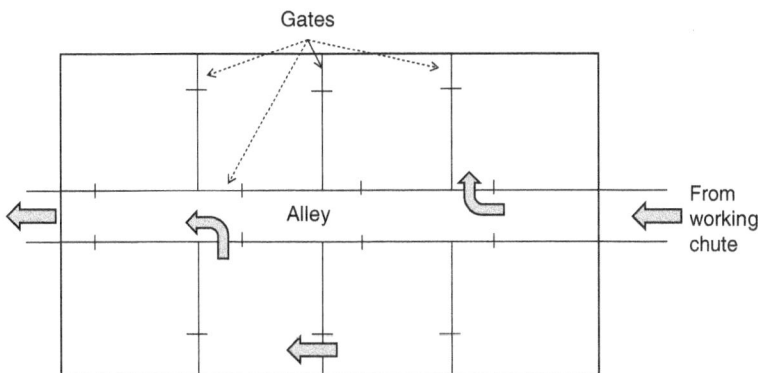

Fig. 10.8. Example sorting pens. The alley in the middle is the same width as the gates so a gate that closes a holding pen can also be fastened to close off the alley. Gates should be hinged to match the most common animal flow and have much greater utility if they can be made to open in multiple directions. Gates between adjacent pens (away from the alley) provide added utility for moving and sorting animals.

that portable equipment be properly anchored while in use, and that there are not many loosely assembled pieces that can add extra noise or visual distractions for the animals and might impede their movement (plywood not securely fastened and banging against panels when cattle push against them, chains used to connect panels with long loose ends that swing, etc.). This concept is true for permanent equipment and facilities as well.

10.3.4 Transportation

In recent years, there has been growing attention paid to welfare of cattle during transport, and several industry organizations have developed guidelines for transporting cattle. The same general principles of effective animal handling and knowledge of behavior apply to loading and transport of cattle; however, there are additional considerations specific to transportation that should be evaluated as a component of

BQA programs. Most of the recommendations on cattle transportation have been developed in Europe, Australia, Canada and the USA. Some specific transportation-related considerations are discussed here.

Loading

Pens and ramps used for loading cattle should minimize the potential for animals to slip. The loading ramp height and slope should be appropriate for the intended truck and trailer height. Facilities that ship cattle routinely have permanent loading/unloading ramps. Operators that do not load cattle often may use portable loading ramps. These are adjustable in regard to height, but simply changing the height automatically changes the slope as well. Grandin (2008) and others have warned against using excessively steep ramps. The slope is usually recommended to be from 10° to 20° (or 11 to 22%).

The loading chute and ramp should have some type of non-slip flooring such as grooved concrete, wire panels, sand, rubber mats or wooden cleats. The loading ramp may use these materials in a continuous slope or be composed of stair steps. Stair steps should have a rise of 100 mm and be 300 mm long, and cleats should be spaced 200 mm apart to match cattle stride (Grandin, 2008). The loading chute/ramp should be the same width as the opening on the truck or trailer for ease of loading. For instance, if trucks have single-file doors entering the trailer, the width of the ramp/chute should match this so that animals load single file and cannot get beside one another, which may cause injury and/or added stress.

Hauling

Effective and timely transportation of cattle is a vital marketing component. Several factors impacting the welfare and economic value of cattle should be considered in their transport; these include animal density, road conditions, driver style and qualifications, vehicle and trailer maintenance and cleanliness, time spent loaded, scheduling and communications, weather conditions, animal variation in size, trailer design, and individual animal differences (see Fig. 10.9). Many of these factors are interrelated, and therefore should be considered collectively. These general considerations are also true if cattle are to be transported by boat or even by airplane, although several specific details are obviously different.

Fig. 10.9. North American cattle truck. The upper photo shows the truck and trailer as it appears, whereas the lower figure shows the layout of the interior eight compartments within the trailer. Having multiple, smaller groups of animals within a trailer provides for better animal welfare as well as prevention of weight shifting during transport. This type of trailer may be referred to as a 'pot belly' as the center section is lower than the front and rear sections.

The density of the animals in the truck is one of the most obvious considerations, and the amount of space needed per animal relates directly to animal size, temperament, stage of production and social interactions. The amount of variation in the group of animals is usually not under the control of the manager; however, the grouping of animals within a trailer or even within compartments of a trailer can be an effective element in minimizing transportation stress. Table 10.4 shows recommended loading densities in Australia for cattle of different sizes.

Transportation varies considerably around the world. In Australia, road trains (two or three trailers per truck, each with two decks) are commonly used to haul cattle. If these trucks were hauling 500 kg cows loaded at the recommended density from Table 10.4, there would be 24 cattle per deck, 48 cattle per trailer, and 144 cattle per truck. Several markets rely on import/export of live cattle overseas, and it is common for large ships holding 10,000–30,000 cattle to be used. These types of ships have impressive interior designs for housing cattle and feed as well as waste management. The same quality assurance principles and guidelines are recommended for ships as for trucks and trailers.

Obviously cattle of different sizes need different amounts of space. Additional considerations than can impact on space requirements are the presence of horns and distance traveled and the corresponding amount of time spent on the truck. Table 10.5 has values from Grandin (2008), which show recommended loading densities of polled versus horned cattle. As all horned cattle are not the same in regard to horn length, direction or size, different types of horned cattle may have different space requirements as well. The guidelines in Table 10.5 should be adequate for animals that have medium to small horns. Cattle that have had their horns 'tipped' (the ends removed) still behave as horned cattle, and Grandin (2008) has recommended that if more than 10% of the cattle are horned or tipped, the shipment of animals should receive the same space allotment as if they all were horned.

The European Commission (2002) has provided guidelines for transport of cattle among EU member nations for cattle hauled long distances and for small calves; these values are provided in Table 10.6. Cattle that are older typically do not lie down during transport in trucks and trailers, whereas young calves are much more prone to lie down while being hauled. Of course if animals are transported for a long enough time, they will lie down due to fatigue. Much more space per animal must be allocated if they are to lie down

comfortably during transport as opposed to remaining standing. This is the main reason cattle shipped by boat are provided pen space with a larger floor area than in trucks, as these trips may last 2 to 14 days.

It is desirable that cattle remain standing during transport in trucks. The loading densities used in industry are based on the assumption that cattle remain standing. If an animal lies down it has a much higher potential of being stepped on or another animal falling on it, potentially causing injury. This illustrates why it is important to separate animals of different sizes during shipment and the need for internal compartments in stock trailers. If a load of cow-calf pairs is to be shipped and the calves are young, they certainly need to be sorted from cows and housed together in separate compartments. If young calves are more prone to lie down during transport and they are housed with larger and/or more aggressive animals, they are more likely to get trampled, and this is exacerbated by long traveling distances.

Table 10.4 Recommended truck loading densities for cattle in Australia.

Mean live weight per animal (kg, lb)	Floor area per animal (m², ft²)	Animals per 12.2 m single-deck trailer
250, 551	0.77, 8.3	38
300, 662	0.86, 9.3	34
350, 772	0.98, 10.5	30
400, 882	1.05, 11.3	28
450, 992	1.13, 12.2	26
500, 1103	1.23, 13.2	24
550, 1213	1.34, 14.4	22
600, 1323	1.47, 15.8	20
650, 1433	1.63, 17.5	18

Taken from Animal Health Australia (2012)

Table 10.5 Recommended truck loading densities for horned and polled cattle.

Mean weight of fed steers or cows, kg (lb)	Horned, tipped, or more than 10% horned or tipped, m² (ft²)	No horns (polled), m² (ft²)
360 (800)	1.01 (10.9)	0.97 (10.4)
454 (1000)	1.20 (12.8)	1.11 (12.0)
545 (1200)	1.42 (15.3)	1.35 (14.5)
635 (1400)	1.76 (19.0)	1.67 (18.0)

From Grandin (2008).

Table 10.6. European recommendations for cattle transportation space and rest.

Cattle, size	Travel duration (hours)	Floor space allowance (m²)	Travel duration (hours)	Floor space allowance (m²)
500 kg	Up to 12 hours	1.35	Over 12 hours	2.03
Calves	First travel period (hours)	First rest (hours)	Second travel period (hours)	Second rest (hours)
Under 6 months	8	6	8	24
Over 6 months	12	6	12	24
	Floor space for young calves of different sizes			
Weight (kg)	Up to 55	55–110	110–250	
Space allowance (m²)	0.3–0.4	0.4–0.7	0.7–0.8	

From European Commission (2002).

Distances traveled where cattle spend less than 12 hours loaded in a trailer are most desirable, and the EC has recommended that calves under 6 months of age not spend more than 8 hours in a loaded truck. In some instances, remaining still in a loaded trailer may be equally or more stressful than being in transit. If trucks are parked in direct sunlight in extremely hot temperatures for long periods, this will probably cause more stress from heat and possibly air quality as opposed to moving, which provides a flow of air. Large commercial cattle trucks are designed to protect animals from environmental elements. Producers that have their own trucks or trailers but which do not have closed roofs, solid sides, etc. need to consider environmental conditions during transit of cattle so that they do not cause unintended added stress from wind chill, lack of ventilation during high temperatures, and other potential problems.

In general, the thought is that cattle learn to associate various activities with resulting outcomes, and if treatment is aversive animals can take more time and/or require more force to move through facilities or respond to handling practices. In the same way, cattle that eat something undesirable, which causes pain or metabolic distress (or to experience undesirable outcomes associated with specific flavors) can have adverse effects on what plants they eat, and/or what areas of pastures they utilize for grazing. It is also likely that animals can perceive the same situation differently, and that some animals are more easily trained than others. Understanding and implementation of these considerations provides the potential for improved understanding and assessment of animal welfare.

10.4 Welfare

The World Organization for Animal Health (OIE) defines animal welfare as follows: 'Animal welfare means how an animal is coping with the conditions in which it lives. An animal is in a good state of welfare if (as indicated by scientific evidence) it is healthy, comfortable, well nourished, safe, able to express innate behaviour, and if it is not suffering from unpleasant states such as pain, fear, and distress.' This definition has been widely adopted by various animal organizations across the world. The European Commission (2012) recently provided descriptions and concepts of five 'freedoms' developed to ensure proper animal welfare as being: (i) freedom from hunger, thirst and malnutrition, (ii) freedom from fear and distress, (iii) freedom from physical and thermal discomfort, (iv) freedom from pain, injury and disease, and (v) freedom to express normal patterns of behavior.

There are many production-based management practices known to cause physical pain or discomfort based on people's perceptions of how animals react to those practices. Obvious examples known to cause stress because they cause pain are dehorning, branding and castration. However, there are other situations or practices that are less obvious in the amount of stress they induce, such as interaction with animal caretakers and transportation. Among the European Commission's five freedoms, several of these can be accomplished completely, but freedom from fear, distress and physical discomfort may be hard to quantify.

As mentioned throughout this chapter, many animals respond quite differently to the same stimuli. If avoidance of momentary pain or distress is a priority, it is possible that animal welfare and well-being could be worse in the overall scheme of things (for instance if we did not restrain and vaccinate animals to avoid distressing them, but they subsequently contracted an infectious disease that severely threatened their health or life). Situations that cause chronic and persistent pain and discomfort to animals are to be avoided. Banning castration or some other painful management technique severely limits the potential for producers to have flexibility in their management systems. Minimizing the pain and discomfort associated with a procedure that is known to be painful and uncomfortable is an entirely different line of thought than completely banning the procedure. If a group of intact males fight more amongst themselves than a group of castrated males, which group has experienced more discomfort? Sometimes the answers to these types of questions are not so clear-cut. As a result, some discussion devoted to the topic of 'stress' and its relationship to behavior is presented here.

10.4.1 Considerations of Stress

Stress might be explained as the condition of mental, emotional and/or physiological strain resulting from unfavorable circumstances. Various concepts brought up throughout this chapter, such as animal temperament, behavior and handling, are definitely stress related because the same conditions may be more or less stressful for individual animals, and the manner in which various animal handling and transportation activities are performed can increase or decrease an animal's stress. Certainly animals cope with a variety of influences on a daily basis, and individual animals can cope with the same set of conditions with differing mechanisms. Several hormones can be involved with various physiological changes associated with stress and are described in Table 10.7.

Table 10.7. Hormones and chemicals involved in stress-related responses.

Name and abbreviation	Category	Action/activity	Site of production	Stimulus for production or release
Corticotropin-releasing hormone (CRH)	Protein-peptide	Stimulates secretion of ACTH	Hypothalamus	Stressful situations
Adrenocorticotropic hormone (ACTH), may also be called corticotropin	Protein-peptide	Stimulates synthesis and secretion of glucocorticoids such as cortisol	Anterior pituitary	Stressful situations
Vasopressin (VP) or antidiuretic hormone (ADH)	Protein-peptide	Maintenance of water, glucose and salts in circulatory system, increases blood pressure	Hypothalamus	Dehydration, certain chemicals such as acetylcholine, and pain
Norepinephrine (NE, may be called noradrenaline)	Amino-acid derived	Increases heart rate and pressure, increases blood flow to muscles	Adrenal medulla and some neurons	Fight-or-flight situations
Epinephrine (EPI, may be called adrenaline)	Amino-acid derived	Increases heart rate, muscle contraction, vasoconstriction of blood vessels, vasodilation of air passages, increases lipolysis and gluconeogenesis	Adrenal medulla and some neurons	Fight-or-flight situations
Cortisol	Steroid-glucocorticoid	Increases gluconeogenesis, suppresses inflammation and immunological responses	Adrenal cortex	Stressful situations
Acetylcholine	Neurotransmitter	Muscle activity, and may be involved in decision making and attention, stimulates release of epinephrine and norepinephrine	Neurons (cells of central and peripheral nervous systems)	Mild stress situations and/or learning processes

10.4.2 Feedlot and Housing Considerations

Cattle in confined facilities can be housed outside in pens or indoors. It is very important that proper space and ventilation be provided for cattle in confinement scenarios. Depending upon the climatic environment, in concert with the type of cattle (genetics and stage of production), feedlot settings may need to provide protection from wind, solar radiation, rain/mud, or dust. This may also present a challenge because structures that provide desirable blockage of wind during winter may be undesirable if they block wind during the heat of summer. In some cases, there may be ways of storing feed (stacks of hay bales, covered silage, etc.) that provides wind blockage during winter, but these supplies can be used before summer. If someone is considering a new livestock facility in areas where it is undesirable for people to live such as near a landfill, etc., housing of animals where they would be continually subjected to other undesirable conditions (smoke, fumes from pollutants, etc.) should be avoided. Also, animals that become ill need extra protection from environmental elements compared to their healthy contemporaries, and special housing or facility space for this use should be provided. Table 10.8 illustrates some guidelines on cattle housing and feeding space allowance.

10.5 Summary

There is no substitute for regular observation of cattle behavior to help understand components of beef cattle production. Understanding of 'normal' and 'abnormal' behaviors in general as well as observation and recognition of individual animal differences for what are 'normal' and 'abnormal' are also very important. Cattle, as most animals, can and will respond to a wide variety of stimuli, including training. Training in regard to proper handling techniques can reduce stress levels for both animals as well as personnel. Design of cattle handling equipment and facilities can complement (or inhibit) animal experiences and the effectiveness of training effort. Behavior of cattle is as important when they are not under the influence of managers as when they are being processed through facilities or being transported. Cattle will modify their behavior in response to conditions that are deemed unfavorable to alleviate stress, such as seeking shade to reduce heat stress or seeking cover and separation from the herd before giving birth. Overall welfare and well-being of cattle are very important to their production efficiency and longevity, and documentation of proper cattle welfare management techniques is likely to become increasingly important for some consumers and/or regulatory agencies.

Table 10.8 Recommended space allowance per animal for housing and feeding cattle.

	Calves, 180–380 kg	Feedlot cattle, 360–545 kg	Bred heifers, 360 kg	Cows, 455 kg	Cows, 590 kg	Bulls, 680 kg
			Open lots with no barn, m²			
Unpaved lots with mound (including mound space)	14.0–28.0	23.2–46.5	23.2–46.5	18.6–46.5	28.0–46.5	46.5
Mound space, 25% slope	1.9–2.3	2.8–3.3	2.8–3.3	3.7–4.2	3.7–4.2	4.7–5.6
Unpaved lot, 4–8% slope, no mound	28.0–55.8	37.2–74.4	37.2–74.4	32.5–74.3	32.5–74.3	74.3
Paved lot, 2–4% slope	3.7–4.7	4.7–5.6	4.7–5.6	5.6–7.0	5.6–7.0	9.3–11.6
			Barns (unheated cold housing), m²			
Open front with dirt lot	1.4–1.9	1.9–2.3	1.9–2.3	1.9–2.3	2.3–2.8	3.7
Enclosed, bedded pack	1.9–2.3	2.8–3.3	2.8–3.3	3.3–3.7	3.7–4.7	4.2–4.7
Enclosed, slotted floor	1.1–1.7	1.7–2.3	1.7–2.3	1.9–2.3	2.0–2.6	2.8
			Feeder space (mm) for different types of animals when fed:			
Once daily	457–559	559–660	559–660	610–762	610–762	762–914
Twice daily	229–279	279–330	279–330	305–381	305–381	–
Free choice grain concentrate	76–102	102–152	102–152	127–152	127–152	–
Self-fed roughage/high forage	229–254	254–279	279–305	305–330	305–330	–

From FASS (2010) Chapter 6.
This table is presented in imperial units in Table A11.

10.6   Study Questions

10.1 If you were to design a cattle feedlot for 250 animals, where they would be grown from 250 kg to 500 kg:

10.1a How much total space (m², hectares, etc.) would you want to allot for the pens, and why?

10.1b Draw a possible layout of this facility (animal pens, alleys, feeding areas, loading area and cattle handling/processing area).

10.2 If you were to evaluate animal temperament for a herd of cattle that have not been scored for this trait in the past, which type(s) of scoring system(s) would you utilize, and why?

10.3 You are negotiating with a cattle hauler to transport 500 kg cows for you. The trip will take approximately 8 hours. This hauler uses double-deck trailers that measure 2.5 m wide by 12 m long for each deck, and he says that he has put 60 cows per trailer in the past. What do you think about what this hauler has told you?

10.4 Why are cattle trailers with multiple compartments per deck preferable to trailers with one single open space per deck?

10.5 If you had to choose between: (a) proper cattle handling techniques of workers but less than ideal facility design/layout, or (b) ideal facility design/layout but poor cattle behavior knowledge and lack of care from workers, which of these two scenarios would you prefer as the manager of the operation and why?

10.6 Would you say that cattle temperament has an optimal intermediate level, or should an extreme be favored in selection of breeding stock? Explain.

10.7 Describe five behaviors of cattle that you would consider abnormal.

10.7   References

Animal Health Australia (2012) *Australian Animal Welfare Standards and Guidelines – Land Transport of Livestock*. Animal Health Australia (AHA), Canberra.

Bailey, D.W. (2004) Management strategies for optimal grazing distribution and use of arid rangelands *Journal of Animal Science* 82(Suppl), E147–E153.

Blackshaw, J.K., Blackshaw, A.W. and McGlone, J.J. (1997) Buller steer syndrome review. *Applied Animal Behaviour Science* 54, 97–108.

Bonsma, J. (1975a) Judging cattle for functional efficiency, Beef Cattle Sci. Handb. 12, 23–36 (as cited by Grandin and Deesing, 2005).

Bonsma, J. (1975b) Crossbreeding for increased cattle production, Beef Cattle Sci. Handb. 12, 37–49 (as cited by Grandin and Deesing, 2005).

Erlinger, L.L., Tolleson, D.R. and Brown, C.J. (1990) Comparison of bite size, biting rate and grazing time of beef heifers from herds distinguished by mature size and rate of maturity. *Journal of Animal Science* 68, 3578–3587.

European Commission (2002) *The Welfare of Animals During Transport (Details for Horses, Pigs, Sheep and Cattle)*. Report of the Scientific Committee on Animal Health and Animal Welfare, Brussels.

Fang, X., Downey, E., Runyan, C.A., Sawyer, J.E., Hairgrove, T.B., Ridpath, J.F., Gill, C.A., Mwangi, W. and Herring, A.D. (2012) Correlations of temperament with titer and hematological responses of crossbred steers challenged with bovine viral diarrhea virus. *Journal of Animal Science* 90(Suppl. 3), 223(Abstr.).

FASS (2010) *Guide for the Care and Use of Agricultural Animals in Research and Teaching, 3rd Edition*, January 2010. Federation of Animal Science Societies, Champaign, IL.

Gauly, M., Mathiak, H., Hoffmann, K., Kraus, M. and Erhardt, G. (2001) Estimating genetic variability in temperamental traits in German Angus and Simmental cattle. *Applied Animal Behaviour Science* 74, 109–119.

Grandin, T. (2000) *Livestock Handling and Transport*, 2nd ed. CABI Publishing, Wallingford, UK.

Grandin, T. (2008) Cattle Transport Guidelines for Meat Packers, Feedlots, and Ranches. Available at: http://www.grandin.com/meat.association.institute.html (accessed 19 December 2011).

Grandin, T. and Deesing, M.J. (2005) Genetics and behavior during handling, restraint, and herding. Available at: http://www.grandin.com/references/cattle.during.handling.html. Accessed on March 15, 2005.

Gubernick, D.J. (1980) Maternal "imprinting" or maternal "labelling" in goats. *Animal Behaviour* 28, 124–129.

Hall, S.J.G. (1989) Chillingham cattle: social and maintenance behaviour in an ungulate that breeds all year round. *Animal Behaviour* 38, 215–225.

Hearnshaw, H. and Morris, C.A. (1984) Genetic and environmental effects on a temperament score in beef cattle. *Australian Journal of Agricultural Research* 35, 723–733.

Kabuga, J.D. and Appiah, P. (1992) A note on the ease of handling and flight distance of *Bos indicus, Bos taurus* and their crossbreds. *Animal Production* 54, 309–311.

Le Neindre, P., Trillat, G., Sapa, J., Ménissier, F., Bonnet, J.N. and Chupin, J.M. (1995) Individual differences in docility in Limousin cattle. *Journal of Animal Science* 73, 2249–2253.

Mitlöhner, F.M., Morrow-Tesch, J.L., Dailey, J.W., and McGlone, J.J. (1999) Altering feeding times for feedlot cattle reduced dust-generating behaviors. *Journal of Animal Science* 77(suppl. 1), 148(Abstr.).

Morris, C.A., Cullen, N.G., Kilgour, R. and Bremner, K.J. (1994) Some genetic factors affecting temperament in *Bos taurus* cattle. *New Zealand Journal of Agricultural Research* 37, 167–175.

Pajor, E.A., Rushen, J. and de Passillé, A.M.B. (2000) Aversion learning techniques to evaluate dairy cattle handling practices. *Applied Animal Behaviour Science* 69, 89–102.

Price, E.O., Thos, J. and G.B. Anderson (1981) Maternal responses of confined beef cattle to single versus twin calves. *Journal of Animal Science* 53, 934–939.

Reinhardt, C., Reinhardt, A. and Reinhardt, V. (1986) Social behaviour and reproductive performance in semi-wild Scottish Highland cattle. *Applied Animal Behaviour Science* 15, 125–136.

Riley, D.G., Chase, Jr., C.C., Coleman, S.W., Randel, R.D. and Olson, T.A. (2004) Assessment of temperament at weaning in calves produced from diallele matings of Angus, Brahman, and Romosinuano. *Journal of Animal Science* 82(Suppl 1.), 6(Abstr.).

Sandlin, B.A., Brown, A.H. JR. Z.B. Johnson, J.A. Hornsby, R.T. Baublits, and B.R. Kutz (2005) Case Study: Postpartum maternal behavior score in six breed groups of beef cattle over twenty-five years. *The Professional Animal Scientist* 21, 13–16.

Sato, S., Sako, S. and Maeda, A. (1991) Social licking patterns in cattle (*Bos taurus*): Influence of environmental and social factors. *Applied Animal Behaviour Science* 32, 3–12.

Schloeth, R. (1958) Le cycle annual et le comportment social du taureau de Camarague. *Mammalia* 22, 121–139 [as cited by Lott, D.F. (1979) Applied ethology in a nomadic cattle culture. *Applied Animal Ethology* 5, 309–319]

Selman, I.E., McEwan, A.D. and E.W. Fisher (1970) Studies on natural suckling in cattle during the first eight hours post partum: I. Behavioural studies (dams). *Animal Behaviour* 18, 276–283.

Shrode, R.R. and Hammack, S.P. (1971) Chute behavior of yearling beef cattle. *Journal of Animal Science* 33(Suppl. 1), 193(Abstr.).

Stookey, J.M. (2001) Buller steer syndrome. University of Saskatchewan Western College of Veterinary Medicine Applied Ethology. Available: http://www.usask.ca/wcvm/herdmed/applied-ethology/articles/bullers.html. (accessed 20 July 2010).

Stricklin, W.R., Heisler, C.E. and Wilson, L.L. (1980) Heritability of temperament in beef cattle. *Journal of Animal Science* 51(Suppl. 1), 109(Abstr.).

Taylor, L.F., Booker, C.W., Jim, G.K., and Guichon, P.T. (1997) Epidemiological investigation of the buller steer syndrome (riding behaviour) in a western Canadian feedlot. *Australian Veterinary Journal* 75, 45–51.

Voisinet, B.D., Grandin, T., Tatum, J.D., O'Connor, S.F. and Struthers, J.J. (1997) Feedlot cattle with calm temperaments have higher average daily gains than cattle with excitable temperaments. *Journal of Animal Science* 75, 892–896.

Wiggers, D.L., McGlone, J.J., Morrow-Tesch, J.L. and Dailey, J.W. (1998) Behavior of feedlot cattle fed once or three times per day. *Journal of Animal Science* 76(suppl. 1), 102(Abstr.).

11 Carcass and Meat

The ultimate value of beef cattle in most settings is utilization of animal body components primarily for food. In dairies, milk production (quantity and quality) takes on a primary purpose; in other scenarios, draft, transportation or ceremonial uses may be primary functions. No matter which of these primary uses exist, the ultimate conversion of cattle into food and other useful components is very important. Sanitation is critical for every step and process when cattle are converted into edible end-products. It is important to have inspection systems to ensure the safety and wholesomeness of beef and associated end-products and by-products for all processes involved. Grading and classification of beef carcasses and associated products based on specific standards relating consistency from consumer expectations and satisfaction are needed for domestic and international trade. This chapter discusses general concepts associated with conversion of cattle into carcasses and associated by-products, food safety, issues related to beef palatability and consumer eating satisfaction, and comparisons of selected beef grading systems used in some countries.

11.1 Conversion of Cattle to Carcasses and End-Products

There are three primary considerations in regard to beef carcasses and the resulting meat: (i) food safety and wholesomeness, (ii) amount of beef yield from the live animal as well as the resulting carcass, and (iii) eating quality and palatability of associated products. The food safety aspect is the most critical, otherwise the yield and quality of condemned carcasses or products and/or processes that harm consumers are irrelevant.

The degree of beef yield from the carcass is primarily a function of amount of muscle and fat, but amount of gut fill, size of digestive tract and internal organs, amount of bone, and amount of water loss also influence the yield percentage. Dressing percent

is a term used to describe the percentage of the live animal weight that ends up as the carcass weight, and is calculated by dividing the carcass weight by the live weight. In some cases the terms 'hot carcass weight' or 'chilled carcass weight' may be used, and these simply refer to when the carcass weight was taken; hot or warm carcass weight refers to the weight before it is placed in a cooler, whereas chilled weight is the weight several hours (such as 24 or 48) after being in a cooler. Typical water loss during the chilling or cooling period is 1–2% for the first 48 hours, but can be 3–6% for carcasses aged 14 days. The steps involved in converting cattle into beef products and by-products are briefly discussed below.

11.1.1 Steps in Conventional Harvest (Slaughter) Process

The general procedure in killing cattle to harvest meat that has been found to be the most efficient from a production standpoint and most appropriate from a food safety angle involve the general steps (in order) of: (i) stunning, (ii) bleeding, (iii) head, feet and hide removal, (iv) evisceration, (v) carcass splitting, (vi) washing and (vii) chilling. These steps are followed in packing plants. For certain scenarios, there may be exceptions or omissions of certain steps. For instance in a 'wet market' situation, where the animals are harvested and the meat is made available immediately after death, there is not likely to be splitting of the carcass, and there will not be a chilling period. The stunning phase may also be quite different or non-existent. In some regions, entire carcasses are transported to local meat markets where individual cuts can be selectively taken from the carcass at the customer's request.

Depending upon the amount of labor and the design and equipment in the facilities, it may only take 20–30 minutes from stunning to cooler. Large commercial beef plants can process one animal per minute or faster within a production line and they may have multiple lines. It is not uncommon for large corporate beef plants to process 4000–5000 cattle per day using two 8-hour work shifts. Many packing plants will have their carcasses graded after a 24- to 48-hour chilling period. Figure 11.1 shows a beef carcass with 10 mm of fat cover.

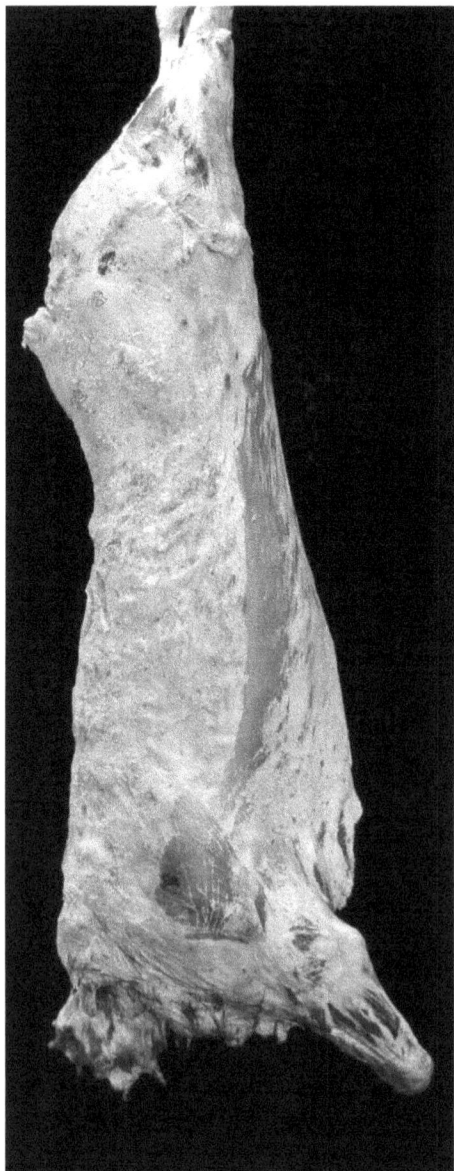

Fig. 11.1. Side view of beef carcass. Photo courtesy of D.B. Griffin.

In large plants, after carcasses are chilled they are graded and then are fabricated into different cuts, and how the carcasses are fabricated and the associated names of the beef cuts can be very different from one country (or region) to the next. Carcasses with similar grades may be separated into their own groups for fabrication. In several countries it is common that whole beef carcasses from smaller plants are shipped daily to local butcher or meat shops. It is common that shoppers will come to these shops, tell the butcher what carcass parts they want, and have the butcher cut off the area(s) of the carcass they want to purchase. It is routine that in these situations the beef carcass components are sold (and consumed) the same day the animals are harvested.

11.1.2 Ritual Beef Slaughter (Kosher and Halal)

Kosher and halal are terms used to describe what is 'fit and proper' to eat for two groups of people, Jews and Muslims, respectively. These terms are used to describe a wide array of foods and beverages including meat. Both kosher and halal requirements are rooted in holy scriptures of these world religions, and the basic requirements for each are briefly discussed here, and heavily rely on the meat science course notes of Savell (2013).

Kosher

Only certain animals are acceptable under kosher requirements, which allow consumption of species such as cattle, sheep, goats, deer and bison, but prohibit consumption of others such as camels, rabbits and pigs. The kosher ritual slaughter is known as shechitah, and the person who performs the slaughter is called a 'shochet', both from the Hebrew root Shin-Chet-Tav, meaning to destroy or kill. The method of slaughter is a quick, deep stroke across the throat with a perfectly sharp blade with no nicks or unevenness. This method is thought to cause unconsciousness within 2 seconds. This process ensures rapid, complete draining of the blood, which is also necessary to render the meat as kosher. The shochet is not simply a butcher; he must be a pious man, well trained in Jewish law, particularly as it relates to kashrut. In many cases a rabbi may be involved or serve as the shochet. Following use of acceptable animals and ritualistic slaughter, animals must be classified as 'glatt'. The term glatt means smooth and refers to the lungs of the animal during inspection. Inspectors known as 'bodeks' examine certain organs from each animal. The lungs of each animal are examined for any adhesion or other defects such as might follow from severe pneumonia. Glatt kosher meat must be soaked and salted with 72 hours of slaughter, because it is not customarily subjected to the extra spraying ('begissing') that can keep meat moist prior to soaking and salting.

Consumption of blood is forbidden, which is why complete bleeding and soaking of meat in salt and water is so important in removing residual blood. Also, the sciatic nerve and its adjoining blood vessels may not be eaten. The process of removing this nerve is time-consuming and not cost-effective, so many processors where kosher markets are not the primary target may simply sell the hindquarters to non-kosher butchers.

Halal

Information for this section was obtained from Savell (2013) and Aidaros (2005). There are two methods of halal slaughter (Dabh and Nahr). Dabh involves severing the trachea, esophagus and jugular veins and is used for sheep, goats, cattle and buffaloes. Nahr involves cutting the blood vessels at the base of the neck and can be used for camels, cattle and buffalo.

Halal slaughter dictates that before slaughter animals should be rested, be well fed and well looked after; animals should not be starved at the point of slaughter. No animal should be slaughtered in front of another animal. Separation between holding and slaughter areas is recommended to ensure that animals awaiting slaughter do not see other animals being bled. The knife should not be sharpened in front of the animal. The knife and cut should not occur vertically but be drawn horizontally across the neck in a single motion. The cut should not be close to the chest or too near to the head and should be a clean cut without tearing of tissues. The cut for bleeding cannot be used to severe the head from the body. The person slaughtering the animals should hold the knife in the right hand and the left hand should be used to hold the animal's head and restrain the animal. Any Moslem having reached puberty is allowed to slaughter after saying the name of Allah and facing Makkah (Mecca). The name of Allah has to be mentioned before or during slaughter and must be said by a member of the Moslem faith. Animals should be provided with enough time to bleed out completely and further preparation, and dressing of the carcass must be delayed until all signs of life and cerebral reflex have disappeared. Violating any of these instructions and recommended procedures leads to the slaughter being rendered non-halal.

Islam prohibits the consumption of pork, strangled animals, animals devoured by wild beasts, blood, meat from animals found dead, fatally beaten animals, horn-butted animals, animals sacrificed to idols, animals slaughtered by people other than people of the scriptures (Christians and Jews), and animals dedicated to any other than God.

11.2 Structure and Composition of Muscle

There are three general types of muscle: (i) skeletal, (ii) cardiac and (iii) smooth. Skeletal muscle is attached directly or indirectly to the skeleton and is responsible for the physical movement in live animals, whereas cardiac muscle maintains the blood flow through actions of the heart, and smooth muscle is responsible for involuntary movements within the digestive and reproductive systems. It is skeletal muscle that makes up the bulk of the beef carcass, and warrants the majority of discussion here.

11.2.1 Skeletal Muscle

There are many different levels of organization with skeletal muscle. The basic unit of contraction for this type of muscle is the sarcomere (the skeletal muscle cell). Sarcomeres (Fig. 11.2) contain both thin and thick myofilaments in an alternating pattern, where thin myofilaments are between thick myofilaments and vice versa. The thin myofilaments partially overlap end-to-end with the thick myofilaments, and the region that is open between thin myofilaments end-to-end is known as the H-zone. Thick filaments are composed primarily of the protein myosin; thin myofibrils are composed mainly of the proteins actin, troponin and tropomysin. The connecting units between sarcomeres are called Z filaments (and this region is termed the Z-line). There are very many sarcomeres (perhaps 8000 or so) which together make up a myofibril; subsequently, many myofibrils (possibly 2500)

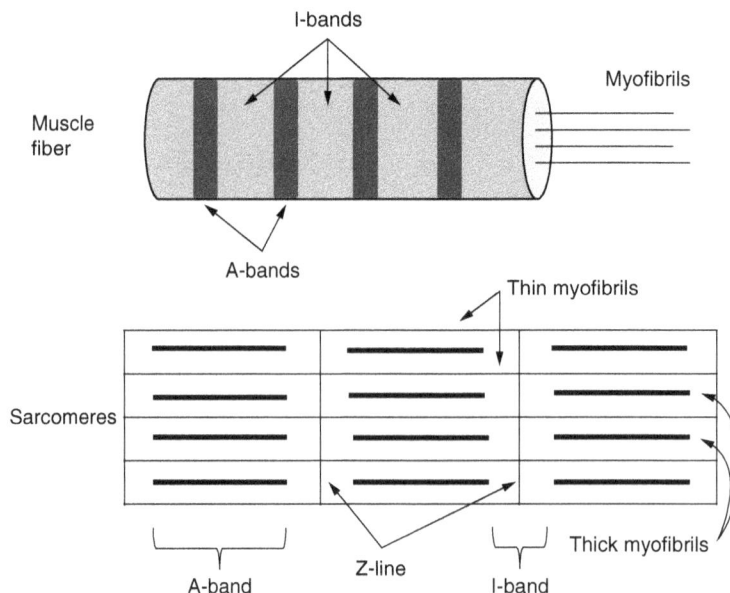

Fig. 11.2. Structures of muscle fiber (above) and sarcomeres (below) as viewed from side. Adapted from Aberle *et al.* (2001).

make up a muscle fiber. A muscle fiber will have a striated appearance with darker and lighter bands. The darker bands are known as A-bands, which represent the area where the thicker myofibrils are found, and the lighter bands are known as I-bands, and these areas are where only the thin myofibrils are found. Muscle fibers are composed of skeletal muscle cells, which are much larger than most cells of the body, and have multiple nuclei (as opposed to a single nucleus in most cell types). Muscle fibers are grouped together to form muscle bundles, and muscle bundles are grouped together to form entire muscles.

Connective tissue is closely associated with skeletal muscles, and there are several types of connective tissues. These can vary from extracellular substances that have anything from a jelly-like to a tough, fibrous consistency. Connective tissue can surround muscle fibers (endomysium), muscle bundles (perimysium) or entire muscles (epimysium) to provide structural support. Very often the extracellular fibers of connective tissue associated with skeletal muscle are comprised of the proteins collagen, elastin and reticulin. Characteristics of the muscle bundles (Fig. 11.3), muscle fibers and associated connective tissues can contribute to toughness of the resulting meat,

and this is why certain muscles and/or cuts may need to be cooked and prepared differently.

11.2.2 Conversion of Muscle to Meat

Information from this section relies greatly on information from Aberle *et al.* (2001), Lawrie (1998), Romans *et al.* (2001) and Swatland (1984), and is presented as a very general overview. For more detailed discussions one of these meat science references or others should be consulted.

As an animal dies, a series of complex biochemical processes can occur, and the conversion of living tissue such as skeletal muscle into meat takes several hours to complete. After the animal is dead, its tissues and cells attempt to keep operating, and energy is being utilized to maintain temperature and organizational structure. Removal of blood depletes the oxygen supply to the muscles, and sets into motion the steps of rigor mortis, or stiffening of the muscles.

Energy is stored in muscle as glycogen, which is made up of thousands of glucose molecules. The energy supply in the muscle (ATP) becomes depleted as ATP breakdown continues due to the enzyme ATP-ase, and this increases inorganic phosphate levels. This in turn stimulates the breakdown of

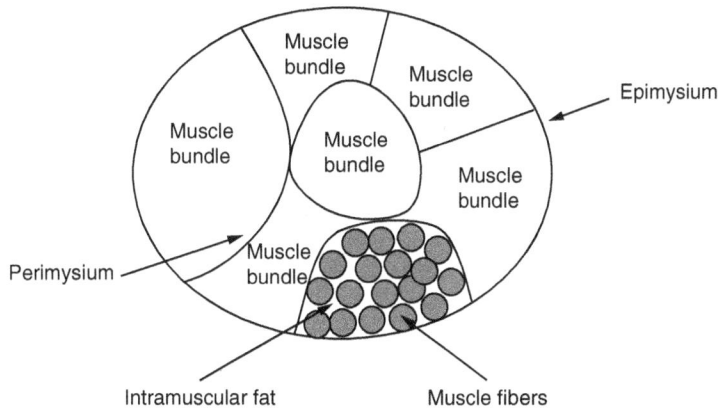

Fig. 11.3. Organizational structure of skeletal muscle. This schematic shows a cross-section through a muscle. Muscle fibers are only shown in one of the seven muscle bundles. Each muscle fiber is surrounded by endomysium connective tissue, whereas perimysium and epimysium are connective tissues surrounding muscle bundles and entire muscles, respectively. Intramuscular fat (marbling) is found within muscles between the muscle fibers, whereas intermuscular (seam) fat is found between muscles.

glycogen to lactic acid. This is known as the anaerobic pathway of glycolysis (the aerobic pathway for glycolysis leads to the Krebs, or citric acid cycle). This lowered level of ATP and resulting subsequent steps causes the pH in the muscle to begin to fall (become more acidic), and the muscles are less able to maintain the structural integrity of proteins. As ATP is the energy source for muscle contraction and relaxation, the components of the muscle fibers (actin and myosin filaments) become locked in place as ATP is depleted, and this is what causes rigor mortis. This process can be seen with the skeletal muscle of animals a few minutes following death as these muscles twitch (this is very obvious with the cheek muscles of the head).

Following death the pH begins to drop from around 7.0 to ultimately end up around 5.5. The rate of decline of ATP is related to the rate of pH decline. As lactic acid builds up, this allows for muscle protein to break down and is related to tenderness of the muscles. The pH and rate of decline is also related to water-holding capacity of the muscle.

11.2.3 Beef Carcass Quality Defects

The word 'quality' can have many different uses and interpretations. In this chapter beef carcass quality is used to refer to the attributes and characteristics of the meat that can impact consumer palatability and value, and where appropriate can also refer to influences on beef carcass quality grades where

those systems exist. There are several potential beef quality defects that can be associated with the muscle biochemistry and/or activity. These can be attributes of individual animals, or environmental conditions. The typical, and desirable, color of cattle skeletal muscle at death is purple, but within 24 hours the skeletal muscle will be bright red. The muscle glycogen level is typically 1.0% at death and drops to 0.1%, and the pH will go from approximately 7.0 to 5.5 or 5.6. However, in some animals the glycogen level may be low to begin with, and this can affect the normal post mortem biochemical processes. A few of the beef muscle quality defects associated with biochemical pathways are individually discussed below.

Improper muscle color

There is a phenomenon that results in 'dark' or 'dark-cutting' beef, where these animals may only have 0.2–0.5% muscle glycogen at death, and although the existing muscle glycogen and ATP will be depleted, there is not enough glycogen to allow desirable levels of lactic acid to accumulate, and therefore the pH may only decrease to 6.3–6.7. The muscle of these animals will be quite dark, and never appear bright red. A condition associated with this phenomenon may be called dark, firm and dry (DFD) beef. On the other hand, if at the time of death there are low glycogen levels and already existing fairly high levels of lactic acid, the pH may decline to undesirable low levels such as 5.0 and below, and this produces muscle with a pale color, undesirable soft texture and extreme water loss (pale, soft and exudative, or PSE).

Cold shortening

Cold shortening is associated with a too rapid cooling of beef carcasses and decreases tenderness of muscle fibers; as a result, beef carcasses are chilled in coolers above freezing (such as 4°C,) but not frozen, or kept in coolers below 0°C immediately after harvest. The reason the phenomenon is called cold shortening is that the muscle fibers will contract more, producing shorter and wider muscle fibers that will require more force to cut or chew. Freezing muscle of cattle that has high levels of ATP is thought to produce very tough beef.

Heat ring

Heat ring is a problem where the outer portion of the longissimus muscle appears darker, with a coarser texture, and may sink down away from the surface area where a beef carcass is ribbed for grading. This is thought to be related to animals not having enough external fat cover and therefore producing different cooling rates within the muscle.

Blood splash

Blood splash is another defect that may be influenced by the stress level and/or activity of cattle before death. There are many capillaries in muscle and fat, and it is thought that if animals are sufficiently stressed, an abrupt increase in blood pressure may cause some of these capillaries to burst and appear as many dark spots in the muscle or fat. Keeping animals as quiet and calm as possible before harvest, as well as bleeding as soon as possible after stunning, are both important to minimize the incidence of blood splash.

Bruising

Bruising can be another potential problem when animals are handled roughly and/or facilities are improperly designed. Bruising causes discoloration of the muscle and will result in affected areas being trimmed off the carcass (and loss of potential carcass weight). Generally, cattle handling and movement strategies that are known to reduce and minimize animal stress (both long term and short term) are recommended to reduce the potential incidence of most of these defects.

11.3 Food Safety and Inspection Systems

Proper inspection systems are critical to ensure food safety for consumers. These typically involve both ante mortem (before death) inspection of animals as well as post mortem (after death) inspection of carcasses, and monitoring of the complete harvest (slaughter), fabrication and distribution processes. The main goals of inspection systems are to prevent human illness from any potential disease from infected or contaminated beef products. This is the role of government-regulated inspection, and full understanding of inspection requirements and implementation are critical for trust-based trade across countries that utilize different systems and languages.

11.3.1 HACCP – Hazard Analysis of Critical Control Points

The roots of HACCP can be traced back to the early years of the US space program, where food safety was a critical responsibility for preparing foods for astronauts, and was first used in processes such as canning and other cooked food applications where food safety failure could have huge consequences (Romans *et al.*, 2001). HACCP is based on the prevention of potential problems as opposed to inspection and the identification of problems after they occur, but it was not until the 1980s that many fresh meat processors began to explore and incorporate HACCP guidelines (Romans *et al.*, 2001).

In general, the seven principles of HACCP (taken from Romans *et al.*, 2001) are (i) a hazard analysis of each process must be carried out to identify and list food safety hazards, which can be biological, chemical or physical, (ii) the critical control points (CCPs) of each process must be identified; CCPs are steps and/or processes at which control can be applied and a food safety hazard eliminated, prevented or reduced to acceptable level, (iii) the critical limits for preventative measures associated with each CCP must be established; a critical limit is the minimum or maximum value to which a process parameter must be controlled, (iv) monitoring requirements for CCPs must be determined; monitoring is integral and consists of measurements to ensure a CCP is within the established critical limit, (v) the HACCP plan must include corrective action to be taken when monitoring indicates a deviation from critical limit at a CCP, (vi) effective record-keeping procedures that document the entire HACCP system must be developed and maintained; the ability to provide current as well as historical, objective and relevant data is needed for regulatory compliance and industry assurances, and (vii) HACCP systems must be systematically verified periodically; this ensures that current procedures are in compliance with the HACCP plan, and ensures that the HACCP plan is reviewed to determine whether modification or revalidation is needed.

In many countries there are regulatory agencies that require HACCP systems to be in place and regularly assessed. However, development of a HACCP plan even when there is no governmental regulatory oversight will help prevent food-borne illnesses and can give companies marketing advantages, either domestically or internationally.

11.3.2   Beef Quality Assurance (BQA) Programs (Best Management Practices, BMP)

Beef quality assurance (BQA) programs (may be referred to as best management practices, BMP) educate producers about how their management and husbandry practices impact the ultimate value of beef carcasses and beef cuts for consumers. Most people involved in cattle production are concerned about stewardship and animal welfare as well as consumer confidence. Unfortunately, there is also a concern about false claims and blame shifting when something undesirable occurs. The goal of BQA programs is simply to get producers to think about the potential implications of what happens throughout the beef cattle production process from pasture to plate. The main issues that come into play are (i) animal handling and welfare, and (ii) animal health and nutritional management.

In regard to animal handling, moving animals and using facilities where there is minimal animal stress and excitement (discussed further in Chapter 10) are always best. This conjures up some obvious examples such as not shouting at animals, not hitting animals, not using obviously harmful facilities with sharp corners. But this can also include not such obvious things such as mixing groups of animals

together prior to harvest, transportation for long distances to processing facilities, air flow patterns within trailers, etc. Rough animal handling and ill designed or maintained equipment and facilities increase the chances of injury to animals and handlers, promote potential for bruising, which obviously lowers carcass value, and may increase the potential to produce dark-cutting beef. Furthermore, consumers are increasingly concerned about the source of their food, and animal welfare considerations are now important criteria for several grocery outlets and food service providers. Many of these stressors can alter blood flow rate, change blood chemistry components, and therefore can impact muscle pH decline and other associated mechanisms associated with conversion of meat muscle and meat aging processes.

11.3.3   Banning of Certain Chemical Compounds

Most countries have some regulatory agency that checks on chemical compounds going into the food chain. A critical component of best management practices is to NEVER use a product or compound that does not have an explicit label clearance for use in cattle. Table 11.1 lists some compounds that are not allowed for use in food animals. Correspondingly, several pharmaceutical products have warning labels for withdrawal time before slaughter; if the label says 21-day withdrawal is required, this means that it is illegal to slaughter animals less than 21 days since the product was administered. This concept is common sense (once it is known and thought about). Cattle producers must view themselves as food producers at all times (this may be for their families or for people on the other side of the world), and it is critical to always read and follow label directions for approved products and to NEVER use products that are not approved. There are agencies in most countries that conduct regular sampling of beef products for a wide variety of chemical residues with traceback systems in place to find and prosecute producers that violate food production laws. Whether this violation is intentional or not, the consequences are the same. Some products are legal for use only in certain countries, and if it is purchased in one country and used in another, the laws for the country where the animals are located and harvested will apply. Most imports and exports of beef are also regularly tested for compounds of concern.

Table 11.1. List of some banned products in food animal species.[a]

Diethylstilbestrol (DES), a non-steroidal synthetic estrogen	Chloramphenicol – an antibiotic
Nitroimidazoles – antibiotic class that includes dimetridazole, metronidazole and ipronidazole	Sulfonamide use in dairy cattle – antibiotic
Nitrofurans – antibiotic class including nitrofurazone and furazolidone	Clenbuterol – beta 2 adrenergic agonist that promotes lean muscle mass with reduced fat deposition
Dipyrone – antipyretic, anti-inflammatory and analgesic compound	Fluoroquinolones – class of antibiotics
Glycopeptides – class of antibiotics	

[a]Based partly on Payne *et al.* (1999) for education of US veterinarians. In some cases products are known carcinogens (cancer causing), in others such as with some antibiotics products may be similar to those used in human medicine, and, some antibiotics may be approved for use in companion animal species but not food animal species. Other countries have similar or modified lists.

Most typically used health products such as vaccines, antibiotics and parasite treatments have established withdrawal times, and producers (and veterinarians) should always read the manufacturer's labels before any product is used.

Many groups have been concerned about the use of pharmaceutical products in food-producing animals, and this interest has particularly increased among wealthier nations. Wilson *et al.* (2003), with the World Bank, performed a thorough review of the use of pharmaceutical products in food animals and stated that except for a small number of isolated cases, it was very hard to link human illness with consumption of veterinary drugs used for animal feed or for animal health protection. They went on to state that even though outbreaks of diseases have been suspected to be the consequence of antibiotic use in animal feed, lack of scientific evidence cannot always prove that the antibiotic is the actual cause of the disease or illness (Wilson *et al.*, 2003). Use of antibiotics in food animal production should be continually scrutinized and researched to better understand human health and antibiotic-resistant pathogens.

11.4 General Factors Affecting Beef Carcass Composition

Numerous studies have evaluated many factors that influence beef carcass composition. The major finding from these studies involving genetics, animal age, nutrition level, growth-promoting products and animal gender are briefly summarized in this section. There are many potential interactions among all of these factors, as well as considerable variability within all of these factors.

11.4.1 Genetic Background

Within *Bos taurus*, there are biological differences between British-based and Continental European-based cattle with regard to tenderness, marbling and muscle deposition that correspond to differences in growth rate, weight at the time of slaughter, and fatness at slaughter. British types seem to have slight advantages in tenderness when other composition factors are held constant, but increased tenderness in several cases when differences in other factors such as fat thickness, time on feed, etc. are not equal. Continental European cattle tend to be taller, have heavier and more muscular carcasses at a constant fat thickness, and require a longer time on high-energy diets to reach a constant fat end-point. The two main differences (related to genetics) between *Bos taurus* and *Bos indicus* cattle relate to tenderness and degree of marbling where *Bos taurus* (particularly British type) has advantages in both traits. Certain genetic types such as Japanese Black (Wagyu) and Korean Hanwoo have extremely high marbling potential, particularly when fed for long periods; other genetic types such as double-muscled breeds like Belgian Blue, Piedmontese, etc. have very little to any marbling.

11.4.2 Physiological Age

Age is associated with palatability. It is usually assessed by skeletal ossification and dentition, but may also be indicated by lean muscle color. Increased age is associated with reduced palatability, particularly in cattle over 42 months old. Differences in tenderness appear to be a function of differences in age. The contractile state of muscle and the amount of connective tissue affect tenderness but are not significantly different between breeds and differ due to animal age and sex.

11.4.3 Nutrition

Higher planes of nutrition will speed up growth rate as well as fat deposition. Type of diet, such as grass-fed vs concentrate-fed, and even type of forage and type of grain, can impact meat color, flavor and fatty acid profile. Differences in marbling are driven by differences in plane of nutrition where excess energy above maintenance is provided, which also increases external fat thickness, but early life influences including fetal programming are thought to influence marbling as well.

11.4.4 Growth-Promoting Hormones and Hormone-Like Compounds

Several compounds such as bovine somatotropin (bST), testosterone, estrogen and progesterone (and their synthetic analogs) and beta-agonists have been shown to increase rate of gain, efficiency of gain and red meat yield by increasing muscle deposition at the expense of fat deposition. Extremely intense use of growth-promoting hormone products may reduce tenderness. Marbling may be reduced as a function of reduced external fat.

11.4.5 Environment

How animals handle stress can affect energy metabolism. This may be due to illness because severity of bovine respiratory disease has been shown to reduce carcass marbling; impacts on tenderness do not seem to be obviously impacted, but have not been intensively studied.

Energy used to conserve and dissipate body heat can affect efficiency of gain and carcass composition when conditions are drastically different from the thermoneutral zone.

Long-term stress is thought to increase the risk of producing 'dark-cutting' beef (color of the lean muscle is much darker than typical and is not the color that consumers are accustomed to).

11.4.6 Gender

Differences exist in bulls vs steers vs heifers of the same age for muscle, fat and bone growth patterns and percentages. In general, the ranking from bulls to steers to heifers are: (a) leanest to fattest, (b) most muscular to least muscular, (c) longest time to shortest time to reach a fat-constant end-point, and (d) lowest marbling to highest marbling.

11.4.7 Live Animal Indications of Beef Carcass Traits

It is easy to evaluate cattle for muscle and fat cover by visual appraisal and practice, and this is discussed in Chapter 6. It is much more difficult to evaluate intramuscular fat based on fat cover as these traits alone have a low degree of relationship. Ultrasound evaluation (discussed in Chapters 7 and 13) can be useful in predicting muscle, fat cover and intramuscular fat. It is virtually impossible to estimate beef tenderness based on any live animal characteristic or ultrasound evaluation, although this has been an active area of research. In some instances bruising may be evident on live animals, but usually it is impossible to know about most bruises and intramuscular lesions from injections by appearance of the cattle.

In almost all instances, production records on the animals can be combined with live animal evaluation to increase the predictive ability related to many carcass traits. Previous management, age of animals, time on feed or forage, genetic background, etc. are useful to know to better estimate both carcass quality and yield components. For instance, general muscle-to-bone ratios are highest in double-muscled cattle, intermediate in non-doubled muscled beef breeds and lowest in dairy breeds. The order of deposition for fat types (Aberle *et al.*, 2001) are mesenteric (surrounding digestive system and internal organs; necessary for many metabolism-related functions), perinepheric/perirenal (surrounding kidneys and provides for their insulation), intermuscular (between individual skeletal muscles, known as seam fat), subcutaneous (deposited between skin and muscle/skeleton, evidenced by body condition score), and intramuscular (within muscles and referred to as marbling, a primary factor in beef quality grading systems).

Historically, it has been expected that the order of fat deposition goes in the order from internal fat needed for fundamental body functions to intramuscular fat being last to be deposited, however studies have shown there can be substantial variation among and within animal types to the degree of fat deposition and its pattern. Fat thickness is related to both yield and quality because a minimum amount is needed to reduce post mortem problems associated with chilling and to produce minimum amounts of marbling; however, it is not necessary to have high levels of external fat to have high levels of marbling, as the general relationship is not precise (Fig. 11.4).

Grading of beef carcasses provides customers/consumers with standards associated with meat palatability and product yield. The term 'quality' here generally refers to beef eating quality (palatability). Several factors may be considered in determining

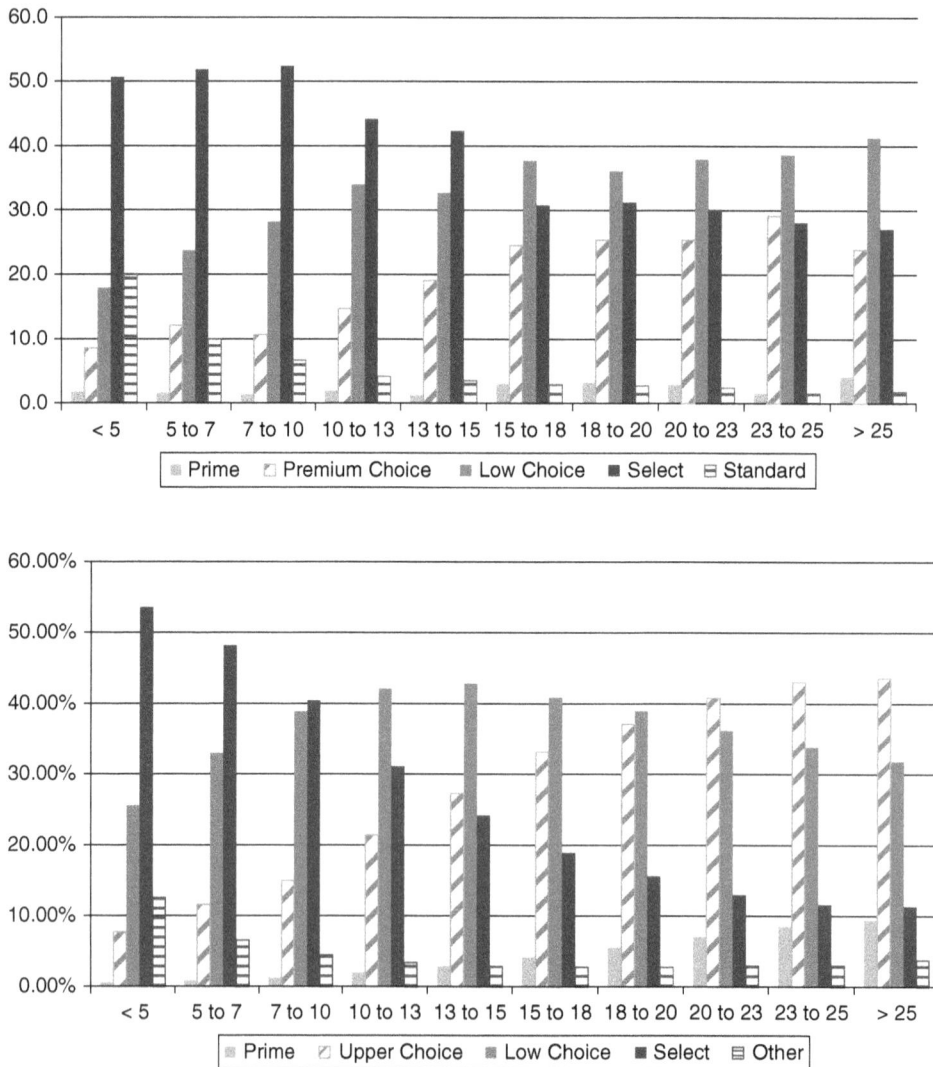

Fig. 11.4. Distribution of US beef carcass quality grades across fat thickness categories (in mm). Data from 2000 (top graph) and 2011 (bottom graph) US National Beef Quality Audits. This illustrates that all quality grade categories can be found across all possible fat levels (although not at the same incidence). Simply using fat cover to predict marbling score is imprecise, particularly across different breeds and feeding regimens. Premium Choice or Upper Choice refer to average and High Choice together, which is the marbling criterion for many premium type programs in North America. Original categories classified in 0.1 inch (2.5 mm) increments; categories here are rounded to nearest mm.

different grades of beef carcasses, and the factors that dictate beef quality and beef yield are not the same (although some may be related) in all markets. The primary factors that influence most beef quality grades are: (i) animal age, (ii) degree of intramuscular fat (also called marbling), and (iii) texture, firmness and color of the muscle. Factors that influence the yield of edible beef (muscle or red meat) from the carcass include amount of external (subcutaneous on live animal) fat, amount of internal fat such as that around internal organs like the kidney and heart, the degree of muscularity, and the carcass weight.

11.5 Factors that Affect Palatability and Consumer Acceptance

Palatability is the general term used to describe eating satisfaction. Components of palatability include tenderness, juiciness, flavor and overall consistency. There are many factors that can impact palatability such as age, type of diet, amount and distribution of fat, sex class, etc. as well as breed and family differences that can influence the palatability of various components. But cooking techniques, seasoning and post mortem aging also affect palatability components, and individual consumer preferences also exist. Moreover, many of these components can interact with one another, and coupled with price of product, also influence a consumer's opinion about palatability. It is beyond the scope of this book to discuss all of these various components. However, a major driver in influencing consumer acceptance in regard to palatability and therefore acceptability revolves around receiving a product that matches their expectations (and not getting something 'worse' than expected). This notion has encouraged many companies to develop and market various beef product lines that differentiate their product from generic or commodity beef.

In many higher-valued lines of beef products, consumers have become increasingly interested in the processes and the history associated with the beef products they purchase. In some cases these interests are driven by eating quality-related concerns, but for others it may be food safety, dietary and health interest, or animal welfare. The interests and abilities to document and market these different types of beef products from a business perspective are discussed in detail in Section 13.1.2.

In many situations, a trained taste panel may be employed to evaluate beef palatability. People on these panels are trained to use a standardized scale such as a 1 to 8 scale for multiple characteristics (such as flavor, juiciness, tenderness, etc.) based on a standardized size of meat such as 1 cm². Tenderness has also been evaluated with objective types of techniques such as shear force (the amount of force required to slice through a piece of meat). Several of these techniques have been used, but one of the oldest and most widely used has been the Warner–Bratzler shear force, where six to eight cores are removed from a steak of standardized diameter (such as 1 cm) and a blade passes cuts through the cross-section of the core and records how much force (as in kg) were required to cut it.

Several post mortem techniques have been used to increase tenderness of beef (or reduce shear force), such as electrical stimulation, amount of aging, use of salt solutions, etc. Beef increases in toughness after death for a few days and then begins to decrease, due to various biochemical processes associated with post mortem rigor mortis and pH change. Substantial variation exists across and within breeds of cattle for various measures of tenderness. Calcium-dependent proteases referred to as calpains influence the post mortem muscle fiber breakdown, and these calpains are in turn inhibited by another class of compounds referred to as calpastatin. As a result, use of calpastatin has been employed as a screening tool for cattle with potentially tough beef in some instances. Also, work conducted in Australia where the carcass is hung by the pelvis as opposed to the Achilles tendon (referred to as Tenderstretch) has also shown some improvement in beef tenderness in some cattle. Tenderness is a much larger concern for whole muscle types of cuts such as steaks and roasts as compared with products that are cut into small or thin pieces, or are ground.

11.5.1 Compositional Aspects of Beef Carcasses

The yield of beef carcasses generally refers to the amount of beef that is provided from the carcass for sale to consumers and is related to its compositional aspects such as percent muscle (or lean), percent fat and percent bone. These can vary substantially across animal body condition score and degree of muscle expression. Beef carcasses are fabricated into wholesale cuts, which in turn are further processed into various retail beef cuts. Retail cuts are much more variable across countries than the wholesale cuts, but all are influenced by regional and/or cultural preferences. Table 11.2 shows some general carcass compositional expectations from cattle, and Fig. 11.5 identifies typical beef wholesale cuts.

The main nutritional advantage to beef as well as other red meats is that it provides a source of high-quality protein. It also provides important minerals such as zinc and iron as well as several B vitamins. The goal of using beef or other meats in the diet is to help provide a well-balanced diet. The taste, as well as the nutritional content, influences consumption of beef products just as it does most foods. In general, many people say they enjoy the taste of beef. Portion control in many scenarios may be much more of a challenge in the diet than the nutritional content of any beef product. However, some more specific nutritional aspects such as fatty acid composition are becoming of more interest to some groups of consumers.

11.5.2 Fat and Fatty Acids in Beef

The total fat (lipid) content of beef is well documented to influence palatability. The types of fat deposited in the body of cattle are not all the same in regard to chemical make-up. There are different fatty acids in beef that can affect aroma, flavor, texture and health impacts. It is counter-intuitive for many people to think that a higher fat type of product may be healthier (when consumed in recommended amounts) compared to a lower fat product, but this can be the case. Much of the information on fatty acid composition in this chapter is based heavily on information from Dr S.B. Smith (Texas A&M University).

Fatty acids are long chains of carbon and hydrogen atoms with a carboxyl (COOH) group attached to one end. The number of carbon atoms and the arrangement of the bonds between carbon atoms dictate the properties and the names of these compounds. Fatty acids are used as fuel sources for many types of cells. Classes of fatty acids are saturated fatty acids (SFAs, no double bonds), monounsaturated fatty acids (MUFAs, one double bond between carbon atoms) or polyunsaturated fatty acids (PUFAs, two or more double bonds). Fatty acid names are described in regard to the number of carbon atoms along with the number of double bonds. The most abundant fatty acid in beef is oleic acid (18:1, MUFA). The SFAs palmitic (16:0) and stearic (18:0) contribute substantially to the overall fatty acid composition of beef and beef fat. Linoleic acid (18:2) contributes very little to beef fat. The MUFA:SFA ratio is related to the melting points of lipids in lean and fat trim of beef and therefore fat softness.

Table 11.2. Percent (%) of cattle live weight and carcass weight expected in roasts, steaks, lean trim, fat trim and bone.

Item	Live weight	Carcass weight
Roasts and steaks	23	39
Lean trim	15	25
Total retail beef	38	64
Fat trim	14	24
Bone	7	12
Total	59	100

Taken from Romans *et al.* (2001) with expectation of 12 mm of 12th rib subcutaneous fat. FAO (2013) suggested lean, fat and bone percentages might be 52%, 32% and 16%, respectively for leaner cattle.

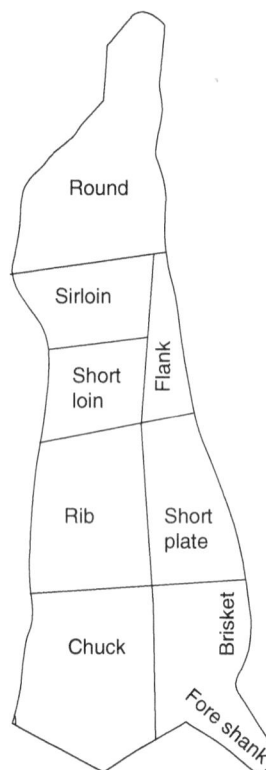

Fig. 11.5. General descriptors for wholesale cuts from beef carcasses. Exact locations of distinction between these cuts may vary across countries or markets. Wholesale cuts are further fabricated into retail cuts, the names of which may vary greatly across countries.

Cattle fed a standard, maize-based finishing diet in the USA produce subcutaneous fat and marbling fat that have consistently low melting points. This situation may be quite different where grains such as barley or wheat are fed as opposed to maize. When these grains are fed in combination with whole cottonseed or rumen-protected cottonseed oil, the melting point of the fat can exceed 45°C (113°F). This fat is very hard because it is very high in SFAs and therefore has a low MUFA:SFA ratio. MUFAs constitute 35–45% of the total fatty acids in beef produced in the USA (St John *et al.*, 1987). Dryden and Marchello (1970) and Westerling and Hedrick (1979) demonstrated that the more oleic acid in beef, the greater the overall palatability. Some portion of the effect of oleic acid on increasing palatability of beef may be due to the fat softness associated with this fatty acid (Perry *et al.*, 1998; Smith *et al.*, 1998). This provides a more fluid mouth feel, which many people perceive as more desirable. The differences in beef fatty acid profiles can vary substantially in forage-fed vs grain-fed animals.

11.5.3 Grain-Fed vs Grass-Fed Beef

There has been much discussion about the various merits of grain-based vs forage-based feeding of cattle in production of beef. Many studies can be found in which there are distinct statistical advantages of one over the other for different traits or characteristics. Both are acceptable, but these approaches can produce different types of products. Like any tool, how these are used dictates the end result, more so than whether or not one or the other is being used. The biggest difference between these two production systems is that cattle require more time to reach the same degree of body fat with forage-based feeding as compared to grain-based. Cattle that are grass fed can become fatter than cattle that are grain fed, but usually cattle that are grain fed are on feed a long enough time that they deposit substantial amounts of body fat.

In general, grass-fed beef tends to be lower in overall fat content than grain-fed beef, and due to differences in fatty acid profile, grass-fed beef also possesses distinct flavor and unique cooking qualities, and has the potential for reduced shelf life. The type of the grain as well type of forage can impact flavor and fatty acid profiles, as can breed type and length of feeding period, and these factors may interact with each other. In a very large Australian study, Johnston *et al.* (2003) found that finishing system of pasture vs feedlot affected all beef quality traits evaluated where the pasture-finished cattle had higher shear force values, lower sensory panel tenderness scores and darker meat color, but geographic region had no impact on these traits in both temperate and tropically adapted cattle. The tropically adapted heifers had lower tenderness scores and darker color than steers, but this effect was not seen in temperate breeds.

In an extensive review of research spanning three decades, Daley *et al.* (2010) stated that grass-fed beef (on a g/g fat basis) has a more desirable SFA lipid profile (more C18:0 cholesterol neutral SFA and less C14:0 and C16:0 cholesterol elevating SFAs) as compared to grain-fed beef, but grain-fed beef consumers may achieve similar intakes of both n-3 and CLA through consumption of higher fat portions with higher overall palatability scores. The major MUFA in beef, oleic acid, has been found to lower LDL-cholesterol without affecting beneficial HDL-cholesterol (Grundy *et al.*, 1988). One of the major SFAs in beef, stearic acid, has been found to have no effect or even to lower serum cholesterol (Bonanome and Grundy, 1988). Additionally, several clinical studies have shown that lean beef, regardless of feeding strategy, can be used interchangeably with fish or skinless chicken to reduce serum cholesterol levels in patients with high cholesterol (Daley *et al.*, 2010). Grain-finished beef and grass-finished beef need to be viewed as choices by consumers and by producers, not as competitors where one is valid and the other is not.

11.6 Comparisons of Beef Carcass Grading Standards and Systems

Many countries have developed and implemented standards for grading of beef carcasses to aid in merchandizing products within their boundaries. Additionally, many private company brands have also been developed that either incorporate existing grading standards or use their own criteria. This section provides general discussion and descriptions of some beef carcass grading systems that have been employed in some countries; consequently, there are probably some standards that have been left out due to lack of English-based descriptions. Many grading systems are based on evaluation of fat and muscle at particular rib locations (6th, 10th or 12th) of carcasses (Fig. 11.6).

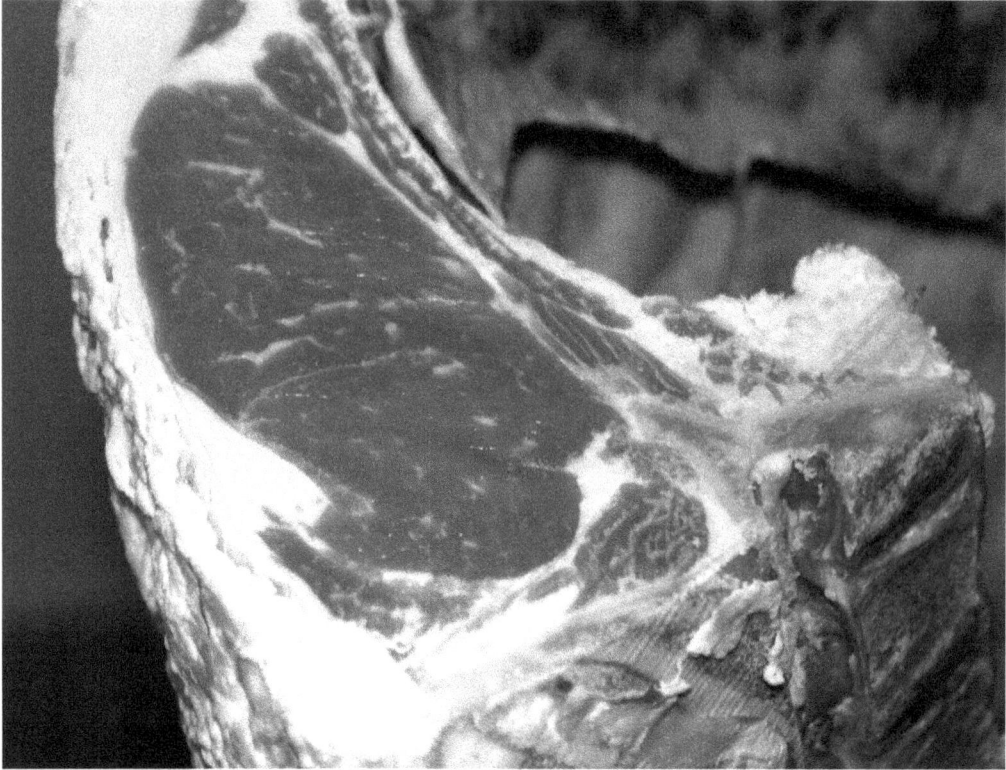

Fig. 11.6. Beef carcass that has been ribbed at the 12th–13th rib interface. Light specks within the longissimus dorsi muscle indicate intramuscular fat (marbling). Evidence of external fat and intermuscular (seam) fat are also evident (upper left). The flat surface (bottom right) indicates where the carcass was split through the vertebrae. Photo courtesy of D.B. Griffin.

11.6.1 USA

As the North American beef industry has been heavily based on grain (primarily maize) feeding of surplus cattle since the 1960s, the US system based on quality and yield grades has been used since then. Quality grade is based primarily on animal maturity and marbling (Fig. 11.7), and yield grade is a 1 to 5 system based on loin eye muscle area (12th rib), hot carcass weight, internal (kidney, pelvic and heart [KPH]) fat, and fat thickness (12th rib) where animals with a higher numerical grade are expected to have lower percentages of meat yield from the carcass. Almost all US feedlot cattle are A maturity (under 30 months of age) at harvest and the vast majority are feedlot finished.

11.6.2 Canada

The carcass quality grading system of the Canadian Beef Grading Agency (www.beefgradingagency.ca)

is summarized in Table 11.3. It incorporates fat thickness, but there is also a yield grade system where three grades are employed. Many cattle harvested in Canada are leaner than US cattle, probably as a reflection of a higher percentage of forage-based finishing.

11.6.3 European Union and EUROP System

The standardized system used throughout the EU member nations is referred to as the EUROP classification system. The two primary components of this classification are conformation, fatness and sex class categories (Fig. 11.8). Conformation is based on an assessment of the overall shape of the carcass and is divided into five main classes: E, U, R, O and P (hence the EUROP acronym), with U, O and P classes subdivided into upper (+) and lower (−) designations.

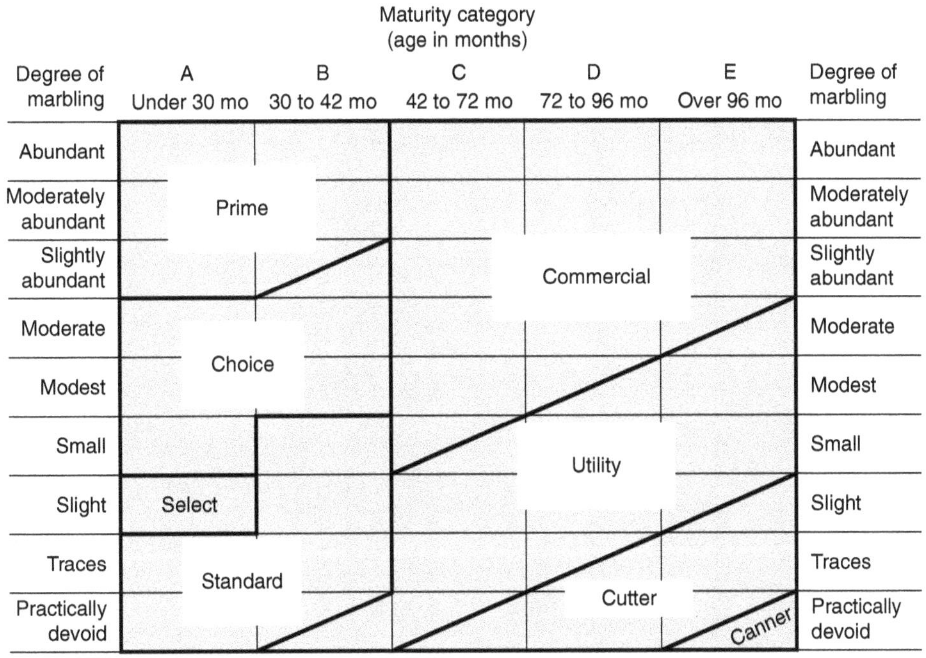

Fig. 11.7. Marbling scores and maturity categories associated with US beef carcass quality grades (USDA, 1997).

Conformation class E describes carcasses of extreme muscle expression (referred to as 'outstanding shape' in some publications), and particularly represents carcass type produced by double-muscled cattle. Conformation class P describes poorly muscled carcasses of 'inferior shape', typically produced by cattle of extreme dairy type. In much of Europe (and many other areas), milk is produced from more dual-purpose types of cattle as opposed to the USA and Canada, where selection for increased milk production has changed the muscular conformation of dairy cattle. In this system, external fatness is based on an assessment in five classes from 1 (very lean) to 5 (very fat), with classes 4 and 5 being subdivided into leaner (L) and fatter (H) categories. Sex class categories are denoted by the letters A (young bull), B (bull), C (steer), D (cow) and E (heifer).

11.6.4 South Africa

The following information was taken from the South African (KwaZulu-Natal) Ministry of Agriculture website (KZNDAE, 2014). A carcass grading system had been in use at South African abattoirs since 1985,

but was replaced by a carcass classification system on 26 June 1992 (Agricultural Product Standards Act, 1990; Act No. 119 of 1990). Numerous surveys conducted in South Africa have shown that consumers (mainly housewives) had little knowledge or understanding of the beef carcass grading system, and that this grading system was largely used by butchers to buy the type of meat they believed their customers wanted. It was also stated that a 'classification' system as opposed to a 'grading' system might be better for marketing to consumers with widely differing preferences as opposed to a grading system that implies some are 'better' than others, or that some are inferior. The classification system is based on six components of (i) carcass weight, (ii) age of the animal, (iii) fat content, (iv) carcass conformation, (v) damage to the carcass, and (vi) designation of older bull carcasses. The components of the South African beef carcass classification system are more fully described in Table 11.4.

11.6.5 Australia

In Australia there are two beef grading schemes, AUS-MEAT and Meat Standards Australia, or MSA.

Table 11.3. Descriptions of Canadian beef quality grades.

Grade	Maturity (age)	Muscling	Rib eye muscle	Marbling	Fat color and texture	Fat thickness
CANADA PRIME	Youthful	Good to excellent with some deficiencies	Firm, bright red	Slightly abundant	Firm, white or amber	≥2 mm
CANADA AAA, AA, A	Youthful	Good to excellent with some deficiencies	Firm, bright red	AAA – small AA – slight A – trace	Firm, white or amber	≥2 mm
B1	Youthful	Good to excellent with some deficiencies	Firm, bright red	No requirement	Firm, white or amber	<2 mm
B2	Youthful	Deficient to excellent	Bright red	No requirement	Yellow	No requirement
B3	Youthful	Deficient to good	Bright red	No requirement	White or amber	No requirement
B4	Youthful	Deficient to excellent	Dark red	No requirement	No requirement	No requirement
D1	Mature	Excellent	No requirement	No requirement	Firm, white or amber	<15 mm
D2	Mature	Medium to excellent	No requirement	No requirement	White to yellow	<15 mm
D3	Mature	Deficient	No requirement	No requirement	No requirement	<15 mm
D4	Mature	Deficient to excellent	No requirement	No requirement	No requirement	≥15 mm
E	Youthful or mature	Pronounced masculinity	No requirement	No requirement	No requirement	No requirement

From Canadian Beef Grading Agency.

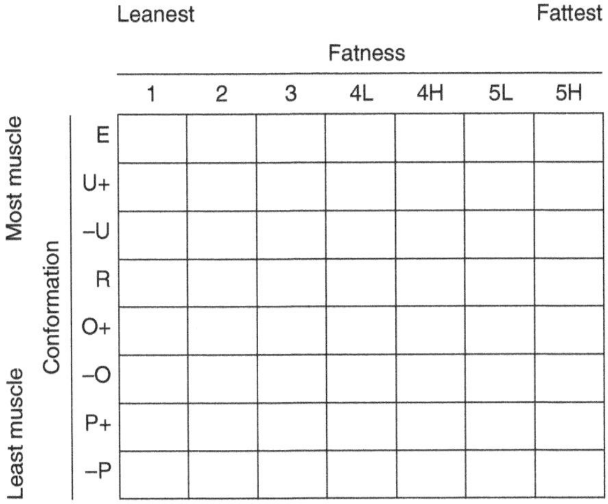

Leanest   Fattest

Fig. 11.8. Description of EUROP beef carcass classification system. Letters E, U, R, O and P designate decreasing degree of muscularity (double-muscled cattle would be E conformation), and numbers 1 through 5 designate increasing degree of fatness. Conformation class is provided first, followed by fatness class (i.e. R4L, –U3, P+5L, etc.). This classification system promotes total meat yield.

The AUS-MEAT system is for classification of grain-fed beef carcasses and uses carcass weights, external fat, dentition, sex class, carcass shape, marbling scores, lean and fat color. The MSA system goes beyond AUS-MEAT and is based on classification of individual cuts of beef with additional factors that are provided in Table 11.5.

11.6.6 Asia

Both Japan and South Korea have detailed beef carcass grading systems where both quality grade and yield grade are assessed. Both employ meat color and fat color scores, but the Japanese system also utilizes meat brightness, and fat luster, texture and firmness scores. The fat scores are a reflection of the type of cattle and the length of feeding period that impacts fatty acid profiles. The quality grade scales in both South Korea and Japan go far beyond those used in the USA (Table 11.6). The percent intramuscular fat in the US quality grade system goes from 1 to 12 or 13%, but the Korean system goes up to 19% intramuscular fat, and the Japanese system goes to 34%. The highest beef marbling scores in Japan are rare, and an A5 grade carcass resulting from marbling score 12 is extremely rare but may be valued at $20,000. China has recently developed and employed a beef carcass quality grading system that is based on animal age and marbling score.

Polkinghorne and Thompson (2010) described research in Australia, Korea, Ireland, USA, Japan and South Africa that showed consumers across diverse cultures and nationalities had a very similar view of beef eating quality. These authors also found that consumers will pay higher prices for better eating quality grades and this generally was not affected by demographic or meat preference traits of consumers. Polkinghorne and Thompson (2010) recommended that an international language on palatability become standardized to help global beef trade. It must be determined across different global regions which types of beef carcasses are the most cost-effective to produce for domestic and international markets; it should be emphasized again here that simply pursuing carcass characteristics without reference to the cost of production, including from the cow herd perspective, may lead to undesirable and non-sustainable outcomes.

11.7 Cattle By-Products

Offal is the term used to describe the parts of the animal's body that do not remain with the carcass. Some are edible (consumed by people) and others are inedible (not consumed by people), but the vast majority of these by-products are quite useful for a

Table 11.4. Beef carcass classification system and components in South Africa.

1. Carcass mass (carcass weight):

The carcass mass is the cold mass (cold dressed mass or chilled carcass weight) of the carcass after refrigeration. The mass is expressed in kilograms. Carcasses are auctioned at controlled abattoirs as cents per kilogram cold carcass mass.

2. Cattle age based on dentition (incisors):

Four age categories are used based on presence of permanent incisors of A (0), AB (1 or 2), B (3 to 6) and C (over 6). Only cattle that do not have any permanent incisors erupted are in A. Carcasses are classed as veal until the first pre-molar erupts (5 to 6 months old), after which the carcass falls in the A age category.

3. Fat:

Carcasses are assigned to one of seven fat classes (0 to 6 scale) by a trained official based on a visual assessment of external carcass fat content and distribution (provided below).

Fat class	Description	Subcutaneous fat (%)	Fat thickness (mm)
0	No fat	Under 3	0
1	Very lean	3.3	<1
2	Lean	4.1	1 to 3
3	Medium	5.2	>3 and ≤5
4	Fat	6.3	>5 and ≤7
5	Overfat	7.3	>7 and ≤10
6	Excessively fat	7.8	>10

4. Conformation:

Trained officials place a conformation score (1 to 5 scale) based on amount and expression of muscle; the conformation classes and descriptions are: 1 (very flat), 2 (flat), 3 (medium), 4 (round) and 5 (very round).

5. Damage to carcass:

No description was provided, but a 3-point scale is reported to be used.

6. Gender:

All bull carcasses except those in A age category are designated with the code M/D.

From KZNDAE (2014).

Table 11.5. Description of Meat Standards Australia (MSA) beef.

- Assuring the eating quality of MSA beef requires standards to be maintained from paddock to plate
- Cattle that meet the MSA requirements are graded at MSA licenced abattoirs. A National Vendor Declaration (NVD) and an MSA vendor declaration, which are checked by the grader and livestock personnel, are sent with the cattle.
- Each carcass is graded by an MSA accredited grader with an eating quality grade assigned for each individual cut.
- Each carcass is identified with a carcass ticket and the following information is recorded in the Data Capture Unit:
- Body number and lot number – cattle from individual vendors will be kept in separate lots
- Carcass weight – important in determining weight for maturity
- Sex – male or female
- Tropical breed content – the hump height is also measured to guarantee the most accurate eating quality grade
- Hanging method – determined as being either Achilles hang or tenderstretch
- Hormonal growth promotants – will affect MSA score obtained for different muscles
- Ossification – measured to determine carcases maturity
- Marbling – using both the MSA and AUS-MEAT measurement systems
- Rib fat – a minimum of 3 mm is required, measured at the AUS-MEAT standard site. Overall fat cover is also assessed including any hide puller damage
- pH and temperature – pH is measured using a pH meter and must be below 5.71. Temperature should be below 12°C according to AUS-MEAT standards
- Meat color – recorded using AUS-MEAT standard meat color chips. Meat colors in the range of 1B to 3 are accepted depending on the abattoir or brand specification Other measurements that do not impact on eating quality can be taken at the customers' request, including:
- Eye muscle area (EMA) – measured in square cm using an AUS-MEAT grid
- Fat color – recorded using AUS-MEAT chips from 0 (white) to 9 (yellow) If the carcass meets all MSA and company specifications, it is eligible to have cuts packed and sold as MSA.

From MLA (2012).

variety of uses. Dog treats can be made from hooves, containers may be made from horns (historically used for storage of gunpowder, etc.), and many ceremonial items and decorations can use various cattle by-products. However, many other non-obvious uses such as some cosmetics, explosives, and numerous leather products are also made in part with inedible cattle by-products. Edible by-products include the organs, and these are typically referred to as variety meats. Since recent bovine spongiform encephalopathy (BSE, or mad cow disease) outbreaks, it is not recommended that brains of other mammals be consumed by humans as the prion organism that causes BSE in cattle is found in the central nervous system of infected cattle (this is a similar concept to scrapie in sheep, and chronic wasting disease in deer and antelope species, which are caused by prions) and the risk of developing the human variant (Creutzfeldt–Jakob disease) of this type of human disease, although quite small, has to be minimized. As a result, brain, spinal cord and major nerves close to the spinal column are not recommended to be consumed, especially from animals over 30 months old. This is also why feeding of ruminant-derived meat and bonemeal to ruminants has been banned in most countries.

One thing associated with the animal that has little value is the contents of the digestive system (although even it can be used for fertilizer); therefore animals are generally withheld from feed for 12–18 hours before being harvested at most commercial processing facilities. Water is typically not withheld prior to harvest as dehydration could add to animal stress as well as impeding exsanguination and hide removal. The hide is the single most valuable cattle by-product in most instances and is a primary commodity for the leather industry. Table 11.7 shows some typical expected percentages and weights of cattle hides and other various by-products.

In some markets, the variety meats may be a less valuable source of income than the carcass, but in other markets the collective value of the variety meats may be equal to or even exceed the value of the carcass. When there are regional and/or cultural differences in demand for these different types of products, this provides much trade potential and the possibility of added value for trade and development of new markets. Some edible cattle by-products (and their principal uses) from Aberle *et al.* (2001) include beef extract (soup and bouillon), bones (gelatin, confectionaries, ice cream, jellied products), blood (sausage component), calf skin

Table 11.6. Comparisons of Korean, US and Japanese beef quality grading systems.

Korea											
Quality grade	3	2		1			1+		1++		
Marbling score	1	2	3	4	5	6	7	8	9		
IMF%[a]	<5	5–7	7–9	9–11	11–13	13–15	15–17	17–19	>19		

USA											
Quality grade	SE	CH-	CHo	CH+	PR-	PRo	PR+				
Marbling score[b]	SL	SM	MD	MT	SLAB	MAB	AB				
IMF%[c]	3.8	5.0	6.7	7.3	10.1	11.1	12.0				

Japan											
Quality grade	1	2	3	4				5			
Marbling score	1	2	3	4	5	6	7	8	9	10	11	12
IMF%[c]	5.1	8.3	10.4	14.2	13.9	16.9	18.6	20.9	23.0	26.1	31.5	34.0

[a]Paek (2007); marbling scores are assessed at the 6th rib in Korean and Japanese carcasses and 12th rib in US carcasses (marbling score will be higher at 6th rib versus 12th rib).
[b]In reference to A maturity (less than 30 months of age) cattle; signs '-', 'o' and '+' represent low, average and high levels within quality grade (i.e. low Choice, etc.); for marbling scores SL = slight, SM = small, MD = modest, MT = moderate, SLAB = slightly abundant, MAB = moderately abundant, AB = abundant; US cattle over 42 months of age cannot be graded as Choice or Prime.
Values are minimum amount anticipated for respective marbling score (BIF, 2010).
[c]Percent IMF values taken from 146 sampled carcasses (Cameron et al., 1994).

Table 11.7. Expected by-product percentages and projected weights from cattle of different biological type.[a]

	Percentage (%)			Weight (kg)		
	English	Holstein	≥50% *indicus*	English	Holstein	≥50% *indicus*
Green hide	8.2	6.5	9.2	45.1	35.8	50.6
Trimmed hide	7.7	6.0	8.7	42.4	33.0	47.9
Fleshed hide	6.7	5.7	7.4	36.9	31.4	40.7
Cured hide	5.9	5.0	6.3	32.5	27.5	34.7
Liver	1.45	1.96	1.26	8.0	10.8	6.9
Heart	0.48	0.50	0.39	2.6	2.8	2.1
Tunic tissue	0.026	0.02	0.024	0.1	0.1	0.1
Green tripe	2.04	2.15	2.00	11.2	11.8	11.0
Scalded tripe	1.24	1.13	1.04	6.8	6.2	5.7
Oxtail	0.26	0.21	0.28	1.4	1.2	1.5
Weasand meat	0.030	0.034	0.032	0.2	0.2	0.2
Pancreas	0.084	0.093	0.086	0.5	0.5	0.5
Sweetbread	0.091	0.104	0.062	0.5	0.6	0.3
Kidney	0.223	0.255	0.215	1.2	1.4	1.2
Whole head	2.35	2.56	2.13	12.9	14.1	11.7
Cheek meat	0.45	0.47	0.35	2.5	2.6	1.9
Head meat	0.20	0.20	0.15	1.1	1.1	0.8
Oxlips	0.097	0.110	0.089	0.5	0.6	0.5
Salivary glands	0.098	0.063	0.083	0.5	0.3	0.5
Skull	1.51	1.77	1.47	8.3	9.7	8.1
Whole tongue	0.78	0.75	0.71	4.3	4.1	3.9
Trimmed tongue	0.39	0.36	0.35	2.1	2.0	1.9
Tongue trim	0.21	0.19	0.23	1.2	1.0	1.3
Tongue meat	0.026	0.051	0.012	0.14	0.28	0.07
Gullet	0.096	0.105	0.078	0.5	0.6	0.4

Continued

Table 11.7. Continued.

	Percentage (%)			Weight (kg)		
	English	Holstein	≥50% *indicus*	English	Holstein	≥50% *indicus*
Tongue bone	0.018	0.018	0.015	0.10	0.10	0.08
Tongue root muscle	0.042	0.029	0.027	0.23	0.16	0.15
Blood	2.48	2.93	2.62	13.6	16.1	14.4
Hooves	1.76	1.78	1.79	9.7	9.8	9.8
Lungs	0.85	1.1	0.92	4.7	6.1	5.1
Small intestine	2.29	2.71	1.99	12.6	14.9	10.9
Spleen	0.20	0.24	0.20	1.1	1.3	1.1
Ear and lip trim	0.24	0.23	0.24	1.3	1.3	1.3
Large intestine	4.31	5.15	4.09	23.7	28.3	22.5
Trachea	0.12	0.28	0.13	0.7	1.5	0.7
Ruffle fat	0.95	1.3	0.85	5.2	7.2	4.7
Final rail trim	0.95	0.24	0.52	5.2	1.3	2.9
Heart bone	0.006	0.005	0.007	0.03	0.03	0.04
Bile	0.066	0.096	0.075	0.4	0.5	0.4

[a]Percentages relative to live weight from 550 kg live weight basis with external 12th rib fat thickness of 10–12 mm. Percentages taken from Terry *et al.* (1990).

trimmings (gelatin, confectionaries, ice cream, jellied products), cheek meat (sausage), esophagus (known as weasand, sausage), fat (shortening, candies, chewing gum), intestines (sausage casings), stomach components (sausage or variety meat), oxtail (soup stock), and the heart, kidneys, liver, spleen/melt, pancreas/sweetbread and tongue as variety meats. Table 11.8 provides nutritional values for some of these variety meats.

11.7.1 Use of Inedible By-Products

There are many uses for various inedible by-products from cattle. Many cattle organs and glands have compounds that can be extracted for human medicine and biomedical uses. Also, several cattle products or compounds serve a variety of industrial and manufacturing processes and uses, and several are listed in Table 11.9. The uses of these inedible by-products plus those of the numerous edible by-products therefore add value to the live animal. The difference in the live weight value of the animal and its respective carcass value is referred to as the drop value and represents the total value of the by-products (edible and inedible). Some markets may report the drop value credit (or drop value price) as a function of the live weight value (or price). The drop value is highly dependent upon access to markets that can utilize the various by-products. The drop credit is typically considerably more at a large packing plant that can merchandise large volumes of

Table 11.8 Nutritional composition of cooked organ meats.[a]

Variety meat	Calories (cal)	Protein (g)	Fat (g)
Brain	111	10.5	7.4
Heart	153	26.3	4.5
Kidney	165	26.3	5.9
Liver	160	21.5	7.6
Lung	104	18.8	2.6
Pancreas	256	29.1	14.6
Spleen	125	23.9	2.6
Thymus	105	18.4	2.9
Tongue	187	26.2	8.3

[a]Based on 100 g portions from young (veal) calf. From Romans *et al.* (2001).

various by-products than at a small abattoir, and may have little to no value at low-volume facilities. As a result, there may be large regional or geographical differences in the drop value that is attainable from similar cattle simply due to location and available facilities, or the acceptance or lack thereof for the use of certain by-products.

11.8 Summary

The ultimate value of cattle is typically dependent upon their carcass and by-product traits. Although the carcass is thought of as the primary end-product of beef cattle, numerous edible and inedible by-products

Table 11.9. Examples of some inedible cattle by-products and uses.

Blood	Collagen (derived from connective tissues and beef hides)
Cell culture laboratories: Bovine serum albumin provides a wide variety of macromolecular proteins, low molecular weight nutrients, carrier proteins for water-insoluble components, and other compounds necessary for *in vitro* growth of cells, such as hormones and attachment factors. Serum adds buffering capacity to the medium and binds or neutralizes toxic components in the growth milieu. Home and industrial uses: Plywood adhesives, fertilizer, foam fire extinguisher, chemical fixer for dyes	Numerous uses: Hemostats, vascular sealants, tissue sealants, orthopedic implant coatings, vascular implant coatings, artificial skin, bone graft substitutes, corneal shields, injectable collagen for plastic surgery, injectable collagen for incontinence treatment, meat casings, food additives, artificial dura maters, dental implants, wound dressings, anti-adhesion barriers, platelet analyser reagents, research reagents, antibiotic wound dressing, lacrimal plugs

Fatty acids (derived from tallows)	Glycerin products (derived from tallows)
General uses: Plastics, tires, candles, crayons, cosmetics, lubricants, soaps, fabric softeners, asphalt emulsifiers, synthetic rubber, linoleum (metallic stearate), PVC (calcium stearate), jet engine lubricants, carrier for pesticides and herbicides, wetting agents, dispersing agents, defoamers, solubilizers, viscosity modifiers Many specific examples (oleic acid, stearic acid, etc.) are used for specific purposes or products	Glycerin derivatives: A wide range of pharmaceuticals including cough syrups and lozenges, tranquilizers, eyewashes, contraceptive jellies and creams, ear drops, poison ivy solutions, solvent for digitalis and intramuscular injection, sclerosing solutions for treatment of varicose veins and hemorrhoids, suppositories, gel capsules Glycerol: Solvent, sweetener, dynamite, cosmetics, liquid soaps, candy, liqueurs, inks, lubricants, antifreeze mixtures, culture nutrients for antibiotics

Tallow (fats and oils derived from meat, bone, hooves, and horns)	Gelatin (derived from collagen)
Various industrial tallows used as lubricants and greases: Top White Tallow, All-Beef Packer Tallow, Extra Fancy Tallow, Fancy Tallow, Bleachable Fancy Tallow, Prime Tallow, Special Tallow, No. 2 Tallow, A Tallow, Choice White Grease, Yellow Grease	Non-food uses include cosmetics, industrial uses, photographic paper, and photograph development as an aid in binding for glues, adhesives and emulsion and, as a binder in pills and suppositories

From Klinkenborg (2001).

are also quite economically important. As with other traits discussed throughout this book, both genetic and non-genetic influences can result in widely different carcass size and composition characteristics. All cattle producers must be mindful of the potential consumer issues that their management methods could impact, including placement of injections and the feeding and administration of only approved feeds and products. Stewardship regarding knowledge and adherence to product label directions and withdrawal times are crucial for immediate and long-term considerations. Food safety is important throughout all steps involved from conversion of live cattle into various edible products, including transportation, storage and food preparation. Many systems exist across the world for both beef inspection for food safety considerations as well as beef grading for marketing considerations. Different types of beef products are desired by various consumers, and the ability to provide consumers with the beef products they want is a crucial component of beef production system sustainability.

11.9 Study Questions

11.1 Should cattle producers care about beef carcass defects in their cattle due to their management practices even though they sell cattle on a live basis? Explain why or why not.

11.2 The live weight of market cattle is 600 kg. If the carcass from this type of animal weighed 354 kg, what would the dressing percent be?

11.3 List six edible by-products from cattle.

11.4 Market cattle weigh 635 kg. If the dressing percent was 61.5%, what would the carcass weight be?

11.5 List six inedible by-products produced from cattle.

11.6 If beef carcasses weigh 290 kg, and the estimated dressing percent was 60%, what was the live weight of these cattle?

11.7 Briefly discuss HACCP and its principles.

11.8 What would be reasonable estimates of weights for the heart, tongue, liver and oxtail from a 600 kg beef animal? What would the expected values be on a 400 kg beef animal?

11.9 Is it likely that cattle of various breeds would have differing yields of carcass beef and/or edible by-products? Discuss why.

11.10 What is meant by the term 'beef quality grade'?

11.11 Is general hygiene of workers in a beef processing plant important to consumers? Explain why or why not.

11.12 If the harvest live weight of cattle is 500 kg, what are reasonable expected weights for the liver, heart, kidney and whole tongues from this type of animal? How much might these values differ from British types of cattle to dairy types?

11.10 References

Aberle, E.D., Forrest, J.C., Gerrard, D.E. and Mills, E.W. (2001) *Principles of Meat Science*, 4th edn. Kendall Hunt Publishing Company.

Aidaros, H. (2005) Global perspectives – the Middle East: Egypt. *Rev. Sci. Tech. Off. Int. Epiz.* 24, 589–596.

BIF (2010) Chapter 3 Animal Evaluation In: *Guidelines for Uniform Beef Improvement Programs*, 9th edn, Beef Improvement Federation. Available: www.beef-improvement.org. (accessed 12 November 2012).

Bonanome, A. and Grundy, S.M. (1988) Effect of dietary stearic acid on plasma cholesterol and lipoprotein levels. *New England Journal of Medicine* 318, 1244–1248.

Cameron, P.J., Zembayashi, M., Lunt, D.K., Mitsuhashi, T., Mitsumoto, M., Ozawa, S. and Smith, S.B. (1994) Relationship between Japanese beef marbling standard and intramuscular lipid in the *M. longissimus thorasis* of Japanese Black and American Wagyu cattle. *Meat Science* 38, 361–364.

Daley, C.A., Abbott, A., Doyle, P.S., Nader, G.A. and Larson, S. (2010) A review of fatty acid profiles and antioxidant content in grass-fed and grain-fed beef. *Nutrition Journal* 9, 10. Available at: http://www.nutritionj.com/content/9/1/10 (accessed 8 December 2012).

Dryden, F.D. and Marchello, J.A. (1970) Influence of total lipid and fatty acid composition upon the palatability of three bovine muscles. *Journal of Animal Science* 31, 36–41.

FAO (2013) Meat cutting and utilization of meat cuts. Food and Agriculture Organization of the United Nations document repository. Available at: http://www.fao.org/docrep/004/t0279e/t0279e05.htm (accessed 13 December 2013).

Grundy, S.M., Florentin, L., Nix, D. and Whelan, M.F. (1988) Comparison of monounsaturated fatty acids and carbohydrates for reducing raised levels of plasma cholesterol in man. *American Journal of Clinical Nutrition* 47, 965–969.

Johnston, D.J., Reverter, A., Ferguson, D.M., Thompson, J.M. and Burrow, H.M. (2003) Genetic and phenotypic characterization of animal, carcass, and meat quality traits from temperate and tropically adapted beef breeds. 3. Meat quality traits. *Australian Journal of Agricultural Research* 54, 135–147.

Klinkenborg, V. (2001) Cow Parts. *Discover Magazine* online. Available at: http://discovermagazine.com/2001/aug/featcow (accessed 14 January 2012).

KZNDAE (2014) Agricultural publications and production guidelines Department of Agriculture and Environmental Affairs, Kwazulu-Natal Province, South Africa. Available at: http://www.kzndae.gov.za/en-us/agriculture/agricpublications/productionguidelines.aspx (accessed 11 May 2014).

Lawrie, R.A. (1998) *Lawrie's Meat Science*, 6th edn. CRC Press, Boca Raton, FL.

MLA (2012) Meat and Livestock Australia web page. Available at: http://www.mla.com.au/Marketing-red-meat/Guaranteeing-eating-quality/MSA-beef/Grading (accessed 16 October 2012).

Paek, B.H. (2007) An overview of beef cattle (Hanwoo) production in Korea, 2nd Korea – U.S. – Japan International Joint Symposium: Producing High Quality Beef for the Asian and Domestic Markets, October 15 & 16, 2007. Texas A&M University College Station, TX, USA.

Payne, M.A., Baynes, R.E., Sundlof, S.E, Craigmill, A., Webb, A.I. and Riviere, J.E. (1999) Drugs prohibited from extra label use in food animals. *Journal of the American Veterinary Medical Association (JAVMA)* 215, 28–32.

Perry, D., Nicholls, P.J. and Thompson, J.M. (1998) The effect of sire breed on the melting point and fatty acid composition of subcutaneous fat in steers. *Journal of Animal Science* 76, 87–95.

Polkinghorne, R.J. and Thompson, J.M. (2010) Meat standards and grading: A world view. *Meat Science* 86, 227–235.

Romans, J.R., Costello, W.J., Carlson, C.W., Greaser, M.L. and K.W. Jones (2001) *The Meat We Eat*, 14th edn. Interstate Publishers, Inc., Danville, IL.

Savell, J.W. (2013) Course notes for ANSC 307 Meat Science, Texas A&M University. Available: http://meat.tamu.edu/ansc-307/ (accessed 13 December 2013).

Smith, S.B., Yang, A., Larsen, T.W. and Tume, R.K. (1998) Positional analysis of triacylglyerols from bovine adipose tissue lipids varying in degree of unsaturation. *Lipids* 33, 197–207.

St John L.C., Young, C.R., Knabe, D.A., Schelling, G.T., Grundy, S.M. and Smith, S.B. (1987) Fatty acid profiles and sensory and carcass traits of tissues from steers and swine fed an elevated monounsaturated fat diet. *Journal of Animal Science* 64, 1441–1447.

Swatland, H.J. (1984) *Structure and Development of Meat Animals*. Prentice-Hall, Inc., Englewood Cliffs, New Jersey USA.

Terry, C.A., Knapp, R.H., Edwards, J.W., Mies, W.L., Savell, J.W. and Cross, H.R. (1990) Yields of by-products from different cattle types. *Journal of Animal Science* 68, 4200–4205.

USDA-Agricultural Marketing Service (1997) *United States Standards for Grades of Carcass Beef*, Washington, DC., USA.

Westerling, D.B. and Hedrick, H.B. (1979) Fatty acid composition of bovine lipids as influenced by diet, sex and anatomical location and relationship to sensory characteristics. *Journal of Animal Science* 48, 1343–1348.

Wilson, J.S., Otsuki, T. and Majumdar, B. (2003) Balancing food safety and risk: Do drug residue limits affect international trade in beef? Paper prepared for presentation at the American Agricultural Economics Association Annual Meeting, Montreal, Canada, 27–30 July 2003.

12 Financial and Economic Aspects

One of the most important goals of beef cattle production is profit (at least that is the assumption the author has used throughout this book). This does not mean that environmental resources, animals or customers are to be exploited, nor does it mean that all potential production segments, or components of the beef production system, are or will be profitable all the time. Three major factors must be kept in mind by cattle producers: (i) cattle are able to utilize roughage-based diets (that humans and monogastric animals cannot) to yield high-quality protein products for people, (ii) cattle are able to utilize lands not suitable for production of many other agricultural crops, and (iii) the ultimate value of cattle and their associated products is a function of the acceptability and demand for their products by consumers.

It has been said by several people that 'you cannot manage what you do not measure', and this concept is true of all aspects of beef cattle production. In many countries, tremendous changes in animal type and size have been seen as a result of increased nutrition and other associated management practices, as well as genetic changes through selection. In many areas of the world our industry has increased production potential on a per animal basis as well as on a per land unit area basis. We have made great 'improvements' in many output-type traits such as growth, animal weight, carcass weight, etc. In an attempt by producers to increase profit, they naturally strive for increased income, which typically is weight-based. However, to truly determine profitability, both income and expense have to be considered:

$$\text{Profit} = \text{Income} - \text{Expenses} \qquad (12.1)$$

This equation is not complex, but obtaining the required information may be quite complicated (or at least need considerable discipline, especially in regard to documentation of expenses). The goal of this chapter is to discuss the assessment of management-related considerations and decisions that have potential to alter profitability for beef cattle producers.

12.1 Cost vs Benefit Comparisons

When faced with new information or a potential change in management, producers must consider whether or not this new approach is worth it. If producers are not positioned or conditioned to evaluate scenarios from a systems-based point of view, or if they have inadequate information, they may make the wrong decision even though they have the right motive.

Consider the example of providing extra nutrition to calves nursing cows. If a cow-calf producer knew that providing extra feed to the calves increased the sale weight of these calves, does that information mean the producer should feed the calves? The first response should be 'it depends', and understanding why 'it depends' is the key to making the right decision (and the right decision will not always be the same for all scenarios).

In order to evaluate the possibilities, managers need to think about the 'benefits' vs the 'costs' when deciding if a management practice is worth the effort. The hard part for many people in making the decision is (i) identifying what the components are for the benefits and/or the costs, and (ii) coming up with realistic values for these benefits and costs to use as their assumptions. The process of thinking about the scenarios and realizing what the components are may be more important than precisely knowing the actual values for the components (but the closer these assumed values are to being the actual values, the better the information producers/managers have to make decisions). For many of the examples in this chapter, the assumptions for costs, prices, animal performance, etc. may be markedly different for many areas of the world; however, this line of logic can be used for any circumstance or scenario.

As an example, consider two herds of beef cattle with varying level of performance for percent calf crop weaned and annual cow expense. Herd A has a 90% calf crop weaned with annual cow expense of $375; Herd B has 98% calf crop weaned with annual cow expense of $475. If both herds were to wean calves that are the same size (average of 226.8 kg, or 500 lb) that are valued at the same price ($2.53/kg, or $1.15/lb): (i) How do you determine which herd is more profitable on a per cow basis? (ii) If Herd B spent an additional $100 per breeding cow to go from 90% to 98% calf crop weaned, was the added expense worth it?

The weight of calf weaned per cow exposed to breeding for Herd A would be 90% of 226.8 kg (500 lb), or 204.1 kg (450 lb); the weight of calf weaned per cow exposed to breeding for Herd B would be 98% of 226.8 kg (500 lb), or 208.7 kg (490 lb). In Herd A $375 was spent per cow, and $517.50 was received per cow (weight of calf weaned per cow exposed [204.1 kg or 450 lb] × $2.53 per kg price or $ 1.15 per lb price), for a profit of $142.50 per cow exposed. In contrast, $475 was spent per cow and $563.50 per cow was received (208.7 kg [490 lb] of calf weaned per cow exposed × $2.53 per kg [$1.15 per lb] price), for a profit of $88.50 per cow exposed in Herd B. Therefore, the answer to the first question above is that Herd A is more profitable per cow by $54 than Herd B. This does not address the overall profitability of these two herds, however, because there may be different numbers of cows, and the total profit (and/or loss) for the operation is also a very important consideration.

If Herd B had 61% more cows that Herd A, the total profit would be the same because the profit per cow was 61% higher in Herd A, and if Herd B for instance had 400 cows, and Herd A had 200 cows, the total profit for Herd B would be $35,400 as compared to $28,500 for Herd A. The profitability for an operation is critical, but the evaluation of profitability on a total amount, or a profit per cow amount, or even a percent return on money invested (or percent return on assets) gives different values, and they have different interpretations. Also, evaluation of only one of these profit measures may lead to different management strategies and outcomes than others, or simultaneous evaluation of multiple measures. Consider two herds where one herd (Herd C) spends $400 per cow exposed per year and has income of $500 per cow exposed per year. The other

(Herd D) spends $1000 per cow exposed per year and has income of $1200 per cow exposed per year. Which herd is more profitable? Herd C has a net return of $100 per cow, whereas Herd D has a net return of $200 per cow. Herd C has a return on investment of 25% ($100 profit relative to $400 expense), whereas Herd D has a return on investment of 20% ($200 profit relative to $1000 expense).

The answer to the second question of whether or not the extra $100 expense per cow would be worth the extra 18.1 kg (40 lb) of calf weaned per cow, resulting in an additional income of $46 per cow is a bit more obvious when the scenario is evaluated in this manner, and the answer would be 'no'. It is not a good decision to add $100 of expense to gain $46 of income. If the added input costs were not considered, or not known, it would not be obvious what the answer to the question would be.

Many producers automatically believe when they perform management practices that result in higher animal value or sale prices that this automatically leads to profit. For instance, preparation of calves for feedlot conditions in North America is referred to as preconditioning (weaned, received recommended vaccinations, trained to eat from feed bunk, etc.), and calves that are documented to be preconditioned typically receive $3–8 per 100 lb ($0.07–0.17 per kg) price premium; however, several studies (Pate and Crockett, 1978; Peterson *et al.*, 1989; Pritchard and Mendez, 1990) have indicated that preconditioning may not be an economically justified practice when feeding costs are high. Many producers may fail to compare cost vs benefits because they have not thought about a break-even scenario.

12.1.1 The Concept of Break-Even Price or Value

The concept of a break-even price or cost of production or investment value is important for producers and managers to determine as a benchmark value. How far away the actual price or cost or value, etc. is from the break-even point determines the level of profit (or loss). Several people have alluded to the concept of determining this value in the simple equation:

$$\text{Future value} = \text{current value} + \text{cost} \atop \text{to obtain future value} \qquad (12.2)$$

Equation (12.2) is simple, with only three components, but determining reliable estimates for these values may be very complicated, particularly for someone who has not attempted this calculation in the past. The concept of using this formula is illustrated with the example below. Again, it is important for the reader not to focus too much on the actual values used as these vary greatly across time and geographical areas, but to focus on the approach used, and consider how it might be applied to other scenarios.

Suppose steers that weigh 317.5 kg (700 lb) are priced at $2.18 per kg ($0.99 per lb). We are trying to decide whether we should retain ownership of these calves and feed them at a feedlot. We guess that they will finish the feeding period at 567 kg (1250 lb). We also guess that the cost of gain in the feedlot will be $1.65 per kg ($0.75 per lb of gain). What is the break-even price on these steers at the end of the feedlot phase? Think about this on a per steer basis.

$$(567 \text{ kg} \times \text{future price}) = (317.5 \text{ kg} \times \$2.18)$$
$$+ (249.5 \text{ kg} \times \$1.65)$$

Or, in imperial units,

$$(1250 \text{ lb} \times \text{future price}) = (700 \text{ lb} \times \$0.99)$$
$$+ (550 \text{ lb} \times \$0.75)$$

The only unknown is the price that the 567 kg (1250 lb) finished steers will sell for (the weight gain and the cost of gain are estimates and are assumed to be known, but these will not truly be determined until the cattle are sold at the end of the feeding period). We know the current value based on knowing the weight of the animals and the market price. This can be determined precisely (but if these are estimates, they are subject to being unknowns as well). The current value is $693. If these steers will weigh 567 kg (1250 lb), that means they must gain 567 – 317.5 or 249.5 kg (1250 – 700 = 550 lb) of weight in the feedlot. Therefore, the cost to obtain the future value is 249.5 kg × $1.65 (550 lb × $0.75) = $412.50. As a result, the future value that equals the current value and the cost of getting to the future value is $693 + $412.50 = $1105.50, and the break-even finished steer price is determined by dividing this value by the future weight ($1105.50 ÷ 567 kg [1250 lb]) and would be $1.9497 per kg ($0.8844 per lb).

By using this simplistic equation, producers and managers can think about what factors impact their levels and costs of production, and what value or price they might expect to get for their efforts and expenses. This same logic can be applied in the purchase considerations for breeding bulls and females.

12.1.2 Purchase Price and Value Considerations for Herd Bulls and Replacement Females

Both commercial cow-calf and seedstock producers must realize that there is a break-even purchase price on bulls and females purchased for breeding (even when the value or concept is unknown to the buyer). Many producers probably do not know what this value is, however. It costs money and time to collect performance data, produce EBVs/EPDs, conduct genetic tests, etc., but it also takes effort to adequately visually evaluate breeding animals for structural soundness, udder conformation, masculinity/femininity and temperament. For bulls purchased for breeding, length of productive life in daughters is also extremely important if replacement heifers are kept, but this is not easy to determine. In females purchased for breeding, their length of productive life is probably the largest single production trait affecting their profit potential (provided exceptional values for other traits do not exist). All of these types of information come at a price. Paying $1000 for a herd sire is not realistic for commercial cow-calf producers, but neither is paying $10,000. The value of the bull is relative to the types and amount of information available for him, but it is also relative to where that bull fits into a specific production system. A young herd sire might be worth $4000 in a particular environment/system, but might only be worth $1500 in another. The number of offspring produced (which includes longevity), the performance/value of his offspring and the salvage value should be what dictates a bull's value, and therefore his break-even purchase price.

Using the same logic as with purchased stocker or feeder calves, the same general formula can be used to calculate the break-even purchase price of bulls and females intended for breeding. The challenge is to accurately incorporate expected levels of production, and prices over time. One of the most important factors that dictate profitability of breeding cattle is the number of progeny they produce, and therefore this is usually dictated by longevity. If they have acceptable progeny and last a long time, they will have substantial value. Breeding animals that produce undesirable offspring are not ideal, but they are better than animals that do not produce many or any progeny.

Table 12.1 gives values of annual net return (income minus expenses) of types of beef females for various different scenarios. The total income is based on the production of calves and the salvage value of the female, and the expenses are based on her purchase price and annual maintenance costs, including depreciation (discussed a bit later in the chapter).

For these input and output assumptions, it can be seen how the longevity of beef cows is required to pay for their initial investment and maintenance. The decrease in reproductive performance is directly responsible for profit potential. The values can be used in a spreadsheet so that varying costs, weights and prices can compare a variety of scenarios.

Figure 12.1 illustrates the concept of production values of herd bulls relative to different prolificacy levels. The production values of bulls are functions of the performance of their progeny and the number of progeny plus their salvage values. Expenses include annual maintenance and their purchase price. The scenarios in Figure 12.1 exclude the bull's purchase price, and therefore these production values can be thought of as the break-even purchase prices of bulls (which are highly variable from –$1000 to over $20,000) relative to the scenario. It may seem outrageous to pay several thousand dollars for a herd bull when other bulls may only cost a few hundred dollars, but think about what information is really needed to determine which type of bull is the best buy for the scenario.

12.1.3 Investment and Financial Analysis Calculations

Cattle managers may not be accustomed to some of the financial terms or considerations associated with these types of analyses. However, a cattle production unit should be viewed and evaluated as a business, and business managers that understand and utilize these types of considerations have advantages over any competitors that do not consider them. Some of the more fundamental aspects of these concepts are discussed here.

The concept of depreciation is the reduction in value of an asset with the passage of time, and usually due to wear and tear. A new car depreciates

from its purchase value, and the rate of depreciation is related to how well and how much it is used. The value of a cow actually increases over time as long as she remains productive, and peaks around 4–7 years, but that is not how depreciation of her replacement cost is calculated. If a heifer is purchased for $1000, the simple depreciation is to take that amount, divide it by the number of years she will be in the herd, and add that amount to her annual costs. The problem is that there is no way of knowing how long she will be in the herd. Very often, the purchase of equipment, vehicles, structures, etc. is depreciated over a set number of years (such as 7 years) or for the length of its life, in what may be called the straight-line method, where the same depreciation amount is deducted each year:

$$\text{Depreciation} = (\text{purchase value} - \text{salvage value})/\text{years used} \quad (12.3)$$

Opportunity cost is the loss of potential gain or value from other alternatives when one particular alternative is chosen, and this is a bit hard to visualize. As a result it may be related to using land resources for other opportunities as opposed to beef cattle production (which may or may not be a realistic situation). Very often opportunity costs are provided in analyses of beef cattle or other agricultural enterprises, but in-depth consideration of this concept is beyond the scope of this chapter.

There are also several types of investment analyses that can be used in beef cow-calf scenarios and aid in decision making regarding replacement breeding cattle on an individual animal or on a group of animals basis. Three specific calculations of payback period, net present value and break-even purchase price may be useful (Meek *et al.*, 1999; Mathews and Short, 2001; Falconer, 2012). There are some free spreadsheet calculators available on the internet regarding these and other analyses targeted towards beef cattle, but like any computer program (garbage in, garbage out concept), useful information for specific scenarios is critical for realistic calculations.

If producers and managers can determine some of these economic values on both the animals they are producing and the animals they are considering for purchase, they will have added information to evaluate their options (i.e. be further empowered to reduce risk, discussed more later in the chapter). Managers may be faced with options where they can produce animals more cheaply than they can

Table 12.1. Average annual net return per cow relative to production assumptions.

No. of years past purchase	Perfect calving record	Calf production scenario				
		Skips one calf 2 years after purchase	Skips one calf 3 years after purchase	Skips two calves (2 and 3 years after purchase)	Skips two calves (2 and 6 years after purchase)	Skips three calves (2, 5 and 8 years after purchase)
1	−975	−975	−975	−975	−975	−975
2	−225	−538	−225	−538	−538	−538
3	−42	−250	−250	−458	−250	−250
4	38	−119	−119	−275	−119	−119
5	81	−44	−44	−169	−44	−169
6	108	4	4	−100	−100	−100
7	127	38	38	−52	−52	−52
8	141	63	63	−16	−16	−94
9	151	81	81	12	12	−57
10	159	97	97	34	34	−29
11	165	109	109	52	52	−5
12	171	119	119	67	67	15
13	175	127	127	79	79	31

Assumptions here include purchase of pregnant heifers for $1050 each, annual cow cost of $400, salvage value of $450, all calves produced weigh 226.8 kg (500 lb) and are valued at $2.76 per kg ($1.25 per lb). Straight-line depreciation [(purchase cost − salvage value)/years] included as annual expense. Calculations based on total income − total expenses / number of years. As salvage value is included for each year past purchase, these can also be viewed as longevity scenarios (the average annual return for 4, 8, etc. years with production assumptions in play to that point). Notice how many years it takes a cow to begin achieving a positive cash flow relative to these different scenarios.

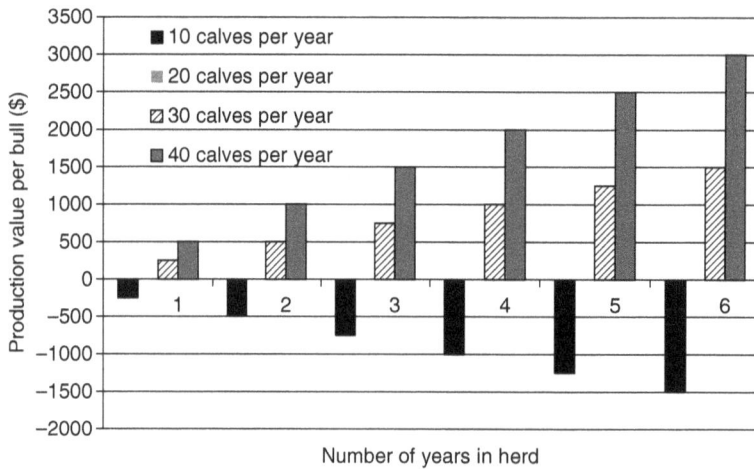

Bulls producing under 20 calves per year produce net losses that accumulate (20 calves per year results in zero loss or gain in value).

The production value can be thought of as the break-even purchase price. The bull with the highest value is $3000 here.

Higher progeny numbers through more cows bred and more years of service result in higher production values.

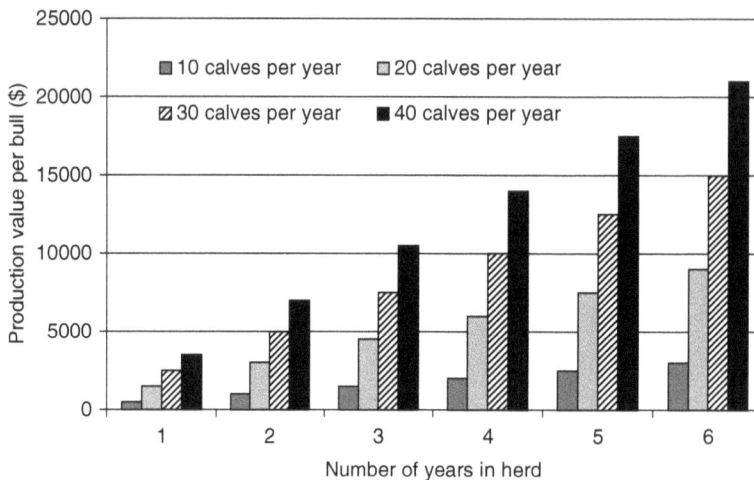

When progeny value is higher, there is increased profit potential of bulls. Several scenarios here result in herd sires with production value over $10,000.

Fig. 12.1. Production values of herd bulls relative to number of progeny, longevity and production potential of herd. Total expenses assumed to be $500 per bull per year; a salvage value of $600 per bull is also assumed. The assumption for annual net return per cow weaning a calf is $25 in the upper graph and $100 in the lower graph.

purchase equivalent ones, or they may find they can purchase replacements more cheaply than they can produce them. This type of knowledge provides the basis for more economically precise decision making.

12.2 Assessment of Expenses and Relationships with Profit

Numerous factors are associated with costs, but these may often go unrecognized if not formally evaluated. Some of these specific input costs are individually discussed here, but not necessarily all

cost categories are covered. This section should give readers an idea about the right questions to ask and the values to obtain or consider that can influence overall profitability.

12.2.1 Consideration of Interest Associated with Input Costs and Loans

One thing that may be overlooked by cattle producers if they need to borrow money from a lending source is the additional expense due to interest on financing a loan. A high-interest loan can erode potential profit in some circumstances, and just because an entity will loan producers money to finance agricultural production opportunities, this does not mean it is the best opportunity for the producers; it is critical to fully understand the calculations and obligations involved. For this example, to calculate interest associated with a production loan, consider the purchase of three cattle and the associated feed and veterinary costs. Three animals are purchased with an average weight of 240 kg (529 lb), and their purchase price is $1.32 per kg ($0.60 per lb). The anticipated feed cost will be $165 per animal, and the anticipated veterinary costs will be $8.50 per animal. A banker has agreed to finance 90% of the total costs of the program (animal and feeding costs combined) for 200 days at 6.0% simple interest. The expected combined weight of the finished animals is 1260 kg (2778 lb). Using this information: (a) calculate the total interest costs for the three animals, (b) calculate the interest cost per kg of finished animal, and (c) calculate the cost of gain (in $ per kg) for this program.

To address these questions, we first need to do some basic calculations. The cost of the feeding program will be $165 + $8.50 = $173.50 per animal, and $173.50 × 3 = $520.50 for all three animals. The purchase costs will be $1.32/kg × 240 kg = $316.80 per animal and $316.80 × 3 = $950.40 total. As a result, the total cost of the entire enterprise will be costs of the feeding program plus the purchase cost of the animals ($950.40 + $520.50), or $1470.90 in this example.

Question (a). The banker will finance 90% of the total costs, so the amount financed will be $1470.90 × 0.90 = $1323.81. Next we must determine the amount of finance charges per day. To do this, we take the amount financed, multiply it by the interest rate, and then divide that amount by 365 because interest rates are expressed on an annual or 365-day basis. This therefore becomes ($1323.81 × 0.06)/365 = $0.2176 per day, but because we will finance this cost for 200 days, we must multiply to get the finance charges for 200 days so that $0.2176 × 200 = $43.52. Equation (12.4) below shows the formula for calculating simple interest based on an annual percentage rate (R) and set number of days:

$$\text{Interest} = \left(\frac{\text{amount financed} \times R}{365} \right) \times \text{number days financed} \quad (12.4)$$

Taking any production loan without understanding the concepts of interest presents a large risk for producers. High-interest loans can quickly erode profit potential from production aspects, and in some cases produce debt for the manager.

Question (b). Because these animals gained 180 kg each, or 540 kg total, and the interest costs were $43.52, we simply divide the interest cost by the weight gain so that $43.52/540 = $0.0806/kg. The values used for this example will deviate significantly across global regions, however the logic and approaches are universal.

Question (c). The total cost of the program (for all three animals) is determined from above as $520.50 for feed plus vet costs plus an additional $43.52 for interest, or $564.02. The total weight gain of the three animals is determined as the final weight of the group (1260 kg) minus the initial weight of the group (240 kg × 3 = 720 kg), or 1260 kg – 720 kg = 540 kg. The cost of the gain is calculated by dividing the cost by the weight gain, or $564.02/540 kg = $1.0445/kg, or $1.04/kg ($0.47 per lb) when rounded off. This value may also be interpreted as $104.45 per 100 kg of gain (or $47.37 per 100 lb). In the USA, costs and market prices are usually expressed as dollars per 100 lb. No matter what units are involved, the concepts of the calculations are the same.

The cost of gain number is important to know because if it is lower than cattle prices, then additional weight provides additional value, and as animals weigh more, they should receive more profit. If the cost of gain is calculated to be close to the animal prices, then caution needs to be used as additional weight gain may reduce potential profit or even compound debt. If no previous information about similar cattle exists to estimate cost of gain (which itself is related to feed intake, feed efficiency, animal health and feed costs), then there will be much potential variation around any estimate used, and therefore more inherent risk associated with the activity.

There can also be different types of contractual arrangements in regard to paying for grazing lands and growing cattle. For instance, some agreements are based upon a set cost relative to the animal's weight gain and others may be based upon a set charge per animal per unit of time. This is more commonly done with growing animals than breeding herds (the same logic would apply to breeding herds; the performance measures relative to contract would differ though). Again the actual values used in the example here are not as important as the logic to make these and other related comparisons. Assume that there are two offers available for a set of growing calves. One option is to pay (or charge, if the perspective is from that of the land owner) a price of $0.6615/kg of gain ($0.30/lb) or pay a set price of $12 per animal per month. The variables that would make one scenario or the other more potentially attractive are the amount of time and the amount of weight gain. If neither of these can be guessed accurately, this adds risk to our decision over which route to take. The adage that 'knowledge is power' is true in all matters related to economic management.

In an effort to help US cow-calf producers evaluate their production and financial measures simultaneously, an area referred to as integrated resource management (IRM) and standardized performance analysis (SPA) was developed. This approach has provided for a standardized way to evaluate economic performance in cow herds. Under SPA analysis, producers must incorporate all expenditures and income associated with the cow-calf production enterprise, and it treats cow-calf production as a stand-alone business, which may or may not be the case. It even accounts for charges of land rent, if the land is owned, and includes depreciation of assets, which many ranchers may not consider, as well as salary for time invested. SPA analyses are based on an accrual accounting basis as opposed to a cash basis. Table 12.2 shows values for herds enrolled in the southwest USA SPA database from 1991 to 2002; as a result, the actual values do not reflect current market situations. However, general trends in other regions of the USA (Dhuyvetter and Langemeier, 2010; FINBIN, 2012) and several countries (Deblitz, 2012) illustrate the same concept that a major driver of profitability is based upon keeping expenses as low as possible. Figure 12.2 illustrates expense categories across profitability levels for some cow-calf operations.

12.3 Interpretation of Variation

This section provides some more detailed discussion regarding variation and its calculation. The concept of variation, variability, uniformity, etc. is complex, and has different meanings for different people and different type of traits. We rely on genetic variation to make selection response, we rely on variation in feed nutritional contents and prices to match supplementation needs and identify the best feed choices, and we rely on variation in consumer preferences to develop and market different products. Variation may be good or bad,

depending upon how it is packaged. In fundamentals of statistical analysis there is a concept referred to as analysis of variance, where variation is partitioned into (or attributed to) different components of between factors vs within factors. This concept is important for managers to recognize and to consider.

In many cases, it is the 'within' variation that robs efficiency as this type of variation is associated with variability within a production unit or management unit. As a result, this within variation could be in reference to within a pasture, within a feedlot pen, within a beef product line, etc. The 'between' variation is the type of variation that differentiates one production unit or management unit from another unit. As a result, between variation can represent differences between pastures, between truckloads of cattle, between feedlot pens, and between different beef products and/or different branded beef programs.

Many biologically and economically important traits follow a normal, or bell curve distribution. The bell curve is centered at the mean of the trait, and the height and width of the curve is dependent upon the standard deviation of that trait. This concept was discussed in Chapter 3 regarding phenotypic, genetic and environmental variation. The conceptual formula to calculate sample variance (s^2) is:

$$s^2 = \frac{\sum (y_i - \bar{y})^2}{n - 1} \qquad (12.5)$$

The standard deviation is the square root of the variance (s). The terms σ^2 and σ may be used to represent variance and standard deviation, respectively (usually used when discussing populations as opposed to samples, but they have the same interpretation). In this formula y_i represents individual observations and $\bar{y}$ represents the mean observations of some trait, and n is the number of observations. As can be seen from the formula, variance is relative to the difference (i.e. deviation) between individual observations and the mean; the greater the differences, the more variation, and vice versa. In a normally distributed trait, approximately 68% of all observations are expected to be within one standard deviation of the mean, approximately 95% of all observations are expected to be within two standard deviations of the mean, and approximately 99.7% of all observations are expected to be within three standard deviations of the mean. For a particular trait, a larger standard deviation indicates a higher degree of variation.

Table 12.2. Southwest US cow-calf standardized performance analysis (SPA) summary of key measures (1991–2002).[a]

	Percent calf crop weaned (%)	Average weaning weight (kg)	Kg calf weaned per cow exposed	Capital investment per cow ($)	Cost of production per cow[b]	Grazing and feed cost per cow	Cost of production ($ per kg wwt)[b]	Return on assets market value (%)	Net income per cow ($)[b]
No. of cows				*Summary by cow herd size category*					
1–49	83.1	228.2	192.4	$6049	$554	$244	$3.44	−2.82	($187)
50–99	83.2	231.4	191.5	$4683	$387	$201	$2.01	2.19	$6
100–199	82.1	233.3	192.4	$3610	$445	$194	$2.32	−0.08	($45)
200–299	81.6	236.8	193.1	$3718	$422	$139	$2.18	−1.24	($34)
300–499	83.2	239.9	198.6	$2955	$413	$141	$2.09	−0.43	($19)
500–999	82.0	242.0	199.4	$3112	$387	$133	$1.90	1.40	$16
1000+	79.6	243.8	192.8	$3520	$334	$124	$1.76	1.16	$29
Average	82.2	236.6	194.6	$3891	$421	$167	$2.25	0.0	($33)
Quartile				*Summary by profit quartile*					
Top 25%	85	243.5	205.9	3917	$322	$134	$1.52	6.6	$140
2nd 25%	83	237.6	195.5	3397	$362	$152	$1.85	2.4	$29
3rd 25%	81	238.1	194.1	3243	$413	$164	$2.16	−1.4	($44)
Low 25%	80	227.2	184.6	5049	$590	$217	$3.42	−7.7	($261)

[a]All production measures are calculated on the basis of cows exposed to breeding. Simple averages were calculated for 424 herds and total of 306,610 cows, with approximately 60 herds in each size group.
[b]Cost of production was full financial pre-tax cost including depreciation per breeding cow and weaned calf.
From McGrann (2003).

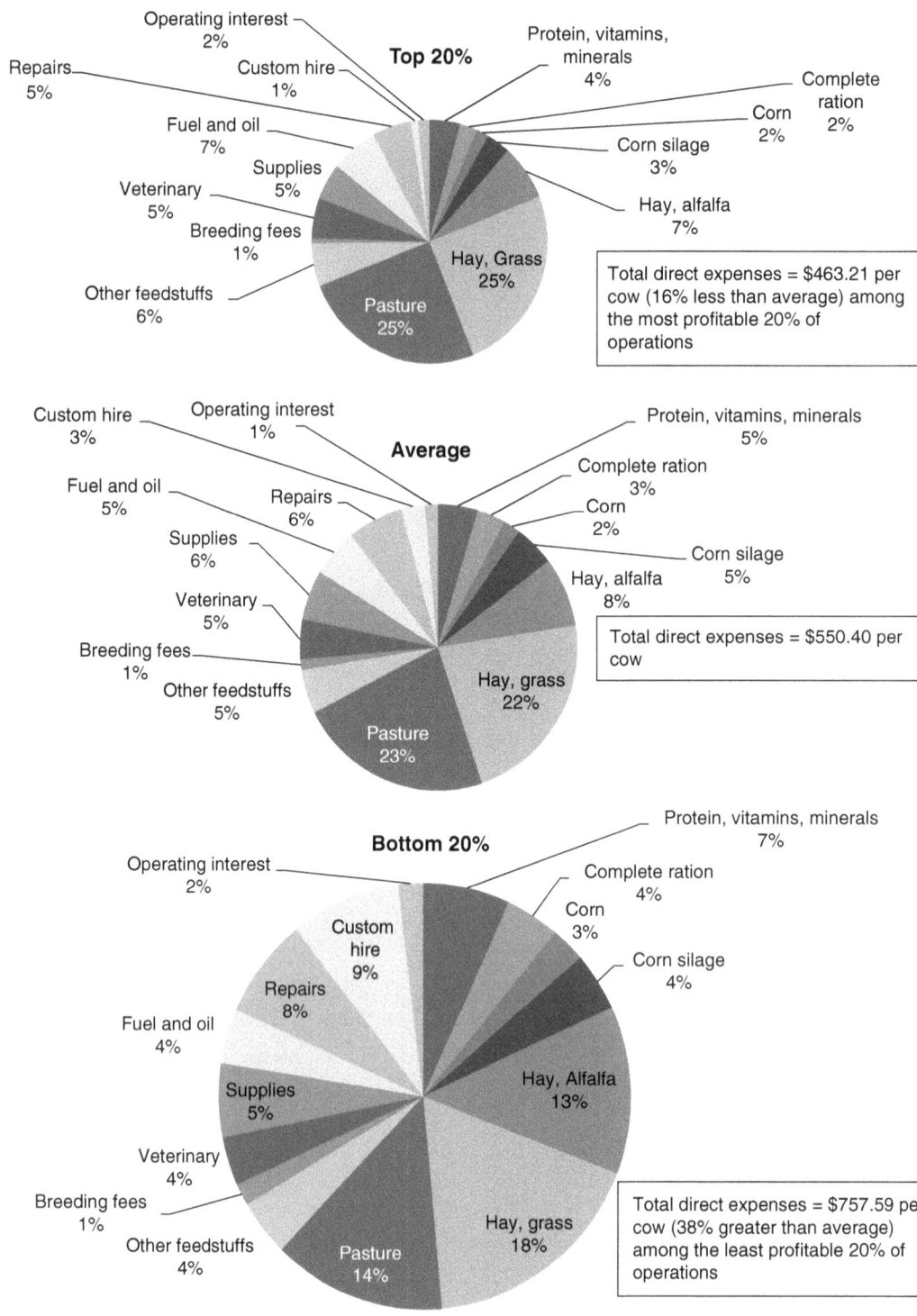

Fig. 12.2. Direct expense category percentages across cow-calf profitability groups in 2012 (FINBIN online database including several states in the USA).

However, it may be misleading to only evaluate the standard deviation when making considerations across traits. For example if the standard deviation for birth weight was 4 kg, and the standard deviation for weight at 12 months of age was 40 kg, there may be the same degree of variability relative to the mean of the trait even though the numbers for the standard deviation values are different. Another measure of variability that helps with this type of consideration is called the coefficient of variation (or CV), which is simply the standard deviation divided by the mean (Equation 12.6). As a result, the CV can then be thought of as the percent of the standard deviation relative to the mean:

$$CV = \frac{s}{\bar{y}} \qquad (12.6)$$

Table 12.3 shows the means, standard deviations and CV values for several traits from southwest US SPA cow herd values. These data, as well as those from other analyses, show that it is typical for more relative variation to exist in expenses and profitability measures than for most output-related traits such as weights and calf production.

12.3.1 Concepts of Variation Associated with Production Considerations

In the manufacturing and service industries, product consistency is a primary 'quality' factor, and usually comes from performing the process exactly the same and/or producing the same thing time after time. This type of variation can be thought of as 'within' variation (i.e. variation within a product or production line). From this standpoint, variation is bad because it produces inconsistency and lack of uniformity within a product or process. However, there is another type of variation that is recognized as being good (although it may not be recognized at first as a type of variation), and that is differences between products or processes. This type of 'between' variation gives rise to different products and different processes. It is good for customers and consumers to have choices, but relative to the choice made, products need to be consistent within that choice. These types of within and between variation need to be thought of by cattle managers because they not only influence end-product types of considerations, they also affect the cost and efficiency of production. Within any cattle operation there are production units (pastures, herds, etc.), and recognition of the variability associated with these units can lead to improved management.

Very often, to produce *similar* performing cows, fed cattle, carcasses and beef products, we need to treat many cattle and carcasses *differently*. Examples of this concept from previous chapters might include (i) sorting cows/heifers into different feeding or management groups to attain a target body condition, (ii) supplemental feeding of steers and heifers differently, (iii) sorting of calves of different ages or frame sizes to be fed or developed for different lengths of time, etc. Another type of variation to consider deals with the concept of

Table 12.3. Means, standard deviations and coefficients of variation from southwest US SPA analyses.

Trait	Average	SD	CV
Pregnancy %	88.5	5.9	7
Calving %	81.3	8.7	10
Weaning %	79.4	8.4	10
Weaning weight (kg)	222.0	33.2	14
Kg calf weaned per cow exposed	177.6	39.5	19
Feed cost per cow (US$)	118.08	78.06	61
Grazing cost per cow (US$)	102.51	65.54	63
Raised feed hectares per exposed female	0.12	0.24	164
Grazing feed hectares per exposed female	8.62	6.84	98
Total investment per breeding cow – costs basis ($US)	2871	4125	128
Percent return on assets – costs basis	1.6	8.33	698
Net income per cow ($US)	9.29	168.16	822

From Southwest SPA data 2012 (Bevers, 2012).

'diversification' within an operation; this is where different groups of animals or types of products can be produced. This concept is discussed further in Section 12.3.2.

12.3.2 Risk Management

Production risk for livestock enterprises can be thought of as the variability in production performance values or output that is linked to variability in financial or economic outcomes. In the context of the definition of risk given above, production risk is related to the variance around expected output rates. Thus, managing production risk is related to the management of variability in output as a means of managing the stability of revenues or net income. Using this base definition, a framework for both the evaluation and management of production risk can be constructed for livestock enterprises. A relatively small proportion of the published research regarding risk management in agriculture centers around production risk management, and of the published studies explicitly focused on production risk, few are associated with livestock production risk (thoughts taken from Sawyer and Sugg, 2009). In general, risk management is important throughout beef cattle production, and is something that producers probably do unconsciously as part of their routine management, but without recognizing what constitutes risk management, its usefulness or potential application to different scenarios may not be recognized. Table 12.4 summarizes categories of risk management strategies and possible examples that can be relevant to beef cattle production operations.

There are many examples of risk management considerations, and they may need to be tailored to fit local conditions. Very often the concept of risk management for cattle producers may be based on the concept of reduction of 'within' variation and increasing of 'between' variation, which usually includes sorting of animals into management and/or outcome groups (separating thin from adequate body condition cows for targeted supplemental feeding, feeding animals longer that have more potential to deposit marbling, etc). However, this concept of risk management can also be related to the manager's sense of business opportunities and the potential for diversification within or across production systems. For instance, in some situations, a ranch may be able to produce and market breeding heifers, mature cows, stocker/grazer calves, hay and/or crops, and ecotourism opportunities. Having cattle that are acceptable to many types of markets and people can also reduce risk as many more potential buyers and purposes will exist as compared to highly specialized types of cattle. Also positioning the operation to rely less on 'artificial' inputs (purchased feed, fertilizer, herbicides, etc.) can reduce economic risk, but typically shifts more responsibility to the cattle and the ability to match cattle potential with natural resources.

Pingpoh *et al.* (2007) reported on the incidence of various risk management strategies used by beef cattle farmers in northwest Cameroon to be: transhumanance – moving animals for grazing of seasonal resources (64%), burning pastures – burning of old plant residue to provide higher quality new growth (58%), keeping traditional cattle species/types – use of adaptation related considerations (100%), mixed cropping–pastoral management – multiple uses of lands and crops (60%), cattle placed in more than one herd – catastrophic events will not wipe out entire group (48%), secondary economic activities – not solely relying on one source/type of income (90%), and joint cattle ownership – animals owned by partners for multiple people with vested interests (36%).

Properly matching cattle types to resources and markets that adequately reward managers for their efforts is a component of overall risk management. The type of cattle production system will dictate, at least in part, what type(s) of animals can be marketed (see Table 12.5).

In some areas, it may be uncommon for producers to sell young cattle such as weaners or stocker/feeder calves. In other areas, it may be uncommon to sell older cattle, except as culled breeding animals. When animals are sold (or attempted to be sold) that are different from the typical expectations of local cattle buyers, there will probably be reduced prices for these 'unusual' animals. In areas where there is a mixture of different cattle uses, there is also probably the potential to market several different cattle types (the term 'type' here does not necessary indicate genetic influence, but it could) for multiple uses. Managers should evaluate the potential to sell multiple 'types' of animals as their overall operational assessment.

12.3.3 Traits of Successful Managers

A US study (Klinefelter, 2007) of farm profitability was conducted from 1980 to 1987, in which the

Table 12.4. Consideration of beef cattle production risk management strategies.

Category	Description	Example
Avoidance	Choosing not to engage in the activity that creates a specific risk exposure; because most ranches are multi-enterprise firms, avoidance of particular production operations may constitute an effective risk management strategy.	It might be the decision of a cow-calf operation not to precondition calves in order to avoid losses associated with mortality or low weight gain during activity.
Mitigation	The variance associated with a production outcome, or resulting from a specific production driver, will be reduced through managerial action; the most likely strategy that many cattle managers will use for production risk management.	Use of supplemental feedstuffs to offset variation in forage quality; mineral supplementation for certain soil profiles; vaccination for known disease threats; early weaning of calves.
Transfer	The source or outcome of production risk can be passed to another agent or entity.	A landowner or operating company may provide acreage and care for cattle on a fixed fee per head basis as opposed to owning cattle; use of insurance policies where particular risk is transferred to the insurer for a fee.
Retention or acceptance	Management accepts the variance in production; may occur when the tolerance for risk relative to exposure is high, or when the expected value of accepting risks is greater than other alternatives.	Managers may accept some risks due to greater than average technical efficiency, knowledge/data from previous experiences, or some other competitive advantage they possess.
Diversification	Exposure to a particular source of production risk exists in one enterprise, and operation of another enterprise that is not exposed to same source helps to offset negative exposure.	Having multiple types of cattle or production system components such as cow-calf and stocker operations; having different groups of cattle to market during year; being able to sell cattle to different buyers or markets.

Taken from Sawyer and Sugg (2009).

author compared characteristics of the top 25% of operations (the most profitable) against the bottom 25% of operations (the least profitable, i.e. lost the most money). The most profitable farms had crop yields that were 5% more and costs per unit of production that were 5% less than the least profitable farms. These two profitability groups had similar machinery and equipment, and they also had similar debt-to-asset ratios. The net income per farm was +$50,000 per year in the most profitable group and –$25,000 per year in the least profitable group. The main differences between these two profitability groups were attributed to planning and monitoring of yearly performance, where the most profitable group was comprised of the best managers for those components.

As mentioned earlier, McGrann (2003) evaluated beef cow-calf operations in the southwest USA (1991–2002) with the summary of key traits previously provided in Table 12.2. Some of the main points from the analysis illustrated traits of effective cow-calf managers needed for increased potential of profitability. These included considerations such that producers must make a business management commitment, and measure and monitor progress toward specific written goals, the ability to execute a business plan seemed to separate successful from non-successful managers, and that most ranches had inadequate inventory and management accounting systems to accurately measure and monitor performance.

Table 12.5. Considerations of marketing flexibility with different types of beef cattle operations.

Type of cattle or operation	Description	Potential vulnerability considerations
Purebred (seedstock) cow herd only	Land base devoted to as many cows as realistic based on carrying capacity; production of breeding animals for other cow-calf operations, calves sold young.	All animals are high in price, and if forced to sell out quickly, only market prices for commercial cattle may be available; only way to reduce stocking rate if needed is to sell high-value breeding animals; early weaning of calves may greatly reduce their value, and/or jeopardize breed society records.
Commercial cow herd only	Land base devoted to as many cows as realistic based on carrying capacity; production of market animals for various markets, but calves sold young.	Reduction of stocking rate if needed will reduce breeding herd size; there are probably more attractive opportunities to sell calves of various ages than purebred calves.
Cow herd with growing animals past weaning	Land base devoted to combination of cows and calves that are grown following weaning (stocker and/or finisher).	If reduction of stocking rate is needed this may be done largely or entirely by selling growing animals without reduction of breeding herd.
Finisher cattle only	Land base devoted to as many cattle as possible that will be finisher market animals.	These cattle are probably owned for long periods of time, maybe up to 2 or 3 years in some cases, so needed reduction in stocking rate could result in cattle being sold before optimal time; can be offset by having multiple groups of different ages.
Finisher and stocker (weaner) cattle	Land base devoted to combination of different age groups of cattle that will be ready to sell/market at different ages.	Similar to the finisher operation described above, multiple times for selling animals of different ages are possible; young cattle can be destined for finishers or sold to others.
Stocker (weaner) cattle only	Land base devoted to young, growing calves that are kept for relatively short lengths of time.	Stocker calves are owned for much shorter periods of time (3–8 months) than finisher cattle and therefore offer potential to sell at regularly scheduled time points even when production conditions deteriorate.

All scenarios and assumptions include having stocking rates properly matched with carrying capacity. Stocker calves here could also refer to replacement heifers.

Stanley and Danko (1996) authored a book titled *The Millionaire Next Door*, where they documented and characterized traits of US millionaire households based on detailed data they collected. A primary theme of this work was that 'typical' millionaires were not what the general public expected. The median prices paid by US millionaires (from 1996) for some items were $399 for business suits, $140 for dress shoes and $235 for wristwatches. These prices were not 'cheap', but they were far from extravagant, which many people might expect. Another main point was that 86% of US millionaires were self-made, meaning that they had not inherited wealth, and most built businesses and/or investments over time (hardly any got rich quickly). Finally, the authors characterized typical US millionaires as (i) frugal, (ii) good at planning, (iii) not afraid to take calculated risk, and (iv) believers in education.

There is a difference between being a low-cost operation and a frugal operation. There are instances where effective managers spend more money to increase profit, and other instances where they reduce costs to increase profit. Having the ability to accurately gauge whether expenses provide adequate benefits is the key to effective business management. In many areas of the world, lack of education regarding realistic agricultural production and lack of access to financing probably inhibit many livestock producers (including the

Table 12.6. Considerations of marketing different types of cattle from commercial cow-calf herds.[a]

Type of animal sold	Description	Options/considerations
Young calves	Calves close to weaning age.	When calf prices are high, and cost of added weight gain is high, it may be best to sell at this point.
Cull breeding animals	Breeding animals that have been removed from herd due to some undesirable issue.	Time of year and available feed resources can influence value fluctuation. Selling bred cows a few months later than selling open cows immediately may be a consideration.
Stocker (weaner) calves	Calves over weaning stress and ready for grazing-based weight gain.	If adequate grazing is available these animals can be held for several weeks to accumulate value, costs relate strongly to forage resources.
Finisher calves	Calves that are older, maybe a year of age, that have already been developed to some extent, but are now ready for finishing through grazing or feedlot.	Holding of these animals depends greatly on feed/forage resources and costs. Timing within and across years may need to be considered.
Breeding age heifers	Heifers that have been developed and are ready to be bred; age(s) depend upon local conditions.	It may be a subtle distinction of heifers used for finishing and breeding. When cattle breeding numbers are low, these females can receive premiums.
Bred heifers	Heifers that have been confirmed pregnant, typically 4–6 months pregnant.	When breeding inventory is down and forage resources are good, bred heifers may command high premiums per animal as compared to open females.
Young cows	May be sold as ready to breed with calf at side, pregnant with calf at side, pregnant without calf; may be 2–5 years old depending upon local conditions.	Young cows sold from breeding herds should be those that are not anticipated to be genetically superior, especially for longevity. Body condition greatly affects value.
Medium age cows	May be sold as ready to breed with calf at side, pregnant with calf at side, pregnant without calf; may be 5–8 years old depending upon local conditions.	Medium age cows sold from breeding herds should be those that are not anticipated to be genetically superior, especially for longevity. Body condition and mouth condition affect value.

[a]These options also exist with purebred/seedstock herds, but the prices received for various purebred animals will probably be the same as commercial animals for many stages or ages. One option that is probably not realistic for most commercial operations is sale of bulls for breeding stock, as individual performance data are usually not recorded, and this information is important for bull buyers.

poverty-stricken), but the concepts for more cost-effective cattle management are true across widely different circumstances.

12.4 Marketing and Related Business Considerations

Much discussion has been presented about how cattle farmers and ranchers produce cattle, but now discussion turns to how they may market their animals. Broad-scale views about general industry structure affect marketing potential. For instance, does the industry need to be more concerned about marketing live animals, marketing beef products, or are both equally important? Is it expected that seedstock vs commercial cow-calf producers view marketing differently? Also, how do public/consumer perceptions of the industry influence existing and emerging marketing opportunities? All of these issues affect current prices and long-term trends. Of course there are also numerous short-term conditions that affect prices (drought conditions, recent rains, current number of animals entering markets, supply of beef, etc.) that producers should be aware of as well. Basic economic theory tells how as supply increases, prices tend to decrease because demand decreases and vice versa.

The broad-scale issues may not be apparent that they affect the prices that farmers and ranchers receive at their local markets. The major marketing considerations for most cattle producers revolve around three primary factors of (i) when they are selling, (ii) what they are selling, and (iii) how they are selling. These three considerations are individually addressed below.

12.4.1 Marketing Timing (When)

When is the time of year or time in the production calendar that the animals are sold. This is typically based upon the timing of certain production practices such as weaning calves at a particular time of year, or the target market sought, but may also be influenced by current market and/or environmental conditions.

12.4.2 Marketing Different Types of Animals and/or Products (What)

What refers to the type(s) of animals that can be sold. For typical commercial cow-calf operations in North America, this is most commonly calves at or shortly after weaning, and culled breeding animals. However, these producers can market a wide array of different types of animals (Table 12.6). Likewise, producers that sell breeding cattle can merchandize different types of animals and/or products at a variety of ages and stages of production, and including semen, embryos, and partial ownership of breeding animals. Some producers may only purchase young cattle to be developed under grazing programs, but these cattle can also be marketed at various ages and stages of development.

12.4.3 Marketing Avenues and Outlets (How)

It must become obvious that some producers may have more potential to increase income, some producers may have more potential to decrease expenses, and some producers may have potential to do both. Only focusing on income or expenses will not guarantee increased profitability; both

have to be evaluated simultaneously. A major potential impact on income revolves around marketing considerations. The types of animals and products that a producer can market as well as the different types of marketing alternatives available to producers can be widely different across geographical regions, within countries and across countries. In North America, Western Europe, and Australia and New Zealand, producers can sell their cattle through auctions, by direct sales, through electronic means, through cooperative efforts, and through market integration, and combinations of these options (Fig. 12.3). The two major distinctions in marketing outlet are public (auction) and private (direct).

An auction is where there are several potential buyers that are all able to bid on animals at the same time, with the highest bid on purchasing the animals. This marketing route can be used for selling young animals, slaughter animals, breeding animals and cull animals. It may happen at a single location such as a ranch or at a commercial, professional facility designed for this, or even on the internet, or through a combination of these options. The advantages of auctions are that they provide for a quick sale, and if there are many potential buyers this is also an advantage. Disadvantages can include risk associated with a small number of potential buyers, the value of an animal may be based on quick, phenotypic assessment, and the seller has less potential to influence price (they are more price takers). If little or no information about the animals being sold is shared with potential buyers, producers are even more limited in their ability to differentiate the value of their animals. In North America, the most common way young cattle (called feeder calves) are sold is through regional,

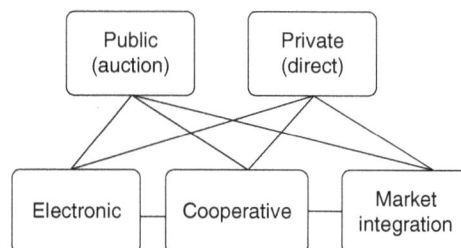

Fig. 12.3. Primary marketing route considerations for cattle and beef producers. Sales are usually classified as either public or private, but can involve all other route possibilities in various combinations.

weekly auction markets where quick, current visual assessment and weight are used to determine value. If producers went to the extra effort and expense to increase the value of their animals, yet this information is not passed on to potential buyers, they may actually reduce their profit by following recommended, desirable management practices.

Direct sales can be used to sell all types of cattle. Advantages of this type of marketing are that the sellers can market their animals more effectively by conveying previous management and/or genetic information to prospective buyers. Disadvantages for the seller with direct marketing are that they have more responsibility to be familiar with current and emerging market conditions, and that they may have to contact many potential buyers individually before the animals are sold. Both auctions and direct sales can occur via electronic sources.

Cooperative marketing is where groups of producers at the same level of the production chain work together to market their products. The term 'horizontal integration' has been used to describe this concept. Several producers might put their calves together in order to sell a complete truck load of cattle, or fill a large pen at a feedlot, etc. which they could not do by themselves. Groups of breeders of a particular breed or group of breeds might decide to sell breeding animals collectively. In some cases, specialized auctions for cattle may happen, where calves of similar age, size, muscle and color are put into lots by people organizing the sale, where the sellers do not get to choose whose cattle theirs are sold with. The biggest advantages of cooperative or group marketing are that it gives access to potential marketing outlets or buyers that individuals would not otherwise have access to, and that it allows potential for more informed marketing to add value. The biggest disadvantage of cooperative marketing is that it forces several producers to work together and there is less potential to differentiate one producer's cattle from another, and there is usually a set standard of management that all producers must adhere to.

Retained ownership is a term that has been used to describe market integration (Fig. 12.4); vertical integration and vertical alignment are also terms that describe this same concept, where animals are owned through multiple production system components. There may be many different possible levels of retained ownership available to producers (selling calves 45–60 days post-weaning, selling calves as yearlings after a forage-based growing phase, selling cattle at the end of the feedlot period, selling cattle as carcasses, etc.), or in some regions there may not be these different options. The concept of value-based marketing of beef began to be used in the USA in the late 1980s and referred to the idea that cattle should be valued based on their actual carcass traits as opposed to on a live basis with carcass performance estimated. The concept of value-based marketing can be applied to all types of cattle, beef or any other product because not all

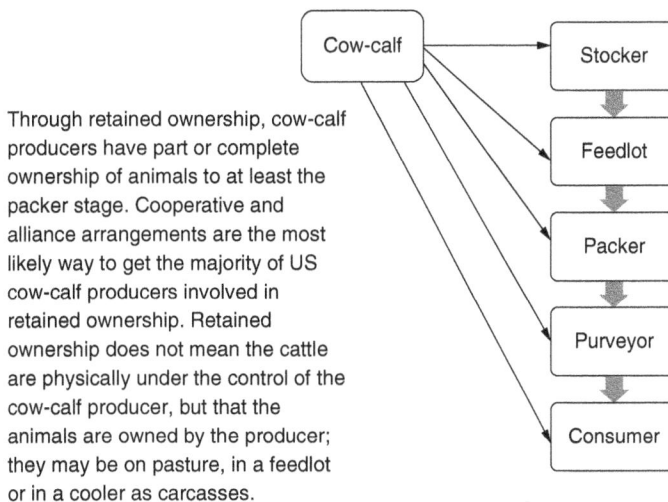

Through retained ownership, cow-calf producers have part or complete ownership of animals to at least the packer stage. Cooperative and alliance arrangements are the most likely way to get the majority of US cow-calf producers involved in retained ownership. Retained ownership does not mean the cattle are physically under the control of the cow-calf producer, but that the animals are owned by the producer; they may be on pasture, in a feedlot or in a cooler as carcasses.

Fig. 12.4. Concept of retaining ownership of cattle past the operation of origin.

customers perceive the same characteristics as having the same value. In some cases, simply knowing the production process or origin of cattle or beef products may add more value than the actual characteristics of the animals or products themselves. This concept is built upon in Chapter 13.

The term 'alliance' began to be used in the US beef industry in conjunction with the discussions on value-based marketing (Sartwelle *et al.*, 2000). Alliances are typically groups of entities or individuals working together to provide for increased market integration (Fig. 12.5). Several people have said that somewhere during the beef production process, someone will usually profit from cattle before they reach the packer. Likewise, it is expected that usually someone will make a profit on cattle by the time they reach the consumer. As a result, market integration provides for ownership across several production segments. The potential to eliminate the 'middleman' can allow for increased profit. It also allows for more marketing to differentiate products through production process and/or origin verification, and marketing cattle or products that are closer to the consumers also provides opportunities for value-based marketing. If cattle are profitable at different production segments, these profits can be accumulated with increased market integration. However, if cattle are not profitable in several segments, there is also potential to

accumulate debt. Of course, as there is increased market integration there must also be a prolonged amount of time until the ultimate value of animals is determined, and this may be a serious disadvantage for some producers from a cash flow perspective.

The corporate structure and vertical integration in the poultry and pork industries in many countries encompasses these concepts, and they have more structure than most beef cattle alliances because they own the breeding animals and dictate genetic decisions and feed the herds and flocks in controlled, grain-based environments. The capital investment required to own a breeding herd of beef cows large enough to supply a large packing plant or beef market on a consistent basis is huge (this is a major point of consideration in Chapter 13). As a result, many beef alliances have targeted beef cow producers that want to work together under organized leadership as opposed to a single-owner concept.

12.4.4 Concepts of Added Value

In general there is increasing potential for value-added considerations throughout many agricultural production systems, including those from cattle (and cattle themselves). Each production segment adds value, but marketing to consumers is the biggest opportunity (Fig. 12.6). The share of the on-farm value of food products is a small share of the total

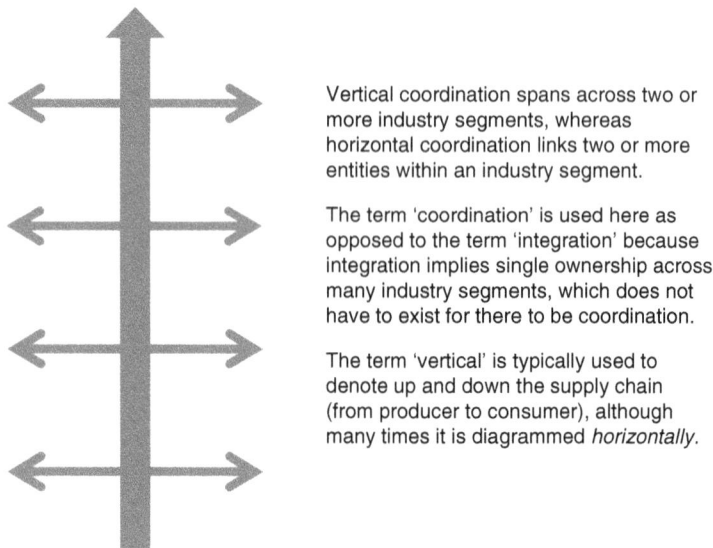

Vertical coordination spans across two or more industry segments, whereas horizontal coordination links two or more entities within an industry segment.

The term 'coordination' is used here as opposed to the term 'integration' because integration implies single ownership across many industry segments, which does not have to exist for there to be coordination.

The term 'vertical' is typically used to denote up and down the supply chain (from producer to consumer), although many times it is diagrammed *horizontally*.

Fig. 12.5. The concepts of vertical coordination and horizontal coordination.

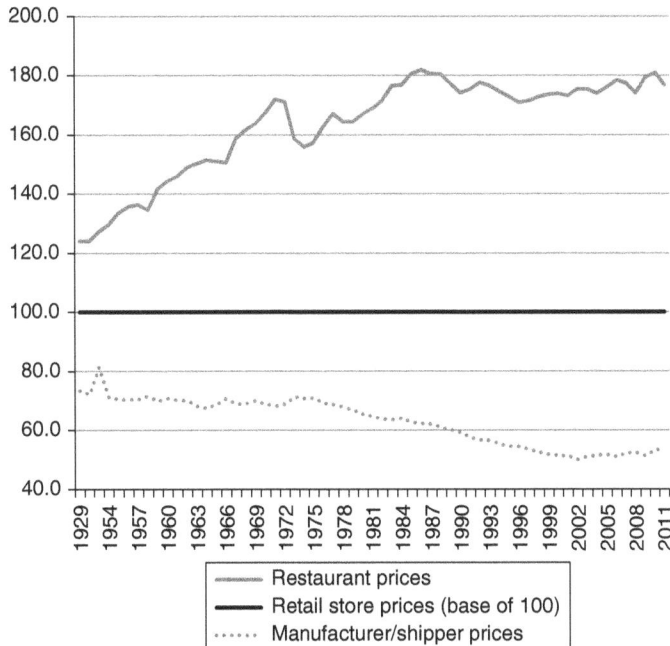

Fig. 12.6. Relative prices of US foods at three components of food production system illustrating value-added concept. Retail food price was set as the base of 100 for each year. From USDA Economic Research Service Food Expenditures online database.

value, and this trend has been ongoing for quite some time, especially in developed markets globally.

In the USA, the percentage of the farm value of foods has continued to shrink as the prices received by farmers have remained much flatter than the prices paid by consumers, especially for value-added products. This has occurred for a variety of reasons. From 1993 to 2008, the farm share of food expenditures continued to decrease, but the trend was very different for food purchased for home consumption as compared to food purchased for away-from-home consumption (as through restaurants, etc.). The farm share remained about 23% during this time for in-home food, but decreased from 10.5% to 4.7% for away-from-home food (Canning, 2011). Figure 12.7 shows percentages of US food production system components relative to food expenditures.

12.4.5 New Business and Marketing Opportunities

Many new products and business models have been successful because they provided part of a solution to somebody else's problem. This has been true with many agricultural by-products that can be fed to ruminants. Managers need to continually assess new opportunities, but they also need to not discontinue something that works just to try something new. There may be numerous situations to provide non-traditional ways to add value to certain livestock processes or products.

12.5 Summary

It is critical that cattle producers think about the cost vs benefits of various management practices and decisions, and that they have the proper measurements to make these assessments. Emphasis on total profitability as opposed to only production output traits (or only inputs) will usually provide for a more balanced and sustainable management approach. Understanding the concepts and measurements of variation provides the potential for more precise management and marketing decisions and the basis of understanding production risk. In all regions, access to markets and current market prices are important for both short-term and long-term

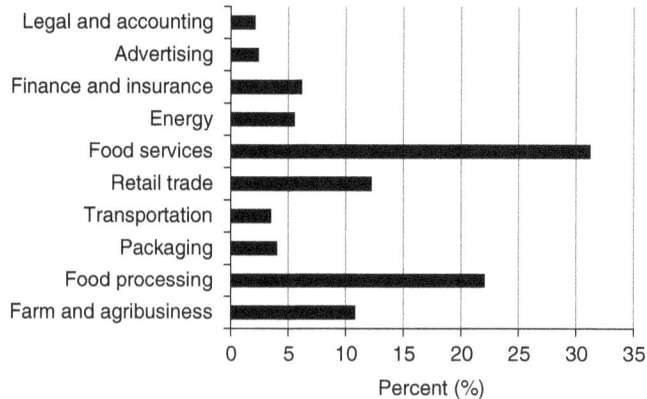

Fig. 12.7. Percentages of 2011 US food expenditures due to system components. From USDA Economic Research Service Food Dollar Series.

decisions. The ability to coordinate types of cattle with environmental conditions and appropriate market options is the foundation for profitable beef cattle production in most regions of the world.

12.6 Study Questions

12.1 An operation has used creep feeding of its calves for added calf growth and weaning weight. For this problem, assume calf prices will be $3.53 per kg.

12.1a If the feed cost $0.44 per kg, each calf consumed 68 kg of feed, and each calf gained an extra 22.7 kg of weight, was the creep feeding a good choice?

12.1b If the feed cost $0.39 per kg, each calf consumed 295 kg of feed, and each calf gained an extra 22.7 kg of weight, was the creep feeding a good choice?

12.2 The baseline level of production in a cow herd has been 88% calf crop weaned with 215.5 kg average weaning weight. Assume that each bull in this herd breeds 25 cows per year, and lasts 4 years. This operation has spent $3000 per bull in the past. If this operation could purchase bulls for $4000 each with resulting percent calf crop weaned being 92% per year and average calf weaning weight of 231.3 kg (with all other expenses staying the same), would this be a financially good thing to do?

12.3 You have a group of stocker steers that average 344.7 kg; current market price for these calves is $2.76 per kg. You could put these calves into a feedlot where they should finish in 5–6 months at 578 kg with an estimated cost of gain of $2.43 per kg.

12.3a What would be the break-even price on these cattle as fed steers coming out of the feedlot (express as $ per kg)?

12.3b If these cattle actually sold as fed steers for $2.78 per kg, how much money would you make or lose per head?

12.3c If you thought the cost of gain would more likely be $2.58 per kg, how would that change your considerations?

12.4 An operation has implemented a rotational grazing system. The kg of calf weaned per cow exposed went from 221.4 kg before the change to 215.5 kg afterwards, but they were able to maintain 15% more cows annually (no other changes in costs or performance), how would the new grazing management compare to the previous management? You can assume $475 per cow per year for expenses and $3.53 per kg for calf prices.

12.7 References

Bevers, S. (2012) Southwest SPA data and summary. Texas Rolling Plains Agricultural Economics Program, Texas A&M AgriLife Extension. Available at: http://agrisk.tamu.edu/. (accessed 10 December 2012).

Canning, P. (2011) A Revised and Expanded Food Dollar Series: A Better Understanding of Our Food Costs USDA-ERS Economic Research Report Number 114.

Deblitz, C. (2012) Costs of production for beef and national cost share structures. AgriBenchmark Working Paper 2012/3. Available at: http://www.agribenchmark.org/. (accessed 9 December 2012).

Dhuyvetter, K., and Langemeier, M. (2010) Differences between high, medium, and low profit cow-calf producers: An analysis of 2004-2008 Kansas Farm Management Association cow-calf enterprise. Kansas State University Report (18 pp.). Available at: www.agmanager.info. (accessed 9 January 2013).

Falconer L. (2012) Economic Tools to Evaluate Herd Liquidation Decisions for Breeding Cattle Texas A&M AgriLife Extension Service. Available at: http://animalscience.tamu.edu/files/2012/04/beef-economic-tools-to-evaluate.pdf. (accessed 14 December 2012).

FINBIN (2012) Farm financial management database, searchable livestock databases. Available at: http://www.finbin.umn.edu/. (accessed 11 December 2012).

Klinefelter, D. (2007) Causes of farm and ranch failures. *Texas Beef Cattle Management Handbook*, Texas A&M AgriLife Extension Service publication # b-1630.

Mathews, Jr., K.H. and Short, S.D. (2001) The beef cow replacement decision. *Journal of Agribusiness* 19, 191–211.

McGrann, J. (2003) Standardized Performance Analysis – SPA – for Southwest Herds – 1991–2002. Extension Report 8/5/03, Texas Cooperative Extension, Texas A&M University System.

Meek, M.S., Whittier, J.C. and Dalsted, N.L. (1999) Estimation of net present value of beef females of various ages and the economic sensitivity of net present value to changes in production. *The Professional Animal Scientist* 15, 46–52.

Pate, F.M. and Crockett, J.R. (1978) Value of pre-conditioning beef calves. Florida Agricultural Experiments Stations Bulletin 799.

Peterson, E.B., Strohbehn, D.R., Ladd, G.W. and Willham, R.L. (1989) The economic viability of pre-conditioning for cow-calf producers. *Journal of Animal Science* 67, 1687–1697.

Pingpoh, D.P., Mbanya, J., Ntam, F. and Malaa, D. (2007) Some risk management practices among the beef cattle farmers of the North West Province of Cameroon: Effect on technology dissemination. 106th seminar of the EAAE, Pro-poor development in low income countries: Food, agriculture, trade, and environment, 25–27 October, 2007 – Montpellier, France, 1–8.

Pritchard, R.H. and Mendez, J.K. (1990) Effects of pre-conditioning on pre- and post-shipment performance of feeder calves. *Journal of Animal Science* 68, 28–34.

Sartwelle, III, J.D., Davis, E.E., Mintert, J. and Borchardt, R. (2000) Beef cattle marketing alliances. Texas Agricultural Extension Service. Pub# L-5356 (RM 1-9.0).

Sawyer, J.E. and Sugg, D. (2009) Production risk management for cow-calf and yearling systems. Lectureship Series Paper 8 pp, King Ranch Institute of Ranch Management, Texas A&M University-Kingsville.

Stanley, T.J. and Danko, W.D. (1998) *The Millionaire Next Door*. Gallery Books.

13 Supply Chain and Integrated Production Considerations

The goal of this chapter is to discuss the procedural and management considerations that tie beef cow herds to the end products of the beef production system (i.e. the complete supply chain). For many scenarios or operations, not all of the industry components may be involved directly in the business (i.e. cow herd owners may derive income primarily from sales of young calves, feedlot owners can buy young calves and sell finished cattle to meat companies, food service companies buy beef cuts and sell cooked products to customers, etc.), but all industry components and their interrelationships are important to consider for both the immediate and long-term sustainability of individual enterprises, and for entire industries. Much of the discussion in this chapter revolves around concepts of production/product quantity and quality, meaning that both the end results and the processes used to obtain the end results are important and affect the beef production chain. Many considerations also build upon managerial and marketing considerations discussed in Chapter 12. The considerations associated with supply chains rely heavily upon different types of production systems. This goal of this chapter is to familiarize the reader with how production levels affect animal output (or off-take) and resulting total beef production, to emphasize the importance of individual animal identification and traceability and to highlight the importance of beef cattle production assurance programs.

13.1 Expanding the View of Beef Cattle Production into the Business Philosophy

Ideas about viewing beef cattle operations as business entities were introduced in Chapter 12, primarily in the context of individual operations. Our discussion is expanded here with further considerations regarding customer satisfaction and variability.

13.1.1 Ideas Associated with Quality Management

Dr Edwards Deming is thought by many to be the founder of expressing concepts associated with quality management for businesses, and his famous '14 points' originally presented in the book *Out of the Crisis* have served many entities as management guidelines. These points were proposed to provide a more efficient workplace, higher profits and increased productivity (Deming Institute, 2012). Many if not all of these points can be utilized or adapted for a variety of beef cattle entities or enterprises. These could pertain to a meat company, a corporate cattle feedlot, a seedstock bull producer, a commercial cow-calf producer, etc.

An important consideration of supply chain management with respect to cattle genetics is do you give customers what they want or what they need? The education component of customers (from bull buyers to beef consumers) cannot be overlooked, and on first glance it appears that we obviously give customers want they want, but we can also influence what they want (and the value they place on it). The ethical considerations of cattle production with respect to broad-scale stewardship of both animal and natural resource aspects cannot go without serious consideration. Some cultures or societies can afford various preferences for products much more than others. Many of these considerations have been woven into beef quality assurance (BQA)/best management practice (BMP) recommendations and formalized programs in many countries.

13.1.2 Beef Supply Chain Considerations

In the business world supply chain management deals with the processes and associated concepts of procurement, manufacturing, distribution and sales of product(s) of a company. This same discussion is relevant when we are discussing the beef industry at various levels (global, international partnerships, national, regional, corporate or individual producer), but the economic importance of specific practices and outcomes may have very different meanings across these industry levels. This section discusses the general supply chain considerations associated with beef production.

The most simplistic beef supply chain model involves in some cases the cattle producer and the consumer, but in commercial-oriented settings the most simplistic model typically involves the cattle producer, the meat packer, meat distributors and consumers (Fig. 13.1). Ownership has typically changed between each of these sectors. If there were ownership and management from animal production throughout the chain to the consumer, this would be called vertical integration (Chapter 12). This has become the corporate structure of the specialized poultry and swine industries in many nations.

In fully integrated poultry and swine production, a corporation owns all the animals and makes all the genetic decisions. The animals are delivered to contract growers. The contract growers provide for the facilities and management of the animals for a specified amount of time. The corporation is responsible for the transportation of the animals to and from the grower, but the grower is responsible for the daily care and management of the animals until they leave their facility. On leaving the grower facility, the animals are transported to a processing plant where they are

harvested and fabricated into meat products. The processing facility may be owned and operated entirely by the corporation, or the corporation may contract with a processing plant for a set amount of days or shifts per week. Corporations may also contract with producers to house breeding animals. Under this model, the contract producers get paid based on the number of animals at the end of the contract and resulting weight gain, so they are getting paid directly for their management, but they are generally not paid directly for consumer satisfaction. This type of arrangement is generally referred to as 'factory farming', because there is intensive management, animals are housed in confined spaces, and there is corporate oversight. Public concerns over confined food animals have led to the development of new markets for non-conventional production, and this line of thought is discussed more toward the end of the chapter. Figure 13.2 compares some production times in food animal systems.

This type of complete, corporate ownership is much more difficult to economically justify in the beef cattle industry because of the enormous amount of capital investment per breeding cow relative to the annual level of production. In many areas of the world, beef cows (and/or other ruminants) are present (and productive) on land resources that have few other potential uses or return on investment because the land is not suitable for farming, and because they can harvest forage resources in ways other food animals cannot.

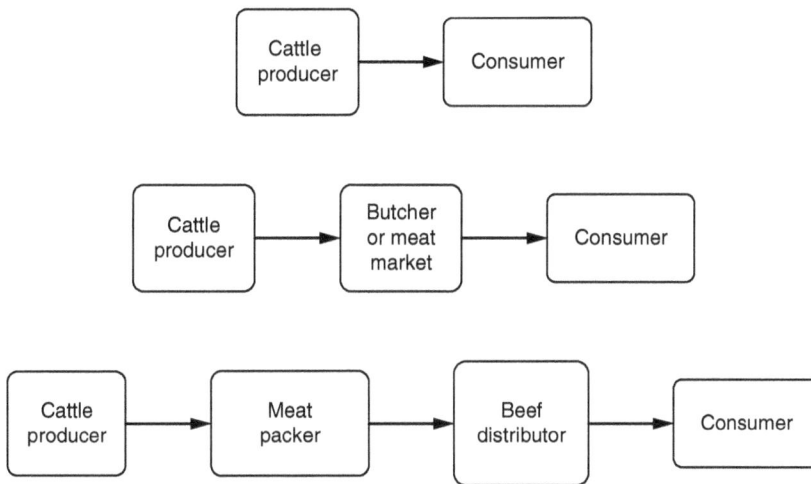

Fig. 13.1. Examples of beef industry supply chains

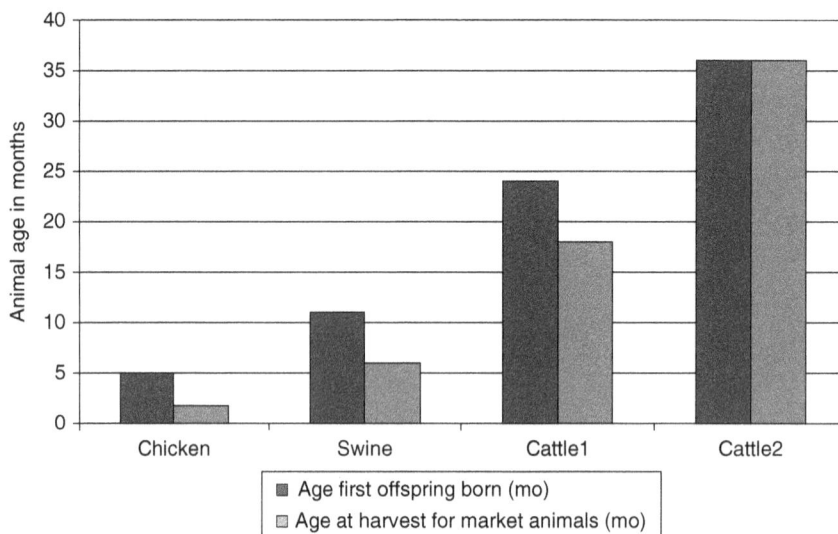

Fig. 13.2. Relative times for first female reproduction and time to market for some farm animal species. Values for Chicken, Swine and Cattle1 are relative to intensive production in developed countries. Cattle2 is relative to forage-fattening in many areas of the world.

An effective beef supply chain management strategy cannot occur without the fundamental knowledge of the biology and production methods influencing beef cattle.

13.2 Animal Numbers and Supply Considerations

In the discussion of beef supply chains it may not be obvious at first that reproduction and animal health in cow herds are primary factors that dictate the number of animals potentially available to provide the beef products. The carrying capacity of the land resources, the amount of land resources, and nutritional capabilities associated with growing cattle are also key components in beef supply chains. For corporate success, a consistent supply of product is critical. Consistency in regard to the availability of the product (i.e. it is available for purchase when the consumer wants it) as well as consistency in regard to the product expectations when purchased (i.e. having the same characteristics with each purchase) are both drivers of product success. We will discuss consistency as related to product quality later in the chapter. For the moment, we concentrate on consistency of supply.

13.2.1 Breeding Herd Aspects

As with the production of beef per animal harvested, production per cow exposed to breeding also impacts animal supply, and therefore beef supply. The percentage of animals based on the breeding cow herd inventory that could be harvested is a function of reproductive ability, as well as animal health and survival. If 80% of the cows exposed to breeding weaned calves, and 90% of the calves survived post-weaning to harvest and breeding age (10% loss of animals post-weaning), this relative herd inventory rate would be calculated as (0.90 × 0.80) = 0.72, or 72%. The percentage of replacement females and males needed to maintain herd size is included in this number. Under a very productive scenario where 90% of animals make it to harvest and breeding age from cows exposed to breeding, and replacement rates of 30% of heifers and 1% of males are needed from each calf crop, over 75% of the breeding cow inventory number can be harvested annually (100 breeding cows would yield 75 animals for harvest) and still maintain its original number. On the other hand with lower reproductive rates, reduced animal health and survival, where there is a combined weaning and harvest/breeding survival rate of 55%, and replacement rates of 52% of heifers and 4.5% of males are retained for breeding, the relative number of animals available for harvest annually would be below 40% (less than 40 animals to harvest from 100 cows). Along these lines of thought, the concept of increased productivity per unit of land area would also allow for increased beef production with the same amount of land, or constant beef production levels with reduced land use.

13.2.2 Concepts of Cow Breeding Herd Size as Related to Supply Chain Considerations

It may not be obvious yet how the aspects of fertility and animal health (discussed in Chapters 8 and 9, respectively) directly impact supply chain potential. The production of 'excess' animals from a beef breeding herd is what allows for (i) the production of animals for market/meat production, and/or (ii) the ability to increase herd size. In situations with very harsh production environments, it may be a huge challenge to simply maintain breeding herd size; if breeding herd size cannot be maintained, there is no potential for sustainability, much less growth.

The principles pertaining to cow herd fertility and survival rates can be applied to estimate herd size maintenance and production of excess animals. The number of breeding females in this line of discussion designates those that are of breeding age desired to become pregnant and wean a calf in the upcoming calving seasons. A major assumption in this exercise is that we are expecting heifers and cows to produce calves as regularly as is realistically possible, so the goal is a live calf per female per year for several years. The specific production environment and many associated considerations will dictate how realistic a goal this is, but developing specific goals is very important to visualize increased efficiency and profitability.

The scenarios in Table 13.1 are not equal in their overall level of desirability or production potential, particularly across widely differing production environments. All these calculations were set up with formulas in a spreadsheet to evaluate multiple scenarios. In these scenarios it is assumed that only reproductively active females from the base herd in one year are allowed to stay in the base herd for the subsequent year, but this may not be realistic or desired in all cases as young cows that do not breed back as 3-year-olds may reproduce successfully for

many subsequent years following a ship. As a result, the approach used here could yield slightly higher herd size numbers depending upon the culling criteria involved. Females that are not reproductively successful need to be removed from breeding herds typically, but the exact level of fertility that is economically justifiable may vary widely across production environments. In these calculations, cows that become pregnant, but fail to wean a calf, are included in the subsequent breeding herd, and this is also not always economically justified; without animal ID it would be impossible to know which if any cows are habitual problems in this regard.

Based on the scenarios described in Table 13.1, the number (or percentage) of heifer calves that need to be kept just to maintain a breeding herd of 100 females varies from 4 to 32 (10% to 100% of the heifer calves). In some cases, the desire to maintain constant breeding herd size is a primary production concern. In other circumstances, it may be desired to grow the number of females in the breeding herd, and this concept is also directly related to fertility, survival and percentage of heifers that can be kept as potential replacements. Table 13.2 provides some example scenarios where different herd growth rates can result, and where different production levels can result in the same annual growth rates.

In some circumstances use of sexed semen may be an appropriate strategy for increasing herd numbers (or for increasing incidence of desired sex of calves, male or female). Some related discussion on cow herd size and production of animals through artificial insemination with sexed semen is provided. The main benefit that use of sexed semen provides is the opportunity to produce excess animals of one sex, but not the ability to produce excess animals in general; production of excess animals is a function of fertility and survival rates. These two concepts must be in order before other types of traits such as growth or carcass aspects can be emphasized. The technology that results in sexed semen and production of highly skewed sex ratios has been proven. Whether or not bulls that are useful for particular situations have sexed semen available is a different but important consideration. For the examples provided in this chapter, it is assumed that a 90% calf sex ratio will result from sexed semen use.

The same considerations are involved in calculating herd growth rates when sexed semen AI is used as when natural service matings are used, however,

Table 13.1. Levels of reproduction and survival that result in constant breeding herd size.

	Column						
A	B	C	D	E	F	G	
Number (or percent) of breeding age females – cow herd base	Rate of breeding cow herd used for subsequent calf crop	Calf crop born rate	Calf survival rate from birth to weaning	Calf crop weaned rate	Rate for female calves kept as replacements	Survival rate from weaning to breeding	Scenario comments
100	0.85	0.85	0.92	0.78	0.38	0.98	Adequate fertility, negligible fetal losses, and good survival rates.
100	0.82	0.82	0.84	0.69	0.55	0.95	Moderate fertility and cow loss, no fetal losses, low calf survival rate to weaning.
100	0.76	0.74	0.80	0.59	1.00	0.80	High calf losses from birth to weaning (20%) and weaning to breeding age (20%).
100	0.85	0.83	0.92	0.76	0.42	0.94	Adequate fertility, good survival rates.
100	0.92	0.88	0.93	0.82	0.20	0.95	High fertility and survival rates.
100	0.88	0.87	0.93	0.81	0.30	0.95	Adequate fertility, good survival rates.
100	0.78	0.75	0.90	0.68	0.70	0.94	Marginal fertility, adequate survival rates.
100	0.94	0.91	0.96	0.87	0.15	0.96	High fertility and survival rates allows for more surplus animals to be produced.
100	0.80	0.72	0.90	0.65	0.67	0.90	Marginal fertility and cow loss due in part to embryo losses, good animal survival past birth.
100	0.73	0.71	0.85	0.60	1.00	0.90	Low fertility, marginal animal survival past birth.
100	0.96	0.94	0.96	0.90	0.10	0.98	Extremely high fertility and animal survival rates.
100	0.68	0.66	0.98	0.65	1.00	0.98	Low fertility with extremely high survival rates; requires 100% of female calves kept for breeding.

These are some scenarios that allow for maintaining a set, constant number of breeding females. Within a scenario, lower levels of performance will result in reduction of the breeding herd size; however, increased performance levels within a scenario allow for potential to increase herd size, and/or to generate more surplus animals. Assumes 50% of calves born are female.

Table 13.2. Considerations of production levels to grow cow breeding herds.

A	B	C	D	E	F	G	H	I	J	K	L
Number females in the breeding herd[a]	Pregnancy rate[a] (0.86 = 86%)	Calf crop born[a]	Calf survival birth to weaning[a]	Calf crop weaned[a]	Female calves (0.50 = 50%)	Rate of the female calves kept as replacements	Replacement survival rate – weaning to breeding	Number of replacement heifers[b]	Number of cows of base herd remaining[c]	Females available for upcoming breeding season	Rate of increase per year in breeding herd[d]
100	0.96	0.94	0.96	0.90	0.50	0.79	0.98	36	96	131	0.31
100	0.92	0.88	0.93	0.82	0.50	1.00	0.95	41	92	131	0.31
100	0.95	0.93	0.96	0.89	0.50	0.70	0.97	31	95	125	0.25
100	0.86	0.84	0.96	0.81	0.50	1.00	0.96	40	86	125	0.25
100	0.88	0.86	0.93	0.80	0.50	0.76	0.98	30	88	118	0.18
100	0.88	0.84	0.90	0.76	0.50	0.88	0.90	33	88	118	0.18
100	0.96	0.93	0.96	0.89	0.50	0.40	0.96	18	96	113	0.13
100	0.83	0.80	0.92	0.74	0.50	0.90	0.92	33	83	113	0.13
100	0.79	0.76	0.84	0.64	0.50	1.00	0.88	32	79	107	0.07
100	0.76	0.73	0.90	0.66	0.50	1.00	0.94	33	76	107	0.07

[a]Based on breeding age females exposed to breeding during previous corresponding breeding season.
[b]Number of replacement heifers = column A x column E x column F x column G x column H rounded to whole number.
[c]Number of base herd cows remaining = column A x column B, assumes automatic annual culling for lack of pregnancy and other issues such as mortality rounded to whole number.
[d]Number of females available for breeding in current year divided by the initial base cow number from previous year, subtracted from 1.
These columns are designed for a spreadsheet format to evaluate various scenarios.

the percentage of the breeding herd that is bred by AI is included in the calculations. If we assume a 60% conception rate from AI matings, and two rounds of AI matings are used, then 60% of the breeding herd will become bred from the first service, and 60% of the remaining 40% will be bred from the second service (0.60 × 0.40 = 0.24) so that a total of 84% of the herd gets bred by AI. If it is assumed that 40% of the herd gets bred per AI service, then with two rounds of AI, there is expected to be 64% (40% on first round + 24% on second round) and so on. Therefore there can be multiple ways to obtain the same percentage of AI conception rate.

An annual breeding herd growth rate that would be very high when utilizing natural service would be around 31%; this would require high levels of fertility and survival as well as close to 100% of female calves becoming replacements. However, use of sexed semen AI could nearly double the annual rate of herd growth (but this also reduces number of males to sell). The doubling of the growth rate can result in considerably less than half the time, as happens with compound interest:

$$Amount\ of\ money\ at\ point$$
$$in\ time = P(1+r)^n \qquad (13.1)$$

where P = principal, r = interest rate for the time period, and n = number of time periods.

There are many factors that influence the annual herd growth rate, but once that has been approximated, it can be used similarly to r in the compound simple interest equation to estimate herd size after n number of years, where the principal investment (P) is the initial base herd size. Below is an example where beginning with 100 females, and an annual growth rate of 15%, the expected breeding herd size 4 years later would be approximately 175 females.

$$100(1 + 0.15)^4 = 174.9\ or\ 175$$

Beginning with 100 breeding females, annual growth rates of 0.05, 0.10, 0.20 and 0.30 for 15 successive years would result in potential breeding herd sizes of 208, 418, 1540 and 5118, respectively. No matter the situation, the value of the excess animals must be considered for annual operation costs and therefore economic sustainability. A herd may have the potential to increase in herd size rapidly, but cannot afford the lost income from low sales of excess cattle (males and/or females). An integral part of supply chain management for a business (and cow herds are businesses and should be viewed in that context) is cash flow as well as overall profitability, and one or the other (or both) may limit or dictate feasibility of various scenarios (see Table 13.3).

13.2.3 Cattle Numbers and Beef Production

Most people are familiar with the concepts of supply and demand as they relate to economics and price. Many different factors obviously affect both of these concepts. In general, as societies have accumulated wealth, there has been a corresponding increase in meat consumption. This is because the price of meat is considerably higher per unit of weight compared to many other foods such as grains and vegetables. Meat either costs more to produce for the same unit of weight as these other foods, or is more costly to acquire if hunting is involved. The supply may be dictated by the demand, or if the

Table 13.3. Performance measures of cattle in South Africa under different management systems.

Trait	Communal	Commercial	Recommended
Average calving %	40	65	85
Pre-weaning calf mortality %	50	4	2
Post-weaning calf mortality %	15	2	2
Calves weaned p.a.	5	16	21
Calves available for sale (after replacements)	2	15	20
Average weight (kg)	150	180	205
Price per kg (Rand)	3.50	8.00	8.25
Potential monetary value (Rand)	1,050	21,600	33,825
Potential monthly income (Rand)	87.50	1,800	2,818

Data from Republic of South Africa Department of Agriculture, Forestry and Fisheries (KZNDAE, 2014).

supply is limited, the demand can exceed the supply and thus make it more costly. Discussion here transitions to animal number considerations relative to beef production output.

Tables 13.4 and 13.5 illustrate examples of animal and land resources needed to supply commercial beef packing plants of various sizes under a forage-based finishing scenario and under a combination grazing cow herd and feedlot supply scenario (relative to scenario assumptions). This illustrates the point that large amounts of cattle and land must be managed to supply these plants.

For a processing plant to be economically viable as a stand-alone industry enterprise, it must make efficient use of its resources (facility infrastructure and labor), and this means having a consistent supply of cattle for daily operations. These tables also make no assumptions in regard to amount of beef produced per carcass, which also affects profit potential; if there is a fixed processing cost per animal, then there can be higher profit margins for processing heavier cattle and carcasses.

None of the values presented in these tables should simply be taken at face value, but should be

Table 13.4. Land capacity needs to supply annual production of beef packing plants of various sizes from a forage-based finishing system.

Plant daily kill capacity	Operating days per year	Number of cattle harvested per year	Cow herd needed to supply	Land area needed for cow herd (ha)	Land area needed for growing and finishing cattle (ha)	Total land area needed (km²)
50	260	13,000	21,667	131,526	26,305	1,578
100	260	26,000	43,333	263,051	52,610	3,157
500	260	130,000	216,667	1,315,257	263,051	15,783
1000	260	260,000	433,333	2,630,514	526,103	31,566
1500	260	390,000	650,000	3,945,771	789,154	47,349
2000	260	520,000	866,667	5,261,028	1,052,206	63,132
3000	260	780,000	1,300,000	7,891,542	1,578,308	94,699
4000	260	1,040,000	1,733,333	10,522,056	2,104,411	126,265
5000	260	1,300,000	2,166,667	13,152,570	2,630,514	157,831

Assumptions here include stocking rate of 6.1 hectares per cow annually for cow herd, and number of animals annually harvested are 60% of breeding cow herd, and stocking rate of 2 hectares of forage per animal harvested. These estimates do not include land area needed for any potential supplements provided to cow herd or animals during growing and finishing phases.

Table 13.5. Feedlot and land capacity needs to supply annual production of beef packing plants of various sizes.

Plant daily kill capacity	Operating days per year	Number of cattle harvested per year	Number of feedlot turns per year	Feedlot capacity needed	Cow herd needed to supply	Land area needed for cow herd (ha)	Land area needed for grain (ha)	Land area for cow herd + grain (km²)
50	260	13,000	2	6,500	21,667	131,526	2,818	1,343
100	260	26,000	2	13,000	43,333	263,051	5,637	2,687
500	260	130,000	2	65,000	216,667	1,315,257	28,184	13,434
1000	260	260,000	2	130,000	433,333	2,630,514	57,795	26,883
1500	260	390,000	2	195,000	650,000	3,945,771	84,552	40,303
2000	260	520,000	2	260,000	866,667	5,261,028	112,736	53,738
3000	260	780,000	2	390,000	1,300,000	7,891,542	169,104	80,606
4000	260	1,040,000	2	520,000	1,733,333	10,522,056	225,473	107,475
5000	260	1,300,000	2	650,000	2,166,667	13,152,570	281,841	134,344

Assumptions for this scenario include 9.1 kg of grain consumed per animal per day during feedlot phase of 180 days, grain yield of farm land of 3048 kg per hectare, stocking rate of 6.1 hectares per cow annually for cow herd, and number of animals annually harvested are 60% of breeding cow herd. These estimates do not include land area needed for other components of feedlot diet such as protein sources or other ingredients, or land area needed for any potential supplements provided to cow herd.

understood in regard to how the assumptions were used to generate the expectations. Readers are encouraged to use a spreadsheet to consider calculations involved to answer several of the study questions at the end of this chapter. What should become apparent is that with the feedlot finishing system, only 11% of the land area is required for finishing cattle as compared to the land area under the forage finishing system; however, the total land use requirement of the feedlot finishing system is 85% that of the forage finishing system due to the large land areas required to maintain the supporting cow herds. This is a very simplistic example, but the analogy holds true across a wide array of situations. And ultimately, what dictates which type of system works best is related to its sustainability, based on economic, ecological and societal components. The thought process of matching a production system to its resources is quite analogous to matching cattle genetics to environmental resources for herd sustainability.

13.3 Animal Identification

Identification of cattle has a long history, much of which was nicely reviewed and discussed by Blacou (2001). Identification of cattle has two distinct levels (that often serve complementary purposes) but should not be confused: (i) identification of ownership, and (ii) identification of individual animals. Identification of ownership (or intended purpose as for sacrifice, etc.) is a much older concept than identification of individuals, which typically has been used to identify pedigree information and is now a key component in traceability considerations for beef supply chain management and/or marketing. Reasons for identification could relate to production management, health management, disease outbreak control, proof of ownership, market advantages or specifications (domestic and/or international), consumer issues/confidence, and supply chain trace-back.

There are many methods of cattle identification, including plastic ear tag, tattoo, fire branding (hide or horn), freeze branding (hide), electronic (RFID ear tag, microchip, bolus, etc.), biological (nose print, retina scan, DNA), neck chains, dewlap tags and ear marking. A major consideration with animal identification is whether or not each method is permanent or temporary. In the strictest sense no method of identification is permanent because hide brands and ear tattoos can be altered, etc. However,

these two distinctions are valid because it is not nearly as easy to alter a 'permanent' ID method as it is a 'temporary' method. Some type of redundancy on each animal is preferred, and written or electronic records need to correspond to values physically on the animals. Convenience of use is critical, especially across different potential owners or managers, as well as across different industry segments. The usefulness of the different types of ID is a critical consideration when evaluating adoption or continued use of various methods. This refers to usefulness for the manager, but should also include consideration of potential usefulness of other possible owners of the animals and other sectors in the industry.

The two most common types of cattle identification in many areas of the world are hide brands and ear tags, and both of these can be used for proof of ownership and individual animal ID; however, the degree of usefulness of both of these methods for these purposes depends upon the specific scenarios in which they are used. Suppose that a seedstock producer wants to identify his or her cattle for both ownership and individual animal record keeping. Branding an individual number on each animal is an option, as is having one or two ear tags in each animal with the same number. Likewise, a holding brand specific to that operation (and recognized by a legal authority) can be placed on the hide, but can also be put on an ear tag. If this producer only used ear tags, the information could be lost from the animal more easily than the hide brand can be altered. Moreover, only using brands will limit the amount of information that can be placed directly on the animal because it is wise to minimize the amount of hide damaged by branding from a quality assurance aspect as well as a potential animal welfare consideration. Additional detailed information can be included on the ear tag, showing items such as the sire and/or dam of the animal, name and/or telephone number of owner, and different colored ear tags can be used to differentiate contemporary groups, breed combination, sire/dam line, etc. In situations where ear tags are not readily available, an ear marking system can be used.

It is always best to have some degree of identification redundancy actually on the animal (see example in Fig. 13.3) in case an issue arises with one type becoming altered or lost, as well as a written or computer database with complete information. This is analogous to having a backup copy of computer

Fig. 13.3. This cow has multiple types of identification. The individual animal number is 660, shown on the left horn and the ear tag. She was born in 2003, indicated on the right horn. The ear tag also shows additional pedigree information. Redundancy of individual animal ID on the animal itself is wise, and the complete information needs to be recorded and stored in some type of database or file.

files (and many other things in life) in that it is best to consider this scenario and have a plan implemented in case it is needed, before it is actually needed. Many contingency plans are developed after an emergency has occurred, which is usually too late to help the immediate situation. The redundancy on the animal should be of two different ID methods (brand and tattoo, brand and ear tag, etc.), not simply two ear tags that could both be lost or cut out at the same time. There are several biological methods of individually identifying cattle such as eye retinal scans, nose prints and DNA, which make excellent permanent records as components of databases; however, these have higher costs and limited utility in regard to immediate (chute-side) sorting of animals when ID is in question.

The specific scenario in which the cattle operation resides may help dictate which type(s) of identification are most useful, particularly in regard to proof of ownership. In situations where animals are commingled, proof of ownership can be crucial; this can be intentional, as with cooperative or communal grazing of animals, or unintended, as when a hurricane or typhoon or other weather disaster destroys fences or forces animals together that normally would be kept separate.

Many managers who are accustomed to dealing with large numbers of animals, but not accustomed to individually identifying animals, may think that management based on individuals is either not realistically possible, or not cost effective. Olvera (2010) studied the concept of using individual cattle weights as compared to the average weight of lots (i.e. pens) of feedlot cattle to apply a weight-based antibiotic upon feedlot arrival. In this study, use of individual animal weights yielded a lower cost per animal compared to using the lot average weight. Cost per animal can be reduced when a product is administered at a lower weight (or a lower dose for the correct weight), but that also introduces risk into the production system as the medication will have reduced efficacy, and this reduced cost may end up being more costly in the bigger picture. More precise management where individual variability can be addressed holds potential for both increased production efficiency as well as a story to market to consumers.

13.4 Traceability Considerations

The concepts of traceability (trace back) are integral to supply chains for a variety of reasons. These include public health and food safety, compliance with regulatory policy and agencies, domestic and international trade, consumer/customer confidence, marketing opportunities, and internal cost/return evaluation. To address many of these issues, several countries have implemented national animal identification programs. Some corporations or companies have implemented animal traceability into their routine operations. Traceability becomes more challenging in supply chains where there are multiple owners of animals. Pendell *et al.* (2010) stated that domestic and international consumers are increasingly demanding traceability associated with meat products, and that this is rapidly becoming the global standard. Cattle producers in countries that do not require identification systems probably need to be proactive and implement some type of individual animal identification system that can be useful to them. At the very least, these producers will have the ability to differentiate the more productive from the least productive animals; these producers may also be able to participate in some type of marketing arrangement or alliance where they will be paid higher prices due to having this type of information. The value of this type of information is important, but is difficult to estimate. Pendell *et al.* (2010), in discussing the fact that there was no mandatory national cattle ID program, said a US domestic retail beef demand increase of less than 1% would pay for animal ID costs (assuming 90% adoption) in the USA, but it

is unknown whether US domestic demand would increase as a result of a mandatory national system.

Certainly many countries have implemented national ID programs to garner or increase international demand for their beef products. Murphy *et al.* (2008) stated that ability to access information stored on one country's database for 'value-added' benefits in another country holds tremendous opportunity for all of North America; this is also probably true across many countries, but this would also increase some risk considerations, at least for the potential of data corruption or loss. Table 13.6 summarizes some aspects of cattle traceability that has been reported in different countries.

One of the key implications of traceability is that it has the potential to increase information flow in the supply chain (Fig. 13.4), theoretically resulting in improved allocation of economic value. Consumer interest in beef labeling cannot be taken for granted (Verbeke and Ward, 2006). Souza-Monteiro and Caswell (2004) reviewed traceability systems and defined traceability as the ability to follow the movement of a food product through specified stage(s) of production, processing and distribution; they also discussed how others (i.e. Golan *et al.* (2003)) characterized traceability systems by additional considerations such as breadth (the amount of information recorded), depth (how far backward and forward traceability was maintained) and precision (ability of the system to pinpoint the original source of a problem).

Table 13.6. Examples of cattle identification and traceability system guidelines.[a]

Country	Premises ID	Individual cattle ID	Group or lot ID	Electronic ID	Animal movement recorded	Animal number retired
Australia	M	M	V	M	M	M
Botswana	V	M	No	M	M	V
Brazil[b]	M	M	V	V	M	M
Canada	V	M	No	M	V	M
EU	M	M	V	V	M	M
Japan	M	M	V	V	M	M
Mexico	V	V	V	V	V	V
Namibia	M	M	V	V	M	M
New Zealand[b]	V	V	V	V	V	V
South Korea[b]	M	M	V	V	M	M
Uruguay	M	M	V	M	M	M
USA[b]	V	V	V	V	V	V

[a]Adapted from Bowling *et al.* (2008).
[b]Program voluntary in these countries; requirements of those who choose to participate. M = mandatory, V = voluntary.

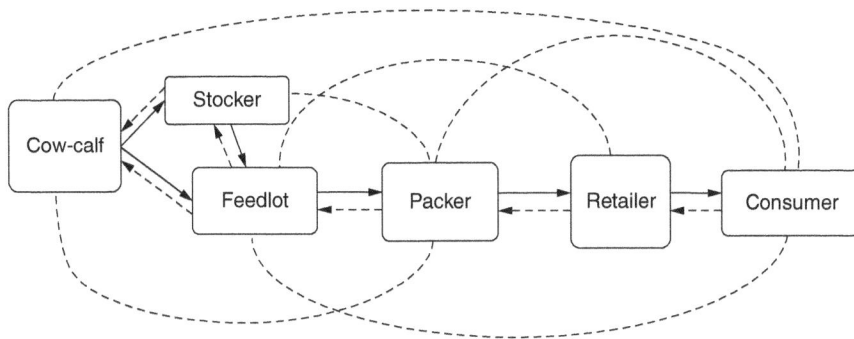

Fig. 13.4. Potential for open information along the supply chain of beef production for North American industry. Animal and product flow represented by solid arrows; traditional information and market signals represented by broken arrows; possible information exchange between sectors represented by broken lines.

The different types of information that are available or conveyed to consumers also have varying degrees of value. Resende-Filho and Buhr (2008) used beef injection-site lesion incidence to simulate the economic value that can be attained through information flow from end-product only, as lesions affect the beef value but are only discovered after the sale, and the supplier can go unrecognized if not specifically tracked. They found that even an imperfect system with a relatively unreliable traceback mechanism could act as a credible deterrent and produce favorable changes by suppliers with appropriate economic value allocation for positive and negative outcomes. The spread of bovine spongiform encephalopathy (BSE) in Europe prompted EU policy makers to implement a mandatory beef traceability and labeling system in 2002. Gracia and Zeballos (2005) evaluated consumers and retailers in Spain and found they both felt that overall consumer confidence was increased with the system, but both these groups also believed that the system had increased beef prices, possibly more than its actual value.

Based on consumer survey data in Belgium, Verbeke and Ward (2006) found interest from consumers appeared low for information directly related to traceability and product identification but was much higher for information such as easily interpretable indications of quality such as certified quality marks or seals of guarantee, as well as for mandatory standard information like expiration date. Verbeke and Ward (2006) also said that traceability in itself has little marketing potential unless accompanied by trustworthy indications of quality. Sugahara (2009) stated that the Japanese people

have been increasingly aware and concerned about food safety issues and described two traceability systems that pertain to beef. The beef traceability system targets domestic beef and records date of birth, sex, type, breeding location and breeding manager; the system of production information disclosure (JAS) is for other beef, pork and farm products and requires assurances regarding animal medicines, agricultural chemicals, fertilizers and detailed types of chemicals used. It has been possible in Japan for several years for consumers to access data about beef products at retail stores directly through smartphone technology based on information provided on labels.

There continues to be new or non-traditional approaches that can be incorporated into traceability systems including information from DNA profile (Dalvit *et al.*, 2007; Orrù *et al.*, 2009) and from isotope profile (Schmidt *et al.*, 2005) to name a couple. The concepts of traceability and beef production assurance are similar, but they are not exactly the same concepts; more traceability considerations are discussed later in the chapter, in Section 13.6.1.

13.5 Development of Beef Supply Chains

Many developing areas of the world are expanding or altering their beef cattle production and supply chains, or are evaluating these considerations. Waldron *et al.* (2010) discussed these considerations directly regarding China, but the considerations pertain to most other areas as well (even if details do not). These authors stated that China has embarked on an agricultural modernization program with far-reaching implications for rural development, food safety and trade. Waldron *et al.* (2010) stated that a major focus has been to build high-value supply chains and large, modern agro-industrial enterprises, but warned that interventionist policies to fast-track the development of high-value supply chains can have perverse outcomes and that a more incremental and facilitative approach to modernization should be pursued based around the development of mid-value supply chains (see Fig. 13.5). Simply taking a system that works in one area of the world and forcing its structure and implementation on another is likely to have very different results with little to no chance of sustainability (a very similar concept to taking highly productive animals with high production potential and transplanting them to a region where they are not adapted). Identification and application of the important concepts that will produce desired results for the specific region are critical. Small increases in efficiency throughout the production system can allow for large increases in overall efficiency and profitability (and likewise accumulating loss of profitability with inefficient processes). This is another reason why it is wise to understand

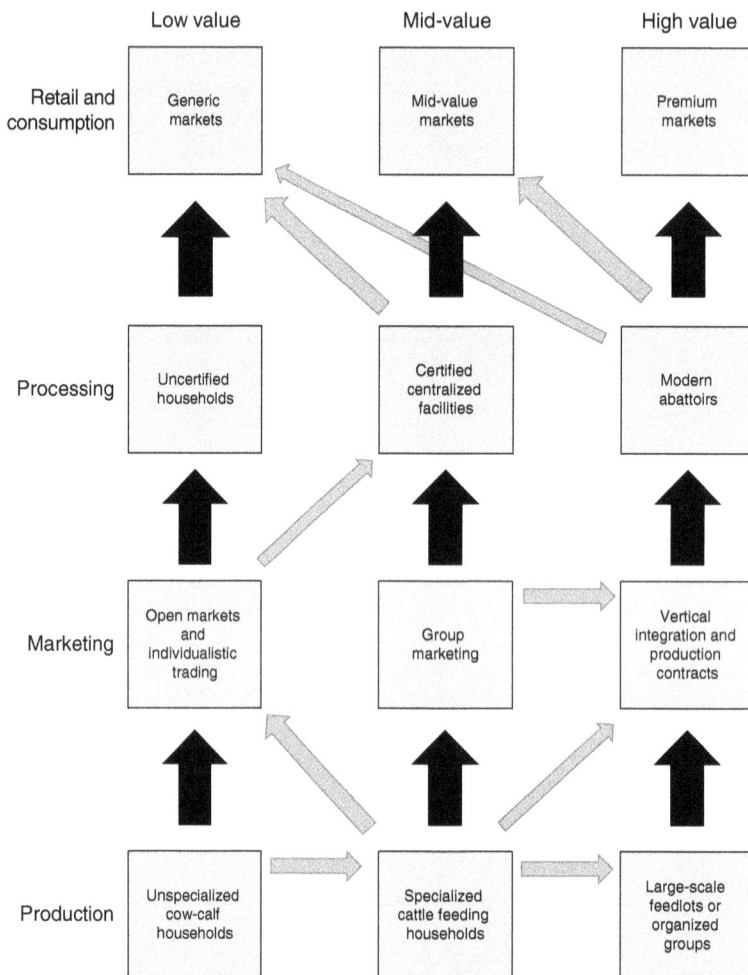

Fig. 13.5. Supply chain considerations in developing areas. The marketing considerations can occur between the production and processing segment, between the processing and retail segment, or both. Graph adapted from Waldron *et al.* (2010).

each incremental process as well as how all the processes collectively work together.

13.5.1 Variation and Uniformity

Several discussions about beef supply chain considerations have revolved around the concept of variation. Certainly, variation in regard to products can be both good and bad. Consumers like to have a choice of products, and in that sense variation is good. This is good for a supplier like a grocery store, good for a manufacturer that offers several products, and good for an industry. Variation is bad, however, when it relates to product inconsistency. These two types of variation can be thought of as across-product (or between-product) and within-product. As discussed in previous chapters, within-production unit variation tends to be a robber of production efficiency, and from the standpoint of product quality as related to consistency, it can be a major problem for consumer satisfaction (Fig. 13.6).

Sources of variation can be of genetic, environmental or combinations thereof in origin. It may be tempting to think that if we can just get rid of the _____ variation (fill in the blank), we will have

a uniform product or process. Most of the time, this line of thinking is too simplistic to be effective. It is true that certain strategies can be used to reduce genetic variation, such as linebreeding, and specific mating strategies (including crossbreeding with terminal crosses). Application of non-uniform practices to genetically similar cattle will produce variable results. Applying the same processes and procedures to non-uniform cattle will also result in non-uniform results. It is from the understanding of potential interactions and the entire production system that should make it clear that to obtain uniformity, different processes and products may have to be incorporated on different cattle types, different climates, different intended consumers, etc. Without a fundamental understanding and consideration of the underlying variation (such as standard deviations, etc.) in conjunction with means for various production traits and their processes, managers have incomplete information as they make decisions.

13.6 Sustainability Considerations

Heitschmidt *et al.* (2004) and others have discussed that for a production system or industry

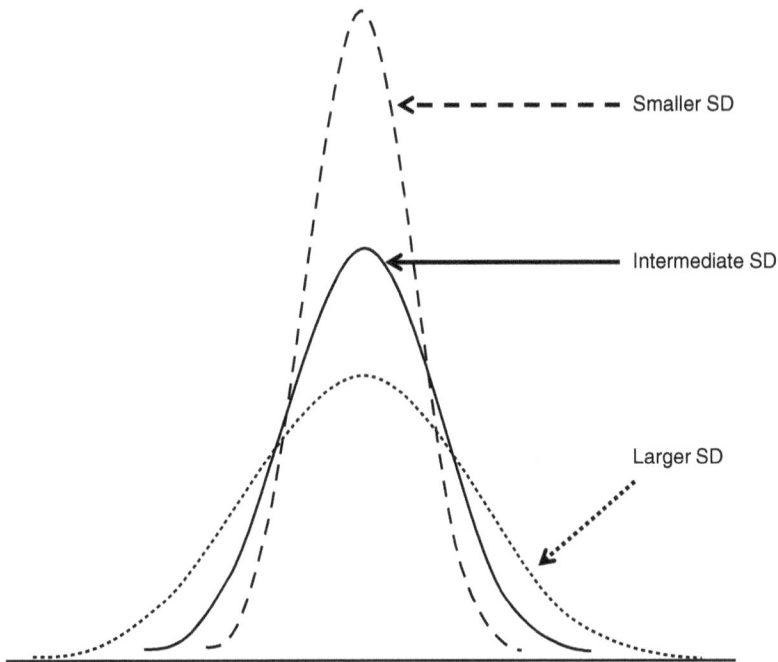

Fig. 13.6. Concept of increasing uniformity with smaller standard deviation.

to be sustainable, there were three important considerations of (i) environmental, (ii) economic, and (iii) societal aspects. These three concepts are inter-related, and producers at all levels within the beef production system should understand and embrace these issues. Producers that are better informed are prepared and able to make better production and marketing decisions. Producers are also often struggling to handle the day-to-day decisions that keep the beef cattle operations functioning; these are continual and time-consuming. If producers are too busy to ever think about 'big picture' concepts and long-term planning, their sustainability is at risk. If producers only think about long-term planning and 'big picture' issues, their short-term sustainability is at risk. Management, success and therefore sustainability require constant information gathering, balancing of priorities, evaluation, and decision making (which of course is true in all industries and endeavors in life). This epitomizes the need for systems-based thinking.

Table 13.7 gives values for operations that were average, the bottom 20% and the upper 20% in regard to profitability for four different US beef cattle production components of cow-calf, backgrounding (which is short-term grazing or feeding to develop or precondition feeder calves), replacement heifer production, and finished cattle production across operations several states (see also Fig. 13.7). As emphasized in Chapter 12, the operations that were able to reduce expenses were the ones that were profitable, not necessarily the operations with the biggest animals or the most weight gain or highest performance.

Operations and production chain components that strive for production and economic efficiency have higher chances of maintaining profitability, and assessment of economic sustainability is critical for all operations and production components. This does not, however, mean that all times or all components will be profitable. To fully understand beef production chains, it must be understood that 'less than average' profitability animals may need to be produced in one component in order to capitalize on their economic superiority in another component (or multiple components). Conversely, animals that are acceptable across all components, but not necessarily 'superior' in any component may be more profitable when the whole production chain is considered than animals that are 'superior' for one component but undesirable in one or multiple other supply chain components. Figure 13.8

illustrates the costs associated with producing 100 kg of carcass beef in the USA, Germany, Brazil and Australia from the AgriBenchmark database (Deblitz, 2012).

13.6.1 Beef Quality Assurance, Beef Production Assurance

Beef quality assurance and related programs have evolved to include best practices revolving around good record keeping, animal health and welfare, and food safety, which ultimately can result in more profits for producers and improved products for consumers. When better-quality animals leave farms and ranches and any other site of production, subsequent elements of the supply chain (including consumers) all benefit. When better-quality beef and associated by-products reach the various customers, they are more confident in the products they are buying, and this increases market acceptance and potential for market growth (BQA, 2012–2013). The US National Cattlemen's Beef Association (NCBA) introduced its BQA program in 1982 to address concerns of avoiding residues in beef (Bredahl *et al.*, 2001). Since then, BQA programs and guidelines have been launched in all sectors of the beef and dairy industries.

The Council of Europe has established guidelines for cattle production assurance considerations entitled *Recommendations Concerning Cattle*, which was adopted by the Standing Committee on 21 October 1988. It has guidelines concerning stockmanship and inspection of cattle, buildings and equipment, management, changes of phenotype and/or genotype, along with special provisions discussed for bulls, cows and heifers, and calves. A quote from the preamble of the *Recommendations Concerning Cattle* is provided below:

Aware that the basic requirements for the health and welfare of livestock consist of good stockmanship, husbandry systems appropriate to the physiological and behavioural needs of the animals, and suitable environmental factors, so that the conditions under which cattle are kept fulfill the need for appropriate nutrition and methods of feeding, freedom of movement, physical comfort, the need to perform normal behaviour in connection with getting up, lying down, resting and sleeping postures, grooming, eating, ruminating, drinking, defecating and urinating, adequate social contact and the need for protection against adverse climatic conditions, injury, infestation and

Table 13.7. Financial and production measures among profitability groups of US beef industry segments.

Trait	Cow-calf			Backgrounding			Replacement heifers			Finished cattle		
	Ave	Low 20%	High 20%	Ave	Low 20%	High 20%	Ave	Low 20%	High 20%	Ave	Low 20%	High 20%
Number of operations in category	368	73	74	107	21	22	90	18	18	89	17	18
Number of animals[a]	125	95	145	192	130	204	49	75	69	324	120	453
Total direct expenses per animal ($)	550.4	757.59	463.21	239.55	244.07	241.69	357.41	533.49	287.38	621.5	978.28	622.43
Total overhead expenses ($)	98.43	156.32	88.08	36.42	34.42	67.96	41.07	55.9	37.23	50.76	81.51	75.96
Total labor and management expenses ($)	69.68	78.04	78.83	15.33	18.94	13.6	26.12	21.08	34.83	21.35	31.22	18.42
Net return per animal (income − expenses, $)[a]	25.34	−368.09	266.02	51.06	−133.32	286.63	47.87	−280.87	315.38	−17.38	−436.93	134.22
Feed cost per animal ($ per kg)[a]	398.98	504.43	346.69	194.79	203.24	182.14	268.64	327.04	235.87	537.5	905.83	527.74
Sale price ($ per kg)[a]	3.32	3.11	3.29	3.00	3.06	3.08	1471.23	1151.00	1466.86	2.72	2.68	2.78
Ave sale weight (kg)	267.1	290.7	281.6	345.1	337.9	307.0	448.1	408.2	438.1	607.3	558.3	605.4
Cost of production ($ per kg)[a]	3.13	4.46	2.31	2.82	3.49	1.92	1294.28	1494.13	1143.85	2.75	3.39	2.56
	Kg weaned per cow exposed			*Kg feed per kg gain*			*Weight gain from purchase to sale (kg)*					
Live weight production per animal	213.2	200.5	235.8	118.4	110.2	63.9	130.2	12.7	129.3	298	269	274
Percent death loss[a]	5.0	4.2	4.1	1.1	1.3	2.3	0.4	0.8	0.8	0.8	0.8	0.7
	Feed cost per kg weaned ($)			*Kg feed per kg gain*			*Feed cost per kg sold ($)*			*Kg feed per kg gain*		
Feeding efficiency related measures	1.87	2.52	1.47	10.97	19.83	7.49	0.60	0.80	0.80	9.97	14.37	8.92

[a]Number of animals and measures for cow-calf operations expressed on per cow exposed basis whereas other segments are expressed per animal; sale prices, sale weights and death loss values for cow-calf operations pertain to calves only. From FINBIN (2013) online searchable database that included several states in USA for 2012 calendar year.

disease or behavioural disorder, as well as other essential needs as may be identified by established experience or scientific knowledge (Council of Europe, 2013).

Meat and Livestock Australia (MLA) has also been instrumental in providing guidelines to its producers and assurances to its consumers regarding production quality assurances, and MLA production assurance considerations for on-farm livestock producers are summarized in Table 13.8.

Even when producers have operations in countries that do not have formal BQA/BMP programs or accreditation opportunities, producers can implement these practices of their own accord, and it may also be an option to have some legal authority verify (through witnessed, signed certificate or affidavit, etc.) processes.

In several countries a national animal identification system has been either the starting point or a crucial component of providing production assurance through traceability. There are several countries that have existing or developing quality/ production assurance guidelines, and the goal of this section has been to provide an overview of the concepts and provide some of the more globally discussed systems.

As mentioned earlier in the chapter, considerations and applications in developing countries may have unexpected or undesirable results when existing supply chain structures from one country are directly imposed in another country, particularly those with different structure, policy and enforcement capabilities. Much of the remaining paragraph relies heavily on thoughts taken from Steinfeld *et al.* (2006). Consumers in affluent countries have a major impact on standard setting, yet the results may affect poor and marginal producers, processors and consumers who do not directly trade in the global market and have very little voice in the standard-setting process. The private sector has an increasing influence, while the impact of the public sector is limited, and policies do not always reflect the needs of the various stakeholders in livestock food chains. Food safety requirements are a major determinant shaping the structure of the livestock sector and associated food chains. One of the main aspects of the structural change process described above is the reduction of ties between production locations and consumers, and this can increase potential for food safety concerns if adequate oversight and regulation is not in place.

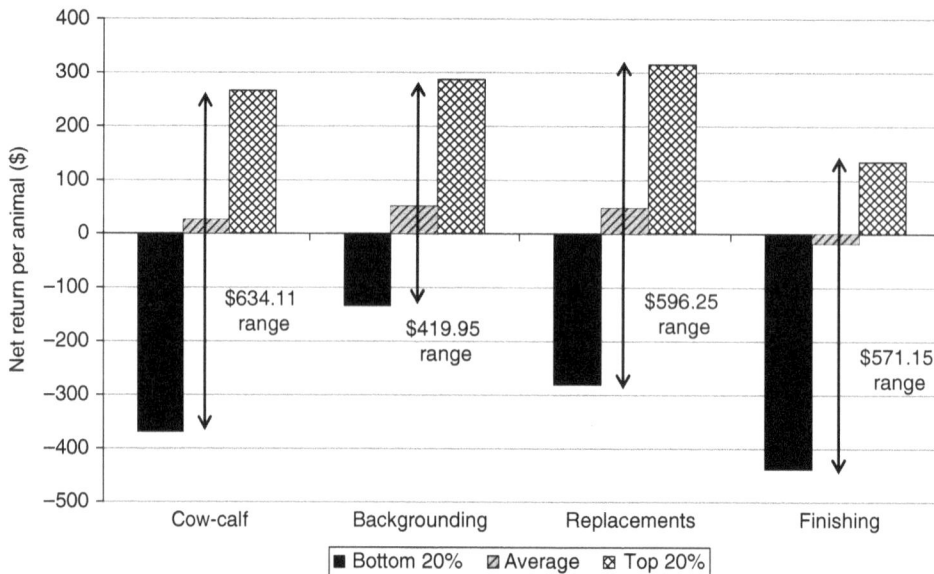

Fig. 13.7. Variation in profit across US beef industry production segments in 2012. Average values per animal with the bottom 20% average, and top 20% profit categories provided within industry components. Production segments/ systems here refer to production of calves nursing cows to weaning at 7 months of age, short-term feeding and growing of weaned calves, development of heifers for breeding cows, and feeding of cattle to 10–12 mm of external fat for carcass production. From FINBIN (2013) online database spanning several states in the USA.

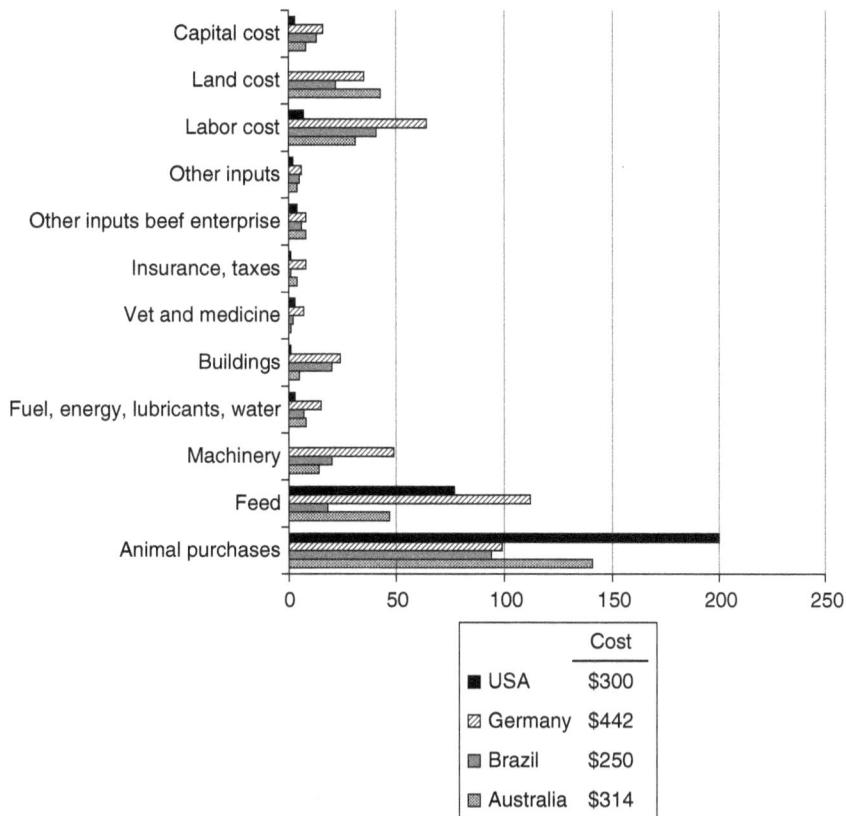

Fig. 13.8. Cost of producing 100 kg of carcass weight. Values are in US$ from 2010 estimates. These are weighted values relative to the enterprises in the AgriBenchmark database. From Deblitz (2012).

Vertical coordination and/or integration, a characteristic of structural change in developing areas, can also therefore increase the importance of food safety considerations.

Steinfeld *et al.* (2006) also discussed that the emergence of multinational food chains and a dramatic rise in the market share of supermarkets and the modern retail sector in many countries have created a number of effects, some of which may affect food safety, including (i) a shift toward cross-border systems, with corporations procuring goods in their different countries of operation, (ii) a shift toward preferred-supplier systems to select producers who meet specific quality and safety standards and lower transaction costs, (iii) a consolidation of production and processing which may have unexpected impacts on safety, and (iv) a shift toward safety and quality standards driven by the private sector.

13.6.2 Concept of Value and Value-Added Beef

The term 'value' is a bit nebulous, as is the term 'quality', because these can mean quite different things to different people, and they can also mean different things across different products for the same person. Yet these are both terms that customers appreciate and relate to, and are therefore widely used in marketing and advertising for many types of products. Lawrence (2004), Smith *et al.* (2006), and Sugahara (2009), as well as several others, have discussed the concept of 'story' beef, where there are growing numbers of consumers in many different types of markets with interest in hearing 'the story' about where their foods come from and how they were processed. As a result, access to certain elements of the production process can be a way to add value for some markets.

Table 13.8. Meat and Livestock Australia Livestock Production Assurance – Level 1, On Farm Food Safety Scheme.

No.	Standard element	Outcomes
1	Property risk assessment	On-farm systems have been implemented to minimize the risk of livestock being exposed to sites that are unacceptably contaminated with organochlorine or other persistent chemicals, or other potential sources of persistent chemicals, and being exposed to sources of potentially injurious physical contaminants in meat intended for human consumption.
2	Safe and responsible animal treatments	On-farm systems have been implemented to ensure that animal treatments are administered in a safe and responsible manner to minimize the risk of chemical residues and physical hazards in livestock intended for human consumption.
3	Fodder crop, grain and pasture treatments and stock foods	On-farm systems have been implemented to manage the exposure of livestock to foods containing unacceptable chemical contamination to minimize the risk of chemical residues in livestock and to eliminate the risk of animal products being fed to ruminant livestock intended for human consumption.
4	Preparation for dispatch of livestock	On-farm systems have been implemented to ensure that the selected livestock are fit for transport and that the risk of stress and contamination of livestock during assembly and transport is minimized.
5	Livestock transactions and movements	A system has been implemented to enable traceability of the current status of all livestock with respect to treatment or exposure to relevant food safety hazards for all livestock movements between livestock production enterprises including to slaughter and live export.

From AUS-MEAT Limited (2012).

However, added value is relative to price, and this is still the largest single influence on most of the purchasing decisions of most customers in many different types of beef markets. A big part of the story currently involves sustainability considerations.

13.6.3 Beef Production and Societal Concerns

In more affluent societies, several societal issues in food supply chains have led consumers to show interest in organic foods, including beef. There are many reasons why some consumers might be interested in organic foods, including health, environment, animal welfare and taste. Some, such as Ilea (2009), have called intensive livestock farming such a problem that more legislation and regulation is needed and that consumers should reduce their demand for animal products. If unnecessary legislation is increased, this in turn increases costs of production as well as prices for consumers, and automatically limits productivity and efficiency increases needed in many areas of the world. Much of the highly visible scrutiny of animal agriculture has been made by those not involved or familiar with agriculture, and certainly not been developed

from a systems-based perspective. For instance, the increased efficiency of production has been ignored in many scenarios comparing livestock environmental impacts (Cooprider et al., 2011).

Smith-Spangler et al. (2012) performed a very thorough investigative review of nutritional and health aspects of organic foods including row crops, fruits and vegetables, as well as poultry and red meat; this study did not find that organic foods were significantly more nutritious or healthier than conventionally produced foods, and both types of foods had similar levels of bacterial presence. Moreover, Capper (2011) reviewed several beef and dairy production systems regarding environmental footprints and made the following points: (i) improved productivity has considerably reduced the carbon footprint of dairy and beef production over the past century, (ii) extensive systems intuitively appear to many people as more environmentally friendly, yet scientific analyses demonstrate that intensive systems reduce resource use, waste output and greenhouse gas emissions per unit of food, and (iii) as livestock production systems continue to make productivity gains, sustainability should be assessed on the basis of environmental, economic and social issues.

Organic foods are much more expensive to produce than conventional foods because they have reduced production efficiency or reduced yield per unit of land to produce, and in many cases have reduced shelf life. Some beef producers could probably manage and market their products through organic programs and receive much higher prices, but the increased prices would have to largely offset the production decreases. Many organic programs have claimed various nutritional advantages over conventional foods, but as more specific scientific analyses have been undertaken, such as reviews like that of Smith-Spangler *et al.* (2012), many of the claims have been found to be greatly exaggerated. If food producers and consumers are truly concerned about the stewardship of environmental resources, this by nature requires a systems-based mode of thinking. People who are after the truth will be able to work together no matter what their background because they have a common goal (to find the truth); this is not the same thing as having an agenda to advance being the primary goal, and if this is the primary goal of various groups, by nature these groups will not be able to work together.

13.7 Summary

The entire production process from beginning to end is referred to as the supply chain. For this book, the supply chain considers the steps and efficiency of production from pregnancy in cows to consumer purchase. For some business models or enterprises the 'consumer' may be the people purchasing edible beef products, but it could also refer to commercial cow-calf producers that purchase breeding animals, feedlot managers or beef packing plant cattle buyers that purchase cattle. Animal identification where production practices are connected over time is an important component of identification of inefficiencies in beef supply chains. Reproductive success in cow herds sets the stage for animal numbers and production potential, and the number of cows in an operation, a region or a nation dictates the total cattle inventory that can be produced from within to supply a population. The ability to supply a variety of beef products to consumers, but where there is uniformity within a product line, are both critical functions related to consumer acceptance and long-term success. Production assurances that can be provided to consumers are also important business

and industry considerations that increase potential for sustainability of beef production systems and supply chains.

13.8 Study Questions

13.1 How many cows would be required to supply a beef processing plant operating 250 days a year if the plant harvests 120 cattle per day? Assume 0.60 of the cows produce animals for harvest.

13.2 List four marketing combinations to sell cattle that would be available for your geographical region, and provide an example of each one.

13.3 How much land would be needed to sustain the cows in Question **13.1** if the carrying capacity was 4 ha per cow per year? What if the carrying capacity was 11 ha per cow per year?

13.4 Identify three strengths and three weaknesses of the beef industry in your geographical region, and briefly state why you believe these are so.

13.5 State whether or not you would recommend that commercial cow-calf producers individually identify their cows, and list three reasons why (or why not).

13.6 If the annual per capita consumption of beef was 20 kg, and the estimated beef yield was 140 kg per carcass, how many cattle would need to be harvested per year to supply a city of 10,000 people? What about a city of 650,000 people?

13.7a In regard to the US, Canadian, Australian and Western European beef industries, discuss whether you think there is more future profit potential in the cattle business due to: (a) change in animal characteristics, or (b) change in marketing strategies.

13.7b Do you think the answer to this question might be different in developing countries? Discuss why or why not.

13.9 References

AUS-MEAT (2012) Livestock Production Assurance – Level 1 On Farm Food Safety *APPROVED STANDARDS* Approved by the Livestock Production Assurance (LPA) Steering Committee 2012 AUS-MEAT Limited, Meat and Livestock Australia.

Blacou, J. (2001) A history of the traceability of animals and animal products. *Rev. Sci. Tech. Off. Int. Epiz.* 20, 420–425.

Bowling, M.B., Pendell, D.L., Morris, D.L., Yoon, Y., Katoh, K., Belk, K.E. and Smith, G.C. (2008) Review: Identification and traceability of cattle in selected countries outside of North America. *The Professional Animal Scientist* 24, 287–294.

BQA (2012–2013) Beef Quality Assurance Home page sponsored by National Cattlemen's Beef Association (NCBA) and the Cattlemen's Beef Board (CBB). Available: http://www.bqa.org/. (last accessed 15 September 2013).

Bredahl, M.E., Northen, J.R., Boecker, A. and Normile, M.A. (2001) Consumer demand sparks the growth of quality assurance schemes in the European food sector, Changing structure of global food consumption and trade. 90–102 USDA Economic Research Service publication WRS-01-1.

Capper, J.L. (2011) Replacing rose-tinted spectacles with a high-powered microscope: The historical versus modern carbon footprint of animal agriculture. *Animal Frontiers* 1, 26–32.

Cooprider, K.L., Mitloehner, F.M., Famula, T.R., Kebreab, E., Zhao, Y. and Van Eenennaam, A.L. (2011) Feedlot efficiency implications on greenhouse gas emissions and sustainability. *Journal of Animal Science* 89, 2643–2656.

Council of Europe (2013) *Recommendations Concerning Cattle*. Available: http://www.coe.int/t/e/legal_affairs/legal_co-operation/biological_safety_and_use_of_animals/farming/Rec%20cattle%20E.asp. (accessed 11 July 2013).

Dalvit, C., De Marchi, M. and Cassandro, M. (2007) Genetic traceability of livestock products: A review. *Meat Science* 77, 437–449.

Deblitz, C. (2012) Costs of production for beef and national cost share structures. AgriBenchmark Working Paper 2012/3. Available: http://www.agribenchmark.org/. (accessed 9 December 2012).

Deming Institute (2012) Available https://www.deming.org/. (accessed 16 November, 2012).

FINBIN (2013) Farm financial management database, searchable livestock databases. Available: http://www.finbin.umn.edu/. (accessed 24 October 2013).

Golan, E.B., Krissoff, F., Kuchler, K., Nelson, G., Price, G. and Calvin, L. (2003) Traceability in the US food supply: Dead end or superhighway. *Choices* 2, 17–20.

Gracia, A. and Zeballos, G. (2005) Attitudes of retailers and consumers toward the EU traceability and labeling system for beef. *Journal of Food Distribution Research* 36, 45–56.

Heitschmidt, R.K., Vermeire, L.T. and Grings, E.E. (2004) Is rangeland agriculture sustainable? *Journal of Animal Science* 82(E. Suppl.), E138–E146.

Ilea, R.C. (2009) Intensive livestock farming: Global trends, increased environmental concerns, and ethical solutions. *Journal of Agricultural and Environmental Ethics* 22, 153–167.

KZNDAE (2014) Agricultural publications and production guidelines Department of Agriculture and Environmental Affairs, Kwazulu-Natal Province, South Africa http://www.kzndae.gov.za/en-us/agriculture/agricpublications/productionguidelines.aspx (accessed 11 May 2014).

Lawrence, J.D. (2004) The cost of meeting consumer demand. In: *Proceedings of the Beef Improvement Federation 36th Annual Research Symposium and Annual Meeting*, pp. 38–41.

Murphy, R.G.L., Pendell, D.L., Morris, D.L., Scanga, J.A., Belk, K.E. and Smith, G.C. (2008) Review: Animal identification systems in North America. *The Professional Animal Scientist* 24, 277–286.

Olvera, I.D. (2010) Statistical and economic implications associated with precision of administering weight-based medication in cattle. M.S. Thesis, Texas A&M University, College Station, TX, USA (December 2010).

Orrù, L., Catillo, G., Napolitano, F., De Matteis, G., Scatà, M.C., Signorelli, F. and Moioli, B. (2009) Characterization of a SNPs panel for meat traceability in six cattle breeds. *Food Control* 20, 856–860.

Pendell, D.T., Brester, G.W., Schroeder, T.C., Dhuyvetter, K.C. and Tonsor, G.T. (2010) Animal identification and tracing in the United States. *American Journal of Agricultural Economics* 92, 927–940.

Resende-Filho, M.A. and Buhr, B.L. (2008) A principal agent model for evaluating the economic value of a traceability system: A case study with injection-site lesion control in fed cattle. *American Journal of Agricultural Economics* 90, 1091–1102.

Schmidt, O., Quilter, J.M., Bahar, B., Moloney, A.P., Scrimgeour, C.M., Begley, I.S. and Monahan, F.J. (2005) Inferring the origin and dietary history of beef from C, N and S stable isotope ratio analysis. *Food Chemistry* 91, 545–549.

Smith, G.C., Savell, J.W., Morgan, J.B. and Lawrence, T.E. (2006) Report of the June-September, 2005 National Beef Quality Audit: A new benchmark for the U.S. beef industry. In: *Proceedings of the Beef Improvement Federation 38th Annual Research Symposium and Annual Meeting*, April 18–21, 2006 Choctaw, Mississippi, pp. 6–11.

Smith-Spangler, C., Brandeau, M.L., Hunter, G.E., Bavinger, J.C., Pearson, M., Eschbach, P.J., Sundaram, V., Liu, H., Schirmer, P., Stave, C., Olkin, I. and Bravata, D.M. (2012) Are organic foods safer or healthier than conventional alternatives?: A systematic review. *Annals of Internal Medicine* 157, 348–366.

Souza-Monteiro, D.M. and Caswell, J.A. (2004) The economics of implementing traceability in beef supply chains: Trends in major producing and trading countries. University of Massachusetts Amherst, Department of Resource Economics Working Paper No. 2004–6.

Steinfeld, H., Wassenaar, T. and Jutzi, S. (2006) Livestock production systems in developing countries: Status, drivers, trends. *Rev. Sci. Tech. Off. Int. Epiz.* 25, 505–516.

Sugahara, K. (2009) In: Li, D. and Chunjiang, Z. (eds) *Computer and Computing Technologies in Agriculture II, Volume 3*, IFIP International Federation for Information Processing, Volume 295. pp. 2293–2301. Springer, Boston.

Verbeke, W. and Ward, R.W. (2006) Consumer interest in information cues denoting quality, traceability and origin: An application of ordered probit models to beef labels. *Food Quality and Preference* 17, 453–467.

Waldron, S., Brown, C. and Longworth, J. (2010) A critique of high-value supply chains as a means of modernising agriculture in China: The case of the beef industry. *Food Policy* 35, 479–487.

14 Global Comparisons of Cattle and Beef

Although there are several examples of particular situations or scenarios from different countries used throughout this book, this chapter is structured to provide some formal comparisons among the beef-producing areas of the world. To adequately accomplish this would probably require an entire book to be devoted to it. However, in this chapter some of the major production-related differences (or similarities) across countries and regions are discussed. There is discussion of physical environment differences related to climate, which hopefully provokes more thought regarding adaptation, genotype by environment interactions, and numerous other supply-related issues in previous chapters. The chapter also has discussion devoted to comparisons of beef production levels, imports and exports, and several societal traits for different areas of the world. Information about the distribution of dairy vs beef cattle and herd size aspects is provided as well. A main point of this chapter is to illustrate some of the primary considerations for global as well as local beef cattle production.

14.1 Global Groups with Interest and Emphasis on Measurement of Agriculture Impacts

Many countries have various agencies related to agricultural policy and production that are charged with monitoring and reporting of data within their boundaries and/or that affect their production and market conditions. Discussion of these types of groups is beyond the scope of this chapter. Comparisons of some traits across countries are needed for a global understanding and appreciation of beef production. For these types of data, groups that evaluate multiple countries simultaneously and their resulting databases and reports are desired. Some major groups and respective websites used for this chapter (and that readers may want to consult for more information) include the World Bank Group (www.worldbank.org),

the Food and Agriculture Organization (FAO, http://www.fao.org) of the United Nations, Codex Alimentarius Commission (http://www.codexalimentarius.org), Consultative Group on International Agricultural Research (CGIAR, www.cgiar.org), International Livestock Research Institute (ILRI), European Commission Eurostat (epp.eurostat.ec.europa.eu/portal/page/portal/eurostat/home), Agri Benchmark (www.agribenchmark.org), United States Department of Agriculture Foreign Agricultural Service (USDA-FAS, www.fas.usda.gov), and Meat and Livestock Australia Limited (MLA, www.mla.com.au). These are not the only groups with useful information; several nations have groups that are excellent at gathering, interpreting and reporting of international data with analyses and reports targeted towards their specific audiences.

14.2 Global Physical Environment

Discussion now moves to target physical and climatic environmental considerations. Most people are probably familiar with the continents of the world (Europe, Africa, Asia, Oceania, North America, South America and Antarctica), shown in Fig. 14.1. All of these are inhabited by people and have cattle, except Antarctica which has environmental conditions that are too harsh for plants and animals. All of these regions have a wide diversity of soil, climate and numerous other resources within their boundaries. Political boundaries separate and define countries. As a result, the political environment as it relates to infrastructure, market regulation, consumer demands, etc. also defines a substantial portion of environmental considerations for beef cattle production and beef product distribution. There can be large variation in physical environment within countries, but a constant political environment provides some degree of production and market consistency. If countries are divided into states and territories, and these in turn are further divided into counties and districts, there is still variability in physical environment expected

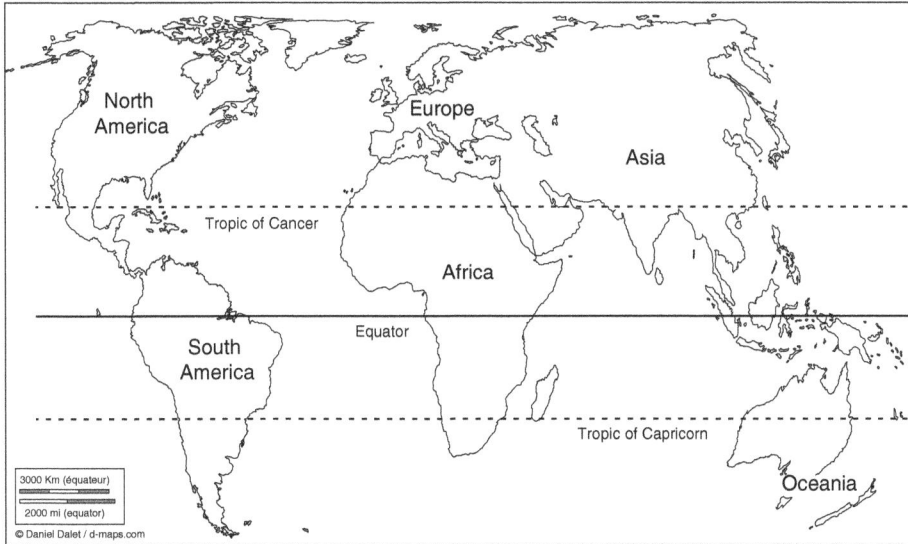

Fig. 14.1. Continents of the world where cattle are located. Map template from d-maps.com.

among and within these units, and, this concept of environmental variation even extends to within farms and ranches in many cases. These aspects of the physical environment relating to precipitation, temperature and ecological zones are discussed in this part of the chapter.

14.2.1 Major Regions of the World

The tropics are adjacent to the equator and bounded by the Tropic of Cancer to the north and the Tropic of Capricorn to the south. Very often tropical conditions indicate high temperatures (frost-free throughout the year) and humidity. The subtropics (or subtropical regions) has a less definitive description but are usually associated with locations immediately north of the Tropic of Cancer and south of the Tropic of Capricorn. Temperate regions are those farther away from the equator and have pronounced mixtures of seasons and mild temperatures, with a defined growing period for many crops bounded by frost-free days. Boreal regions are further removed from the equator and are cold most of the year. Boreal regions are more prevalent in North America, Europe and Asia than South America, Africa or Oceania. Polar regions are those closest to the North and South Poles and have very short growing seasons, if any at all. These conditions are not entirely dictated by distance from the equator as subtropical conditions

may exist near the equator at elevations of 1500 m. Many temperate areas of the world may have higher ambient temperatures during summer months than some subtropical or tropical regions, but the high temperatures do not persist across time. Discussion of different regions of the world must pay adequate attention to the physical environment. Different approaches have been taken to study this concept, and description of the physical, climatic and biological aspects can be extremely detailed. Here, a brief description of the global ecological zones (GEZs) used by the FAO of the United Nations (UN) is provided to give the reader some background and context (see Fig. 14.2).

Regions of the world can be thought of generally as tropical, subtropical, temperate, boreal and polar, respectively, as location moves from the equator toward the North and South Poles. As the earth tilts on its axis throughout the year, regions farther from the equator experience more variation in day length, temperature and other associated concepts. Temperature is also related to elevation (vertical distance from sea level). Precipitation amounts and patterns also help dictate local climatic conditions. The classifications of the 20 GEZs of the world for temperature, elevation and precipitation are found in Table 14.1. Making a broad classification for any area overlooks local variation, but the classification of these global ecological regions are based on the predominant

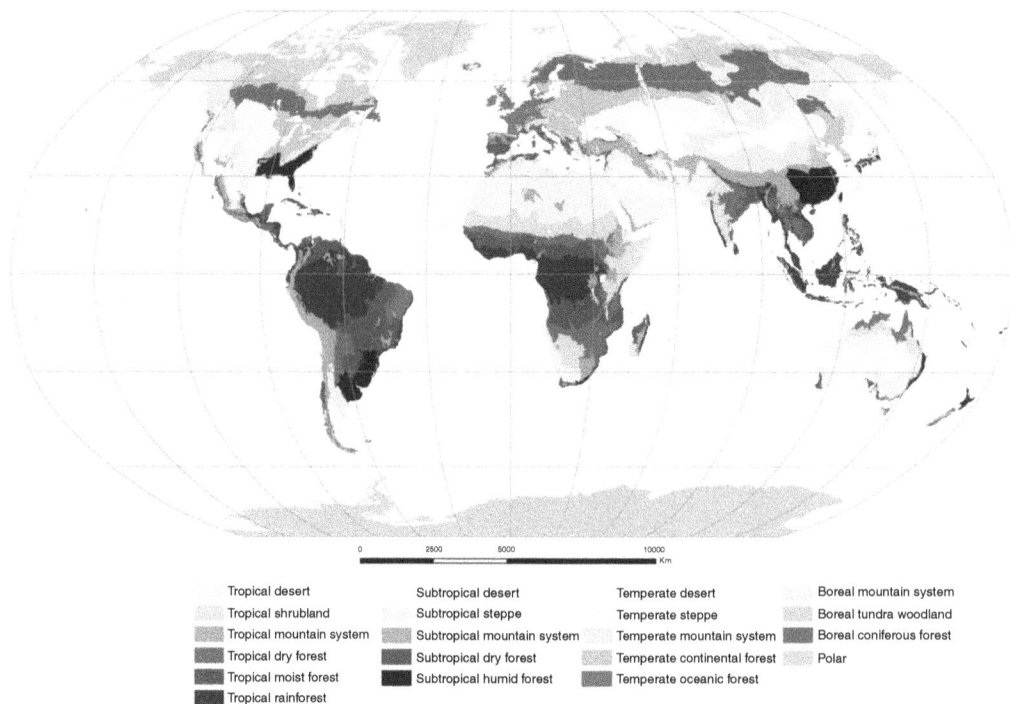

Fig. 14.2. Global ecological zones of the world. From Food and Agriculture Organization of the United Nations (2012) Global ecological zones for FAO forest reporting: 2010 update available at: http://foris.fao.org/static/data/fra2010/ecozones2010.jpg (reproduced with permission).

vegetation types, which in turn are a result of temperature, humidity, rainfall, growing season, soil types, etc. Knowing this type of information provides insight into which types of plants, animals and management strategies may work, as well as indicating many options that would be expected not to work (or be sustainable). This concept should be obvious from use of non-tropically adapted cattle in tropical areas (or vice versa), but applies to many other facets of society such as use of locally adapted lawn and yard plants that suit the environment and can thrive without expensive, artificial inputs. Table 14.2 illustrates the amount of land area across global regions that fall under the FAO GEZ designations.

There can be tremendous variation in geophysical characteristics within a country, and even more so within a continent, and very often there may be more variation within countries or continents than between these entities for these types of factors. However, within countries there are consistent policies and regulations that automatically provide some level of constancy. This is probably

not the case within continental areas, except for the case of the European Union. Nonetheless, the following section gives some broad overviews of beef production considerations across continents, and then some comparisons across countries, with emphasis on some of the more numerically important beef producing and consuming nations of the world.

Detailed discussion of climatic ecological zones is beyond the scope of this chapter, but the relative distributions of these are shown in Fig. 14.3. The reader is encouraged to study world maps and the FAO GEZ maps and related information to not only understand where the individual countries and continental areas are located, but to see the ecological diversity (and similarity) across global regions. Understanding the GEZ concept relates directly to the concept of adaptation of livestock, and the manager's coordination of genetic and production environment resources. It should be recognized that there are similar ecological zones across the world where similar types of cattle can be utilized, and that placement of cattle that are not adapted to particular

Table 14.1. Global ecological zones according to temperature and precipitation.

Climate name	Climate description	Global ecological zone (GEZ)	GEZ description
Tropical	All months without frost, in marine areas mean temperature over 18°C	Tropical rainforest Tropical moist forest Tropical dry forest Tropical shrubland Tropical desert Tropical mountain systems	Wet: 0–3 months dry, during winter Wet/dry: 3–5 months dry, during winter Dry/wet: 5–8 months dry, during winter Semi-arid: evaporation > precipitation Arid: All months dry Approximate > 1000 m altitude (local variations)
Subtropical	Eight months or more where mean is over 10°C	Subtropical humid forest Subtropical dry forest Subtropical steppe Subtropical desert Subtropical mountain systems	Humid: no dry season Seasonally dry: winter rains, dry summer Semi-arid: evaporation > precipitation Arid: all months dry Approximate > 800–1000 m altitude
Temperate	Four to eight months with mean temperature over 10°C	Temperate oceanic forest Temperate continental forest Temperate steppe Temperate desert Temperate mountain systems	Oceanic climate: coldest month over 0°C Continental climate: coldest month under 0°C Semi-arid: evaporation > precipitation Arid: all months dry Approximate > 800 m altitude
Boreal	Up to 3 months with mean over 10°C	Boreal coniferous forest Boreal tundra woodland Boreal mountain systems	Vegetation physiognomy: coniferous dense forest dominant Vegetation physiognomy: woodland and sparse forest dominant Approximate > 600 m altitude
Polar	All months are below 10°C	Polar	Very small to zero plant growth potential

From FAO (2012). Names of the GEZs reflect the dominant zonal vegetation. A dry month is one where the total precipitation (mm) ≤ twice the mean temperature (°C). Readers are encouraged to consult this reference for examples in various global regions.

ecological zones will lead to incompatibilities. Just because the ecological zone is similar does not automatically mean the production environment is the same, however, or that the production outputs and inputs will automatically be the same. It is the first step to consider when bringing cattle into new areas.

14.3 Beef Production Comparisons across Global Regions

Discussion now moves from physical environmental aspects to comparisons of cattle and associated production measures. Many of these statistics are available on individual country bases (the exception mainly being EU member statistics in some cases). Considerations for global beef production need to begin with the discussion of beef cattle numbers, as the number of animals and their relative degree of productivity and profitability dictate the amount and demand of beef produced from the animals. Table 14.3 lists the

top 10 regions pertaining to live cattle numbers, and Table 14.4 the top 10 regions pertaining to beef production, consumption and international trade.

The types of data in Tables 14.3 and 14.4 have been typically reported for many years by various national and international groups. Over time and in accordance with ever-changing political, economic, societal and environmental issues, these values continually change. Also, there can be widely different underlying production aspects that yield similar total numbers, and similar measures of several components of production that in combination with other factors result in widely different total production statistics. Tables 14.5 through 14.7 provide data and bases for some broader considerations regarding some of the potential underlying differences behind some of the gross cattle numbers and overall beef production measures.

As discussed in Chapter 13, the total number of animals as well as their relative degree of productivity

Table 14.2. Distribution of global ecological zones across regions of the world (in 1000 km²).

Global ecological zone	South America	North America	Europe	Asia	Oceania	Africa	Former USSR	Total
Tropical rainforest	6631.2	440.9	0.0	3009.4	481.3	4017.7	0.0	14,580.5
Tropical moist deciduous forest	4302.3	678.0	0.0	1379.5	26.3	4661.2	0.0	11,047.3
Tropical dry forest	1681.6	226.0	0.0	1426.6	468.0	3669.5	0.0	7471.7
Tropical shrubland	103.0	2.1	0.0	1167.1	1063.4	5977.9	0.0	8313.6
Tropical desert	137.6	0.0	0.0	2704.5	0.0	8737.7	0.0	11,579.8
Tropical mountain systems	1886.5	259.1	0.0	834.9	71.2	1473.2	0.0	4525.0
Subtropical humid forest	1199.9	1068.5	0.0	2047.9	281.2	85.1	0.0	4682.6
Subtropical dry forest	100.5	87.0	816.4	129.0	123.3	334.8	0.0	1591.1
Subtropical steppe	639.7	1167.3	5.0	1180.3	1461.1	456.7	0.0	4910.1
Subtropical desert	0.0	1083.9	0.0	1446.3	4139.2	0.0	0.0	6669.4
Subtropical mountain systems	238.2	592.3	149.2	3459.6	0.0	412.4	7.0	4858.6
Temperate oceanic forest	259.1	40.1	1287.1	0.0	218.5	0.0	0.0	1804.9
Temperate continental forest	0.0	2023.8	2906.7	1253.1	0.0	0.0	761.7	6945.3
Temperate steppe	498.3	2121.8	955.5	1115.6	0.0	0.0	1202.7	5893.9
Temperate desert	0.0	742.9	150.3	2181.9	0.0	0.0	2321.9	5397.1
Temperate mountain systems	76.9	1976.8	605.4	3604.8	193.6	0.0	768.8	7226.4
Boreal coniferous forest	0.0	2187.0	2195.7	157.4	0.0	0.0	3924.3	8646.5
Boreal tundra woodland	0.0	2566.4	87.8	0.0	0.0	0.0	1296.9	3951.1
Boreal mountain systems	0.0	1182.3	457.6	109.2	0.0	0.0	4550.1	6299.3
Polar	0.0	3220.4	422.6	0.0	0.0	0.0	1768.3	5411.3
Water	0.0	405.0	0.1	15.6	0.0	158.1	120.2	699.1
No data	8.2	0.3	3.4	0.1	1.8	9.6	0.0	23.4
Total	17,763.3	22,072.0	10,043.0	27,223.1	8528.9	29,993.9	16,722.0	132,346.1

From FAO (2012). Names of the GEZs reflect the dominant zonal vegetation. Data are presented in thousands of square kilometers (e.g. South America has 6.631 million km² of tropical rainforest, etc.).

will dictate total production potential. There is usually more potential for production (or product) per animal when there is a more intensive production system (which may or may not dictate profit), and there can be wide ranges in beef produced relative to the same animal inventory. Several nations import much more beef than they export, and others export much more than they import. When nations are heavily invested in import and export markets, global issues and currency values heavily dictate their local beef prices. Among these countries there are wide differences in environmental resource use, population totals as well as density, and rate of productivity increase. There are also very wide differences in wealth and consumer spending.

It is important for consumers and producers to be informed and aware of the role of agriculture in their local geography and economy as well as the nation as a whole. Agriculture holds great potential in many areas of the world to reduce poverty and to aid in sustainable management of natural resources. These types of values are provided for these countries to indicate to the reader

the large degree of variability that exists across global areas. It seems that many people that have not studied or traveled extensively may think the rest of their country or world may be similar to their circumstances, and an open mind is critical for learning. Over the years, many agricultural development programs appear to have been simply applying a technique that works in one area without thinking about how the information or technology might work in a different area of the world.

Another production aspect across global regions deals with the types of cows that contribute to potential beef production. As beef is the meat produced from cattle, not all beef comes from beef cattle (see Fig. 14.4). In some cases it is unclear how much beef reported from some areas of the world is meat from buffalo or other cattle-related species. No matter which country is considered, at least some of the beef comes from dairy cattle, and in some examples, all beef may come from dairy cattle as opposed to what most consider beef breeds.

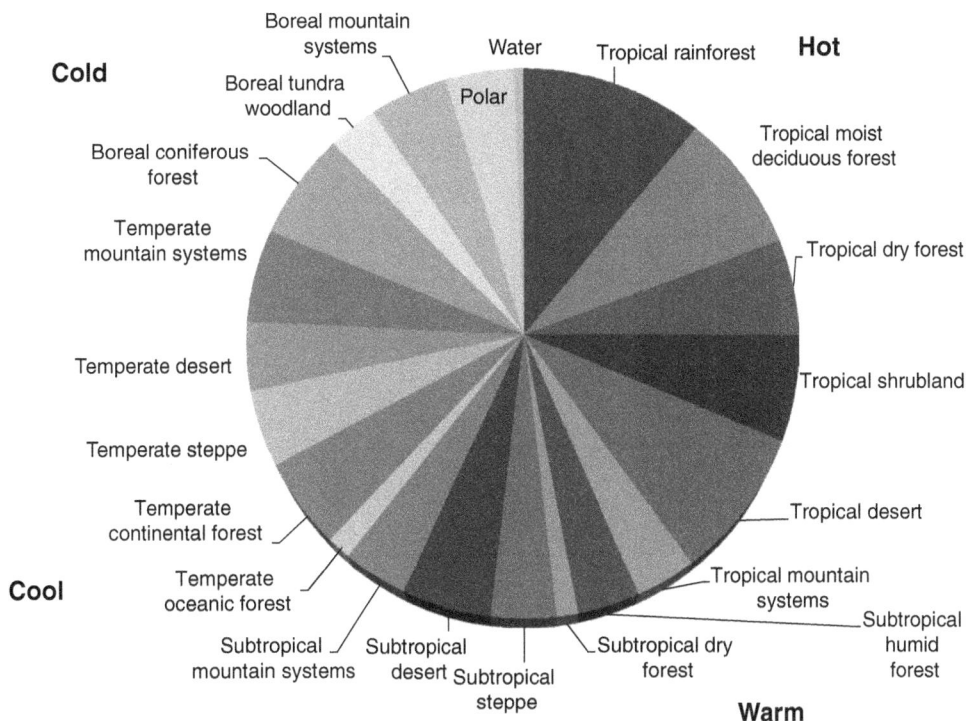

Fig. 14.3. Relative distribution of ecological zones of the world. Data from FAO.

14.3.1 Global Trends and the Livestock Revolution

In the context of discussion about reduction of poverty and the accumulation of wealth, historically, when people have more money, they tend to transition from a low-meat diet (as meat is expensive) to a higher level of meat consumption. For geographical, cultural or religious reasons, the types of meat may or may not change, but the level of consumption tends to increase. Hopefully from the previous chapters and discussion in this book it is apparent that any extreme level of anything is probably detrimental. Numerous research trials have shown that well-balanced diets that provide adequate nutrients (not deficiencies or excesses) result in improved overall health. This is true with cattle and other food animals (which is one good reason that all producers care about animal welfare), and it is true with people. Table 14.8 provides some previous and expected values for milk and meat products in different areas of the world over time, and Fig. 14.5 shows varying levels of beef relative to other meats in the diet for some selected countries.

Figure 14.6 shows the relative livestock production index values that have been reported over time in selected countries representing different areas of the world. Several developing countries have increased their research emphasis and funding on food production during the time frame represented in this figure. Research and development funds are continually needed to increase production efficiency and resource management in all types of countries and cultures. The ability of a population to effectively produce food and have strong trade with other populations (within and between villages, districts, provinces/states and nations) will always remain important. The ability to keep food costs low provides many potential health advantages, and education of consumers regarding nutrition and health from science-based research and public service is critical for societal improvement.

Increased production of beef is expected worldwide in the near future, as is increased global trade in beef products (Fig. 14.7). Over the next few years, it is likely that there will be re-ranking of several leading countries for beef production and trade. Increased production efficiency in many

Table 14.3. Rankings for top 10 global regions for cattle numbers and international cattle trade for 2012.

Rank	Calf crop Country	Number[a]	Total cattle Country	Number[a]	Exports Country	Number[b]	Imports Country	Number[b]
1	India	63.4	India	323.7	Mexico	1539	USA	2256
2	Brazil	49.7	Brazil	197.6	Canada	825	Venezuela	616
3	China	41.0	China	104.3	EU-27	769	Russia	138
4	USA	34.3	USA	90.8	Australia	620	China	115
5	EU-27	29.3	EU-27	86.2	Brazil	512	Egypt	95
6	Argentina	13.8	Argentina	49.6	Colombia	299	Canada	56
7	Australia	10.0	Colombia	30.9	USA	191	Japan	14
8	Russia	6.9	Australia	28.5	Uruguay	75	Mexico	10
9	Mexico	6.8	Mexico	20.1	New Zealand	42	Ukraine	3
10	Colombia	5.1	Russia	19.7	China	28	Belarus	2
	World	280.8	World	1019.3	World	4878	World	3112

[a]Values in million animals.
[b]Values in thousands of animals.
From USDA-FAS (2013); EU-27 is the collective 27 nations of the European Union.

Table 14.4. Rankings for top 10 global regions for beef consumption, production and international trade for 2012.

Rank	Domestic consumption Country	Amount	Total production Country	Amount	Exports Country	Amount	Imports Country	Amount
1	USA	11,744	USA	11,855	Brazil	1524	Russia	1023
2	Brazil	7845	Brazil	9307	India	1411	USA	1007
3	EU-27	7806	EU-27	7765	Australia	1407	Japan	737
4	China	5597	China	5540	USA	1114	S. Korea	370
5	Argentina	2458	India	3460	New Zealand	517	EU-27	348
6	Russia	2395	Argentina	2620	Uruguay	355	Canada	301
7	India	2049	Australia	2152	Canada	335	Egypt	250
8	Mexico	1835	Mexico	1820	EU-27	307	Hong Kong	241
9	Pakistan	1367	Pakistan	1400	Paraguay	251	Venezuela	220
10	Japan	1255	Russia	1380	Mexico	200	Mexico	215
	World	55,513	World	57,170	World	8324	World	6683

From USDA-FAS (2013). Values reported in million kg and are expressed on carcass weight (bone-in) basis; EU-27 refers to the collective 27 nations of the European Union.

developing nations along with low labor costs will provide increased opportunities for both domestic and international markets.

14.4 Livestock Production Systems

As introduced in Chapter 1, Seré and Steinfeld (1996) classified livestock production systems due to soil moisture, daily temperature, how animals were managed and growing season. The type of livestock production system greatly impacts optimal cattle genetic types, realistic (meaning profitable) management and end-products (and by-products). Detailed description of the numerous example production systems is beyond the scope of this chapter, but readers are encouraged to consult Seré and Steinfeld (1996) for examples in many global regions. As discussed earlier, these systems may be used in conjunction with other species or livestock and/or solely in livestock, mixed livestock and farming systems. The specific management and genetic resources used in the same beef cattle production system may be highly variable and should be matched to local physical environments and markets. Operations that have multiple types of animals (through use of multiple systems) which they can market have more potential flexibility to alter animal numbers (and therefore stocking rates)

Table 14.5. Summary of some selected countries for beef production and international trade traits.

Country	Beginning cattle (1000 head)	Beginning beef cow numbers (1000 head)	Beef production (million kg)	Beef per animal of inventory (kg)[a]	Beef imports (million kg)	Imports relative to production (%)[b]	Beef exports (million kg)	Exports relative to production (%)[b]
Argentina	49,597	20,000	2,620	52.8	2	0.1	164	6.3
Australia	28,506	13,865	2,152	75.5	12	0.6	1,407	65.4
Belarus	4,247	0	235	55.3	3	1.3	155	66.0
Brazil	197,550	52,669	9,307	47.1	62	0.7	1,524	16.4
Canada	12,215	3,997	1,075	88.0	301	28.0	335	31.2
China	104,346	46,200	5,540	53.1	99	1.8	42	0.8
Colombia	30,910	9,700	900	29.1	2	0.2	12	1.3
Egypt	6,175	0	280	45.3	250	89.3	0	0.0
EU-27	86,196	12,161	7,765	90.1	348	4.5	307	4.0
India	323,700	0	3,460	10.7	0	0.0	1,411	40.8
Japan	4,172	648	519	124.4	737	142.0	1	0.2
Korea, South	3,354	1,249	312	93.0	370	118.6	1	0.3
Mexico	20,090	6,900	1,820	90.6	215	11.8	200	11.0
New Zealand	10,021	1,053	625	62.4	10	1.6	517	82.7
Russia	19,695	310	1,380	70.1	1,023	74.1	8	0.6
Ukraine	4,426	35	365	82.5	5	1.4	23	6.3
USA	90,769	30,158	11,855	130.6	1,007	8.5	1,114	9.4
Uruguay	11,232	4,100	520	46.3	0	0.0	355	68.3
Venezuela	12,090	0	365	30.2	220	60.3	0	0.0

From USDA Foreign Agricultural Service (FAS) Production, Supply and Distribution (PSD) online database for 2012. Reported in metric tons (tonnes) which refers to 1000 kg or 2205 lb.
[a]Ratio of total beef production (carcass or bone-in basis) divided by beginning cattle numbers.
[b]Ratio of imports or exports divided by beef production.

and to target current market preferences (for instance if yearling heifers are being grown, they can be managed and marketed for potential finisher animals or for replacement heifers depending upon where there is stronger demand). The type of system does not dictate the genetic type of cattle desired; however, the physical environment and market conditions should be primary influences on the genetic type(s) of animals.

Production of cattle that are or will be desirable to multiple markets or potential buyers provides more flexibility than cattle that are highly specialized for production traits. Cattle that are doubled-muscled may be highly favored in many European markets where a high degree of muscle expression and low external fat are favored and there is no evaluation of intramuscular fat. This type of cattle would be highly disadvantageous where larger herds are used (as dystocia is much more common) and feed resources are variable and external fat cover in cows is critical for reproduction. Cattle that have extremely high potential for marbling under Japanese or Korean production systems and markets have lower value under

short-term feeding strategies in North American systems, or under most EU markets that do not evaluate marbling. High production potential cattle in many challenging environments of the world are less desirable in pure form than either indigenous types or when crossed with indigenous types. Some examples of production information from various cattle production systems are provided below.

Peel *et al.* (2011) reviewed the beef cattle production situation in Mexico (Table 14.9). Here, VacaNorte refers to cattle of the north that are produced with the intention of sales into the USA. Use of British and Continental crosses are common. Beef-only and dual-purpose production throughout Mexico may be either semi-intensive or traditional, both relying to varying degrees on zebu influence. In the traditional production system, crop residues (usually from contiguous farming activities) are the primary means by which cattle survive the annual dry season. Criollo production has an emphasis on family milk production and draft use. Only occasionally do cull animals provide meat for home consumption. In many cases,

Table 14.6. Summary of some land and agricultural statistics for selected countries.

Region	Country name	Total land area (km²)	Land area in forests (%)	Land area in agricultural production (%)	Land area in permanent cropland (%)	Cereal yield (kg/ha)	Food production index[a]	Livestock production index[a]
Africa	Kenya	569,140	6.1	48.2	1.1	1,514.3	123.1	123.4
	Nigeria	910,770	9.5	83.7	3.5	1,330.9	104.0	123.0
	South Africa	1,213,090	7.6	79.4	0.3	4,024.0	116.1	130.8
Asia	China	9,327,490	22.5	55.7	1.6	5,705.5	124.0	116.2
	India	2,973,190	23.1	60.5	4.1	2,883.3	129.3	123.5
	Indonesia	1,811,570	51.7	30.1	11.0	4,886.1	124.6	124.2
	Japan	364,500	68.6	12.5	0.8	4,910.7	92.7	99.1
Oceania	Australia	7,682,300	19.3	53.3	0.1	2,097.2	108.3	98.7
Europe	France	547,660	29.2	53.1	1.9	6,859.1	99.2	97.5
	Germany	348,570	31.8	48.0	0.6	6,460.7	105.3	110.6
	UK	241,930	11.9	70.9	0.2	6,984.6	104.1	104.8
North America	Canada	9,093,510	34.1	6.9	0.5	3,527.1	109.3	99.9
	Mexico	1,943,950	33.3	53.1	1.4	3,240.6	105.0	112.0
	USA	9,147,420	33.3	45.0	0.3	6,818.0	105.0	105.2
South America	Argentina	2,736,690	10.7	53.9	0.4	4,672.3	114.1	97.0
	Brazil	8,459,420	61.2	32.5	0.8	4,037.7	127.0	118.2
	Paraguay	397,300	43.8	52.8	0.2	3,480.0	159.1	137.8
World	World	129,709,895	31.0	37.5	1.2	3,708.2	116.6	111.8

[a]World Bank (2013); index relative to 2004–2006 basis.
1 square km = 0.3861 square mile; 1000 kg/ha = 892.4 lb/ac.

Criollo production occurs where crop farming is the primary enterprise (Peel *et al.*, 2010).

Otte and Chilonda (2002) provided values for sub-Saharan Africa (Table 14.10) according to production system type. Pastoral classification here refers to communal and transitory production style, whereas ranching classification is similar to the term of commercial used in other publications.

14.4.1 Herd Size Considerations and Comparisons

The types of traits that are emphasized in small and very small herds have been reported to be very similar to traits emphasized in larger and more specialized cattle operations (for instance Jain and Muladno, 2009, in South Asia). Scholtz *et al.* (2008) reported for South Africa that the reasons for choosing herd bulls was similar across commercial (profit and performance emphasis private herds), communal (herds and situations typical of native peoples) and emerging (transitioning from communal to commercial emphasis) systems in that all three considered similar traits of importance such as performance, conformation, temperament, size, availability, color and horns;

however, actual management was drastically different in that uncontrolled matings occurred 98%, 63% and 11%, respectively, of the time for communal, emerging and commercial producers. Further, 22% of the matings in commercial herds occurred through AI, but was 6.3% in emerging and 0.1% in communal situations.

There are many reasons why beef herds may be of certain animal numbers. Hopefully no matter what the situation, the number of animals is properly matched to the environmental resources. Management may be more complicated in smaller beef herds if the smaller number of cattle is associated with other livestock species and farming activities. If someone has 100 cows, but that is the entire agricultural activity, that scenario would probably be much easier to manage than another operation that has 35 beef cows, 20 sheep, 10 sows and production of one or more crops. As a result, simply comparing herd sizes is an overly simplistic way to make many speculations about households or operations without some understanding of the overall production system, and the context of environmental constraints. Figure 14.8 shows some herd size and industry considerations for South Africa.

Table 14.7. Some population and societal aspects of selected countries.

Country name	Population estimated in 2013 (million)[a]	Population density (people per km²)	% of rural population with improved water access in 2010[b]	Rural population as % of total[b]	Percent of population in labor sectors (%)[a]			Per capita expenditure (US$)[c]	Expenditures on food as % of total[c]	Expenditures per capita in 2000 US$[b]
					Ag	Industry	Service			
Kenya	44.0	77	52	76.0	75	25 combined		668	41.3	370
Nigeria	174.5	192	43	50.4	70	10	20	765	39.6	
South Africa	48.6	40	79	38.0	9	26	65	4652	19.4	2566
China	1349.6	145	85	49.5	35	30	36	2134	21.3	949
India	1220.8	411	90	68.7	53	19	28	892	26.3	524
Indonesia	251.2	139	74	49.3	39	22	48			671
Japan	127.3	349	100	8.9	4	26	70			22,907
Australia	22.3	3	100	10.8	4	21	75	36,292	10.7	15,699
France	66.0	120	100	14.3	4	24	72	24,576	13.3	13,504
Germany	81.1	233	100	26.1	2	25	74	23,937	11.1	14,341
UK	63.4	262	100	20.4	1	18	80	23,728	9.4	18,193
Canada	34.6	4	99	19.3	2	22	76	27,632	9.7	16,209
Mexico	116.2	60	91	21.9	14	23	63	6819	22.7	4506
USA	316.7	35	94	17.6	1	20	79	33,575	6.7	27,175
Argentina	42.6	16		7.5	5	23	72	6125	21.2	
Brazil	201.0	24	85	15.4	16	13	71	7573	15.9	3194
Paraguay	6.6	17	66	38.1	27	19	55			1391
World	7095.2	55	80.5	48.0	36	22	41			3695

[a]CIA World Fact Book online.
[b]World Bank online databases.
[c]USDA Economic Research Service (ERS) online databases.
1 square km = 0.3861 square miles.

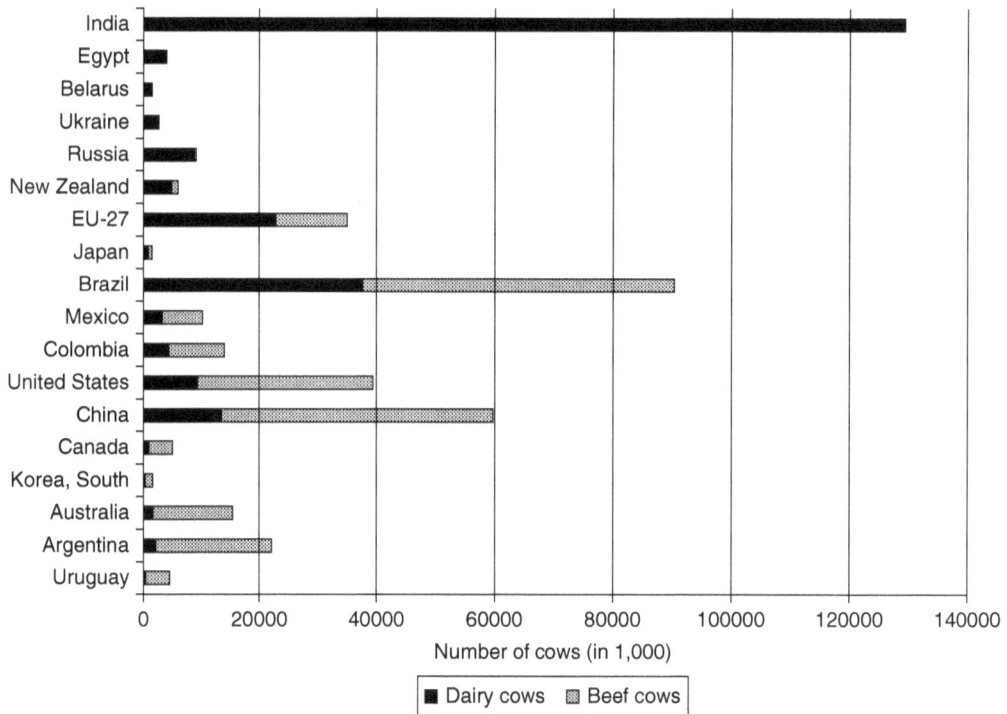

Fig. 14.4. Distribution of cow numbers within some regions of the world. Data are from 2012 market year from USDA-FAS (2013). Values reported are in thousands (i.e. India has 129 million dairy cows, etc.).

Herds involved in communal and emerging sectors of the South African cattle industry are dominated by small herds, as is the case in many regions. Scholtz *et al.* (2008) reported average herd size ranging from 12 to 40 cattle across all provinces except the Northern Cape, where the average herd size was 231. Table 14.11 provides some example herd sizes reported in some countries, and Table 14.12 provides herd size considerations for many European countries. In all instances there is a small percentage of operations that are responsible for large percentages of cattle, but what is considered small or large varies considerably across countries.

Herd size alone does not dictate profitability or production on a per animal basis. There are, however, direct advantages of spreading fixed costs over more animals with larger operations, resulting in lower fixed cost per animal (review Chapter 12 for more detailed discussion if needed). Groups of small herds may be able to work together and accomplish many of the same advantages (and possibly even more) than individual large herds that involve the same number of animals.

Herd size and production system considerations are often highly related. Table 14.13 summarizes data in the international Agri Benchmark (http://www.agribenchmark.org/home.html) dataset and provides some example beef production system considerations across several countries. There are wide differences in these operations regarding herd size, feedstuffs, target markets and expected animal performance. Based on discussions in Chapters 12 and 13, it should be remembered that no set level of production, type of feeding, type of production system, etc. alone dictates profitability. Table 14.14 simply provides some general beef cattle management practices that are typical in some of the different countries discussed in this chapter. Inclusion or exclusion of countries from this table is only a reflection of reported data or first-hand knowledge of the author and does not signify any type of recommendation (or lack thereof).

The type of beef cattle production system is also strongly related to the expected level of animal performance, input considerations and environmental impact. It is beyond the scope of this chapter

Table 14.8. Annual per capita consumption patterns for milk, meat and beef across time.

	1964/66	1974/76	1984/86	1994/96	1997/99	2015	2030
Milk and dairy (kg, whole milk equivalent basis)							
World	74	75	78	77	78	83	90
Developing countries	28	30	37	42	45	55	66
Sub-Saharan Africa	28	28	32	29	29	31	34
Near East and North Africa	69	72	83	71	72	81	90
Latin America and Caribbean	80	93	94	106	110	125	140
South Asia	37	38	51	62	68	88	107
East Asia	4	4	6	10	10	14	18
Industrialized countries	186	191	212	212	212	217	221
Meat (kg, carcass weight equivalent basis)							
World	24.2	27.4	30.7	34.6	36.4	41.3	45.3
Developing countries	10.2	11.4	15.5	22.7	25.5	31.6	36.7
excl. China	11	12.1	14.5	17.5	18.2	22.7	28
excl. China and Brazil	10.1	11	13.1	14.9	15.5	19.8	25.1
Sub-Saharan Africa	9.9	9.6	10.2	9.3	9.4	10.9	13.4
Near East and North Africa	11.9	13.8	20.4	19.7	21.2	28.6	35
Latin America and Caribbean	31.7	35.6	39.7	50.1	53.8	65.3	76.6
excl. Brazil	34.1	37.5	39.6	42.4	45.4	56.4	67.7
South Asia	3.9	3.9	4.4	5.4	5.3	7.6	11.7
East Asia	8.7	10	16.9	31.7	37.7	50	58.5
excl. China	9.4	10.9	14.7	21.9	22.7	31	40.9
Industrialized countries	61.5	73.5	80.7	86.2	88.2	95.7	100.1
Bovine meat (kg, carcass weight equivalent basis)							
World	10	11	10.5	9.8	9.8	10.1	10.6
Developing countries	4.2	4.3	4.8	5.7	6.1	7.1	8.1

Data from Bruinsma (2003).

(and book) to discuss environmental impacts in great detail. However, all operations in future years will probably need to document their environmental impacts (be that carbon footprint or other similar types of measurements) as population growth and competition for environmental resources increase. Table 14.15 summarizes some environmental types of measurements that have been reported for some different beef-producing regions of the world. In the literature describing environmental impact both the types of measurements used as well as their interpretations have not been entirely straightforward for producers to understand. Several whole-systems types of analyses have shown that grain feeding can help reduce total greenhouse gas emissions due to improved feed efficiency of diet as well as reduced feeding time per animal. All these types of comparisons are highly dependent upon their assumptions, which should be carefully studied. It is likely that in the near future many grazing components of beef cattle production, particularly where permanent pastures are used, may be able to receive some type of credit

for being carbon sinks (resources that incorporate carbon into the environment rather that contributing carbon to the environment). As a result, beef cattle operations that can calculate their environmental impacts may have more business opportunities, and they will be better prepared if they start to be the object of environmental scrutiny.

14.5 Summary

There are many fundamental concepts of beef cattle production that can be applied to beef production systems globally. However, the specific strategies of management, use of genetic and environmental resources, and expected outcomes can vary tremendously, and the same production method that produces desirable results in one region may produce disastrous results in another. There is no single desired herd size, output per animal or off-take per land unit that should be a universal target for beef production. Producers and managers need to monitor their resources and their efficiency of production (biologically,

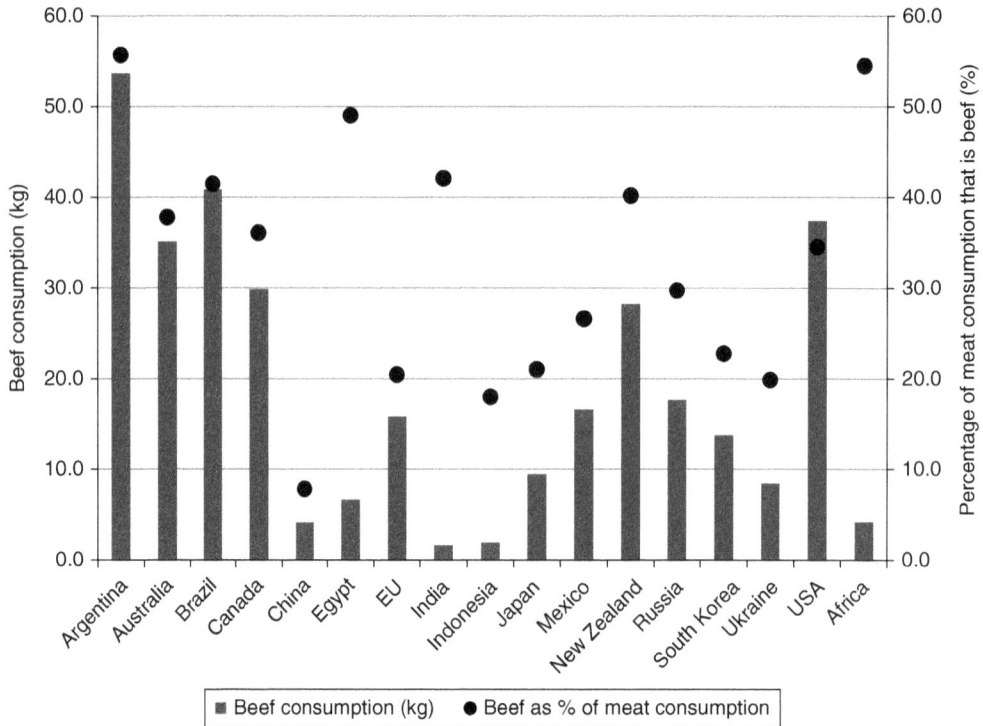

Fig. 14.5. Per capita beef consumption (bars, left axis) and percentage of dietary meat consumption that is beef (dots, right axis) for selected countries and regions of the world. Data from FAPRI (2012).

economically and environmentally) to determine optimal strategies and methods for their specific scenarios. This chapter has provided some values for resources, production values and societal traits to give the reader more of a global comparison and well-rounded understanding of world-wide beef production.

14.6 Study Questions

14.1 Briefly explain the concepts associated with tropical, subtropical, temperate and boreal environments.

14.2 What might be some expected characteristics of steppe as compared to mountain ecological systems?

14.3 Based on 2012 values, list five of the top 10 countries or entities for (a) total cattle numbers and (b) cattle exports.

14.4 Based on 2012 values, list five of the top 10 countries or entities for (a) total beef production and (b) beef exports.

14.5 Generally, what has occurred to meat consumption in societies when wealth has increased from low to medium levels?

14.6 Explain some different considerations for beef cattle production systems that are solely livestock based as compared to those that are combination livestock and farming.

14.7 List three countries that have less than 30% of their cow inventory as beef cows, and list three countries that have over 70% of their cow inventory as beef cows.

14.8 Why have livestock production index values increased more in developing countries than in developed countries from 2000 to 2010?

14.9 What types of factors could influence the level of beef production (from animal as well as carcass standpoints) from a fixed number of breeding cows (i.e. herd size of 100, herd size of 1000, etc.)?

14.10 Is it realistic to have the same beef cattle production practices and considerations in all countries of the world? Explain why or why not. What about across all regions within countries?

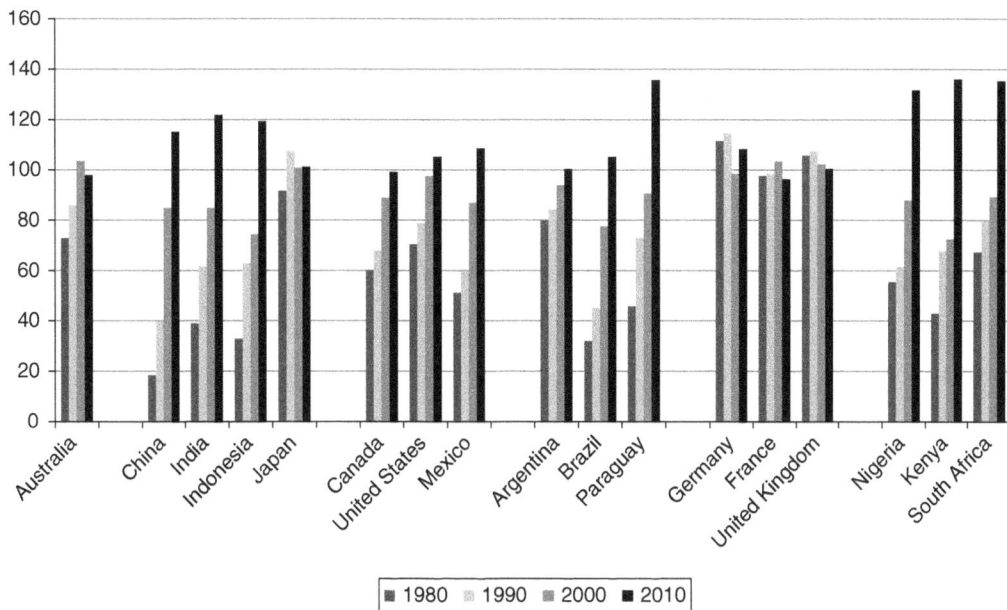

Fig. 14.6. Livestock production index 1980–2010. Livestock production index (LPI) includes production of meat and milk from all sources, products such as cheese, eggs, honey, raw silk, wool and hides/skins where the years 2004–2006 provide the base of 100. The LPI value over time is a function of how much change has occurred relative to the base years of 2004–2006. Values are specific within country and cannot be compared across countries (except to see how much change has occurred within the respective countries). More developed countries have changed the least across this time span as compared to developing countries. Several European countries have lower values for 2004–2006 as compared to earlier years. Data downloaded from World Bank (2013) website.

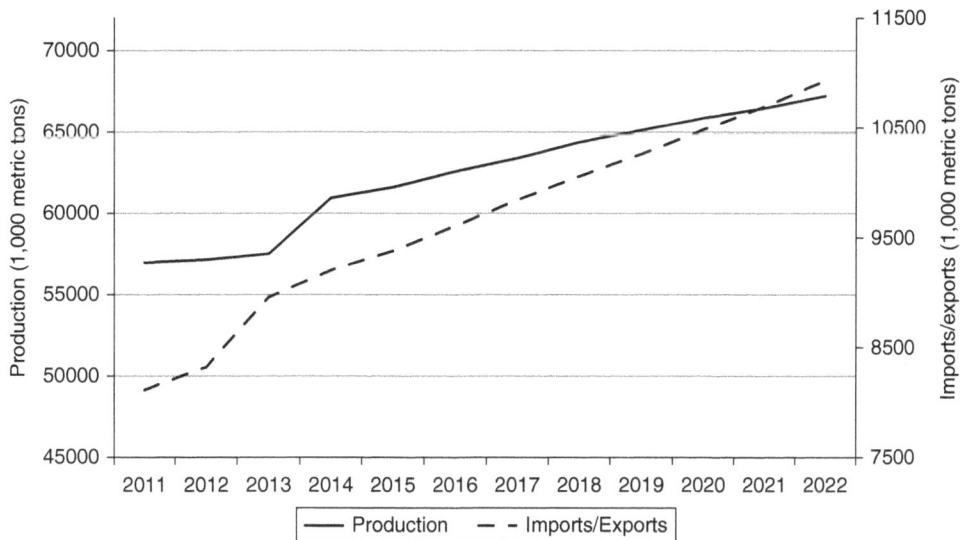

Fig. 14.7. Predicted global beef production (left axis) and international trade (right axis). Data from USDA Economics Research Service. Imports and exports are predicted to be the same amount per annum. Global beef production is expected to increase 17 to 18% while global international trade is expected to increase about 35%.

Table 14.9. Characteristics of Mexican cow-calf production systems.

	VacaNorte	Semi-intensive	Traditional	Criollo
Cow mature weight (kg)	450	435	415	400
Milking ability	Medium/high	Medium	Medium/low	Low
WWT-steers (kg)	180	160	140	120
WWT-heifers (kg)	162	144	126	108
Average WWT (kg)	171	152	133	114
Calving %	70 to 80	60 to 70	50 to 60	40 to 50
Calf death loss %	4	5	8	10
Weaning percent	71	61	47	35
Culling percent	10	10	8	7
Health management	High	Medium/high	Medium/low	Low
Supplementation	Good	Medium	Salt/mineral	Salt/none
Pasture management	Good	Good/medium	None	None

From Peel *et al.* (2011). WWT = weaning weight at 6–8 months of age. Use of cows to provide household milk is common.

Table 14.10. Fertility, mortality and cow size in sub-Saharan Africa across production system types.

	Pastoral		Mixed farming livestock			Ranching
	Arid	Semi-arid	Semi-arid	Sub-humid	Humid	Sub-humid
Calf crop born (%)	61.0	60.5	58.2	60.0	57.4	76.2
Calf mortality (%)	23.1	22.3	20.7	22.3	21.1	10.1
Cow mortality	8.2	7.6	6.2	6.4	4.2	6.2
Age first calving (months)	49	47	47.4	48.4	39.4	41
Weight of mature cow (kg)	246	251	239	256	205	309

From Otte and Chilonda (2002).

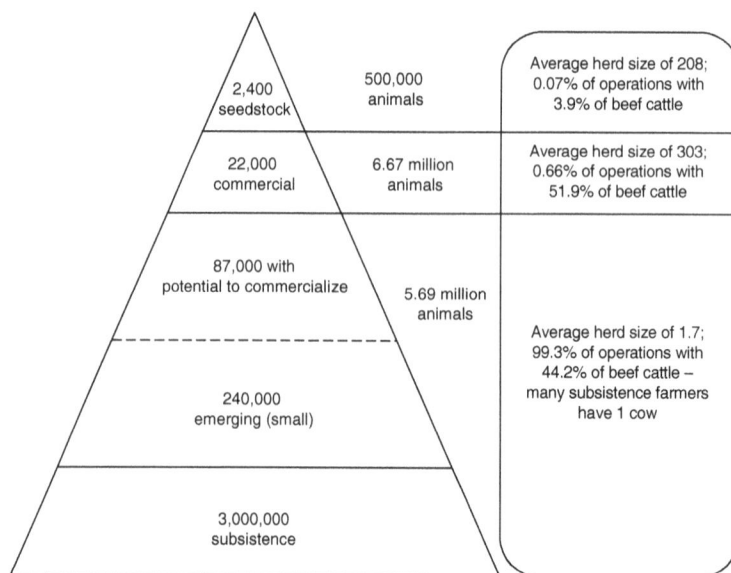

Fig. 14.8. Industry structure for beef and dual-purpose cattle in South Africa. Number of operations/households by sector within pyramid, number of cattle outside pyramid. Adapted from Sholtz *et al.* (2008). Calculations for averages and distributions based on reported values.

Table 14.11. Beef herd size distributions as reported in some different countries.

China Shandong Province 2010 (Waldron and Brown, 2014) – percent of cattle produced (turned off) by herd size category

1 to 9	10 to 49	50 to 99	100 to 499	500 to 999	≥1,000
50%	19%	12%	11%	4%	3%

Argentina (Arelovich et al., 2011) – percent distributions of herd sizes and cattle numbers

	<100	101–250	251–500	501–1000	1001– 5000	>5000
Operations (%)	45.6	23.2	14.8	9.7	6.4	0.3
Cattle (%)	10.8	18.0	2.2	24.2	37.1	7.8

Australian (Bell et al., 2011) – percent distributions of herds due to herd size and sales

	<200	200–800	800–1600	1600–5400	>5400
Northern Australia					
Operations (%)	37.5	31.3	14.0	13.4	3.9
Sales (%)	4	12	13	30	41
Southern Australia					
Operations (%)	56.7	36.7	4.9	1.6	0.1
Sales (%)	18	41	15	13	13

USA (NASS, 2013) – percent distributions of beef herds based on number of cows

	1–49	50–99	100–499	≥500
Operations (%)	79.7	10.8	8.7	0.8
Beef cows (%)	27.7	17.2	38.4	16.7

Table 14.12. Cattle numbers and operations in European countries in 2010.

Country	Total cattle herds			Dairy cows and herds			Other cows[a] and herds			% <10	% >100
	Animals[b]	Holdings	Average	Animals[b]	Holdings	Average	Animals[b]	Holdings	Average		
Austria	2.024	71,940	28.1	0.540	47,740	11.3	0.264	44,400	6.0	83.2	0.0
Belgium	2.593	24,950	103.9	0.521	11,400	45.7	0.528	17,280	30.5	29.5	5.3
Bulgaria	0.586	95,870	6.1	0.334	85,950	3.9	0.020	1,110	17.9	52.3	1.8
Croatia[c]	0.497	47,340	10.5	0.212	41,340	5.1	0.017	3,200	5.2	88.8	0.3
Cyprus	0.053	280	190.8	0.021	200	102.6	0.001	90	15.9	55.6	0.0
Czech Republic	1.329	10,080	131.8	0.381	3,100	122.9	0.169	5,580	30.2	52.0	7.5
Denmark	1.571	13,580	115.7	0.568	4,250	133.7	0.101	8,410	12.0	61.0	0.6
Estonia	0.241	4,620	52.2	0.096	3,520	27.3	0.013	1,120	11.2	68.8	0.9
Finland	0.926	15,640	59.2	0.289	11,910	24.3	0.055	2,260	24.5	32.3	2.2
France	19.506	199,620	97.7	3.720	82,600	45.0	4.136	126,300	32.8	28.9	4.8
Germany	12.535	144,850	86.5	4.165	89,760	46.4	0.665	41,190	16.2	59.4	2.1
Greece	0.652	16,790	38.8	0.131	5,780	22.7	0.137	5,700	24.0	45.8	3.2
Hungary	0.707	19,120	37.0	0.245	11,370	21.6	0.071	2,470	28.7	59.5	6.9
Iceland[c]	0.075	910	82.3	0.026	720	35.7	0.002	140	15.6	57.1	NR
Ireland	6.607	111,000	59.5	1.071	18,460	58.0	1.158	80,070	14.5	49.2	0.3
Italy	5.953	125,880	47.3	1.832	52,130	35.2	0.507	31,240	16.2	54.8	1.6
Latvia	0.394	35,100	11.2	0.166	30,050	5.5	0.018	2,500	7.0	80.8	0.4
Lithuania	0.739	93,050	7.9	0.353	85,020	4.1	0.017	3,440	5.1	89.5	0.3
Luxembourg	0.199	1,460	136.2	0.045	810	55.6	0.032	1,120	29.0	29.5	3.6

Continued

Table 14.12. Continued.

Country	Total cattle herds Animals[b]	Holdings	Average	Dairy cows and herds Animals[b]	Holdings	Average	Other cows[a] and herds Animals[b]	Holdings	Average	% <10	% >100
Malta	0.016	290	54.1	0.007	140	48.1	0.000	30	6.3	66.7	NR
Montenegro[c]	0.080	24,620	3.3	0.051	23,780	2.2	0.001	430	2.3	97.7	NR
Netherlands	3.975	32,830	121.1	1.479	19,810	74.6	0.115	11,610	9.9	70.1	0.5
Norway[c]	0.875	16,900	51.7	0.238	11,130	21.4	0.070	5,200	13.4	49.4	0.2
Poland	5.742	514,120	11.2	2.506	425,780	5.9	0.140	27,110	5.2	88.3	0.3
Portugal	1.430	50,040	28.6	0.278	10,450	26.6	0.442	23,930	18.5	74.3	5.1
Romania	1.990	728,020	2.7	1.151	624,990	1.8	0.034	18,730	1.8	98.5	0.1
Slovakia	0.465	9,310	49.9	0.154	6,300	24.5	0.046	1,490	30.8	59.7	10.1
Slovenia	0.472	36,120	13.1	0.108	10,950	9.9	0.067	20,290	3.3	96.3	0.0
Spain	5.841	111,840	52.2	0.910	29,460	30.9	1.855	74,770	24.8	48.4	4.7
Sweden	1.537	21,590	71.2	0.348	5,620	61.9	0.197	12,190	16.2	50.9	1.2
Switzerland[c]	1.592	41,100	38.7	0.589	32,150	18.3	0.111	11,940	9.3	64.9	0.0
United Kingdom	10.064	85,760	117.3	1.843	23,540	78.3	1.647	59,720	27.6	38.4	4.6

[a]Classification of cows other than dairy assumed to be beef cows, with percentages of herds less than 10 cows and over 100 cows indicated.
[b]Animal numbers in millions. Data from European Commission Eurostat searchable databases.
[c]Non-EU members.
NR = not reported.

Table 14.13. Examples of selected beef finishing systems in different areas of world.

Farm name	No. and type beef cattle sold per year	Main feed sources	Age at start (days)	Finishing period (days)	Average daily weight gain (g/day)	Final live weight (kg)	Dressing percent (%)
AT-7	7 steers	Pasture + grass silage	240	540	704	700	53
AT-30	30 bulls	Maize silage + grains	100	403	1390	705	57
DE-190	120 bulls and 70 feeders	Maize silage + grains	50	437	1291	649	57
DE-240	240 bulls	Maize silage + grains	50	473	1255	673	58
DE-280	280 bulls	Maize silage + grains	60	514	1154	680	60
DE-360	282 bulls and 80 steers	Grass and maize silage + grains	180	360–500	920–1236	620–685	52–57
FR-45	31 bulls and 16 cows	Grass and maize silage + hay + grains	244	265	1566	695	59
FR-90A	90 bulls	Maize silage + grains	274	310–315	1250–1349	673–710	58–61
FR-90B	90 bulls	Maize silage + grains	7	547–557	1110–1122	667–685	54–56
ES-6950	3808 bulls and 3128 heifers	Straw + concentrates + grains	20	313–323	1327–1428	497–528	54–55
IE-75	75 steers	Pasture + grass silage + concentrates	563	365	548	675	54
CZ-160	160 bulls	Grass + maize silage	28	730	836	656	56
CZ-780	780 bulls	Hay + grains	28–345	365–612	805–922	620	54
HU-80	80 bulls	Maize silage + grains	230	230	1304	525	56
HU-440	440 bulls	Maize silage + grains	95	429	933	520	53
PL-12	7 bulls and 5 heifers	Pasture + grass silage + hay + grains	15	535	860	520	56
PL-30	20 bulls and 9 heifers	Pasture + grass and maize silage + grains	15	535	879	530	54

Continued

Table 14.13. Continued.

Farm name	No. and type beef cattle sold per year	Main feed sources	Age at start (days)	Finishing period (days)	Average daily weight gain (g/day)	Final live weight (kg)	Dressing percent (%)
US-7200	7195 steers	Grains + alfalfa hay	265	191	1444	578	61
AR-1000	1000 steers	Pasture + hay	180	463–546	540–549	400–450	58
AR-1300	1300 steers	Pasture + hay + maize stubble + grains	210–255	365–450	549–603	390–425	59
AR-2700	2061 steers and 648 heifers	Pasture + hay + maize stubble + grains	210	365–540	500–644	405–410	59–60
BR-180	180 steers	Pasture	240	1095	319	490	53
BR-500	500 steers	Pasture	210	945	347	480	53
UY-880	880 steers	Pasture + hay + maize stubble	210	527–645	450–550	440	54
AU-1100	922 steers and 184 heifers	Pasture + grains	210	224	964	486	54
NA-125	80 steers and 44 heifers	Pasture	240	690	355	530	57
PK-3	3 bulls	Freshly cut green grains + cottonseed	120	330	463	300	50
PK-50	50 bulls	Freshly cut green grains + concentrates	600–780	180	778	460	50

From Deblitz *et al.* (2004). Farm name relative to country and number of animals. AT = Austria, DE = Germany, FR = France, IE = Ireland, CZ = Czech Republic, HU = Hungary, PL = Poland, US = United States, AR = Argentina, BR = Brazil, UY = Uruguay, AU = Australia, NA = Namibia, PK = Pakistan.

Table 14.14. Description of some selected beef cattle production systems for animal size and age.

Region, cattle/ system	Birth/ young calves	Castration	Weaning	Market or fattening start	Time of fattening	Market
Japan						
Japanese Black steers	30 kg	2–3 months	4 months	10 months at 290 kg	20 months	30 months at 725 kg
Holstein steers	50 kg	2–3 months	3–4 months	6 months at 260 kg	16 months	22 months at 780 kg
Brazil						
Super early maturing	28–38 kg	No	7–8 months, 170–190 kg	8 months at 240 kg	120 days in feedlot	12 months at 420 kg
Early maturing	28–38 kg	No	7–8 months, 170–190 kg	Reared in grazing systems until 18–24 months age and 350 kg then sent to feedlot		
Pasture grazing	28–38 kg	No	7–8 months, 170–190 kg	Kept on variable quality pastures entire time before harvest		30–42 months at 450–500 kg
China	28–38 kg	No	6 months, 120–135 kg	Sold from small household operations at 12–24 months of age following feeding period		
Australia						
Feedlot finishing	28–40 kg	2–3 months or 6–8 months	6–8 months, 200–275 kg	6–8 months, 200–275 kg	4–8 months	14–16 months at 500–600 kg
Grass finishing	28–40 kg	2–3 mo or 6–8 mo	6–8 months, 200–275 kg	6–8 months, 200–275 kg	12–24 months	24–30 months at 550–650 kg

Continued

Table 14.14. Continued.

Region, cattle/ system	Birth/ young calves	Castration	Weaning	Market or fattening start	Time of fattening	Market
Europe	30–45 kg	Not typically	2–8 months, 100–300 kg	2–9 months, 100–300 kg	3–24 months	400–700 kg
USA – beef calves	30–40 kg	2–3 mo or 6–8 mo	6–8 months, 200–275 kg	Calf-fed (6–8 months), Yearling (12 months)	6–8 months, 4–6 months	14–15 months at 560 kg, 17–18 months at 615 kg

Table 14.15. Comparisons of beef production relative to methane production for major beef exporting countries.

	Country									
	EU	Australia	Brazil	United States	Canada	New Zealand	Argentina	India	Uruguay	Paraguay
Beef exports (million tonnes per year)										
Average (2001–2006)	2.317	1.249	1.229	0.695	0.505	0.492	0.409	0.359	0.288	0.106
National herd size (total cattle inventory in millions)										
1988	108.2	21.9	139.6	99.6	10.8	8.1	52.3	200.7	10.3	7.8
2007	89.7	28.4	207.2	97.0	14.2	9.7	50.8	177.8	12.0	10.0
Methane emissions per year (kg per animal)										
Kg/animal	58.0	60.4	55.2	52.5	52.5	60.4	55.2	26.9	55.2	55.2
Methane emissions per country (million tonnes per year)										
1988	6.273	1.320	7.706	5.230	0.565	0.487	2.885	5.397	0.570	0.429
2007	5.204	1.715	11.436	5.093	0.743	0.583	2.801	4.784	0.662	0.552
Change/year (%)	−1.01	1.31	2.12	−0.14	1.62	1.13	−0.36	−0.76	1.35	1.36
Beef produced per country (million tonnes per year)										
1988	9.335	1.588	4.050	10.879	0.947	0.572	2.506	1.094	0.329	0.131
2007	8.106	2.261	7.900	12.044	1.279	0.632	2.830	1.282	0.570	0.220
Change/ year (%)	−1.11	1.72	4.01	0.71	2.92	1.12	0.13	0.52	3.03	0.97
Methane emissions relative to beef production (kg methane/kg beef)										
1988	0.672	0.831	1.903	0.481	0.596	0.851	1.151	4.934	1.735	3.27
2007	0.642	0.759	1.448	0.423	0.581	0.922	0.99	3.731	1.162	2.509
Change/year (%)	0.1	−0.39	−1.82	−0.84	−1.26	0.02	−0.49	−1.27	−1.63	0.39
% change 1988–2007	1.85	−7.23	−29.41	−14.82	−21.34	0.31	−8.87	−21.57	−26.79	7.65

From Millen *et al.* (2011). Databases of herd size and quantity of beef produced, obtained 1988 and 2007, were collected from FAO (2009), and methane production per animal was obtained from the Intergovernmental Panel on Climate Change (IPCC, 2006) according to the authors.
This table is presented in imperial units in Table A12.

14.7 References

Arelovich, H.M., Bravo, R.D. and Martínez, M.F. (2011) Development, characteristics, and trends for beef cattle production in Argentina. *Animal Frontiers* 1, 37–45.

Bell, A.W., Charmley, E., Hunter, R.A. and Archer, J.A. (2011) The Australasian beef industries–Challenges and opportunities in the 21st century. *Animal Frontiers* 1, 10–19.

Bruinsma J., ed. (2003) *World Agriculture: towards 2015/2030. An FAO perspective*. Rome, Food and Agriculture Organization of the United Nations/ London, Earthscan.

Deblitz, C., Charry, A.A. and Parton, K.A. (2004) Beef farming systems across the world: an expert assessment from an international co-operative research project (IFCN) Extension. *Farming Systems Journal* 1, 1–14.

FAO (2012) Global ecological zones for FAO forest reporting: 2010 Update. Forest Resources Assessment Working Paper 179, 52 pp. Food and Agriculture Organization of the United Nations, Rome.

Jain, A.K. and Muladno, M. (2009) Selection criteria and breeding objectives in improvement of productivity of cattle and buffaloes. In: Selection and breeding of cattle in Asia: Strategies and criteria for improved breeding. IAEA, Vienna, Austria.

Millen, D.D., Pacheco, R.D.L., Meyer, P.M., Rodrigues, P.H.M. and Arrigoni, M.D.B. (2011) Current outlook and future perspectives of beef production in Brazil. *Animal Frontiers* 1, 46–52.

NASS (2013) Searchable databases of National Agricultural Statistics Service, United States Department of Agriculture. Available: *http://www.nass.usda.gov/.* (accessed 16 October 2013).

Otte, M.J. and Chilonda, P. (2002) Cattle and small ruminant production systems in sub-Saharan Africa: A systematic review. Food and Agriculture Organization of the United Nations, Rome.

Peel, D.S., Mathews, Jr., K.H. and Johnson, R.J. (2011) Trade, the expanding Mexican beef industry, and feedlot and stocker cattle production in Mexico. USDA Economics Research Service (ERS) report LDP-M-206-01.

Scholtz, M.M., Bester, J., Mamabolo, J.M. and Ramsay, K.A. (2008) Results of the national cattle survey undertaken in South Africa, with emphasis on beef. *Appl. Anim. Husb. Rural Develop.* 1, 1–9. Available: www.sasas.co.za/aahrd/.

Seré, C. and Steinfeld, H. (1996) World Livestock Production Systems, Current Status, Issues and Trends. FAO Animal Production and Health Paper No. 127. Food and Agriculture Organization of the United Nations, Rome.

USDA-FAS (2013) United States Department of Agriculture Foreign Agricultural Service home page. Available http://*www.fas.usda.gov/. (last accessed 22 November 2013).*

Waldron, S. and Brown, C. (2014) Chinese and southeast Asian cattle production, Chapter 7. In: Cottle, D. and Kahn, L. (eds) *Beef Cattle Production and Trade.* CSIRO Publishing and Meat & Livestock Australia, Brisbane, QLD.

Appendix

Table A1. Ordinal (day of the year) calendar (non-leap years).

Day	Jan	Feb	Mar	Apr	May	Jun	Jul	Aug	Sep	Oct	Nov	Dec	Day
1	1	32	60	91	121	152	182	213	244	274	305	335	1
2	2	33	61	92	122	153	183	214	245	275	306	336	2
3	3	34	62	93	123	154	184	215	246	276	307	337	3
4	4	35	63	94	124	155	185	216	247	277	308	338	4
5	5	36	64	95	125	156	186	217	248	278	309	339	5
6	6	37	65	96	126	157	187	218	249	279	310	340	6
7	7	38	66	97	127	158	188	219	250	280	311	341	7
8	8	39	67	98	128	159	189	220	251	281	312	342	8
9	9	40	68	99	129	160	190	221	252	282	313	343	9
10	10	41	69	100	130	161	191	222	253	283	314	344	10
11	11	42	70	101	131	162	192	223	254	284	315	345	11
12	12	43	71	102	132	163	193	224	255	285	316	346	12
13	13	44	72	103	133	164	194	225	256	286	317	347	13
14	14	45	73	104	134	165	195	226	257	287	318	348	14
15	15	46	74	105	135	166	196	227	258	288	319	349	15
16	16	47	75	106	136	167	197	228	259	289	320	350	16
17	17	48	76	107	137	168	198	229	260	290	321	351	17
18	18	49	77	108	138	169	199	230	261	291	322	352	18
19	19	50	78	109	139	170	200	231	262	292	323	353	19
20	20	51	79	110	140	171	201	232	263	293	324	354	20
21	21	52	80	111	141	172	202	233	264	294	325	355	21
22	22	53	81	112	142	173	203	234	265	295	326	356	22
23	23	54	82	113	143	174	204	235	266	296	327	357	23
24	24	55	83	114	144	175	205	236	267	297	328	358	24
25	25	56	84	115	145	176	206	237	268	298	329	359	25
26	26	57	85	116	146	177	207	238	269	299	330	360	26
27	27	58	86	117	147	178	208	239	270	300	331	361	27
28	28	59	87	118	148	179	209	240	271	301	332	362	28
29	29		88	119	149	180	210	241	272	302	333	363	29
30	30		89	120	150	181	211	242	273	303	334	364	30
31	31		90		151		212	243		304		365	31

The day of the month along the left or right columns corresponds to the day of the year in the body of the table. For example, 14 August is day 226 of the year, etc. For leap years (2012, 2016, etc.) 29 February would be day 60 and subsequent dates would be shifted by 1 day where 31 December would be day 366. The age of a calf can easily be calculated by subtracting the days of the year corresponding to the dates in question (i.e. a calf born on 25 February (day 56) and weaned on 22 September (day 265) is 265 – 56 = 209 days of age at weaning).

Table A2. Metric to imperial unit conversions.

Distance	Temperature
1 cm = 0.394 in = 10 mm 1 in = 2.54 cm = 25.4 mm	C = 5/9 * (F − 32) F = 9/5 * (C + 32)
1 m = 3.281 ft 1 ft = 0.3048 m = 12 in	38.6 C = 101.5 F 38.9 C = 102.0 F 39.4 C = 103.0 F
1 km = 3281 ft = 0.6214 mi 1 mi = 1609.344 m = 1.609 km = 5280 ft	40.0 C = 104.0 F 40.6 C = 105.0 F

Area	Mass	
1 sq m = 10.765 sq ft 1 sq ft = 0.0929 sq m	1 g = 0.03527 oz 1 oz = 28.35 g	
1 sq cm = 0.155 sq in 1 sq in = 6.4516 sq cm	1 kg = 2.205 lb = 35.28 oz 1 lb = 0.4535 kg = 453.5 g = 16 oz	
1 hectare = 2.471 ac = 10,000 sq m = 107,650 sq ft 1 acre = 0.4047 ha = 43,560 sq ft = 4047 sq m	50 kg = 110 lb 400 kg = 882 lb	100 lb = 45.4 kg 900 lb = 408 kg
	450 kg = 992 lb	1000 lb = 454 kg
1 sq km = 0.3861 sq mi = 100 ha 1 sq mi = 2.589 sq km = 640 ac	500 kg = 1102 lb 550 kg = 1213 lb 600 kg = 1323 lb 650 kg = 1433 lb	1100 lb = 499 kg 1200 lb = 544 kg 1300 lb = 590 kg 1400 lb = 635 kg

Volume
1 US fl oz = 0.0296 l 1 liter = 0.264 US gal 1 US gal = 3.785 l = 128 US fl oz 1 bushel = 36.4 l 1 cubic ft = 0.0283 cubic m 1 cubic m = 14.046 cubic ft

Table A3. Nutrient requirements of beef cows during gestation and lactation.

Gestating cows, middle third of pregnancy

Weight (lb)	Expected calf BWT (lb)	DM intake (lb/d)	% of BW	TDN (%DM)	NEm (Mcal/lb)	CP (%DM)	Ca (%DM)	P (%DM)	TDN (lb)	NEm (Mcal)	CP (lb)	Ca (lb)	P (lb)
900	63	17	1.9	50	0.44	7.1	0.17	0.14	8.3	7.3	1.2	0.028	0.023
1000	69	18	1.8	50	0.44	7.1	0.17	0.14	9.0	7.9	1.3	0.031	0.025
1100	75	19	1.8	50	0.44	7.1	0.17	0.14	9.7	8.5	1.4	0.034	0.028
1200	80	21	1.7	50	0.44	7.1	0.18	0.15	10.3	9.1	1.5	0.037	0.030
1300	86	22	1.7	50	0.44	7.1	0.18	0.15	11.0	9.7	1.6	0.040	0.033
1400	91	23	1.7	50	0.44	7.1	0.19	0.15	11.6	10.2	1.6	0.043	0.035
1500	96	25	1.6	50	0.44	7.1	0.19	0.15	12.2	10.8	1.7	0.046	0.038

Gestating cows, last third of pregnancy

Weight (lb)	Expected calf BWT (lb)	DM intake (lb/d)	% of BW	TDN (%DM)	NEm (Mcal/lb)	CP (%DM)	Ca (%DM)	P (%DM)	TDN (lb)	NEm (Mcal)	CP (lb)	Ca (lb)	P (lb)
900	63	19	2.1	54	0.50	7.9	0.25	0.16	10.3	9.6	1.5	0.047	0.030
1000	69	21	2.1	54	0.50	7.9	0.25	0.16	11.2	10.4	1.6	0.052	0.034
1100	75	22	2.0	54	0.50	7.9	0.25	0.16	12.1	11.2	1.8	0.057	0.037
1200	80	24	2.0	54	0.50	7.9	0.26	0.17	12.9	12.0	1.9	0.061	0.040
1300	86	25	2.0	54	0.50	7.9	0.26	0.17	13.7	12.8	2.0	0.066	0.043
1400	91	27	1.9	54	0.50	7.9	0.26	0.17	14.5	13.5	2.1	0.071	0.046
1500	96	28	1.9	54	0.50	7.9	0.27	0.17	15.3	14.2	2.2	0.075	0.049

Lactating cows

Weight (lb)	Peak milk (lb/d)	DM intake (lb)	% of BW	TDN (%DM)	NEm (Mcal/lb)	CP (%DM)	Ca (%DM)	P (%DM)	TDN (lb)	NEm (Mcal)	CP (lb)	Ca (lb)	P (lb)
900	10	22	2.5	56	0.53	8.7	0.24	0.17	12.4	11.7	1.9	0.052	0.037
	15	24	2.7	57	0.55	9.6	0.27	0.18	13.7	13.3	2.3	0.065	0.044
	20	26	2.9	59	0.58	10.4	0.30	0.20	15.3	14.9	2.7	0.077	0.051
1000	10	24	2.4	55	0.52	8.5	0.23	0.17	13.0	12.3	2.0	0.055	0.039
	15	26	2.6	57	0.55	9.4	0.27	0.18	14.5	14.0	2.4	0.068	0.047
	20	27	2.7	59	0.57	10.2	0.29	0.20	16.0	15.6	2.8	0.080	0.054
1100	15	27	2.5	57	0.54	9.2	0.26	0.18	15.3	14.6	2.5	0.071	0.049
	20	29	2.6	58	0.56	10.0	0.29	0.19	16.8	16.3	2.9	0.083	0.056
	25	31	2.8	59	0.58	10.6	0.31	0.21	18.2	17.9	3.3	0.095	0.064
1200	15	29	2.4	57	0.54	9.0	0.26	0.18	16.1	15.3	2.6	0.074	0.051
	20	30	2.5	58	0.56	9.8	0.28	0.19	17.6	16.9	3.0	0.086	0.059
	25	32	2.7	59	0.58	10.5	0.31	0.21	19.0	18.6	3.4	0.098	0.066

Continued

Table A3. Continued.

Weight (lb)	Peak milk (lb/d)	DM intake (lb)	% of BW					Lactating cows						
1300	15	30	2.3	56	0.53	8.9	0.26	0.18	16.8	16.0	2.7	0.077	0.054
	20	32	2.4	57	0.55	9.6	0.28	0.19	18.1	17.6	3.1	0.089	0.061
	25	34	2.6	59	0.57	10.3	0.30	0.20	19.7	19.2	3.4	0.102	0.069
	20	33	2.4	57	0.55	9.5	0.28	0.19	18.9	18.2	3.1	0.092	0.064
1400	25	35	2.5	59	0.57	10.1	0.30	0.20	20.5	19.8	3.5	0.105	0.071
	30	37	2.6	59	0.58	10.6	0.32	0.21	21.8	21.5	3.9	0.117	0.078
1500	20	35	2.3	57	0.55	9.3	0.28	0.19	19.7	18.8	3.2	0.095	0.066
	25	37	2.4	58	0.56	9.9	0.30	0.20	21.2	20.5	3.6	0.108	0.073
	30	38	2.6	59	0.58	10.5	0.31	0.21	22.6	22.1	4.0	0.120	0.081

Original values from Beef Cattle NRC and adapted from Oklahoma State University Beef Cattle Manual 4th edn.

Table A4. Nutrient requirements for growing yearling cattle with anticipated 499 kg live weight at 10 to 13 mm of backfat.

Body weight (kg)	ADG (kg)	DMI (kg/day)	Diet nutrient density						Daily nutrient need per animal					
			TDN (%DM)	NEm (MJ/kg)	NEg (MJ/kg)	CP (%DM)	Ca (%DM)	P (%DM)	TDN (kg)	NEm (MJ)	Neg (MJ)	CP (kg)	Ca (g)	P (g)
274	0.3	7.4	50	4.15	1.85	7.2	0.22	0.13	3.7	21.8	4.2	0.5	18	9
	0.9	7.8	60	5.63	3.23	10.0	0.36	0.19	4.7	21.8	12.6	0.8	27	14
	1.3	7.7	70	7.01	4.43	12.7	0.49	0.24	5.4	21.8	20.1	1.0	36	18
	1.6	7.2	80	8.30	5.63	15.3	0.61	0.29	5.8	21.8	25.5	1.1	45	23
	1.8	6.7	90	9.59	6.64	17.8	0.72	0.34	6.0	21.8	29.3	1.2	50	23
299	0.3	7.9	50	4.15	1.85	7.1	0.21	0.13	4.0	23.0	4.6	0.5	18	9
	0.9	8.3	60	5.63	3.23	9.7	0.34	0.18	5.0	23.0	13.4	0.8	27	14
	1.3	8.2	70	7.01	4.43	12.3	0.45	0.23	5.7	23.0	21.3	1.0	36	18
	1.6	7.7	80	8.30	5.63	14.7	0.56	0.27	6.2	23.0	27.2	1.1	45	23
	1.8	7.1	90	9.59	6.64	17.1	0.66	0.32	6.4	23.0	31.0	1.2	45	23
324	0.3	8.4	50	4.15	1.85	6.9	0.20	0.13	4.2	24.7	4.6	0.6	18	9
	0.9	8.9	60	5.63	3.23	9.2	0.32	0.17	5.4	24.7	14.6	0.8	27	14
	1.3	8.7	70	7.01	4.43	11.5	0.42	0.21	6.1	24.7	23.0	1.0	36	18
	1.6	8.2	80	8.30	5.63	13.7	0.52	0.26	6.6	24.7	28.9	1.1	41	23
	1.8	7.6	90	9.59	6.64	15.9	0.61	0.30	6.8	24.7	33.1	1.2	45	23
349	0.3	8.9	50	4.15	1.85	6.8	0.20	0.12	4.4	25.9	5.0	0.6	18	9
	0.9	9.4	60	5.54	3.23	8.8	0.30	0.16	5.6	25.9	15.1	0.8	27	14
	1.3	9.2	70	7.01	4.43	10.9	0.39	0.20	6.4	25.9	24.3	1.0	36	18
	1.6	8.7	80	8.30	5.63	12.9	0.48	0.24	6.9	25.9	30.5	1.1	41	23
	1.8	8.0	90	9.59	6.64	14.8	0.56	0.28	7.2	25.9	34.7	1.2	45	23
374	0.3	9.3	50	4.15	1.85	6.6	0.19	0.12	4.7	27.6	5.4	0.6	18	14
	0.9	9.9	60	5.63	3.23	8.4	0.28	0.16	5.9	27.6	25.5	1.0	36	18
	1.3	9.7	70	7.01	4.43	10.3	0.37	0.19	6.8	27.6	25.5	1.0	36	18
	1.6	9.1	80	8.30	5.63	12.1	0.44	0.23	7.3	27.6	32.2	1.1	41	23
	1.8	8.4	90	9.59	6.64	13.9	0.52	0.26	7.6	27.6	36.8	1.2	45	23
399	0.3	9.8	50	4.15	1.85	6.5	0.19	0.12	4.9	28.9	5.4	0.6	18	14
	0.9	10.4	60	5.63	3.23	8.1	0.27	0.15	6.2	28.9	16.7	0.9	27	14
	1.3	10.2	70	7.01	4.43	9.8	0.34	0.18	7.1	28.9	26.8	1.0	36	18
	1.6	9.6	80	8.30	5.63	11.4	0.42	0.22	7.7	28.9	33.9	1.1	41	23
	1.8	8.8	90	9.59	6.64	13.1	0.48	0.25	8.0	28.9	38.5	1.2	41	23

Original values from Beef Cattle NRC and adapted from Oklahoma State University Beef Cattle Manual.

Table A5. Nutrient requirements for growing yearling cattle with anticipated 1100 lb live weight with 0.4 to 0.5 inches of backfat.

Body weight (lb)	ADG (lb)	DMI (lb/day)	Diet nutrient density						Daily nutrient need per animal					
			TDN (%DM)	NEm (Mcal/lb)	NEg (Mcal/lb)	CP (%DM)	Ca (%DM)	P (%DM)	TDN (lb)	NEm (Mcal)	NEg (Mcal)	CP (lb)	Ca (lb)	P (lb)
605	0.7	16.3	50	0.45	0.20	7.2	0.22	0.13	8.2	5.2	1.0	1.2	0.04	0.02
	1.9	17.3	60	0.61	0.35	10.0	0.36	0.19	10.4	5.2	3.0	1.7	0.06	0.03
	2.9	16.9	70	0.76	0.48	12.7	0.49	0.24	11.8	5.2	4.8	2.2	0.08	0.04
	3.6	15.9	80	0.90	0.61	15.3	0.61	0.29	12.7	5.2	6.1	2.4	0.10	0.05
	4.0	14.7	90	1.04	0.72	17.8	0.72	0.34	13.2	5.2	7.0	2.6	0.11	0.05
660	0.7	17.5	50	0.45	0.20	7.1	0.21	0.13	8.8	5.5	1.1	1.2	0.04	0.02
	1.9	18.4	60	0.61	0.35	9.7	0.34	0.18	11.0	5.5	3.2	1.8	0.06	0.03
	2.9	18.0	70	0.76	0.48	12.3	0.45	0.23	12.6	5.5	5.1	2.2	0.08	0.04
	3.6	17.0	80	0.90	0.61	14.7	0.56	0.27	13.6	5.5	6.5	2.5	0.10	0.05
	4.0	15.7	90	1.04	0.72	17.1	0.66	0.32	14.1	5.5	7.4	2.7	0.10	0.05
715	0.7	18.5	50	0.45	0.20	6.9	0.20	0.13	9.3	5.9	1.1	1.3	0.04	0.02
	1.9	19.6	60	0.61	0.35	9.2	0.32	0.17	11.8	5.9	3.5	1.8	0.06	0.03
	2.9	19.1	70	0.76	0.48	11.5	0.42	0.21	13.4	5.9	5.5	2.2	0.08	0.04
	3.6	18.1	80	0.90	0.61	13.7	0.52	0.26	14.5	5.9	6.9	2.5	0.09	0.05
	4.0	16.7	90	1.04	0.72	15.9	0.61	0.30	15.0	5.9	7.9	2.7	0.10	0.05
770	0.7	19.6	50	0.45	0.20	6.8	0.20	0.12	9.8	6.2	1.2	1.3	0.04	0.02
	1.9	20.7	60	0.60	0.35	8.8	0.30	0.16	12.4	6.2	3.6	1.8	0.06	0.03
	2.9	20.2	70	0.76	0.48	10.9	0.39	0.20	14.1	6.2	5.8	2.2	0.08	0.04
	3.6	19.1	80	0.90	0.61	12.9	0.48	0.24	15.3	6.2	7.3	2.5	0.09	0.05
	4.0	17.6	90	1.04	0.72	14.8	0.56	0.28	15.8	6.2	8.3	2.6	0.10	0.05
825	0.7	20.6	50	0.45	0.20	6.6	0.19	0.12	10.3	6.6	1.3	1.4	0.04	0.03
	1.9	21.8	60	0.61	0.35	8.4	0.28	0.16	13.1	6.6	6.1	2.2	0.08	0.04
	2.9	21.3	70	0.76	0.48	10.3	0.37	0.19	14.9	6.6	6.1	2.2	0.08	0.04
	3.6	20.1	80	0.90	0.61	12.1	0.44	0.23	16.1	6.6	7.7	2.4	0.09	0.05
	4.0	18.6	90	1.04	0.72	13.9	0.52	0.26	16.7	6.6	8.8	2.6	0.10	0.05
880	0.7	21.7	50	0.45	0.2	6.5	0.19	0.12	10.9	6.9	1.3	1.4	0.04	0.03
	1.9	22.9	60	0.61	0.35	8.1	0.27	0.15	13.7	6.9	4.0	1.9	0.06	0.03
	2.9	22.4	70	0.76	0.48	9.8	0.34	0.18	15.7	6.9	6.4	2.2	0.08	0.04
	3.6	21.1	80	0.9	0.61	11.4	0.42	0.22	16.9	6.9	8.1	2.4	0.09	0.05
	4.0	19.5	90	1.04	0.72	13.1	0.48	0.25	17.6	6.9	9.2	2.6	0.09	0.05

Original values from Beef Cattle NRC and adapted from Oklahoma State University Beef Cattle Manual.

Table A6. Nutrient requirements for growing and mature bulls with expected 771 kg mature weight.

Weight (kg)	ADG (kg)	DMI (kg/day)	Diet nutrient density						Daily nutrient need per animal					
			TDN (%DM)	NEm (MJ/kg)	NEg (MJ/kg)	CP (%DM)	Ca (%DM)	P (%DM)	TDN (kg)	NEm (MJ)	NEg (MJ)	CP (kg)	Ca (g)	P (g)
408	0.2	10.0	50	4.15	1.85	7.0	0.16	0.11	5.0	33.5	3.8	0.7	18	14
	0.7	10.4	60	5.63	3.23	7.3	0.23	0.14	6.3	33.5	14.6	0.8	23	14
	1.1	10.4	70	7.01	4.43	8.8	0.30	0.16	7.2	33.5	24.3	0.9	32	18
	1.4	10.0	80	8.30	5.63	10.2	0.36	0.19	7.8	33.5	31.8	1.0	36	18
454	0.2	10.9	50	4.15	1.85	7.0	0.16	0.11	5.4	36.4	4.2	0.8	18	14
	0.7	11.3	60	5.63	3.23	7.0	0.22	0.13	6.8	36.4	15.9	0.8	27	14
	1.1	11.3	70	7.01	4.43	8.1	0.27	0.15	7.8	36.4	26.4	0.9	32	18
	1.4	10.4	80	8.30	5.63	9.3	0.32	0.18	8.4	36.4	34.3	1.0	32	18
499	0.2	11.8	50	4.15	1.85	7.0	0.16	0.11	5.8	39.3	4.2	0.8	18	14
	0.7	12.2	60	5.63	3.23	7.0	0.20	0.13	7.3	39.3	17.2	0.9	27	14
	1.1	11.8	70	7.01	4.43	7.5	0.25	0.14	8.4	39.3	28.5	0.9	32	18
	1.4	11.3	80	8.30	5.63	8.6	0.29	0.16	9.0	39.3	36.8	1.0	32	18
544	0.2	12.2	50	4.15	1.85	7.0	0.16	0.11	6.2	41.8	4.6	0.9	18	14
	0.7	13.2	60	5.63	3.23	7.0	0.19	0.12	7.8	41.8	18.4	0.9	27	18
	1.1	12.7	70	7.01	4.43	7.1	0.23	0.14	8.9	41.8	30.1	0.9	27	18
	1.4	12.2	80	8.30	5.63	7.9	0.26	0.15	9.7	41.8	39.3	1.0	32	18
590	0.2	13.2	50	4.15	1.85	7.0	0.16	0.11	6.6	44.4	5.0	0.9	23	14
	0.7	14.1	60	5.63	3.23	7.0	0.19	0.12	8.3	44.4	19.2	1.0	27	18
635	0.2	14.1	50	4.15	1.85	7.0	0.16	0.11	7.0	46.9	5.0	1.0	23	18
	0.7	14.5	60	5.63	3.23	7.0	0.18	0.12	8.8	46.9	20.5	1.0	27	18
680	0.2	14.5	50	4.15	1.85	7.0	0.16	0.11	7.3	49.4	5.4	1.0	23	18
	0.7	15.4	60	5.63	3.23	7.0	0.17	0.12	9.3	49.4	21.3	1.1	27	18
726	0.2	15.4	50	4.15	1.85	7.0	0.16	0.12	7.7	51.9	5.9	1.1	23	18
	0.7	16.3	60	5.63	3.23	7.0	0.16	0.11	9.8	51.9	22.6	1.1	27	18
771	0.0	15.0	46	3.60	0.00	7.0	0.16	0.12	6.8	54.4	0.0	1.0	23	18
	0.2	16.3	50	4.15	1.85	7.0	0.16	0.12	8.1	54.4	5.9	1.1	27	18

Original values from Beef Cattle NRC and adapted from Oklahoma State University Beef Cattle Manual.

Table A7. Nutrient requirements for growing and mature bulls with expected 1700 lb mature weight.

Body weight (lb)	ADG (lb)	DMI (lb/day)	Diet nutrient density						Daily nutrient need per animal					
			TDN (%DM)	NEm (Mcal/lb)	NEg (Mcal/lb)	CP (%DM)	Ca (%DM)	P (%DM)	TDN (lb)	NEm (Mcal)	Neg (Mcal)	CP (lb)	Ca (lb)	P (lb)
900	0.4	22	50	0.45	0.20	7.0	0.16	0.11	11.0	8.0	0.9	1.5	0.04	0.03
	1.6	23	60	0.61	0.35	7.3	0.23	0.14	14.0	8.0	3.5	1.7	0.05	0.03
	2.5	23	70	0.76	0.48	8.8	0.30	0.16	15.9	8.0	5.8	2.0	0.07	0.04
	3.1	22	80	0.90	0.61	10.2	0.36	0.19	17.2	8.0	7.6	2.2	0.08	0.04
1000	0.4	24	50	0.45	0.20	7.0	0.16	0.11	11.9	8.7	1.0	1.7	0.04	0.03
	1.6	25	60	0.61	0.35	7.0	0.22	0.13	15.1	8.7	3.8	1.8	0.06	0.03
	2.5	25	70	0.76	0.48	8.1	0.27	0.15	17.2	8.7	6.3	2.0	0.07	0.04
	3.1	23	80	0.90	0.61	9.3	0.32	0.18	18.6	8.7	8.2	2.2	0.07	0.04
1100	0.4	26	50	0.45	0.20	7.0	0.16	0.11	12.8	9.4	1.0	1.8	0.04	0.03
	1.6	27	60	0.61	0.35	7.0	0.20	0.13	16.2	9.4	4.1	1.9	0.06	0.03
	2.5	26	70	0.76	0.48	7.5	0.25	0.14	18.5	9.4	6.8	2.0	0.07	0.04
	3.1	25	80	0.90	0.61	8.6	0.29	0.16	19.9	9.4	8.8	2.1	0.07	0.04
1200	0.4	27	50	0.45	0.20	7.0	0.16	0.11	13.7	10.0	1.1	1.9	0.04	0.03
	1.6	29	60	0.61	0.35	7.0	0.19	0.12	17.3	10.0	4.4	2.0	0.06	0.04
	2.5	28	70	0.76	0.48	7.1	0.23	0.14	19.7	10.0	7.2	2.0	0.06	0.04
	3.1	27	80	0.90	0.61	7.9	0.26	0.15	21.3	10.0	9.4	2.1	0.07	0.04
1300	0.4	29	50	0.45	0.20	7.0	0.16	0.11	14.5	10.6	1.2	2.0	0.05	0.03
	1.6	31	60	0.61	0.35	7.0	0.19	0.12	18.4	10.6	4.6	2.2	0.06	0.04
1400	0.4	31	50	0.45	0.20	7.0	0.16	0.11	15.4	11.2	1.2	2.2	0.05	0.04
	1.6	32	60	0.61	0.35	7.0	0.18	0.12	19.4	11.2	4.9	2.3	0.06	0.04
1500	0.4	32	50	0.45	0.20	7.0	0.16	0.11	16.2	11.8	1.3	2.3	0.05	0.04
	1.6	34	60	0.61	0.35	7.0	0.17	0.12	20.5	11.8	5.1	2.4	0.06	0.04
1600	0.4	34	50	0.45	0.20	7.0	0.16	0.11	17.0	12.4	1.4	2.4	0.05	0.04
	1.6	36	60	0.61	0.35	7.0	0.16	0.11	21.5	12.4	5.4	2.5	0.06	0.04
1700	0.0	33	46	0.39	0.00	7.0	0.16	0.12	15.1	13.0	0.0	2.3	0.05	0.04
	0.4	36	50	0.45	0.20	7.0	0.16	0.12	17.8	13.0	1.4	2.5	0.06	0.04

Original values from Beef Cattle NRC and adapted from Oklahoma State University Beef Cattle Manual.

Table A8. Example US feed comparisons for relative cost per pound of TDN and per pound of CP.

Feed	Cost ($) per 50 lb bag	TDN%	CP%	Cost ($) per ton	Cost ($) per lb feed	Cost ($) per lb TDN	Cost ($) per lb CP
			Milled feeds				
Show-ring feed	10.55	0.67	0.12	422	0.211	0.315	1.758
Feedlot finisher	8.20	0.71	0.13	328	0.164	0.231	1.262
Bull grower	6.90	0.63	0.10	276	0.138	0.220	1.380
		Cubes (pellets) for cow herds – can be fed on ground surface					
High energy cube (14%)	7.90	0.69	0.14	316	0.158	0.229	1.129
Range breeder cube (20%)	9.55	0.68	0.20	382	0.191	0.281	0.955
38% HP range cube	12.50	0.69	0.38	500	0.25	0.362	0.658
		Hay purchased in 1200 lb round bales (cost is per bale)					
Alfalfa	115.00	0.65	0.19	191.67	0.096	0.147	0.504
High-quality Bermuda hay	65.00	0.58	0.14	108.33	0.054	0.093	0.387
Low-quality Bermuda hay	50.00	0.50	0.08	83.33	0.042	0.083	0.521
Neighbor's low-quality hay	40.00	0.45	0.05	66.67	0.033	0.074	0.667
		Small square bales of hay purchased at local feed store					
Alfalfa hay (80 lb)	12.00	0.65	0.19	300.00	0.150	0.231	0.789
HQ Bermuda hay (70 lb)	8.00	0.58	0.14	228.57	0.114	0.197	0.816

Prices relative to October 2013. Many feed purchased in bulk can be obtained for $10 to $20 per ton cheaper than in 50 lb bags. Individual commodities can be obtained cheaper when purchased in larger amounts (per truck load, per train car, etc.).

Table A9. Cattle frame score based on age (months) and hip height (in).[a]

	Males[b]							Females					
	Frame score[c]							Frame score[c]					
Age (mo)	3.0	4.0	5.0	6.0	7.0	8.0	Age (mo)	3.0	4.0	5.0	6.0	7.0	8.0
5	37.5	39.5	41.6	43.6	45.6	47.7	5	37.2	39.3	41.3	43.4	45.5	47.5
6	38.8	40.8	42.9	44.9	46.9	48.9	6	38.2	40.3	42.3	44.4	46.5	48.5
7	40.0	42.1	44.1	46.1	48.1	50.1	7	39.2	41.2	43.3	45.3	47.4	49.4
8	41.2	43.2	45.2	47.2	49.3	51.3	8	40.1	42.1	44.1	46.2	48.2	50.2
9	42.3	44.3	46.3	48.3	50.3	52.3	9	40.9	42.9	44.9	47.0	49.0	51.0
10	43.3	45.3	47.3	49.3	51.3	53.3	10	41.6	43.7	45.7	47.7	49.7	51.7
11	44.2	46.2	48.2	50.2	52.2	54.2	11	42.3	44.3	46.4	48.4	50.4	52.4
12	45.0	47.0	49.0	51.0	53.0	55.0	12	43.0	45.0	47.0	49.0	51.0	53.0
13	45.8	47.8	49.8	51.8	53.8	55.8	13	43.6	45.5	47.5	49.5	51.5	53.5
14	46.5	48.5	50.4	52.4	54.4	56.4	14	44.1	46.1	48.0	50.0	52.0	54.0
15	47.1	49.1	51.1	53.0	55.0	57.0	15	44.5	46.5	48.5	50.5	52.4	54.4
16	47.6	49.6	51.6	53.6	55.6	57.5	16	44.9	46.9	48.9	50.8	52.8	54.8
17	48.1	50.1	52.0	54.0	56.0	58.0	17	45.3	47.2	49.2	51.1	53.1	55.1
18	48.5	50.5	52.4	54.4	56.4	58.4	18	45.6	47.5	49.5	51.4	53.4	55.3
19	48.8	50.8	52.7	54.7	56.7	58.7	19	45.8	47.7	49.7	51.6	53.6	55.5
20	49.1	51.0	53.0	55.0	56.9	58.9	20	46.0	47.9	49.8	51.8	53.7	55.6
21	49.2	51.2	53.2	55.1	57.1	59.1	21	46.1	48.0	50.0	51.9	53.8	55.7
Maturity[d]	52.3	54.1	55.9	58.0	60.0	62.0	Maturity[d]	48.2	50.0	52.0	53.9	55.8	57.5
Finished steer weight (lb)[e]	1000	1100	1200	1300	1400	1500	Finished heifer weight (lb)[e]	900	1000	1100	1200	1300	1400
Mature bull weight (lb)[f,g]	1570	1730	1890	2050	2200	2360	Mature cow weight (lb)[f]	1000	1100	1200	1300	1400	1500
Expected carcass weight[h]	600	660	720	780	840	900	Expected carcass weight[h]	540	600	660	720	780	840

Frame score equation for ages of 5 to 21 months = 0.4878 × hht − 0.0289 × age in days + 0.00001947 × age in days[b] + 0.0000334 × hht × age in days − 11.548

Frame score equation for ages of 5 to 21 months = 0.4723 × hht − 0.0239 × age in days + 0.0000146 × age in days[b] + 0.0000759 × hht × age in days − 11.7086

[a] Adapted from Beef Improvement Federation (BIF, 2010) and Hammack (2009).
[b] Steers continue to grow long bone growth longer than bulls and will be 0.5 to 1.0 in taller at 18 to 21 months.
[c] USDA Feeder Calf Frame grade of Medium is frame score of 4.0 to 5.5.
[d] If calved first at 2 years of age, add 1.0 in if calved first at 3 years.
[e] At 0.4 to 0.5 in of fat cover over last two ribs.
[f] At 12 months of age, bulls weigh 50% to 60% of mature weight under non-extreme conditions.
[g] At body condition score of 5 (moderate body fatness) where 1 = extremely thin and 9 = obese; cow weights vary 7 to 8% per condition score and as much as ± 10% for extremes in muscle thickness.
[h] Assuming 60% dressing percent and average muscle of 1.8 sq in of longissimus muscle area per 100 lb of carcass weight and 0.4 to 0.5 in of fat thickness at 12th–13th rib interface.

Table A10. Guidelines for environmental conditions of heat and cold stress in cattle.

Temperature (°F) for heat stress

RH[a]	82	83	84	85	86	87	88	89	90	91	92	93	94	95	96	97	98	99	100
90	91	95	98	102	105	109	113	117	122	126	131	136	141	147	152	158	164	170	176
85	90	93	96	99	102	106	110	113	117	122	126	130	135	140	145	150	155	161	167
80	89	91	94	97	100	103	106	110	113	117	121	125	129	134	138	143	148	153	158
75	88	90	92	95	97	100	103	106	109	113	116	120	124	128	132	136	141	145	150
70	86	88	90	93	95	98	100	103	106	109	112	116	119	123	126	130	134	138	143
65	85	87	89	91	93	95	98	100	103	105	108	111	114	118	121	125	128	132	136
60	84	86	88	89	91	93	95	97	100	102	105	107	110	113	116	119	123	126	129
55	84	85	86	88	89	91	93	95	97	99	101	104	106	109	112	114	117	120	124
50	83	84	85	86	88	89	91	93	95	97	99	101	103	105	108	110	113	115	118
45	82	83	84	85	87	88	89	91	92	94	96	98	100	102	104	106	109	111	114
40	81	82	83	84	85	87	88	89	91	92	94	95	97	99	101	103	105	107	109
35	81	82	83	84	85	86	87	88	89	90	92	93	95	96	98	100	102	104	106
30	80	81	82	83	84	85	86	87	88	89	90	92	93	94	96	97	99	101	102
25	80	81	82	82	83	84	85	86	87	88	89	90	91	93	94	95	97	98	100

Temperature (°F) for cold stress

WS[a]	-10	-8	-6	-4	-2	0	2	4	6	8	10	12	14	16	18	20	22	24	26
25	-59	-56	-53	-50	-47	-44	-41	-38	-35	-32	-29	-26	-23	-20	-17	-14	-12	-9	-6
23	-57	-54	-51	-48	-45	-42	-39	-36	-33	-30	-28	-25	-22	-19	-16	-13	-10	-7	-4
21	-54	-51	-49	-46	-43	-40	-37	-34	-31	-28	-26	-23	-20	-17	-14	-11	-8	-5	-3
19	-52	-49	-46	-43	-40	-37	-35	-32	-29	-26	-23	-21	-18	-15	-12	-9	-6	-4	-1
17	-48	-46	-43	-40	-37	-35	-32	-29	-26	-24	-21	-18	-15	-13	-10	-7	-4	-2	1
15	-45	-42	-39	-37	-34	-31	-29	-26	-23	-21	-18	-15	-13	-10	-7	-5	-2	1	4
13	-41	-38	-36	-33	-30	-28	-25	-23	-20	-17	-15	-12	-9	-7	-4	-2	1	4	6
11	-36	-33	-31	-30	-26	-23	-21	-18	-16	-13	-11	-8	-6	-3	-1	1	4	7	9
9	-30	-28	-26	-23	-21	-18	-16	-14	-11	-9	-6	-4	-2	1	3	6	8	10	13
7	-24	-21	-19	-17	-15	-12	-10	-8	-5	-3	-1	1	4	6	8	10	13	15	17
5	-15	-13	-11	-9	-7	-5	-3	0	2	4	6	8	10	12	14	16	18	21	23
3	-4	-2	0	2	4	6	7	9	11	13	15	17	19	21	22	24	26	28	30

[a]RH = relative humidity (%); WS = wind speed in miles/hour. Shaded areas represent potential danger conditions. Ambient temperatures between 32 and 82°F generally present no or minimal environmental stresses to *Bos taurus* breeds. The range with no to minimal stress for *Bos indicus* breeds is likely 46 to 94°F.

Table A11. Recommended space allowance per animal for housing and feeding cattle.

	Calves, 397 to 838 lb	Feedlot cattle, 794 to 1202 lb	Bred heifers, 794 lb	Cows, 1003 lb	Cows, 1300 lb	Bulls, 1500 lb
	Open lots with no barn, sq ft					
Unpaved lots with mound (including mound space)	14.0–28.0	23.2–46.5	23.2–46.5	18.6–46.5	28.0–46.5	46.5
Mound space, 25% slope	20.5–24.8	30.1–35.5	30.1–35.5	39.8–45.2	39.8–45.2	50.6–60.3
Unpaved lot, 4 to 8% slope, no mound	28.0–55.8	37.2–74.4	37.2–74.4	32.5–74.3	32.5–74.3	74.3
Paved lot, 2 to 4% slope	39.8–50.6	50.6–60.3	50.6–60.3	60.3–75.4	60.3–75.4	100.1–124.9
	Barns (unheated cold housing), sq ft					
Open front with dirt lot	15.1–20.5	20.5–24.8	20.5–24.8	20.5–24.8	24.8–30.1	39.8
Enclosed, bedded pack	20.5–24.8	30.1–35.5	30.1–35.5	35.5–39.8	39.8–50.6	45.2–50.6
Enclosed, slotted floor	11.8–18.3	18.3–24.8	18.3–24.8	20.5–24.8	21.5–28.0	30.1
	Feeder space (inches) per animal when fed:					
Once daily	18 to 22	22 to 26	22 to 26	24 to 30	24 to 30	30 to 36
Twice daily	9 to 11	11 to 13	11 to 13	12 to 15	12 to 15	–
Free choice grain concentrate	3 to 4	4 to 6	4 to 6	5 to 6	5 to 6	–
Self-fed roughage/ high forage	9 to 10	10 to 11	11 to 12	12 to 13	12 to 13	–

Table A12. Comparisons of beef production relative to methane production for major beef exporting countries.

	EU	Australia	Brazil	United States	Canada	New Zealand	Argentina	India	Uruguay	Paraguay
						Country				
Average (2001–2006)	5,108.0	2,754.2	2,710.0	1,531.4	1,113.7	1,085.4	901.7	791.0	636.0	233.5
Beef exports (million lb per year)										
National herd size (total cattle inventory in millions)										
1988	108.2	21.9	139.6	99.6	10.8	8.1	52.3	200.7	10.3	7.8
2007	89.7	28.4	207.2	97.0	14.2	9.7	50.8	177.8	12.0	10.0
Kg/animal	127.9	133.2	121.7	115.8	115.8	133.2	121.7	59.3	121.7	121.7
Methane emissions per year (lb per animal)										
Methane emissions per country (million lb year)										
1988	13,832.1	2,910.1	16,991.4	11,532.5	1,245.2	1,073.2	6,360.5	11,901.5	1,257.4	946.9
2007	11,475.7	3,782.4	25,215.9	11,229.3	1,638.6	1,285.2	6,177.1	10,548.5	1,460.6	1,217.2
Change/yr (%)	–1.01	1.31	2.12	–0.14	1.62	1.13	–0.36	–0.76	1.35	1.36
Beef produced per country (million lb per year)										
1988	20,583.0	3,500.9	8,330.3	23,988.2	2,089.0	1,260.4	5,526.8	2,412.2	724.7	289.5
2007	17,873.5	4,985.5	17,419.5	26,557.7	2,819.3	1,394.4	6,240.2	2,827.6	1,256.9	485.1
Change/yr (%)	–1.11	1.72	4.01	0.71	2.92	1.12	0.13	0.52	3.03	0.97
Methane emissions relative to beef production (lb methane/lb beef)										
1988	0.672	0.831	1.903	0.481	0.596	0.851	1.151	4.934	1.735	3.27
2007	0.642	0.759	1.448	0.423	0.581	0.922	0.99	3.731	1.162	2.509
Change/yr (%)	0.1	–0.39	–1.82	–0.84	–1.26	0.02	–0.49	–1.27	–1.63	0.39
% change 88–07	1.85	–7.23	–29.41	–14.82	–21.34	0.31	–8.87	–21.57	–26.79	7.65

From Millen *et al.* (2011). Databases of herd size and quantity of beef produced, obtained 1988 and 2007, were collected from FAO (2009), and methane production per animal was obtained from the Intergovernmental Panel on Climate Change (IPCC; 2006) according to the authors.

Index

www.ingramcontent.com/pod-product-compliance
Lightning Source LLC
Chambersburg PA
CBHW050104220326
41598CB00043B/7382